Human Neurophysiology

A student text

Oliver Holmes
DSc, MSc, MB, BS, MRCS, LRCP, BA
Senior Lecturer in Physiology
University of Glasgow
Scotland

CHAPMAN & HALL MEDICAL
London · Glasgow · New York · Tokyo · Melbourne · Madras

Published by Chapman & Hall, 2–6 Boundary Row, London SE1 8HN

Chapman & Hall, 2–6 Boundary Row, London SE1 8HN, UK

Blackie Academic & Professional, Wester Cleddens Road, Bishopbriggs, Glasgow G64 2NZ, UK

Chapman & Hall Inc., 29 West 35th Street, New York NY10001, USA

Chapman & Hall Japan, Thomson Publishing Japan, Hirakawacho Nemoto Building, 6F, 1–7–11 Hirakawa-cho, Chiyoda-ku, Tokyo 102, Japan

Chapman & Hall Australia, Thomas Nelson Australia, 102 Dodds Street, South Melbourne, Victoria 3205, Australia

Chapman & Hall India, R. Seshadri, 32 Second Main Road, CIT East, Madras 600 035, India

First edition 1990
Second edition 1993

Typeset in 10/12 Times by Mews Photosetting, Beckenham, Kent
Printed in Great Britain at the University Press, Cambridge

ISBN 0 412 48920 1

A catalogue record for this book is available from the British Library

Library of Congress Cataloging-in-Publication data

Holmes, Oliver.
Human neurophysiology : a student text / Oliver Holmes. – 2nd ed.
p. cm.
Includes bibliographical references and index.
ISBN 0–412–48920–1 (alk. paper)
1. Neurophysiology. I. Title.
[DNLM: 1. Nervous System–physiology. WL 102 H752h 1993]
QP361.H65 1993
612.8–dc20
DNLM/DLC
for Library of Congress 93-25261
CIP

∞ Printed on permanent acid-free text paper, manufactured in accordance with the proposed ANSI/NISO Z 39.48-199X and ANSI Z 39.48-1984

Human Neurophysiology

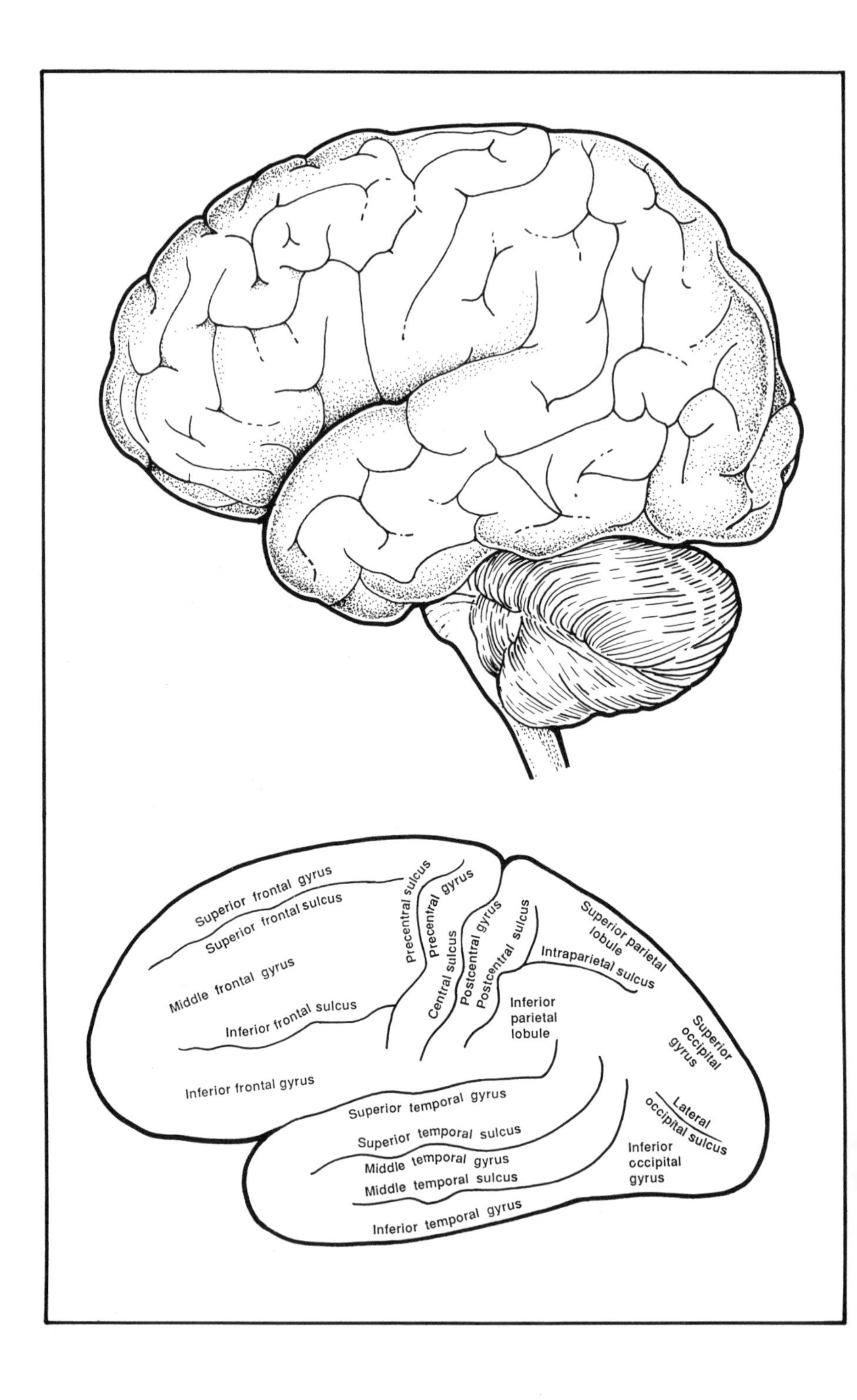
Superior frontal gyrus
Superior frontal sulcus
Middle frontal gyrus
Inferior frontal sulcus
Inferior frontal gyrus
Precentral sulcus
Precentral gyrus
Central sulcus
Postcentral gyrus
Postcentral sulcus
Superior parietal lobule
Intraparietal sulcus
Inferior parietal lobule
Superior occipital gyrus
Lateral occipital sulcus
Inferior occipital gyrus
Superior temporal gyrus
Superior temporal sulcus
Middle temporal gyrus
Middle temporal sulcus
Inferior temporal gyrus

Contents

Preface to second edition xi
Preface to first edition xiii
Acknowledgements xv

1 THE NERVOUS SYSTEM, SENSATION AND MOVEMENT 1

1.1 Overview of the nervous system 1
1.2 Sensation 2
1.3 Motor activity and reflexes 9

2 NERVE 13

2.1 Introduction 13
2.2 The resting potential 13
2.3 Action potential 20
2.4 Compound action potential 27
2.5 Myelination 32
2.6 Appendix A: Cell volume and the sodium pump 35
2.7 Appendix B: Determinants of the membrane potential 36

3 THE PHYSIOLOGY OF RECEPTORS: CUTANEOUS SENSATION 39

3.1 Receptors 39
3.2 Cutaneous sensation 48
Further reading 52

4 THE NEURONE, NEUROMUSCULAR TRANSMISSION AND SYNAPTIC TRANSMISSION 53

4.1 Morphology of nerve cells 53
4.2 Junctional transmission 55
4.3 Synaptic transmission (neuro-neuronal transmission) 60
4.4 The two types of neuronal membrane 67

4.5 Appendix A: Neuromuscular transmission as an exceptional synaptic mechanism 69
4.6 Appendix B: Closing of membrane channels as a synaptic mechanism 70
Further reading 71

5 NEURONAL INTERACTIONS 73

5.1 Synaptic integration 73
5.2 The orthodromic inhibitory pathway 75
5.3 The recurrent inhibitory pathway 77
5.4 Presynaptic inhibition 81
5.5 Surround (lateral) inhibition 84
5.6 Chemical neurotransmitters in the spinal cord 90
5.7 Appendix 91
Further reading 92

6 INTERRUPTION OF PERIPHERAL NERVE AND OF THE SPINAL CORD: SPINAL PATHWAYS 93

6.1 Introduction 93
6.2 Section of a peripheral nerve 93
6.3 Lesions of the central nervous system 98
6.4 Spinal transection 100
6.5 Central nervous lesions: release of lower centres 107
6.6 Central afferent pathways 108
6.7 Descending pathways 111
6.8 Spinal hemisection (Brown-Séquard syndrome) 112
6.9 Appendix A: The reaction of the neurone to axotomy 114

7 PROPRIOCEPTION 115

7.1 Joint and tendon receptors 115
7.2 The receptors of skeletal muscle 117
7.3 Activation of γ motoneurones 123
7.4 Voluntary movement 124
7.5 Which types of proprioceptive activity can be consciously perceived? 129
7.6 The different types of spindle receptor 130
7.7 Appendix: The mammalian muscle spindle in more detail 131

8 SPINAL MECHANISMS 135

8.1 Perception and sensation 135
8.2 Spinal reflexes 141
8.3 The axon reflex 147
8.4 The sequence of recruitment of motor units 149

8.5 Spinal reflexes elicited by stimulation of receptors in skeletal muscle 150
8.6 The contribution of the spinal cord to the organization of complex movements 152
8.7 Appendix: The definition of reflex response time 155
Further reading 156

9 THE SPECIAL SENSES 157

9.1 Introduction 157
9.2 The eye 157
9.3 The ear 168
9.4 Olfaction 176
9.5 Taste 178

10 REFLEXES OF BALANCE 179

10.1 Standing and walking 179
10.2 Righting reactions 181
10.3 Neck reflexes 185
10.4 Synthesis of vestibular and neck reflexes 189
Further reading 195

11 BRAIN STEM MECHANISMS 197

11.1 Introduction 197
11.2 Brain damage 198
11.3 Brain stem reflexes 203
11.4 Decerebrate rigidity 205
11.5 The brain stem reticular formation 209
11.6 Appendix: The Glasgow coma scale 214
Further reading 215

12 THE BASAL GANGLIA AND CEREBELLUM 217

12.1 The supraspinal control of movement 217
12.2 Basal ganglia 219
12.3 The cerebellum 228
12.4 Appendix A: Nigral insufficiency and rigidity 241
12.5 Appendix B: The cerebellum and decerebrate rigidity 241
Further reading 241

13 THE CEREBRAL HEMISPHERES: GENERAL FEATURES, LOCALIZATION OF FUNCTION, THE LIMBIC SYSTEM 243

13.1 The structure of the cerebral hemispheres 243

13.2 The discovery of the motor functions of the precentral gyrus 250
13.3 Decussation in the central nervous system 253
13.4 Limbic system 254
13.5 The hippocampus and memory 257
13.6 The physiological assessment of neural function 259
Further reading 262

14 SEEING AND HEARING 263

14.1 Vision and eye movements 263
14.2 The control of eye movements 263
14.3 The visual pathway 264
14.4 Auditory cortex 272

15 THE CEREBRAL HEMISPHERES: MOTOR FUNCTIONS 273

15.1 The motor functions of the cerebral cortex 273
15.2 Reflex time and reaction time 276
15.3 'Long loop' or transcortical reflexes and voluntary movement 277
15.4 Status of the precentral motor strip 280
15.5 The neural control of skilled human movement 281
15.6 Appendix 282
Further reading 283

16 THE CEREBRAL HEMISPHERES: ASSOCIATION AREAS, HEMISPHERIC ASYMMETRIES, SPEECH, SPLIT BRAIN, MEMORY 285

16.1 Association areas 285
16.2 Hemispheric asymmetry 287
16.3 Speech 288
16.4 The split brain in man 293
16.5 Conditioned reflexes, memory and learning 296
Further reading 299

17 NEUROTRANSMITTER CHEMICALS IN THE BRAIN 301

17.1 Neurotransmitters in general 301
17.2 The postsynaptic receptors for amino acid neurotransmitters 307
17.3 Long-term potentiation 314
17.4 Excitotoxins 315
Further reading 316

18 PAIN **317**

18.1 The sensation of pain 317
18.2 Nociceptive mechanisms 321
18.3 Pathological pains 327
18.4 Appendix A: Primary and secondary hyperalgesia 331
18.5 Appendix B: The gate theory of pain 331
Further reading 331

19 AUTONOMIC NERVOUS SYSTEM **333**

19.1 Peripheral autonomic nervous system 333
19.2 Central autonomic control 341
19.3 Feedback loops 343
19.4 Appendix: The anatomy of the cranial parasympathetic outflow 344

20 THE CEREBRAL ENVIRONMENT **345**

20.1 The constancy of the cerebral environment 345
20.2 Cerebral blood flow 348
20.3 Intracranial pressure 351
20.4 The measurement of cerebral blood flow and metabolism: brain 'imaging' 352
20.5 Pharmacological protection against cerebral ischaemia 356

21 NEUROANATOMY **357**

21.1 The brain and its coverings 357
21.2 The spinal cord 359
Further reading 367

22 SELF-TEST QUESTIONS AND ANSWERS **369**

Glossary 437
References 451
Index 459

Preface to second edition

For this second edition, sections of the book have been extensively revised. These are: cutaneous sensation, spinal mechanisms, the central processing of visual information, hemispheric asymmetries, and neuroanatomy. There is a new chapter on central neurotransmitters. There are several new diagrams, and many of those from the first edition have been redrawn.

In writing the second edition, I have had invaluable advice and criticism from many sources. Using the book with students has afforded me the opportunity of improving explanations that the students found to be inadequate. Comments from colleagues and reviewers have prompted many improvements. In particular I would thank Dr T. D. M. Roberts, who has made many useful suggestions for improvement, and Dr A. B. Chatt, who provided me with review material for the section on signal processing of cutaneous afferent information. My wife patiently supported me during the writing of the book and forgave my neglect of family duties; my sons Alexander and Michael provided valuable secretarial assistance. Finally, I wish to express warm appreciation for the help and encouragement given by Mr P. Remes and his colleagues at Chapman & Hall.

Oliver Holmes
December 1992

Preface to first edition

This textbook of *Human Neurophysiology* is intended for preclinical medical students and other university students studying physiology courses of equivalent standard. In writing the book, I have tried to produce a logical and sequential text starting with the membrane physiology of neurones, proceeding through neuronal interactions to the functioning of the central nervous system as a whole. There is a coverage of sensory physiology and of neuroanatomy sufficient to allow the reader to understand the neurophysiology. I have included examples of the experimental observations on which the subject is based and have included key references, so that the inquisitive can read more widely. There are allusions to clinical applications and also anecdotal 'asides' relating the subject-matter to the everyday experience of the reader, because I have found these to be of interest to students. The book has been kept to a size that can reasonably be read by a junior undergraduate student.

The text is fully illustrated with original line diagrams. There is a substantial section of self-assessment material including structured exercises and multiple choice questions enabling readers to monitor their own progress. A glossary of terms is provided to help readers to understand the jargon of the subject. An informal style of writing has been adopted with the hope of conveying the enthusiasm that the lecturer can achieve in the lecture theatre. This book will have succeeded if it fosters the interest and involvement of the reader with this the most fascinating branch of human physiology.

Oliver Holmes
January 1990.

Acknowledgements

For the figures which have been derived or directly reproduced from books and original articles, the author wishes to thank all the individuals and publishers who have kindly given permission for use of this material. The figures were drawn by Mr Ian Ramsden, of the Department of Medical Illustration in the University of Glasgow.

Figure 1.3d is an original record of the impulse activity in a single afferent nerve fibre from a pulmonary stretch receptor in the anaesthetized rabbit. The response to pulmonary inflation is shown; Figure 2.4c is redrawn from Figure 31 in the book *The conduction of the nervous impulse* (1965) by A.L. Hodgkin, Liverpool University Press; Figure 2.6 is based on oscilloscope records from a class experiment in the Physiology Department, University of Glasgow; Figure 2.9a is derived from Figure 6 in the book *Electrical Signs of Nervous Activity* by J. Erlanger and H.S. Gasser (1937) University of Pennsylvania Press; Figure 2.9b is derived from Chart 8, page 347 of the article 'Numbers and contraction-values of individual motor-units examined in some muscles of the limb, by J.C. Eccles and C.S. Sherrington, *Proceedings of the Royal Society B*, 1930, **106**, 327–356; Figure 2.12 is derived from Figure 5 of the article 'A theory of the effects of fibre size in medullated nerve' by W.A.H. Rushton, *Journal of Physiology*, 1951, **115**, 101–122, by Permission from Dr J.J.B. Jack, Hon. Treasurer of the Physiological Society, Department of Physiology, Parks Road, Oxford, OX1 3PT; Figure 3.6 is derived from *The senses* edited by H.B. Barlow and J.D. Mollon (1982), Cambridge University Press; Figures 7.7b and 7.9 are original oscilloscope records of M and H responses; Figure 9.11 is derived from Figure 1 in the article 'Studies on the morphology of the sensory regions of the vestibular apparatus' by H.H. Lindeman, *Ergebn. Anatomie Entwickll.-Gesch.*, 1969, **42**, 4–113; Figure 10.2 is derived from Figures 4.4 and 4.18 in *The basal ganglia and posture* by J. Purdon Martin (1967) Churchill Livingstone; Figure 10.5 is redrawn from Figure 60 in the book *Korperstellung* by R. Magnus (1924), the 6th volume in the series, *Monographien aus dem Gesamtgebiet der Physiologie der Pflanzen und der Tiere*, Berlin, Verlag von Julius Springer, Springer-Verlag GmbH and Co; Figure 13.2 is derived from Figures 344 and 345 in Volume II of the book *Histologie du Systeme Nerveux de l'Homme et des Vertebres* (1956) S. Ramon Y Cajal, the Consejo Superior de Investigaciones Cientificas, Instituto Ramon Y Cajal, Madrid; Figure 13.3 is redrawn from Figure 22 in the book *The cerebral cortex of man: A clinical study of localisation of function*, by W. Penfield and T. Rasmussen (1950), London, Macmillan Ltd; Figure 13.6 is from original records; Figure 14.2b is derived from the article 'The visual cortex of the brain', pp. 148–56 in '*Perception: Mechanisms and Models*'. Readings from *Scientific American*; Figure 16.2 is derived from Figure 4 of the article 'Perception of bilateral chimeric figures following hemispheric deconnexion' by J.L. Levy, C. Trevarthen and R.W. Sperry (1972) *Brain*, **95**, 61–78. Figure 21.1a is derived from Figure 822 in the book '*Gray's Anatomy*, 25th edn. (1944), edited by T.B. Johnston and J. Whillis, published by Longmans, Green and Co., London, New York and Toronto, with permission from Churchill Livingston publishing Company Ltd; Figure 17.2 is derived from Figure 49 in the book '*Conybeare's Textbook of Medicine*' 16th edn. (1975), edited by W.N. Mann, published by Churchill Livingston Edinburgh.

The nervous system, sensation and movement

1

1.1 OVERVIEW OF THE NERVOUS SYSTEM

The nervous system is the central computing system of the body. At a mechanistic level, it collects information about the world around us and about the body itself, computes an appropriate response and organizes the execution of that response. We have some understanding of how it operates in performing this function. It is also the seat of such phenomena as intellect and emotion. Here our understanding of the relationship between neuronal activity and behavioural manifestation is almost totally lacking.

It is customary to divide the nervous system into central and peripheral parts. The **central nervous system** consists of the brain and the spinal cord. The **peripheral nervous system** includes the nerves of the body which convey information between the central nervous system and peripheral structures such as muscle and sensory receptors, to be described shortly.

The relatively large size and complexity of certain parts of the brain are the features that most dramatically divide the human from the most complex non-human species. In particular, the association areas of the cerebral cortex, described in Chapter 16, show a remarkable expansion in the human brain compared with the brains of non-human species.

The neurone

The nervous system consists of the nerve cells or **neurones**, surrounded by a framework of supporting cells, as shown in Figure 1.1. Neurones are characterized by long extensions from the cell body. These are called **dendrites** and **axons**, to be described later. The function of these extensions is to collect and disseminate information. The cell body (also called the **soma**) contains the nucleus, which is the site of the hereditary material of the cell. The cell body is the metabolic factory of the whole neurone. The cytoplasm

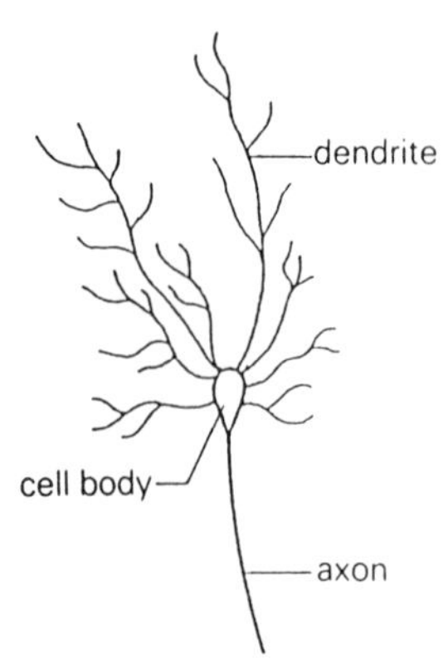

Figure 1.1 Diagram of a neurone.

around the nucleus contains dense rough endoplasmic reticulum, responsible for protein synthesis. It is here that structural proteins for the whole of the cell, including the axon and dendrites, are synthesized. Since the cell body must provide structural proteins for these extensive processes as well as for itself, the rough endoplasmic reticulum is much more conspicuous in the neuronal cell body than in most non-neural cells. Specialized structures needed for neurotransmission are also synthesized in the cell body; we shall be considering these later.

1.2 SENSATION

If it is to be able to react to changes in the world around, the central nervous system needs to be informed about them. The changes are sensed by specialized structures called **sensory receptors**. The body contains millions of receptors. Each senses changes in its own environment and encodes the information to provide information for the central nervous system. This involves changing the type of energy to which the receptor is sensitive (the **stimulus**) into electrical energy, which is the currency of the nervous system. The information is then transmitted to the central nervous system along sensory nerve fibres. Sensory nerve fibres are also called **afferent** nerve fibres, because they conduct impulses towards the central nervous system.

For any sensor, animate or inanimate, stimulation of the receptor can only occur if there is transfer of energy between the stimulus and the receptor. This energy may come from the environment, for instance light energy from a light bulb impinging on a photosensitive receptor in the retina. Alternatively, the energy may be derived from the organism itself; when you put your hand out to feel an object, the energy which deforms the touch receptors in the skin of the hand is derived from the muscular effort of the exploring limb. Since all receptors generate electrical signals, the nervous system cannot determine the nature of the stimulus energy from this electrical signal

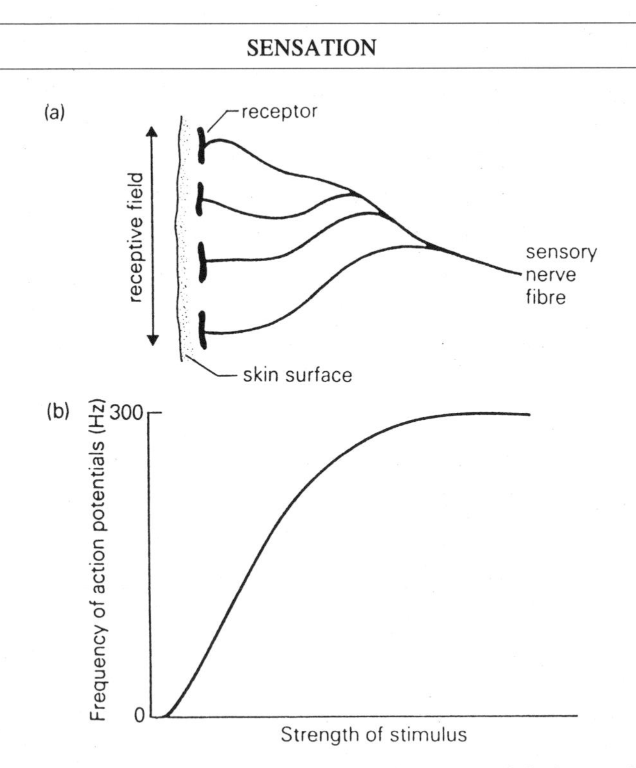

Figure 1.2 (a) Diagram of a sensory unit. (b) The frequency of discharge of a sensory nerve fibre as a function of the strength of the stimulus.

itself; the central nervous system infers this information from the channel along which the information travels. The different types of receptor are connected to different areas of the brain.

The electrical disturbance which the nervous system uses to transmit information along nerve fibres over long distances (for instance from a toe to the spinal cord) and at high speed (up to 100 m/s) is the **action potential**, alternatively called the nerve impulse. The action potential is **all-or-none**. That is to say, every action potential in a nerve fibre is the same amplitude, and the action potential does not change as it travels along the nerve fibre. The fibre cannot transmit information about the intensity of stimulation by altering the amplitude of action potential. Instead, the intensity is signalled by the frequency of action potentials; a high frequency of impulses signals an intense stimulus.

Specificity

Any one receptor is specifically sensitive to one type of stimulus. For instance, the touch receptors in the skin are very sensitive to touch (i.e. mechanical deformation), whereas temperature receptors sense the temperature of the skin. The **modality** is the name that we use to describe the type of stimulus to which a receptor is sensitive.

In most cases, the mechanism of this specificity at the level of membrane physiology is not understood. In a few instances, we have some understanding of the mechanism. For instance, retinal receptors in the eye contain pigment molecules which, when they absorb light, change their configuration, thus influencing the electrical properties of the membrane of the cell in which they lie.

The human has receptors sensitive to a wide range of energies but not to energies such as infra-red, ultraviolet, X-rays, radiowaves and so forth. Our perception of the world is limited to what our receptors and neurones can tell us. Other animals detect stimuli to which we are insensitive. Bees have receptors that can detect the plane of polarized light and may use this information to determine the direction of the sun on cloudy days. Electric eels send out pulses of electricity, and by detecting the disturbances in the electromagnetic field created can detect the presence of obstacles. Birds can detect the earth's magnetic field and use this to guide them during migration. Pit vipers, so called because of pits in their heads housing receptors sensitive to infra-red, use these detectors to sense the direction of warm-blooded prey.

The sensory unit

The receptor is connected to the peripheral end of a sensory nerve fibre. In most cases, several such nerve fibres join together, as shown in Figure 1.2a, to form a single nerve fibre. This then projects to the spinal cord. All the receptors that transmit information along such a common line are a functional unit, which we call the **sensory unit**. The nerve fibre thus gathers information from the area over which these receptors are distributed. For instance, one nerve fibre is responsible for pressure sensation for a patch of the skin. This patch is called the **receptive field** of the sensory unit.

Strength of stimulus

A stimulus that is too weak to initiate impulses in the sensory nerve fibres is said to be **subthreshold**. At stimulus strengths above threshold, there is a range over which increases in stimulus strength are signalled by increases in the frequency of nerve impulses. However there is a limit above which the frequency of nerve firing cannot go. This relationship is shown in Figure 1.2b.

As the stimulus becomes stronger, in addition to causing a rise in firing rate of one sensory nerve fibre, the effect of the stimulus may spread to other sensory units. This effect is called **recruitment** of nerve fibres and is another mechanism for conveying the intensity of a stimulus to the central nervous system.

Adaptation

For different types of receptor, a stimulus that is applied and maintained results in different patterns of impulses in the afferent nerve fibres. In some cases, there is an initial burst of impulses after which the discharge rate falls to a much lower level or ceases altogether (Figure 1.3a) This is the phenomenon of **adaptation**, a decline in intensity of response during sustained stimulation at constant intensity. An extreme instance is when the stimulus elicits one isolated impulse in the afferent nerve fibre at the onset of stimulation. The receptor is said to be 'rapidly adapting'. In this case, the frequency of impulses in the sensory nerve fibre is a poor reproduction of the time course of the stimulus and the central nervous system receives a distorted version of the outside world.

Other types of receptor show no adaptation (Figure 1.3b); the pattern of impulses in their sensory nerve fibres is an accurate reflection of the time-course of stimulation of the receptor. Yet other receptors show an intermediate pattern of behaviour, with frequent impulses in the sensory fibre at the onset of a maintained stimulus and a subsequent decline to a lower resting rate (Figure 1.3c). Whereas Figures 1.3a–c are drawings, Figure 1.3d is from a real experiment in which impulses in a fine sensory nerve containing a single active nerve fibre were recorded. In electrophysiological experiments there is always unwanted background interference and this makes the baseline, before the

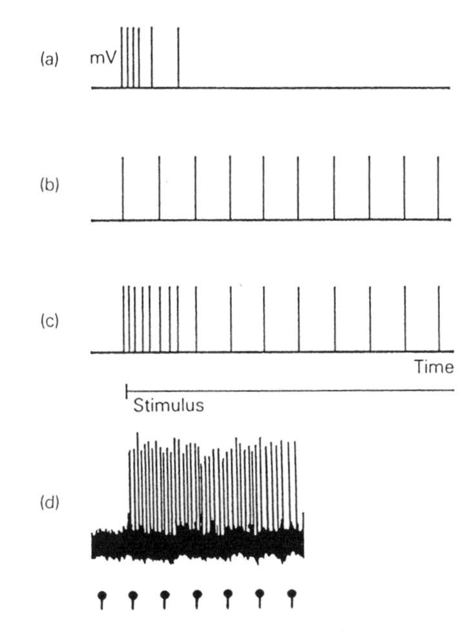

Figure 1.3 Sketches of impulses in sensory nerve fibres from (a) a rapidly-adapting receptor (b) a non-adapting receptor, and (c) a slowly-adapting receptor. (d) Original record of the afferent discharge from a slowly-adapting receptor. Time: 10Hz.

onset of the stimulus, thick and wavy. Since action potentials arise from this 'noisy' background, they do not all reach exactly the same peak voltage. Rapidly-adapting receptors signal changes in the stimulus and not its absolute magnitude. The temperature receptors in the skin are of this type and you can demonstrate this by performing a simple experiment on yourself.

Experiment

Take three bowls of water. Fill them with cold, lukewarm and hot water respectively. Put one hand in the cold water and the other in the hot water. Wait for a minute or two and then put both hands in the bowl of lukewarm water. The lukewarm water feels quite different to the two hands, the previously hot hand signalling it as being cold and the previously cold hand signalling it as being hot.

The sensitivity of receptors

Some receptors are extremely sensitive. For instance, the olfactory receptors in the nasal mucous membrane are so sensitive that four molecules of a stimulating substance can create the sensation of smell. The eye, if fully dark adapted, can detect a flash which is as weak as a candle at 17 miles distance; this corresponds to about four photons entering the eye.

At the other extreme, a hard blow to the eye causes one to see 'stars'. Here, photic receptors are stimulated in the absence of light falling on them; the large mechanical stimulus excites photic receptors. This is a particular case of a general feature of receptors; a large amount of a different sort of energy can excite a receptor not tuned to that particular type of energy. Even a nerve trunk can be stimulated by a strong blow, this being the origin of the sensation when one hits the 'funny bone' (ulnar nerve).

Sensitivity of different parts of the body

Different parts of the body show differing degrees of sensitivity to stimuli. The temperature sensitivity of the tongue is less than that in the skin. It is a common experience that one can drink tea which is too hot for a finger to tolerate. Some areas of the body entirely lack certain modalities of sensation. Brain tissue is insensitive to pain. Viscera are insensitive to cutting, although very sensitive to distension. In the cornea, the only modality of sensation that can be appreciated is pain. Although the cornea is very sensitive to mechanical stimuli, a person cannot distinguish between touch and pain there. Table 1.1 compares pain and touch thresholds at various sites. A stimulus subthreshold for touch at the finger is perceived as painful when applied to the cornea.

Table 1.1 Threshold for sensation (g/mm^2)

	Pain	*Touch*
Cornea	0.2	0.2
Back of hand	100	12
Fingertip	300	3

The diversity of receptors

There is a wide diversity of sensory receptors and of the information that they convey. Some receptors are sensitive to mechanical deformation. Others are sensitive to the chemical composition of the fluid which bathes them. Yet others are sensitive to temperature. The siting of a receptor is also important. A mechanoreceptor lying in a tendon will signal the tension in the muscle to which the tendon is attached. A similar receptor lying in the wall of an artery will signal the hydrostatic pressure of blood inside the artery.

To give some idea of the diversity of sensory receptors, a classification may be useful; here we classify receptors into five groups.

(a) **Cutaneous receptors**, ideally situated to detect changes in the external environment such as hot and cold.
(b) **Proprioceptors** (literally, receptors for feeling oneself) are receptors that give information about the body itself. They are mechanoreceptors that lie in muscles, tendons, ligaments, interosseous membranes, and joint capsules. Their activity is linked with the state of contraction of muscles and the position of joints. They therefore signal the relative positions of the different parts of the body.
(c) **Visceral mechanoreceptors**. 'Viscus' literally means 'hollow'. Anatomically it is the term applied to hollow organs such as the stomach. This group of receptors senses stretching of the walls of the gut, the bladder or the lungs. There are also receptors placed at strategic points in the cardiovascular system to signal blood pressure. They are called **baroreceptors**; they are found in the atria of the heart (the low pressure baroreceptors), and in the carotid sinus and aortic arch (the high pressure baroreceptors).
(d) **Chemoreceptors** are sensitive to the composition of fluid that bathes them. Examples are the taste-buds in the tongue (which sense the chemical composition of the saliva in which they are bathed), the chemoreceptors in the aortic bodies (which sense the partial pressure of oxygen and the hydrogen ion concentration of the blood

that perfuses them), and the osmoreceptors of the hypothalamus in the brain (which sense the osmotic strength of the blood that perfuses them).

(e) **Special senses**, which are housed in the head. These are: the nose for olfactory sense, the taste buds of the tongue for taste, the eyes for vision, the ears for hearing, and the vestibular apparatus for balance and orientation sense.

What do receptors sense?

A thermometer indicates the temperature of its mercury. It gives an indication of the temperature of its surroundings if it comes into thermal equilibrium with those surroundings. A thermometer in interstellar space indicates its own temperature and gives no information about its environment, since this is empty and therefore has no temperature. A cold receptor is stimulated when the receptor itself is cooled. A receptor of this type in the skin signals the temperature of the skin because the receptor is at the same temperature as the skin in which it lies. A deformation receptor is stimulated when it is deformed. Its anatomical siting determines whether it signals information about tension in a muscle, pressure inside a viscus, etc. A light receptor responds to photons of light falling on it.

These obvious statements illustrate the general fact that a receptor generates a signal when energy to which it is sensitive falls on it. Information about the outside world, as in vision, depends on the organism collecting the appropriate energy and beaming it onto receptors in a way which embodies the information sought. In the case of the eye, the eyeball has evolved with the appropriate physical properties to provide an image on the retina of the outside world, and it is on this image that the organism operates. This raises the interesting philosophical point that we can only be aware of what our sense organs tell us. An example is afforded by Figure 5.9, in which our perception is of a slight darkening where horizontal and vertical white lines cross each other. We can of course use a light meter roving across the figure to convince ourselves that the intensity of light does not change at the crossings. But we must still use our senses to make the measurements. It is possible that our senses incorporate a consistent bias such that our whole image of the outside world is distorted, and of this we can never be aware. This, however, makes the assumption that things may exist of which we cannot be unaware. Certain philosophers hold the view that it is fundamentally wrong to compare our image of the world with the world outside because the world outside only exists in so far as we are aware of it.

1.3 MOTOR ACTIVITY AND REFLEXES

On the basis of the information that it receives, the central nervous system elaborates an appropriate response. The only way in which the central nervous system can express itself, whether in the loftiest flight of the philosopher or the deepest emotions of the composer, is in terms of contractions of muscle and secretion of glands. The co-ordination of this activity is therefore of paramount importance in central nervous system functioning.

Introduction to reflexes

Information entering the central nervous system from the receptors has various effects. One is that the ingoing signal may elicit reflex muscular movement. The name 'reflex' comes from the fact that the incoming signal is reflected by the central nervous system out again along the nerve fibres supplying muscles, the so-called motor nerve fibres. Reflex activity involves central mechanisms in the cord which recognize the input and formulate the command message.

Innervation of skeletal muscle

Movement is produced by contraction of muscles commanded by motor nerve fibres from the brain or spinal cord to the muscles. The cell bodies of fibres that innervate skeletal muscle are called **motoneurones** (Eccles 1968, p 2); they lie in the anterior horn of the grey matter of the cord or, in the case of the cranial nerves, in an analogous site in the brain. The axon from each motoneurone projects along a peripheral nerve; here it is called a motor nerve fibre, or an efferent fibre (efferent because it is conducting impulses away from the central nervous system). This nerve fibre projects to a muscle; within the muscle it subdivides and supplies a group of muscle fibres.

Each muscle fibre is innervated by a single motor nerve axon. The motoneurone and the family of muscle fibres that it innervates are known as the **motor unit** (Figure 1.4), a concept similar to the sensory unit described above. In healthy people, one impulse in the motor nerve fibre causes all the muscle fibres which it innervates to contract. The motor unit is the smallest unit of muscular contraction that the central nervous system can control. The muscle fibre is simply a subunit. Its size is limited by factors such as the need for diffusion of oxygen from the interstitial fluid to the centre of the fibre, and diffusion of the products of metabolism in the opposite direction.

A train of impulses passing along a motor nerve cause a state of prolonged contraction (a so-called **tetanic contraction**) in the muscle fibres

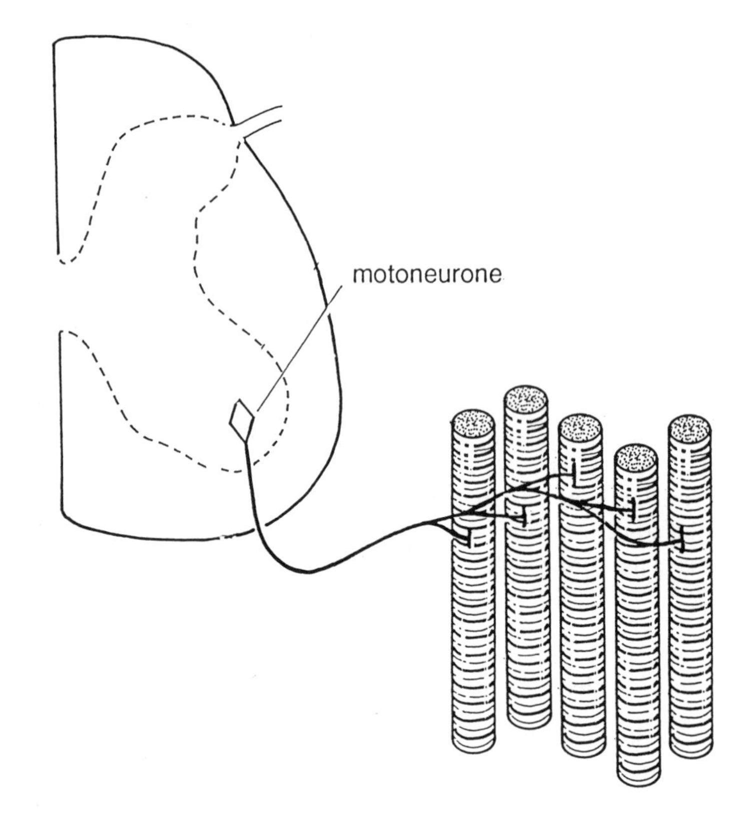

Figure 1.4 A motor unit.

which it innervates. The tension developed is several-fold greater than that of the twitch response to an isolated action potential in the motor nerve.

As with sensory nerve fibres, the all-or-none nature of the action potential precludes amplitude modulation of the action potential as a mechanism for varying muscular effort. The nervous system commands a graded increase of muscular effort by two principal mechanisms, the frequency of discharge of each motoneurone, and recruitment of other motor units to add to the effective tension.

Since the central nervous system controls the whole of a motor unit through a single motor nerve fibre, the size of the motor units limits the degree of control of muscular effort. The number of muscle fibres in a motor unit is relatively large (up to 400) for muscles performing strong coarse movements; an example is the gastrocnemius muscle (the calf muscle) which must sometimes support much of the weight of the body. The number is low in muscles whose contraction must be finely controlled; the extrinsic ocular muscles which move the eyeball are of this type and, for them, a motor unit is typically four muscle fibres.

The motoneurones supplying a single muscle lie together in a column in the spinal cord. This group of neurones is called the **motoneurone pool** for that muscle. A co-ordinated movement involves the excitation of the motoneurone pools of several muscles in the correct sequence. The central nervous system must also recruit, from within each motoneurone pool, the number of motoneurones appropriate for the movement. Figure 1.5 shows the anatomical layout. The motoneurones of different motoneurone pools intermingle to some extent and Figure 1.5 is an idealization which embodies the physiological concept. The motoneurones are particularly numerous at the cord levels corresponding to the limbs; the cord shows swellings at these two levels.

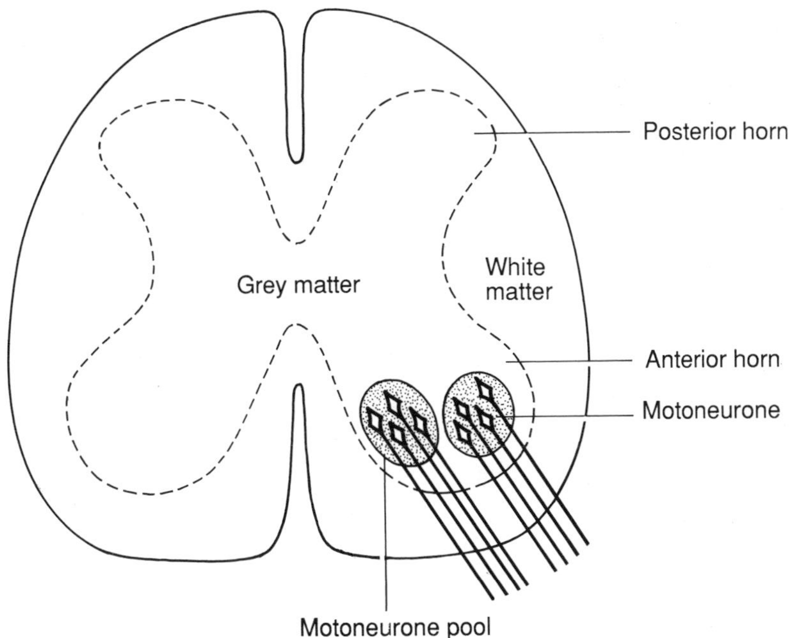

Figure 1.5 Transverse section of the spinal cord. Grey matter (grey because it contains many nerve cell bodies) and white matter (white because it contains only nerve fibre tracts) are shown. In the anterior horn of the grey matter are shown motoneurones belonging to two motoneurone pools.

The production of movement

A movement is produced by a complex programme of contraction of certain muscles and relaxation of others. If the movement is repetitive, such as walking or scratching, the central nervous system must programme a complicated temporal sequence of contractions and relaxations. Movement is produced by muscle shortening. The muscle most directly concerned with a particular movement is called the **prime mover**. Other muscles that assist in producing the same movement are called **agonists**. The joint at which movement is occurring is stabilized and the articular surfaces are prevented from falling apart by continuous activity in all the muscles acting around the joint. This continuous activity in postural muscles is muscle tone or **tonus**. In skeletal muscles (i.e. the voluntary muscles such as those that we use to produce movements of a limb) tonus is due to nerve impulses at low frequency in the motor nerve fibres innervating the muscles. As the articular surfaces move on each other, the distribution of tonus in these supporting muscles is altered. This is part of the **postural** component of movement. For large movements, the other postural components involve changes in whole body posture to prevent the body from falling over. If the tonus in muscles that oppose the movement (the **antagonists**) remain unaltered the system would be inefficient, with the prime mover working against an unnecessary load. This is avoided by spinal cord circuitry which arranges that, when a prime mover contracts, its antagonists relax. For instance, when the forearm is flexed at the elbow joint, there is a reflex relaxation (i.e. a decrease in tonus) in muscles whose contraction would cause the elbow joint to extend. The neural circuitry for this reciprocal relationship between muscles with opposite effects is described in Chapter 5.

Final common path

The motoneurone is a focus point onto which impinge excitatory and inhibitory influences arriving by various routes. All neural influences, whether reflex in origin or from higher centres, converge onto the motoneurones whose discharge causes contraction of muscles. The motoneurone acts as a funnel through which all other influences must exert their effects. The motoneurone stands at the ultimate end of all the integrative niceties of the central nervous system. Its paramount position is recognized by its designation as the **final common path**.

Nerve

2

2.1 INTRODUCTION

The nerve fibre (or nerve axon) has evolved for carrying information very quickly and over long distances from one part of the body to another. Before the evolution of this mechanism, receptors, primitive neuronal integrating circuits, and effectors were constrained to lie close to each other. This limited the overall size of the animal. The acquisition of a mechanism for long distance and rapid communication liberated the animal from this constraint. The central nervous system then developed as the integrating system, far away from many of the sensors on which it relies for information and far away from the muscles which obey its commands. The mechanism for propagation of information along nerve fibres is the nerve impulse. This mechanism depends on the existence of a resting potential across the nerve membrane.

2.2 THE RESTING POTENTIAL

Most of the cells of the body, excitable or not, have a potential difference across their cell or plasma membranes, the inside of the cell being just under 0.1 V negative to the outside. This is called the **membrane potential**. The way in which it arises is described next. There are two factors to be considered, diffusion and electrical forces.

Concentration gradient and diffusion

If dye is dropped into a glass of water, the dye gradually spreads out into the surrounding water, even if the water is not stirred. This is because all molecules are in a constant state of random movement due to their thermal energy; the process is called **diffusion**. The random nature of the movement implies that movement is equally likely in all directions. Because there are more molecules in regions where the concentration is high than where it is low, there is a net movement from regions of high to regions of low concentration. When the dye has spread uniformly throughout the solution, although the dye molecules move individually just as before, there is no further net movement of dye, because there are no concentration gradients. In summary, a **concentration gradient** (or **chemical gradient**, as it is also called) causes net movement from regions of high to regions of low concentration.

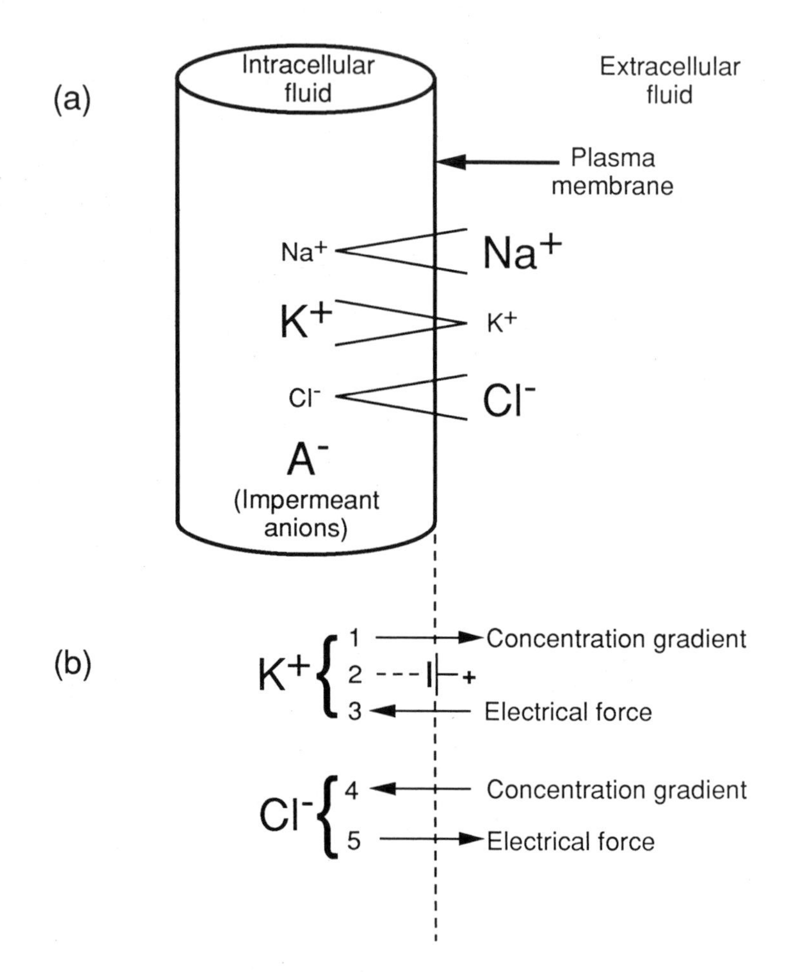

Figure 2.1 The origin of the resting potential. (a) A segment of an axon, with the plasma membrane separating intracellular and extracellular fluid. The relative concentrations of the principal ions in intracellular and extracellular fluids are indicated, large and small letters indicating high and low concentrations respectively. For each of the inorganic ions, the direction of the concentration gradient is indicated by < or >. (b) Physicochemical factors influencing the movement of potassium and chloride ions across the membrane. Reference is made in the text to the numbers.

Similar considerations apply to the movement of chemicals across the cell membrane. For chemicals that can cross the membrane, diffusion results in movement of particles across the membrane, just as in a bulk solution. Each chemical diffuses independently of the others and this allows each to be considered separately. For any chemical that permeates the membrane and that is present in the solutions on either side of the membrane, this chemical will diffuse in both directions across the membrane. If the total amount moving in one direction differs from that moving in the other direction, we say that there is a **net** movement of that chemical across the membrane. We now apply these ideas to the nerve axon.

The intracellular and extracellular fluids, which are separated by the cell membrane, are of different composition. Figure 2.1a indicates the relative concentrations of the principal ions inside and outside the nerve fibre. In the intracellular fluid, there are high concentrations of potassium ions and of large organic ions such as amino acids and proteins. The organic ions are negatively charged and, being large, they cannot cross the cell membrane. In the extracellular fluid, there are high concentrations of sodium ions and chloride ions.

There is an important physicochemical condition that applies to any aqueous solution. In any solution, the total number of positive charges always equals the total number of negative charges, i.e. the net charge is zero. This is the principle of **electrical neutrality**. It allows a rough check on the concentrations given above. In the intracellular fluid, the high concentration of positively charged potassium ions is balanced largely by the high concentration of negatively charged organic ions. In the extracellular fluid, sodium ions are principally balanced by chloride ions.

Across the membrane there is a concentration gradient for each of the four solutes. If there are channels in the membrane allowing a solute to pass, the solute will move to and fro across the membrane due to diffusion.

The origin of the resting potential

The solutes are considered in turn, starting with organic anions. Because of their large size, these cannot cross the membrane and this explains why organic anions do not flow out. The sodium ions are small; despite this, they do not cross the membrane with ease and those that do enter are pumped out again, as described in a later section. The result is that the sodium concentration gradient across the membrane is maintained.

Compared with its low permeability to sodium, the membrane is relatively permeable to chloride ions and even more permeable to potassium ions. The resting potential depends primarily on these potassium ions. Since the potassium concentration is much greater in intracellular than in extracellular fluid, these ions tend to move out of the axoplasm (Figure 2.1b, step 1). Each potassium ion carries with it its positive charge. Net outward movement leaves the inside depleted of positive charge (i.e. negative) whilst increasing the outside positive charge. A membrane potential develops, the inside being negative with respect to the outside (Figure 2.1b, step 2). This in turn exerts an electrical force on the potassium ions. Since positive repels positive, the force on the positively charged potassium ions is directed inwards (Figure 2.1b, step 3). The electrical force counteracts the effect of diffusion and there is no net movement of potassium ions. This explains why the concentration gradient across the membrane for potassium is not dissipated in spite of the high membrane permeabilty to potassium ions.

A similar situation pertains for chloride ions. These ions also cross the membrane with relative ease. The concentration gradient is inward, favouring net movement into the cell (Figure 2.1b, step 4). The electrical force is outward, since chloride is negatively charged (Figure 2.1b, step 5). For chloride, as for potassium, the flows produced by concentration and electrical effects are in opposite directions and the concentration gradient across the membrane is maintained despite a high membrane permeability.

In summary, the different intracellular and extracellular concentrations of ions, together with the differing membrane permeabilities to these ions, results in a resting potential across the membrane. The membrane functions as a battery.

Sodium ions

For these ions, both the concentration gradient and the electrical force are directed from extracellular towards intracellular fluid, but there is no net inward flux of sodium. Does this mean that the membrane is absolutely impermeable to sodium? Or could some other mechanism lead to similar end results? To answer these questions, an analogy may help us; it consists of a flat bottomed boat, floating in water but with some water in the bilges, as in Figure 2.2.

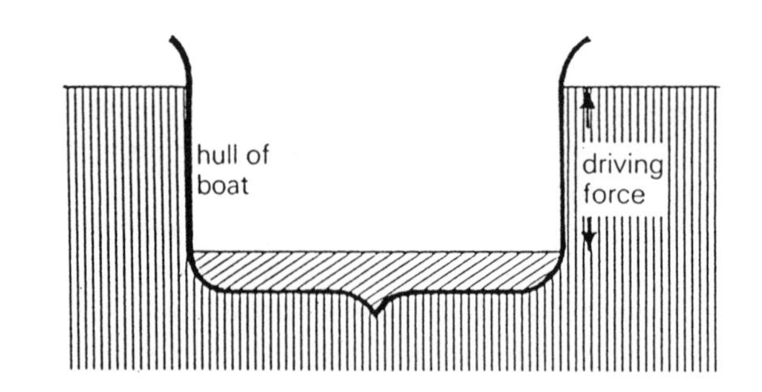

Figure 2.2 Analogy of the electrochemical gradient for sodium ions.

Inside the boat, the level of water is low; outside it is high. This difference in level represents the force tending to drive the sodium ions into the axoplasm. The fact that the height outside is above that inside indicates that the gradient is inwards. So long as the hull has no leaks that allow more water to enter, the boat floats. It will float even if the hull does have holes in it but, in this case, the fluid entering has to be baled out. Such a baling operation will keep the fluid level inside down to an acceptable level. Baling requires energy, because fluid has to be lifted from the bottom of the boat up to the level of the fluid outside. So the boat can be kept floating, at the expense of continued energy expenditure. The cell membrane is leaky for sodium and so sodium enters at a low rate. This sodium entry is counterbalanced by the so-called **sodium pump** which is continually pumping sodium out. Since the pump is working against an electrochemical gradient, it must use energy derived from metabolism.

How would we tell whether the boat's hull were water-tight with no fluid entering or whether it were leaky, with a pump keeping the fluid level low in the bilges? One way would be to put some dye into the outside fluid, and see if it appeared in the inside fluid. The analogous procedure for cells is to add to the extracellular fluid some radioactively labelled sodium and, after a delay, collect intracellular fluid. Analysis of this fluid in a Geiger counter reveals that radioactive sodium does penetrate the cell membrane. Another test is to poison metabolic processes. There is then a net entry of sodium and the sodium concentration gradient runs down. The cells swell and burst. The mechanism is described in the appendix to this chapter.

This dependency of the survival of cells on metabolic activity was recognized during World War II, when blood transfusion was used widely for the first time to save soldiers and civilians who had become exsanguinated as a result of injury. Blood stored with an anticoagulant has a relatively short life because the red cells swell and burst. It was then discovered that the addition of glucose solution considerably prolonged the survival of blood. The glucose provides a source of energy and enables the cell to pump out sodium that leaks in. This is an example of how an understanding of physiological processes can be of practical value in medicine.

Sick cells swell

If the metabolic pump is impaired or poisoned, the inward flux of sodium ions is not counterbalanced. The presence of extra osmotically active particles in the cytoplasm results in swelling, and eventually bursting, of cells. The appendix of this chapter gives a fuller account for the interested reader.

The Nernst equation and equilibrium potential

We have seen that ions are subject to electrical as well as chemical forces. If a membrane is specifically permeable to only one ion in the solutions that it separates, the net flow of that ion from high concentration to low is quickly halted because of the build-up of a voltage gradient across the membrane. The electrical force exactly balances the chemical force. Although there is no net flux, to-and-fro movement continues indefinitely. The membrane potential in this situation is called the **equilibrium potential** for that ion. The relation between the equilibrium potential and the concentrations is given by the Nernst equation.

$$\text{mV} = \frac{61}{\text{n}} \log \frac{\text{extracellular concentration}}{\text{intracellular concentration}}$$

The membrane potential in millivolts (mV in the equation) is expressed as the intracellular voltage with respect to the extracellular voltage. n is the charge of the ion, +1 for sodium and potassium, −1 for chloride. The constant 61 applies at 37°C. The Nernst equation shows that it is the ratio of concentrations across the membrane, not the absolute concentrations, which determines the equilibrium potential. For instance, for a univalent positively charged ion whose concentration in the intracellular fluid is 10 times that in the extracellular fluid, the ratio:

$$\frac{\text{extracellular concentration}}{\text{intracellular concentration}}$$

is 0.1. Since log 0.1 is −1, the equilibrium potential is −61 mV. When this calculation is repeated with the concentrations of potassium ions actually existing in the intracellular and extracellular fluid of nerve, the answer is an equilibrium potential of typically −90 mV.

There is an inconsistency in the above account. On the one hand, the principle of electrical neutrality demands that, in any solution, the numbers of positive and negative charges are equal. On the other hand, we have seen that if a membrane is permeable only to potassium, there is a net flux of potassium ions from the solution containing a high concentration to that containing a low concentration until the appropriate membrane potential builds up. The paradox is easily resolved. The transmembrane transfer of potassium necessary to charge the membrane capacitance to the resting potential is so tiny that no measurable change in concentration of potassium in the fluids on either side of the membrane can be detected. For all practical purposes the equality of numbers of positive and negative charges in a solution is true.

Electrochemical gradient

As we have seen, the equilibrium potential for a particular ion is the membrane potential that exactly balances the chemical gradient for that ion. A concept related to equilibrium potential is that of the **electrochemical gradient** for an ion; this is defined as the difference between the actual membrane potential and the equilibrium potential for that ion. It is the overall force tending to drive the ion across the membrane. In the nerve fibre at rest, this gradient is small for potassium (some 10 mV tending to drive potassium out of the axon) but large for sodium (some 135 mV tending to drive sodium inwards). The membrane at rest is much more permeable to potassium than to sodium; hence the resting potential is close to the potassium equilibrium potential whilst sodium has little effect on the membrane potential at rest.

The giant axon at rest

The pioneering work of Hodgkin and Huxley on the physicochemical basis of the nerve membrane potential was performed on invertebrate giant axons. These nerve fibres are so named because they are up to 1 mm in diameter, far larger than vertebrate nerve fibres, which seldom exceed 20 μm in diameter. It is possible to introduce a microelectrode, which is insulated except at its tip, into the axoplasm of giant axons and to use this to record the voltage across the membrane. It is also possible to extrude axoplasm from giant axons by means of a minute version of a garden roller and to analyse its chemical composition. This, together with a knowledge of the composition of the fluid in which the nerve is immersed, allows calculation of the equilibrium potential for the various ions. This information is shown on the left of Figure 2.4b.

The equilibrium potential for potassium is around −90 mV. The equilibrium potential for chloride is rather smaller, because the concentration ratio is less. For sodium, the equilibrium potential is positive, since the concentration gradient is in the opposite direction from that for potassium.

When the potassium conductance is much greater than the conductances for all other ions the membrane potential is close to the equilibrium potential for potassium. In reality, the ratio of potassium conductance to the other conductances is not so extreme; permeability studies show that, with the nerve at rest, the membrane is highly permeable to potassium ions, rather less permeable to chloride ions, and far less permeable to sodium ions. The membrane potential (typically −80 mV) is about 10 mV less negative than E_{K^+} (typically −90 mV); this difference of 10 mV is due mainly to chloride (E_{Cl^-} = 70mV) with sodium having very little effect (E_{Na^+} = +45mV).

Although the permeability of the membrane to potassium is high by comparison with its permeability to sodium, its permeability is much less than that of a film of water equal in thickness to the cell membrane. Such a film of water is about 10^9 times more permeable than the cell membrane at rest. At the peak of the action potential, the membrane is approximately 10^7 times less permeable than water.

Further explanation of the determinants of the membrane potential is contained in Appendix B of this chapter.

2.3 ACTION POTENTIAL

Physiologically, action potentials are generated either by sensory receptors or by activation from other neurones. Artificially, an action potential can be generated in a nerve at will by applying a depolarizing electrical stimulus (a depolarizing stimulus is one that reduces the magnitude of the resting potential). An artificially evoked action potential is shown diagrammatically in Figure 2.3. This depolarization influences macromolecular proteins in the cell membrane which control channels across the membrane through which sodium ions can flow, these are called **sodium channels**. Depolarization of the membrane causes these channels to open.

The opening of the sodium channels allows sodium to cross the membrane. Since the large electrochemical gradient for sodium is directed inwards, there is a net influx of sodium ions. Each sodium ion carries with it its positive

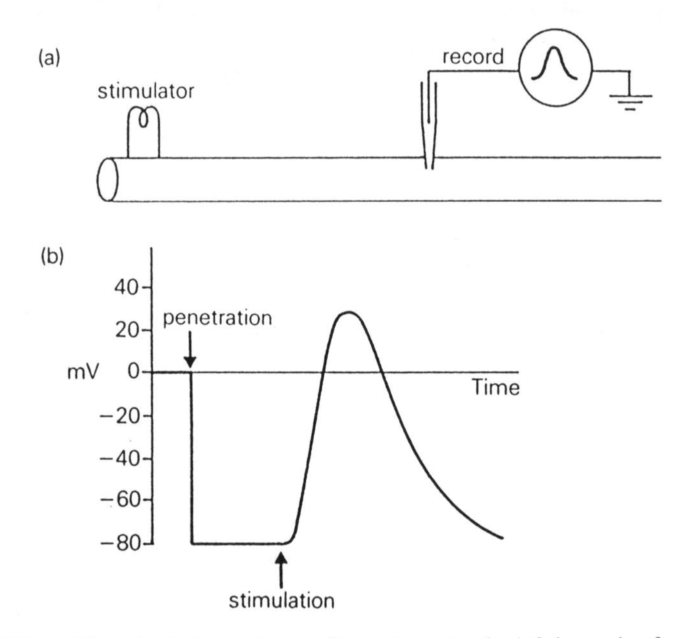

Figure 2.3 (a) The stimulating and recording set-up. On the left is a pair of electrodes applied to the surface of the fibre for delivering stimuli. Towards the right is a recording microelectrode, shown as having penetrated the nerve membrane. (b) The voltage recorded from the microelectrode. At the start of the record the electrode lay outside the nerve membrane. At the time indicated by the arrow on the left, the tip of the electrode penetrated the membrane. At the time indicated by the second arrow, an electrical stimulus was applied through the stimulating electrodes and the action potential was initiated.

charge. This depolarizes the membrane even further and promotes the opening of more sodium channels. This positive feedback loop, summarized in Figure 2.4a, is the mechanism of the upstroke of the action potential. The sodium conductance is increased so much during the upstroke of the action potential that it exceeds the potassium conductance by around 100-fold (the potassium conductance at this stage is at a level similar to that in resting nerve). When the sodium conductance greatly exceeds that for all other ions, the membrane potential moves close to the sodium equilibrium potential; this is the situation at the peak of the action potential (Figure 2.4b). The peak of the action potential is within 10 mV of the sodium equilibrium potential.

The opening of sodium channels is a transient response, lasting less than 0.5 ms; at around the time of the peak of the action potential, the sodium channels close again. At this stage, the electrochemical gradient for potassium is huge and directed outwards. Depolarization causes a delayed increase in potassium conductance. With a large driving force and a high membrane conductance, potassium flows rapidly outwards. The positive charges carried out leave the axoplasmic voltage more negative. The resting potential is thus restored. The permeability changes are shown in Figure 2.4c.

The net entry of sodium ions during the action potential causes a tiny rise in the internal concentration of sodium. After the passage of many action potentials, the maintenance of a low internal sodium concentration demands that excess internal sodium must be pumped out again. This is achieved by increased activity of the sodium pump.

Threshold

In any circumstance in which the membrane of a nerve fibre is depolarized sufficiently (except when it is refractory, see below), an action potential is initiated. The value which the membrane potential must reach before an action potential is initiated is called the **threshold**; this is shown in Figure 2.4b. For a peripheral nerve fibre, it is typically −65 mV. The value of the threshold is sensitive to the composition of the interstitial fluid, the number of action potentials which the axon has recently conducted, and many other factors.

Accommodation

Normally, in order to initiate a nerve impulse with an electrical stimulus, one uses a short sharp shock. If instead a nerve fibre is rapidly depolarized to threshold and the depolarization is maintained, a train of action potentials is produced. The frequency of firing is greatest at the start of the depolarization and thereafter declines to a lower steady level. This decline in frequency

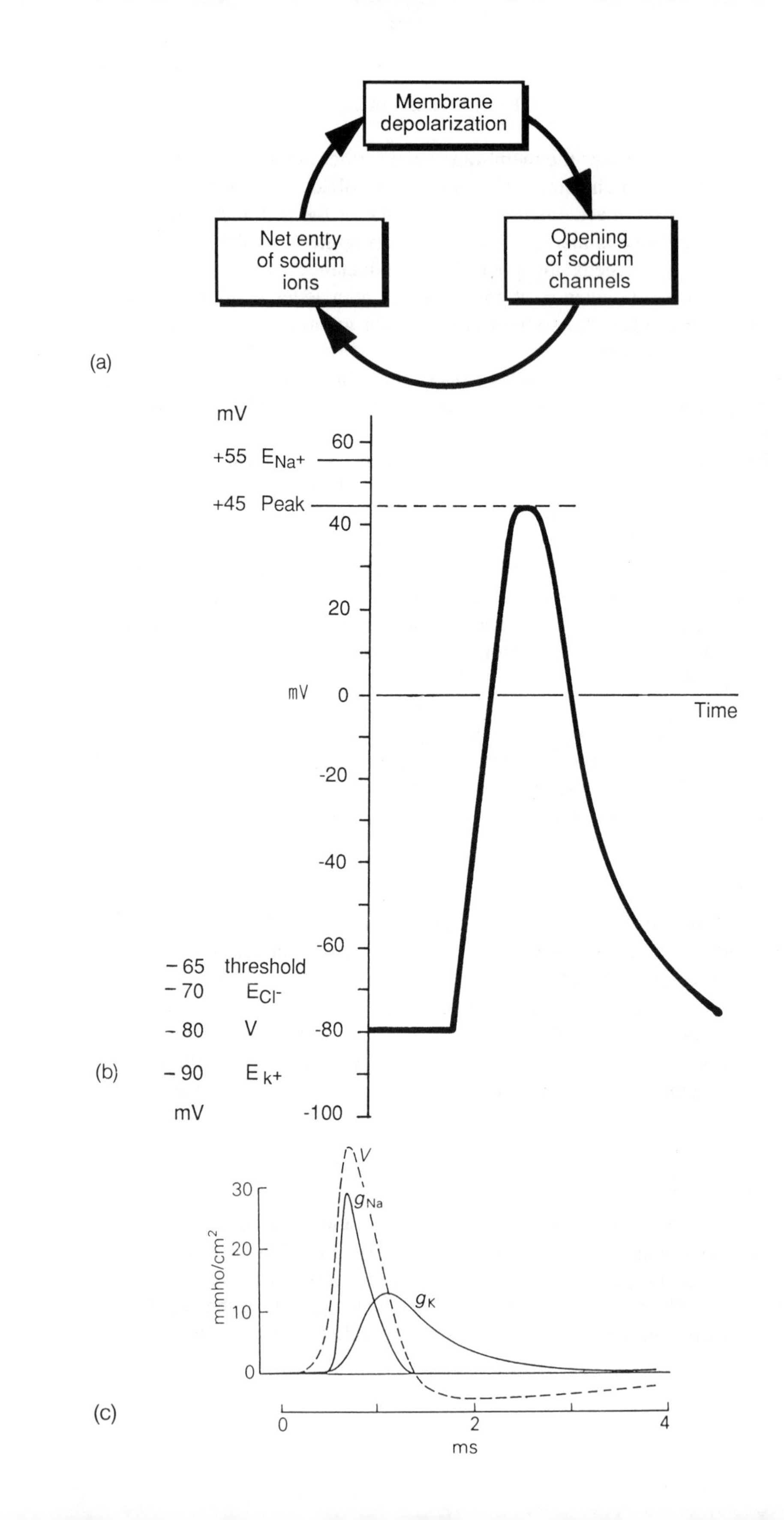

Membrane depolarization
Opening of sodium channels
Net entry of sodium ions
(a)
mV
60
+55 ENa+
+45 Peak
40
20
mV 0
Time
-20
-40
-60
− 65 threshold
− 70 ECl-
− 80 V
-80
(b)
− 90 E k+
mV
-100
V
30
gNa
20
mmho/cm2
10
gK
0
(c)
0
2
4
ms

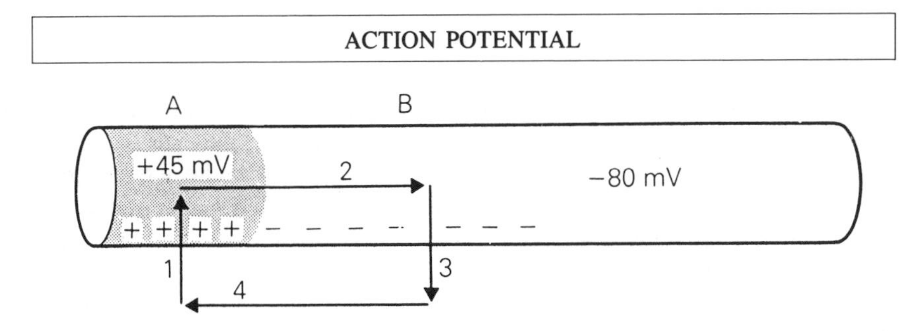

Figure 2.5 The local circuit mechanism for the propagation of the action potential. The shading represents the region of membrane involved in the action potential. The arrows represent flow of current. The nerve impulse is travelling from left to right. See text for explanation.

of impulses is called 'accommodation'. A related observation is that if, instead of rapidly depolarizing the nerve fibre, one gradually depolarizes it, the magnitude of depolarization needed to bring the fibre to threshold is higher than with the rapidly imposed depolarization. Indeed if the rate of depolarization is slow enough, no impulses are initiated.

For the same reason, a sinusoidal current of very low frequency is ineffective in stimulating a nerve because the rate of change of stimulus intensity is too low. A sinusoidal current of very high frequency fails to stimulate because each half cycle is too brief and is cancelled by the next half cycle of current which flows in the opposite direction. The most effective, and therefore the most dangerous, form of electrical stimulation is provided by the domestic power suply at 50 Hz.

Propagation

In our consideration of the mechanism of the action potential, we have confined our attention to events in a small length of nerve. We must now extend our view to the whole length of a nerve fibre, part of which is involved in the action potential mechanism and the rest of which is resting.

As shown in Figure 2.5, there is flow of current between the active region A and the nearby resting region B; the reason for this is now explained. The concentration gradient for sodium ions, together with the high sodium conductance in active membrane, provides a battery which generates active

Figure 2.4 (a) Diagrammatic representation of the changes causing the upstroke of the action potential. An arrow between a pair of variables indicates that an increase in the first causes an increase in the second. (b) Graph of membrane potential changes during an action potential. To the left of the *y* axis are shown equilibrium potentials for sodium, chloride and potassium, together with V, the membrane potential at rest. (c) Time course of permeability changes for sodium and potassium.

inward membrane current at A, shown by arrow 1. In addition to generating the action potential at A, this active membrane current also causes current to flow longitudinally in the axoplasm along the axon from the active region to the neighbouring inactive region B. The internal positivity at A (generated by the inward movement of sodium ions across the active membrane) and the internal negativity (the resting potential) provide an intracellular voltage gradient that causes current to flow through the resistive path provided by the intracellular fluid; the direction of flow of this intracellular fluid is from positive to negative i.e. from A to B, as shown by arrow 2. This current is carried principally by potassium ions. Such a current reduces the internal negativity at B. The circuit is completed by outward membrane current in region B (arrow 3) and current flowing back through the extracellular fluid (arrow 4).

The crucial feature of this local circuit is that the axoplasmic current (arrow 2) reduces the membrane potential in the length of axon just ahead of the active region. As a result, sodium channels in the membrane ahead of the active region open, and region B is triggered to generate its own action potential. This mechanism occurs continuously, not step-wise as the foregoing account might erroneously suggest; the impulse propagates continuously along the axon as a travelling wave. The local circuit is an essential part of the mechanism of propagation of the nerve impulse.

Refractory period

The leading edge of the action potential propagates. Similar local currents are also generated at the trailing edge, but these do not initiate an action potential. The sodium channels of membrane which has recently conducted an action potential are temporarily inactivated and another action potential cannot be conducted for about 1 ms. The time during which the membrane is inexcitable is called the **refractory period**. The refractory period is subdivided into two phases. The **absolute** refractory period occurs immediately after an impulse; during this phase a stimulus, no matter how strong, cannot initiate an action potential. There then occurs the **relative** refractory period, during which the nerve is excitable, but only if a stronger than normal stimulus is delivered.

The phenomenon can be demonstrated on a whole nerve. The electrical records shown in Figure 2.6 were made from a frog nerve dissected out and laid on stimulating and recording electrodes. The experiment was performed at room temperature. Record b is the response to a single stimulus and shows the action potentials in all the fibres in the nerve. Figure 2.6c shows responses to a pair of stimuli with an inter-stimulus interval long enough for all the nerve fibres to have recovered excitability. Figure 2.6d shows seven superimposed responses to pairs of stimuli with varying intervals between them. The responses to the first of each pair of stimuli occur at the same time on the

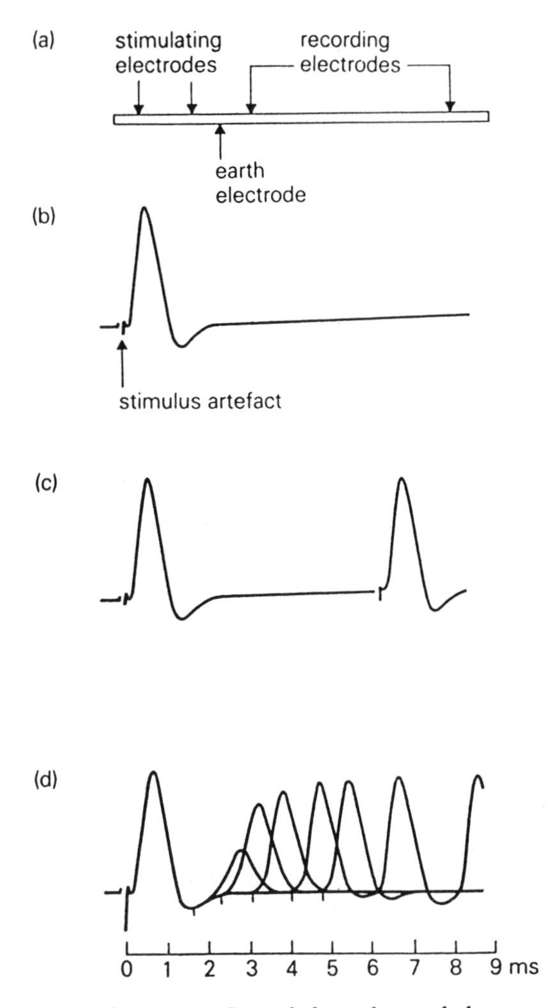

Figure 2.6 (a) Diagram of the set up. Several electrodes touch the nerve. The earth electrode reduces spread of stimulating current from the stimulating to the recording electrodes and so keeps stimulus artefact within acceptable limits. (b) Response to a single stimulus. The instant of stimulation is indicated by the stimulus artefact. The external currents due to the action potentials of all the nerve fibres in the nerve trunk add together to give the compound action potential. The upward deflection indicates that the left-hand recording electrode has become negative to the right hand one. The upward going peak is followed by a small trough; the action potential is slightly diphasic (see below). (c) Two stimuli with a delay of 6 ms between them. The response to the second is the same as that to the first stimulus. (d) A series of eight superimposed traces, with the delay between the pair of stimuli consecutively reduced. At short delays, the compound action potential is of lower voltage because many of the nerve fibres in the trunk are refractory and do not respond to the second stimulus. Frog sciatic nerve. Conduction distance 8 mm. Temperature 24°C. Stimuli 0.1 ms, 15 V; this was adequate for a stimulus to excite all the fibres. Original experiment.

trace; these responses are so consistent that they appear as a single hump, the first in this record. The responses to the second stimuli are seen as humps which are small if the second stimulus occurs very early after the first and which grow as the interval is increased. A small hump indicates that action potentials were occurring in only a few of the nerve fibres; the remainder of the fibres were refractory. If a similar experiment were performed on a single nerve fibre, then the second stimulus would elicit either a full-blown action potential or none at all.

All-or-none

Once the action potential mechanism in a nerve fibre has been initiated, its positive feedback nature ensures that it is all-or-none. Every impulse that a nerve conducts is the same voltage and waveform. As a consequence, the nerve must transmit information coded as the frequency of occurrence of impulses, as described in Chapter 1.

Calcium

The concentration of ionized calcium inside the axon is far less (by a factor of about 10^{-4}) than its concentration in the extracellular fluid (about 1.2 mmol/l). The nerve membrane is relatively impermeable to calcium and any calcium which enters is extruded by a sodium–calcium exchange mechanism. This acts rather like a special swing door which allows sodium to enter and, using the energy which this provides, expels calcium. The metabolic energy for calcium extrusion is thus ultimately derived from the sodium pump mechanism.

If the concentration of ionized calcium in the extracellular fluid falls, the recorded membrane potential does not alter significantly, but less depolarization is needed to reach the threshold. Nerve fibres become hyperexcitable and may fire off action potentials spontaneously. In motor nerves, this results in involuntary contraction of muscles, a condition known as **hypocalcaemic tetany**. This can occur, for instance, as a result of hyperventilation (over-breathing). Conversely, an increase in the concentration of ionized calcium causes the threshold to rise; the membrane is stabilized.

The effect of hypocalcaemia arises from an effect of calcium ions on the nerve membrane. The outer aspect of the membrane has negative charges that occur in pairs, matching the two positive charges on a calcium ion, as shown diagrammatically in Figure 2.7. Extracellular fluid of a normal composition provides sufficient calcium ions to provide a blanket to cover these membrane charges so that the potential just outside the membrane is similar in value to that in the bulk of the extracellular fluid. If the concentration of free calcium ion in the extracellular fluid falls, however, some of the negative charges are unmasked and a voltage gradient exists close to the

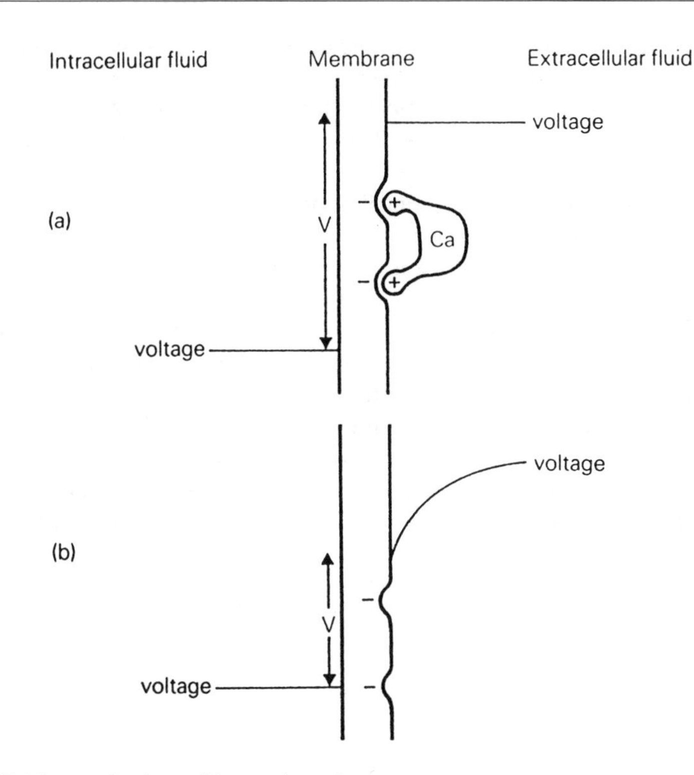

Figure 2.7 The mechanism of hypocalcaemic tetany.

membrane, as shown in Figure 2.7b. The potential across the membrane is actually decreased. This potential is nearer than usual to threshold and so the nerve is hyperexcitable. However, the fact that this sharp voltage gradient is very close to the outer aspect of the membrane means that the potential detected by electrodes between the bulk of the extracellular fluid and the intracellular fluid is unaltered.

2.4 COMPOUND ACTION POTENTIAL

Figure 2.5 showed that some action potential current flows through the extracellular fluid (Figure 2.5, arrow 4). This generates a voltage field in the extracellular fluid which can be recorded as a voltage deflection between a pair of electrodes placed on the surface of the nerve. An action potential in a single fibre results in an extracellular voltage which is much smaller than the action potential recorded across the membrane of the nerve. The voltage of the extracellular fluid near the membrane generating an action potential is negative because positive charges (sodium ions) are leaving the extracellular fluid to enter the axoplasm.

When a nerve trunk is electrically stimulated, action potentials are initiated in the constituent nerve fibres. The extracellular currents generated in the different nerve fibres are additive. An action potential recorded from the surface of the nerve trunk contains contributions from the action potentials in each of the active fibres. The action potential recorded in this way is called a **compound action potential** since it is compounded of contributions from many individual nerve fibres.

The set-up for recording the compound action potential is shown in Figure 2.8a. When an electrical stimulus is applied through the stimulating electrodes, action potentials propagate along the fibres and, as they pass the first recording electrode, a voltage change is recorded. The first recording electrode becomes negative to the second. Most physiologists arrange for this to be recorded as an upward deflection. If the nerve has been damaged between the two recording electrodes so that the nerve fibres cannot conduct to the far end of

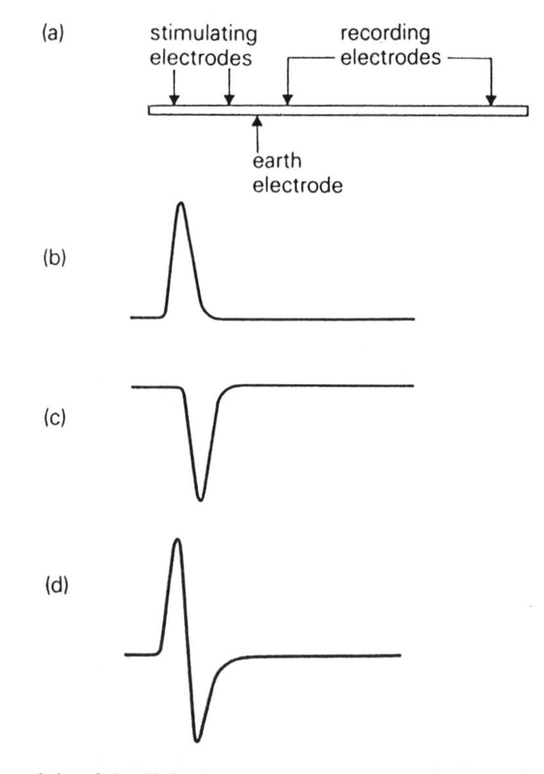

Figure 2.8 The origin of the diphasic action potential. (a) The recording set-up. (b) An action potential recorded as it passes the first recording electrode. (c) The action potential as it passes the second recording electrode. (d) The difference in voltage recorded between the two recording electrodes.

the nerve, the action potential record is a single deflection, as shown in Figure 2.8b. It is called **monophasic**. If nerve conduction along the whole length of the nerve is normal, the action potentials propagate past both recording electrodes. As these action potentials pass the second recording electrode, this electrode becomes negative to the first. This is the opposite polarity from that in Figure 2.8b. It is shown in Figure 2.8c and is slightly delayed because of the longer conduction distance. The physiological record obtained from a nerve which conducts along its whole length is the sum of the upper two traces (Figure 2.8d). The record yielded in this situation is called the **diphasic action potential**, because it consists of two phases of opposite polarity.

Diameter and conduction velocity of nerve fibres; compound action potential

In humans nerve fibres range in diameter from about 0.5 to 20 μm. Thicker fibres conduct more rapidly than thin fibres. When all the nerve fibres in a nerve are stimulated simultaneously by an electric shock applied at one point, the action potentials in the different fibres propagate away from that point at different velocities, like runners in a race. Recording the compound action potential recorded from the nerve at a distance from the site of stimulation is rather like observing the runners as they pass the end post. The runners arrive in a sequence determined by their speeds, the fastest arriving first. Similarly, at the recording electrode, the fast-conducting nerves give an early deflection at the recording electrode and the slower fibres give later deflection. Gasser and Erlanger, working in the early days of electronic instrumentation, were the first to discover that the compound action potential of a peripheral nerve in the frog shows several discrete peaks. These indicate that there are groupings of nerve fibres. For convenience, Gasser and Erlanger labelled the major peaks A, B, and C according to their conduction velocity; peak A is subdivided into α, β and γ, as shown in Figure 2.9a. Each peak contains nerve fibres with particular functions. The A α and γ peaks include efferent nerve fibres to two different sorts of muscle fibre, as we shall see later. Not every nerve exhibits every peak. Indeed, the B peak, which is contributed by autonomic preganglionic nerve fibres (see later), is not present in mammalian peripheral nerves because the layout of the autonomic nervous system is different from that of the frog.

A corollary of this observation of discrete peaks in the compound action potential was that the diameters of the axons in a nerve are not uniformly distributed between the smallest and the largest. Peaks in the histogram can be identified, but this requires special techniques. The first demonstration of such peaks was by Eccles and Sherrington (1930). They confined their attention to the efferent nerve fibres in a nerve supplying a muscle and showed two peaks in the frequency histogram of axon diameter, as shown in Figure

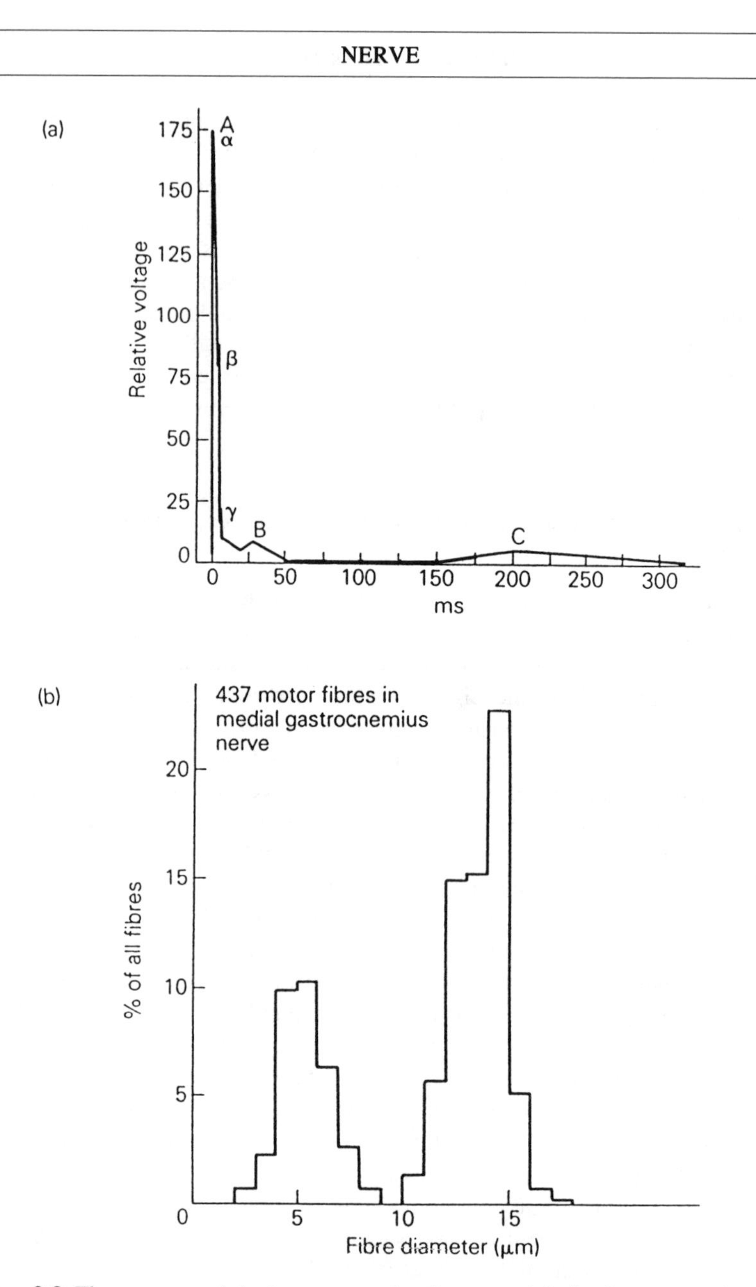

Figure 2.9 The upper graph is the compound action potential of a frog nerve; the conduction distance is large. The lower graph is the frequency distribution of nerve fibres as a function of fibre diameter.

2.9b. These peaks correspond to the group A α and A γ peaks, two groups of efferent fibres which will feature prominently in Chapter 7.

Classification of afferent nerve fibres from receptors in muscle

A nerve supplying a skeletal muscle contains both sensory and motor fibres. In describing the composition of such a mixed motor nerve, Lloyd and Chang (1948) found it useful to describe these different fibres with different nomenclatures (Figure 2.10). They retained Erlanger's A, B, C classification for the motor nerve fibres. For the sensory fibres they introduced a different system using roman numerals I, II, and III. This nomenclature has been retained by muscle physiologists, who classify these afferent nerve fibres according to the siting and properties of the receptors which they innervate. For instance, Groups Ia and II afferents come from different types of receptor within the muscle, whereas Group Ib afferents come from receptors in the tendon. For each group there is a spread in the histogram of conduction velocities, and the histograms for the different groups of afferents so defined overlap. Thus it is possible to have two afferents, one in Group I and the other in Group II, with the same conduction velocity. On the whole, however, the Group I afferents conduct more rapidly than the Group II afferents. Groups I, II, and III correspond roughly in conduction velocity to the motor groups A α, β, γ, and δ respectively, as shown in Figure 2.10. For afferent nerve fibres other than those from skeletal muscle receptors,

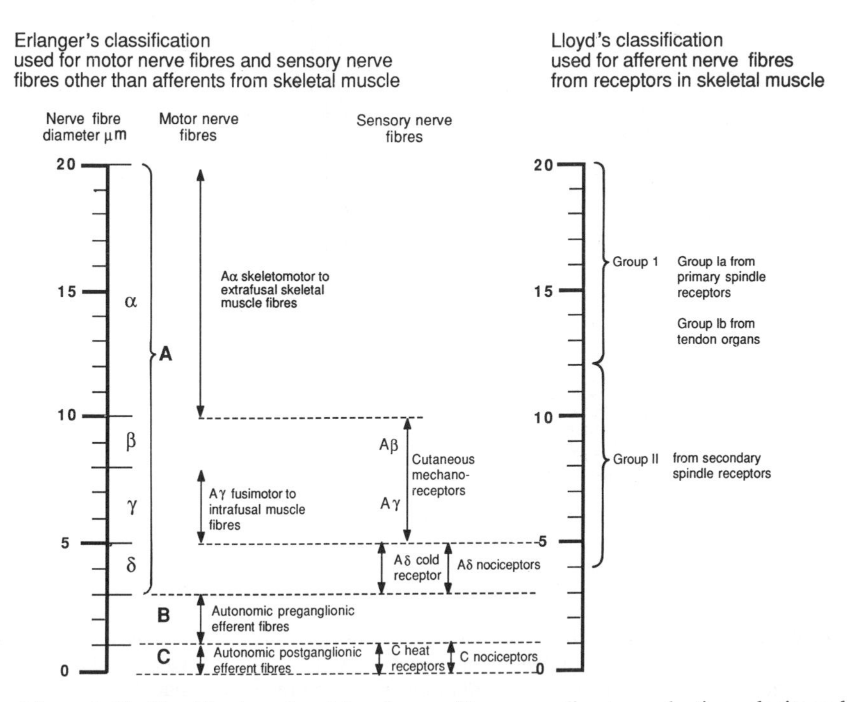

Figure 2.10 Classification of peripheral nerve fibres according to conduction velocity and function.

the A, B, C classification is usually used, although sometimes the Group classification is still to be found. This double nomenclature tends to confuse the uninitiated.

2.5 MYELINATION

As in other tissues, the specialized cells of nervous tissue are surrounded by a framework of supporting cells. In peripheral nerves, these cells are called **Schwann cells**. Each nerve fibre has Schwann cells at intervals along its length.

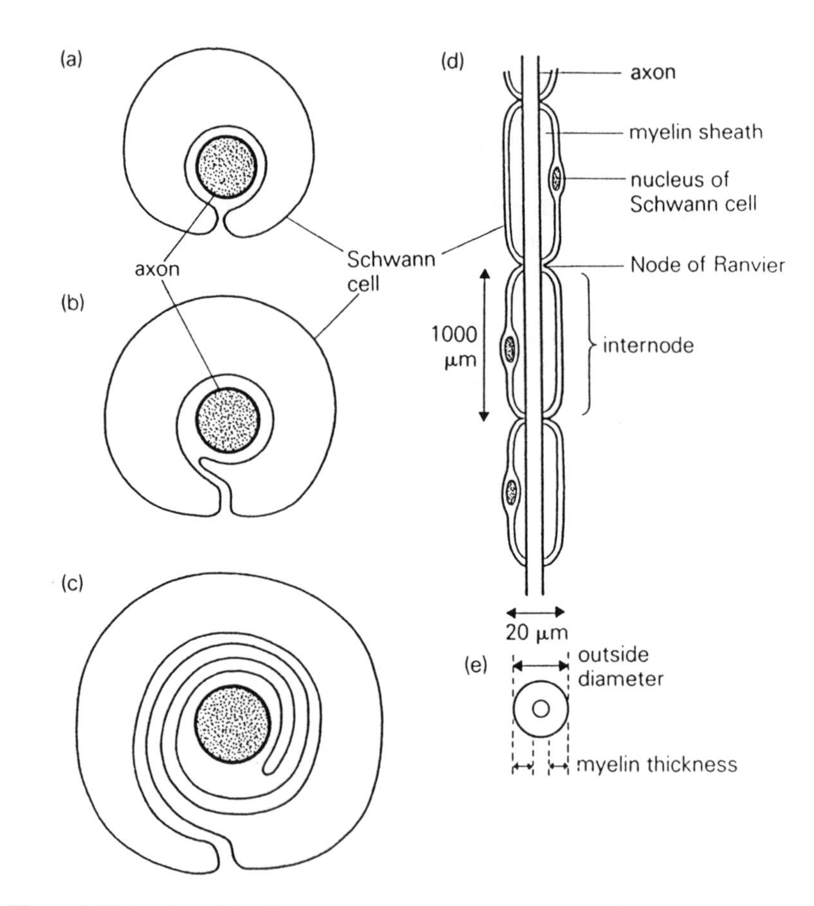

Figure 2.11 (a), (b) and (c) The process of myelination of a nerve fibre; transverse sections are shown. (d) A nerve fibre in longitudinal section. As shown on the scale, the diameter of the fibre is about 18 μm and the length of the internodal region is about 1000 μm. The diagram greatly exaggerates the diameter to length ratio. (e) shows a cross-section of the nerve fibre to indicate how the relative amount of myelin is expressed.

For all but the smallest nerve fibres, these Schwann cells form a thick **myelin sheath** surrounding the axon. Myelin consists of fatty material which is a good electrical insulator. This is important in insulating the axoplasm of the nerve fibre from the extracellular fluid outside the myelin. The myelin sheath consists of a spiral of many layers. This spiral consists of the cell membrane of the Schwann cell and is generated by a mutual rotation of the Schwann cell and the axon, as shown in Figure 2.11a–c. Between adjacent Schwann cells is a gap in the myelin sheath. This is called the **Node of Ranvier** (see Figure 2.11d). The length of nerve fibre covered by one Schwann cell is called an **internode**. The length of the internode is between 0.5 and 2 mm, being greater for thicker nerve fibres.

Saltatory conduction

In myelinated nerve, the action potential is generated at the nodes. Because of the insulating properties of the myelin, the action potential jumps from node to node without involving the membrane underlying myelin (Huxley and Stampfli, 1949). This mode of conduction is called **saltatory**, which means 'jumping'. The speed of conduction is approximately proportional to the outside diameter of the nerve fibre.

The myelin sheath and the axon are both important in the spread of current along the nerve fibre. On the one hand, thick myelin improves electrical insulation and hence reduces internodal losses of current across the cell membrane. On the other hand, a thick axon reduces core resistance and thus improves the longitudinal spread of current. For a nerve fibre of given external diameter, a compromise must be found between having thick myelin and a thick axon. The relative thickness of myelin is expressed as the ratio of the thickness of myelin to the overall thickness of the axon (myelin thickness/outside diameter, Figure 2.11e). Rushton (1951) calculated the effect of this on conduction velocity. He found that, for nerve fibres of the A group, the ratio giving the highest velocity is about 0.7; for those of the B group, it is about 0.5. The thickness of myelin actually found in mammalian nerves is that which would pertain if the nerves evolved to conduct as fast as possible for the size of fibre (Rushton, 1951). A fibres are thickly myelinated and B fibres are thinly myelinated. In A fibres the speed of conduction in m/s at body temperature is approximately six times the nerve fibre diameter in μm. In B fibres, the factor is rather less.

Another advantage of myelination is that, since ionic currents are confined to the nodal membrane, the amount of metabolic energy required for each nerve impulse is proportionally less per unit length of nerve fibre than that for an unmyelinated fibre of similar diameter. Moreover, the ratio of active surface membrane to internal volume is decreased by confining activity to the nodes. This helps in rapid firing since a large internal volume results in small changes in axoplasmic concentration when ions enter or leave.

The smallest fibres in the body are called unmyelinated, although they do have a single thin layer of myelin around them for their entire length. Up to 10 unmyelinated nerve fibres are accommodated within the cytoplasm of a single Schwann cell. The whole length of an unmyelinated nerve takes part in the action potential process. This is called **continuous conduction**. In unmyelinated fibres, the conduction velocity is approximately proportional to the square root of the diameter. If a mammal only possessed unmyelinated nerve fibres, a nerve fibre of 1 cm diameter would be required in order to achieve the conduction velocity of 100 m/s (the speed of conduction of the fastest conducting nerve fibres in man). Nerves would be impossibly bulky. Thus myelination is essential for the nervous system, a rapidly acting system with many lines of communication.

Figure 2.12 shows the theoretical relationship between conduction velocity and diameter for myelinated and unmyelinated fibres. The curves cross at a diameter of about 1 μm. Above this point, myelinated fibres of a given diameter conduct more quickly than unmyelinated and below this point the converse is true. In man, there are few peripheral nerve fibres with diameters in the range of 1–2.5 μm. Nerve fibres with diameters below 1 μm are unmyelinated whereas those with diameters above 2.5 μm are myelinated. The presence or absence of myelin thus appears to reflect the survival value of maximization of conduction velocity.

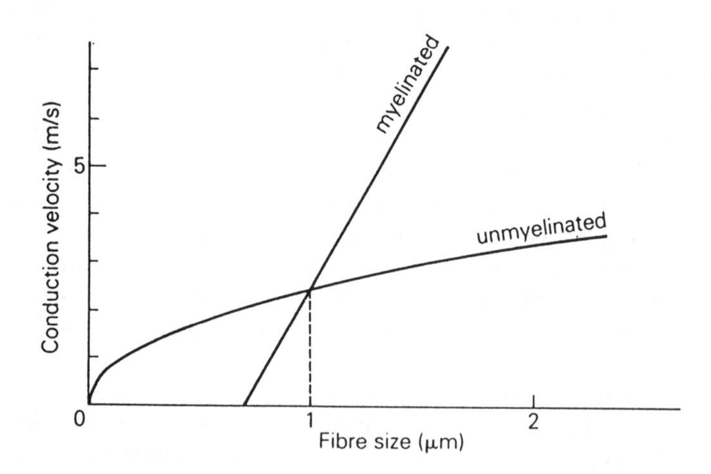

Figure 2.12 Theoretical relationship between conduction velocity and diameter for myelinated and unmyelinated nerve fibres.

The size of nerve fibres in the central nervous system

In the central nervous system, myelinated fibres are on the whole rather thinner than their counterparts in peripheral nerves. If the temperature throughout the body were constant, the conduction velocity would be slower in the central nervous system than in the periphery. The smaller size of the nerves in the central nervous system is connected with the fact that space is at a premium in the central nervous system. This smallness of central nerve fibres is not such a limitation as it might appear. The central nervous system is at core temperature, so a nerve therein always conducts at the same velocity. Nerves in the periphery are subject to the vagaries of shell temperature changes. If a limb is cold, the conduction velocity in the limb nerves falls drastically. Conduction velocity has a temperature coefficient of around 2 for a temperature change of 10°C.

An analogous situation is a central sorting office of the Post Office which has to rely on letters being collected and delivered from outlying districts by van. In the sorting office itself, the rate of sorting is dependable and can be strictly timetabled. Collecting and delivering can be drastically slowed by uncontrollable factors, such as cold weather. It is therefore more efficient to have very fast vans (the peripheral nerves) and to operate at a relatively low, but reliable rate in the sorting office (the central nervous system).

2.6 APPENDIX A: CELL VOLUME AND THE SODIUM PUMP

Since the cell membrane is very delicate and can withstand no hydrostatic pressure difference, any difference in osmotic pressure of the intracellular fluid from that of the extracellular fluid results in a net movement of water across the cell membrane. At equilibrium the total osmotic pressure of the intracellular fluid equals that of the extracellular fluid.

The intracellular fluid contains organic anions, large ions which cannot cross the plasma membrane. The extracellular fluid does not contain significant amounts of such large ions. The membrane is permeable to all the external solutes, although to different extents for different ions. The concentration gradient for sodium is maintained by a pumping mechanism. If this pumping, which is dependent on metabolic energy, is poisoned, then passive movements of permeant ions tend to equalize the concentrations of these ions on the two sides of the membrane. Sodium and chloride move in; potassium moves out. The osmotic pressure contributed to the intracellular fluid by the permeant ions moves towards a value equal to that for the external fluid. However, the osmotic pressure of the internal impermeant chemicals is added to that of the permeant ones, giving the internal fluid an excess of osmotic activity over the external fluid. As a result, there is a net inward diffusion gradient for water. Water enters the cells and they swell. This dilutes the internal fluid, but the permeant solutes will continue to flow and equalize their concentrations on the two sides of the membrane. There is, therefore, always an excess of osmolarity in the internal fluid by comparison with the external fluid. Swelling continues until the cell membrane starts to stretch, at which stage it bursts.

In life, cells do not burst because of the sodium extrusion mechanism. The high concentration of sodium ions in the interstitial fluid balances osmotically the internal impermeant organic anions and the cell adopts an equilibrium volume. However when cells are metabolically very active, they swell. When net sodium influx temporarily exceeds sodium extrusion, cells swell. They return to their previous volume when concentrations of metabolites and ions are returned to their resting values.

2.7 APPENDIX B: DETERMINANTS OF THE MEMBRANE POTENTIAL

Figure 2.13 is a circuit representing the cell membrane. For each ion, there is a corresponding battery with an electromotive force set to the equilibrium potential for that ion. This is represented by V_{K+} for potassium, V_{Cl-} for chloride and V_{Na+} for sodium. The membrane conductance for the ion is represented by a series conductance and is labelled g_{K+} for potassium etc. Permeability studies show that, with the nerve at rest, the membrane is relatively highly permeable to potassium, rather less permeable to chloride and far less permeable to sodium. The conductance for potassium predominates so that the potassium battery dominates the membrane potential. The other ions contribute according to the voltage of their batteries and the magnitude of their conductances. The low permeability to sodium results in sodium having little effect on the resting potential.

Let us consider some particular situations.

Case 1. The potassium conductance g_{K+} is much higher than the other conductances. The negative pole of the potassium battery is then effectively directly in contact with the intracellular fluid. The other batteries then pass current through their respective conductances but this is cancelled by current from the potassium battery, since this battery has a short circuit connection with the inside of the axon. Therefore the membrane potential in this situation is very close to V_{K+}, the voltage of the potassium battery. This represents the resting potential.

Case 2. As for case 1, except that **the sodium conductance is much higher than the other conductances**. By a similar line of reasoning, the membrane potential is close to V_{Na+}, the voltage of the sodium battery. This represents the situation at the peak of the action potential.

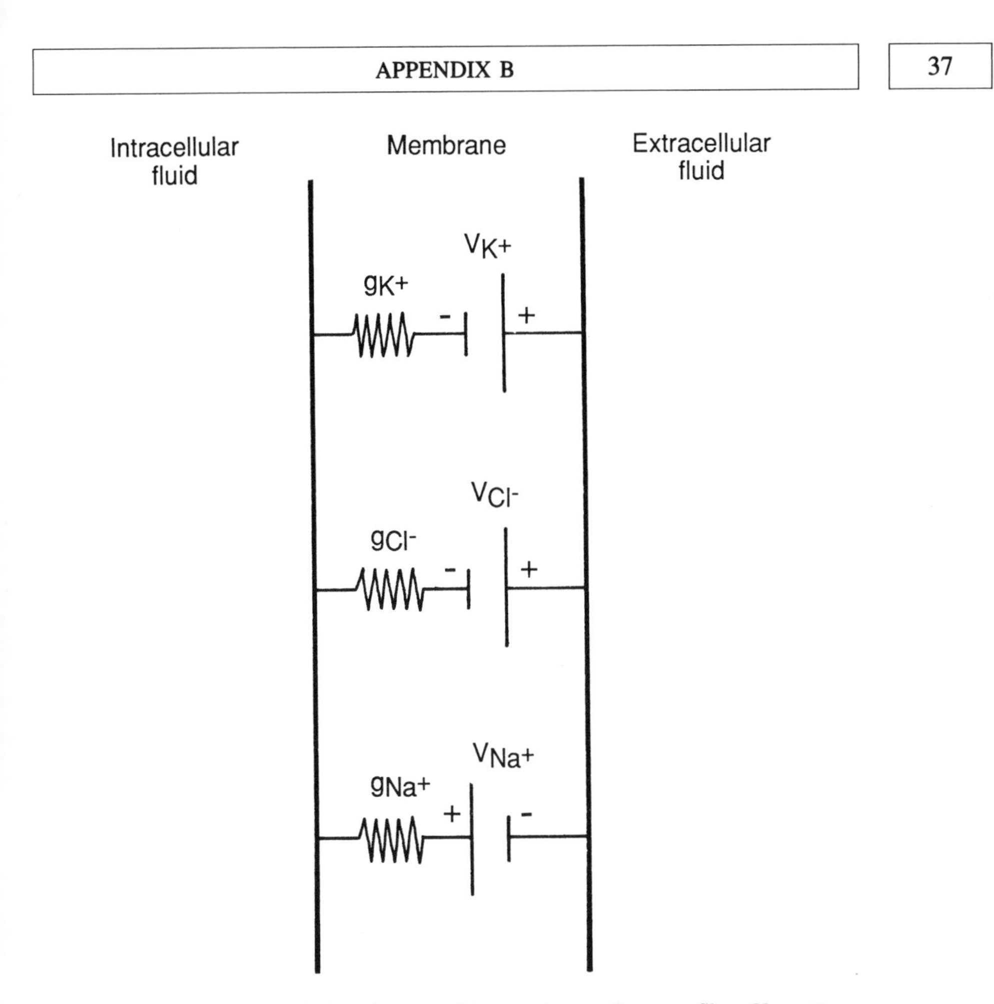

Figure 2.13 Electrical model of an element of the membrane of a nerve fibre. V_{K^+}, V_{Cl^-} and V_{Na^+} represent the driving forces for potassium, chloride and sodium ions across the membrane; g_{K^+}, g_{Cl^-} and g_{Na^+} represent the corresponding membrane conductances for these ions.

Case 3. We consider the nerve membrane at rest, with a relatively high potassium conductance and a low sodium conductance; the membrane potential is close to the potassium equilibrium potential. If now the potassium conductance falls with no change in sodium conductance, the relative contribution of the sodium conductance is increased and the membrane potential moves towards the sodium equilibrium potential. This shows that the same change in membrane potential can be achieved either by the sodium conductance rising or by the potassium conductance falling.

The physiology of receptors: cutaneous sensation

3

3.1 RECEPTORS

The structure of receptors

In all **somaesthetic** (i.e. general body sensory) receptors, the receptor element is the unmyelinated ending of the sensory nerve fibre itself. The receptor potential which we shall be considering shortly is generated in this unmyelinated ending and is conducted **electrotonically** (i.e. passively) to depolarize the nerve fibre membrane, which is capable of initiating a nerve impulse. There is no synapse between receptor and afferent nerve fibre. This contrasts with the situation for the special senses, where, as a rule, the receptor is a modified epithelial cell separate from the end of the afferent nerve fibre. Here, functional connection between the receptor cell and the nerve fibre is by means of release by the receptor of a specific transmitter chemical that interacts with receptor sites on the membrane of the nerve fibre. An example of the arrangement, in inner ear receptors, is shown in Figure 9.9. This illustrates the general arrangement for receptors for taste, vision, hearing, and balance; the exception is olfaction, where the receptor is continuity with the afferent nerve fibre.

An elaborate structure is not essential for receptor specificity. Evidence for this comes from the observation that although one can detect hot and cold, pressure, and other modalities in the skin of the ear lobe, histological examination of this region reveals that the only sensory nerve endings are free nerve endings; there are no anatomically specialized endings.

The energy to which the receptor is sensitive is changed into an electrical signal at the nerve ending. The changing of energy from one form to another is called **transduction**. Rather little is known about the mechanism of transduction and how the membrane of the nerve ending in a stretch receptor, for instance, is depolarized by a small mechanical deformation whereas the membrane of the nerve ending of a cold receptor, which appears exactly the same, is depolarized by cold and not by mechanical deformation.

The receptor potential

Unlike the main length of the afferent nerve fibre, which carries information in digital form as action potentials, the tip of the sensory nerve fibre exhibits different electrical properties. In response to energy to which the receptor is sensitive, the sensory nerve ending generates a potential called the **generator potential** or **receptor potential**. This was first realized by Katz (1950) in the muscle spindle. This is a specialized structure found in skeletal muscles and will be described in Chapter 7. Katz called this potential the **spindle potential**.

The receptor potential is graded

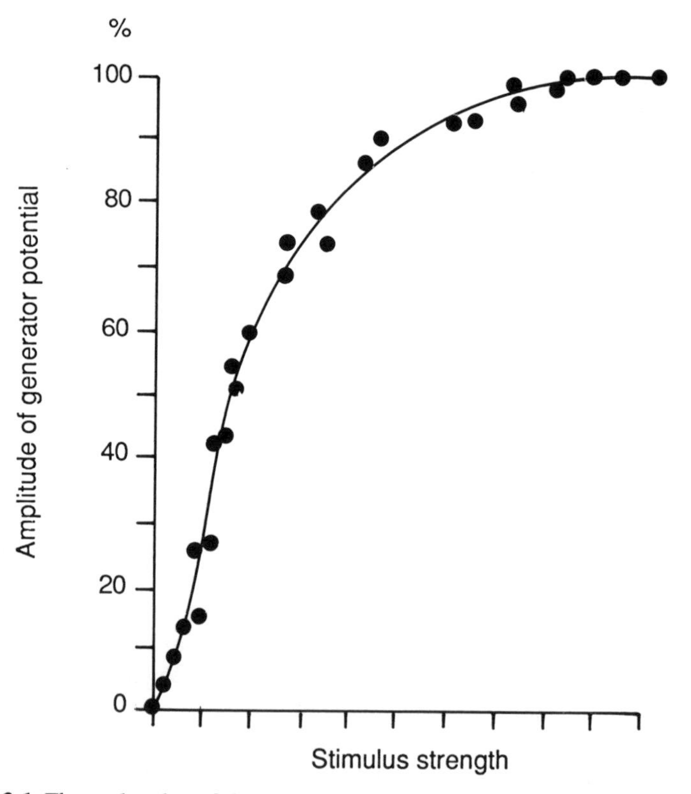

Figure 3.1 The peak value of the receptor potential as a function of stimulus strength.

Much of our knowledge about the physiology of receptors is due to the work of Loewenstein. In 1962, he showed that the receptor potential is **graded**. An increase in stimulus strength causes an increase in the amplitude of the receptor potential. As shown in Figure 3.1, the relationship between strength and response is almost linear in the middle region, with a flattening at either end. Such a relationship is said to be **sigmoid**, because it is the shape of the Greek letter sigma (the Greek **s**).

The receptor potential is localized

Using microelectrodes, Loewenstein (1962) showed that the generator response evoked in the terminal by a small stimulating probe declines exponentially with distance. The receptor potential is thus localized; it does not propagate and, unlike the action potential, it cannot be recorded at a distance from the receptor.

The receptor potential initiates impulses in its sensory nerve fibre

A schematic representation of receptor potentials from a slowly adapting stretch receptor in response to deformations of various strengths is shown in Figure 3.2. It has not proved possible to date to record intracellularly from the naked nerve ending of mammals, because this is so thin and delicate. Certain invertebrate receptors are large enough for intracellular recording, and it is from such records that Figure 3.2 is constructed. Figure 3.2a shows the set-up as if it were possible to record intracellularly, and Figure 3.2b shows the electrical responses obtained from such an electrode. In Figure 3.2b, traces 1 and 2, the response to stimulation consists of a potential which approximately follows the time course of the deformation. Since this receptor potential has a time course which is similar to that of the mechanical deformation, we call the receptor potential an **analogue** signal. In trace 3, the receptor potential reaches threshold and a single action potential is initiated.

Figure 3.2c shows responses from the afferent nerve fibre at a distance from the receptor. There are now only action potentials; there are no receptor potentials, because these are localized to the nerve ending. When weak stimuli are given, as in traces 1 and 2, no action potential is initiated and the record for the sensory nerve fibre is flat. Trace 3 shows a single action potential, corresponding to the action potential arising from the crest of the receptor potential in the nerve ending. Traces 4 and 5 show responses to stronger stimuli which elicit progressively more action potentials.

If the stimulus does not initiate an action potential, it does not influence the central nervous system. Such a stimulus is called **subthreshold**. Examples are shown in Figure 3.2, traces 1 and 2. Although the receptor generates a localized potential, the central nervous system can know nothing of this

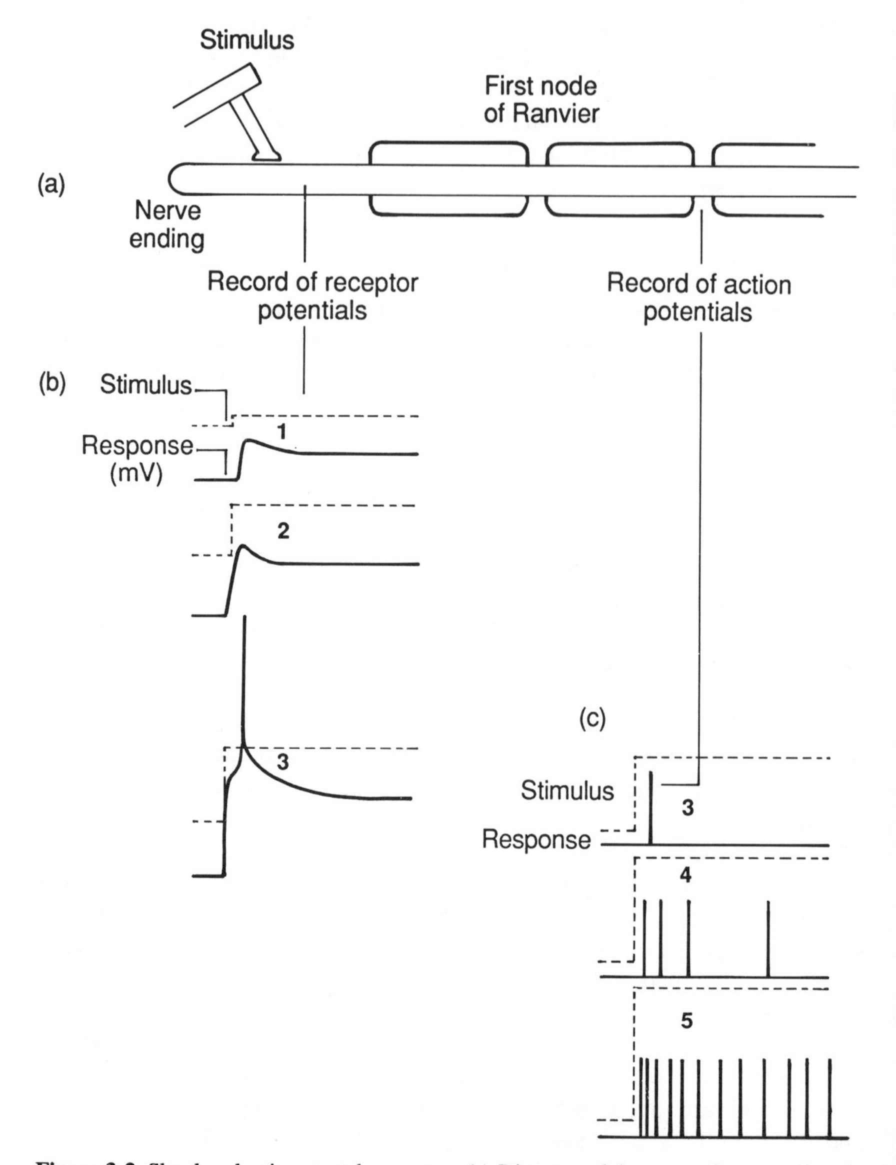

Figure 3.2 Slowly adapting stretch receptor. (a) Diagram of the set-up for recording the membrane potential of a stretch receptor. (b) Records from the receptor of responses to stretches of different strengths. Stretches are shown dashed and responses as full lines. Records 1 and 2 show pure receptor potentials. Record 3 shows the receptor potential with an action potential rising from it; the reason that the action potential is seen here is that this action potential is generated at the first Node of Ranvier and is conducted back passively to the recording electrode in the unmyelinated nerve ending. (c) Action potential records from the sensory nerve fibre connecting with the stretch receptor.

Table 3.1 Comparison of transmission and conduction

	Transmission	*Conduction (action potential)*
Structures	Receptor, synapse, motor end plate	Nerve and muscle fibres
Name of potential	Generator potential synaptic potential end plate potential	Action potential
Characteristics	Graded, localized	All-or-none, propagates over long distances
	Analogue, unidirectional	Digital bidirectional

because no nerve impulses are initiated. The stimulus which just excites an action potential, trace 3 in Figure 3.2c, is called **threshold**. A stronger stimulus is called **suprathreshold**. In the nerve fibre, the information is carried as a pattern of all-or-none impulses, a quite different type of coding from analogue coding of the receptor potential. Table 3.1 summarizes the differences between junctional transmission and conduction of the action potential.

The ionic mechanism of the receptor potential

The receptor potential consists of an opening of membrane gates that allow the passage of all small inorganic ions. The ions present in significant amounts are sodium, potassium, and chloride ions. The receptor potential is a short-circuit of the membrane. The equilibrium potential for the receptor potential (i.e. the membrane potential at which opening of these membrane channels causes no membrane current to flow) is close to zero (it is, in fact, about -10 mV). The mechanism of the receptor potential is therefore different from the action potential, which is due to opening of membrane channels specifically for sodium ions.

To return to the sensory receptor, the nerve impulse is initiated not in the naked nerve ending itself but at the first node of Ranvier (see Figure 3.2a). The receptor potential spreads passively to the first node of Ranvier and depolarizes the nerve membrane at this node. This depolarization opens the voltage-dependent sodium gates and, if the threshold for the generation of an action potential is reached, the action potential mechanism is triggered and propagates along the whole length of the nerve by the mechanism which has already been described in Chapter 2.

Encapsulated endings

While most somaesthetic receptors consist of the free nerve ending alone, a few are encapsulated. The capsule presumably enhances the specificity and sensitivity of the receptor; it is certainly responsible in part for the response characteristics of these receptors, as we shall see. Encapsulated receptors are found in the regions of skin which are highly sensitive to the touch; in humans, the palmar aspects of the fingers are particularly sensitive and this skin is richly endowed with encapsulated receptors. The skin of the palms of the hands, the soles of the feet, and the external genital organs also house encapsulated receptors. Figure 3.3 shows the structure of two such receptors.

Figure 3.3a shows the structure of endings known as **Meissner** corpuscles. These are particularly common at the tips of the fingers. Each lies just beneath the epidermis in the dermal papillae (small finger-like projections of the dermis into the epidermis). The corpuscle consists of flattened cells arranged in lamellae. The lamellar cells appear to be modified Schwann cells. The terminals of the nerve wind their way among the lamellae to end in small

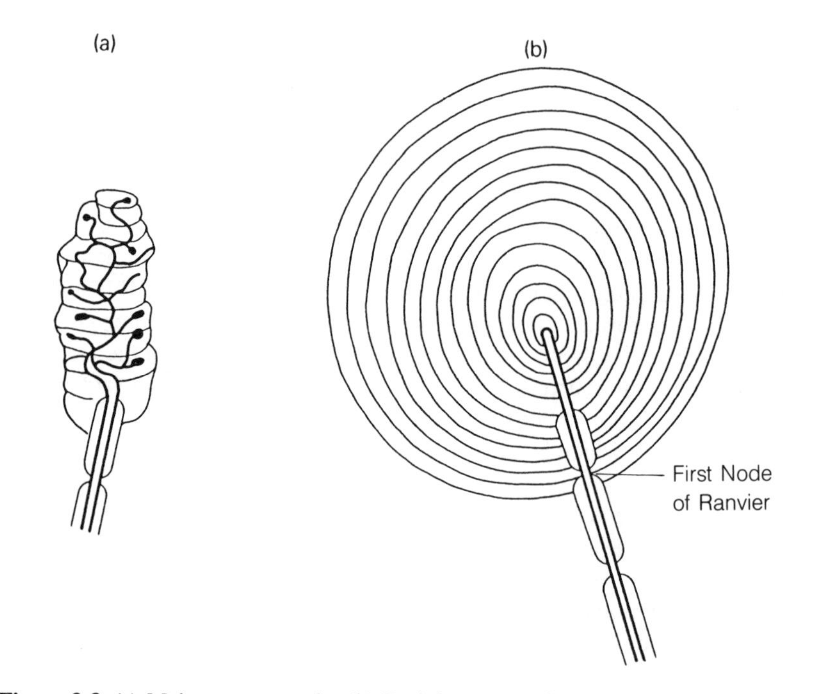

Figure 3.3 (a) Meissner corpuscle. (b) Pacinian corpuscle.

globules. Figure 3.3b shows the structure of the Pacinian corpuscle, which is 1–4 mm in diameter, much larger than the Meissner corpuscle. It lies in the deeper layers of the skin.

The Pacinian corpuscle

The Pacinian corpuscle consists of a series of concentric lamellae of flattened cells, rather like an onion. The lamellae are separated by tissue fluid spaces. At the centre is the nerve ending. In addition to their occurrence in specialized areas of skin, Pacinian corpuscles are found in the mesentery and in the pancreas. Here, they are responsible for the sensation elicited by low frequency vibrations like the banging of a big drum. Animals of the cat family are richly endowed with cutaneous Pacinian corpuscles in the paw pads. A suggested function is that, as the animal crouches to spring for its prey, it can sense vibrations of the ground produced by movement of the prey.

Adaptation due to the physical properties of the Pacinian corpuscle

It is worth reiterating at this point that 'adaptation' (Section 1.2) is the term applied to the decline in frequency of impulses in the sensory nerve fibre during sustained stimulation of its receptor at a constant intensity. Some classes of receptor adapt rapidly; others adapt slowly. For rapidly adapting receptors, the pattern of impulses in the afferent nerve is a poor reproduction of the time course of the stimulus. These receptors signal changes in intensity of energy rather than indicating its absolute intensity. By comparison with adaptation, the term 'accommodation' (Section 2.3) is applied to the observation that there is a decline in frequency of impulses during a maintained depolarization of the axon itself. For a slowly adapting receptor, the adaptation is partly due to accommodation of the sensory nerve fibre. For rapidly adapting receptors, the properties of the receptor itself play the dominant role in adaptation.

Encapsulated endings are rapidly adapting receptors, and this is largely due to the properties of the capsule, as illustrated in Figure 3.4. Figure 3.4a shows the time course of a deformation which is applied suddenly and, after being maintained for a time, is withdrawn. The first sketch in Figure 3.4b shows the corpuscle at rest. In the second sketch, showing the situation immediately after the application of deformation, the mechanical deformation is transmitted to the transducer part of the receptor. A receptor potential is generated (Figure 3.4c) and two action potentials are initiated (Figure 3.4d). In the third sketch, when the deformation is maintained, fluid squeezed between the lamellae has time to move away and the inner lamellae are no longer deformed. No further action potentials are initiated. In the fourth sketch, when the deforming force is removed and the outside lamellae spring back to their original form, the fluid between the lamellae has been redistributed

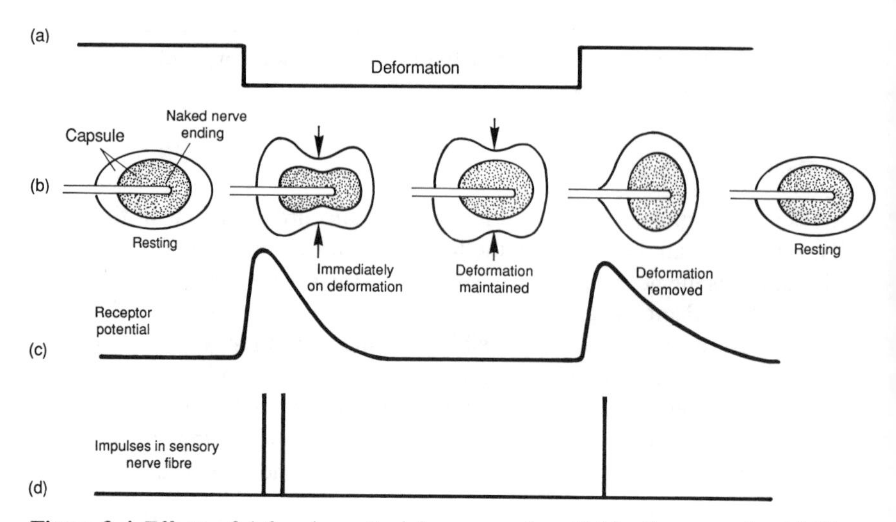

Figure 3.4 Effects of deforming a Pacinian corpuscle – highly diagrammatic. (a) Time course of deformation. (b) Sketches of the receptor at rest, immediately after onset of deformation, after the deforming force has been maintained for some time, immediately after removing the deforming force and finally back in the resting state. (c) Receptor potential. (d) Action potentials on the sensory nerve fibre connecting the receptor to the spinal cord.

and the inner lamella is deformed anew. Another action potential is generated. Finally, the last sketch shows that the status quo is re-established as the fluid oozes back to its resting position.

Evidence for the contribution of the capsule to the adaptation of the receptor comes from the observation that if the capsule is removed, a deforming force applied directly to the nerve ending elicits a prolonged receptor potential and a longer train of impulses in the sensory nerve fibre.

Sensory perception

A stimulus just strong enough to be detected by the subject is said to be at perceptive threshold. For some types of receptor, the perceptive threshold is set by the sensitivity of the receptors themselves; once an afferent nerve impulse is initiated, the stimulus is perceived. For other types of receptors, many afferent nerve impulses are required before the perceptive threshold is reached.

This topic has been studied in elegant experiments with the technique known as **microneurography**, a technique developed by Vallbo (1986). A micro-electrode is inserted percutaneously in the forearm of a human subject. The position of the electrode is adjusted until it picks up action potentials from a single afferent nerve fibre within the nerve. It is then possible to map the receptive field of this afferent fibre and to study the stimulus response characteristics of the receptor.

The sensory receptors of the skin of the hand will be described shortly. Receptors sensitive to indentation of the skin have been partitioned into two groups according to their response characteristics. The group with the lower threshold respond to an indentation of 10 μm; they are rapidly adapting. The other group consists of receptors having a higher threshold; they are slowly adapting.

A weak stimulus just sufficient to elicit a single action potential in a fast adapting receptor is perceived by the subject. When the stimulus is set at a low level, so that it only occasionally elicits an action potential, the subject reports perception of the stimulus on those occasions when an action potential is generated and no perception when no action potential is stimulated. Conversely, when the single nerve fibre is stimulated through an electrode previously used to record action potential activity, the subject perceives the stimulus. For these receptors the neural and perception threshold curves are identical.

Stronger stimuli excite slowly adapting receptors. The innervation of the skin is punctate; the receptive fields of neighbouring quickly adapting sensory units do not always overlap and it is possible to stimulate regions of skin to excite only the slowly adapting, and not the quickly adapting, receptors. When this is done, the perceptive threshold is found to be more than 10 times as great as the threshold for generating nerve impulses. Direct stimulation of the nerve fibre shows that about 20 action potentials are needed before the subject reports perception of sensation. For this type of receptor, the perceptive threshold is set by mechanisms within the central nervous system and not by the receptor.

The overall conclusion of these observations is that, for some types of receptor, the threshold for perception is set by the sensitivity of the receptor whereas for others, it is set by the central nervous system.

Fine tactile discrimination in different areas of the body

Fine tactile discrimination is the ability to discriminate the position of touch stimuli to the skin. One way of testing the fineness of tactile discrimination is to measure the **spatial discrimination**, which means the ability to discriminate two points on the skin as being separate. Two points on the skin are touched, and the task is to determine how close these two points can be for the subject still to recognize them as being separate. You can demonstrate for yourself that the spatial discrimination for touch depends on the region of skin which you test.

Experiment

Ask a friend to act as a subject. Use a pair of callipers with blunt ends. Set the separation of the ends of the callipers to a measured value. Without allowing the subject to see the callipers, touch the skin on the palm aspect of the middle finger with the callipers and ask whether the subject feels one or two points. Vary the separation of the points of the callipers and test your subject several times to determine the least separation which the subject can reliably recognize as two separate points. Note this value. Repeat this measurement on the outer aspect of the upper arm and note the value.

The outcome of this experiment on a typical subject is that two-point discrimination is much better developed in the finger tips, which we use all the time for exploring our environment, than for the upper arm. This difference in spatial resolution is reflected in the representation of the integument (skin surface) in the somatosensory area of the cerebral cortex, as described in Chapter 13.

3.2 CUTANEOUS SENSATION

Having considered the physiology of receptor mechanisms, we now turn to sensation in the skin. There are several types of receptor responsive to mechanical deformation, receptors sensitive to heat and to cold, and receptors signalling damage to the skin.

Axonal composition of cutaneous nerves

There are myelinated and unmyelinated sensory fibres innervating the skin. The thickest fibres are around 14 μm in diameter – considerably thinner than the largest fibres innervating receptors in skeletal muscle. Different types of cutaneous receptor are innervated by nerve fibres of different diameters. For instance, the larger myelinated fibres, above 5 μm in diameter, predominantly innervate mechanoreceptors, whereas heat receptors are innervated by unmyelinated fibres.

Cutaneous receptors

Each receptor is specifically sensitive to a particular type of natural stimulus. The afferent nerve fibre which innervates it is thus a labelled line. The action potentials in different nerve fibres are similar and do not themselves pass on information about the nature of the receptor from which they originated. It is the siting of the nerve fibre which holds the information about the

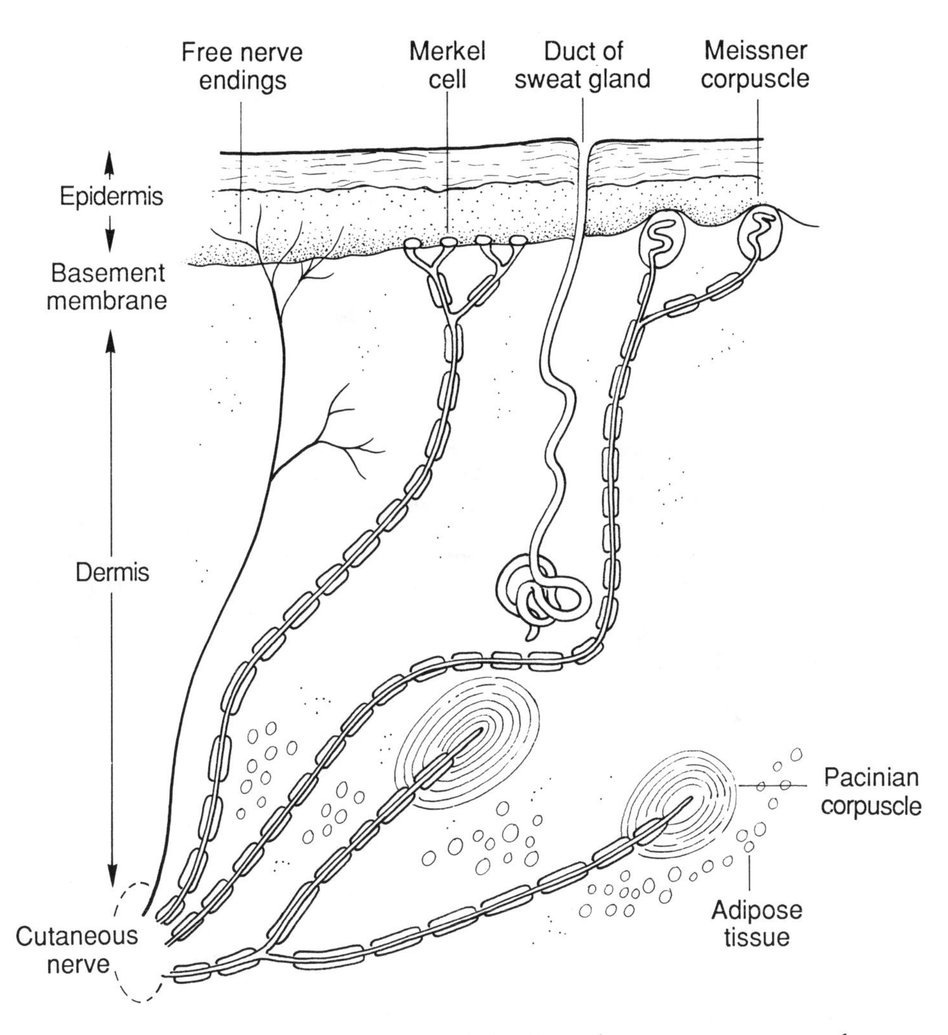

Figure 3.5 Stylized diagram of a section through the skin to show sensory receptors and nerve fibres.

nature of the stimulus. Activity in a particular nerve fibre projects to targets in the central nervous system and this targeting contains the information about the nature of the stimulus.

Three categories of cutaneous receptor can be distinguished: mechanoreceptors, thermoreceptors, and nociceptors. They are shown in Figure 3.5.

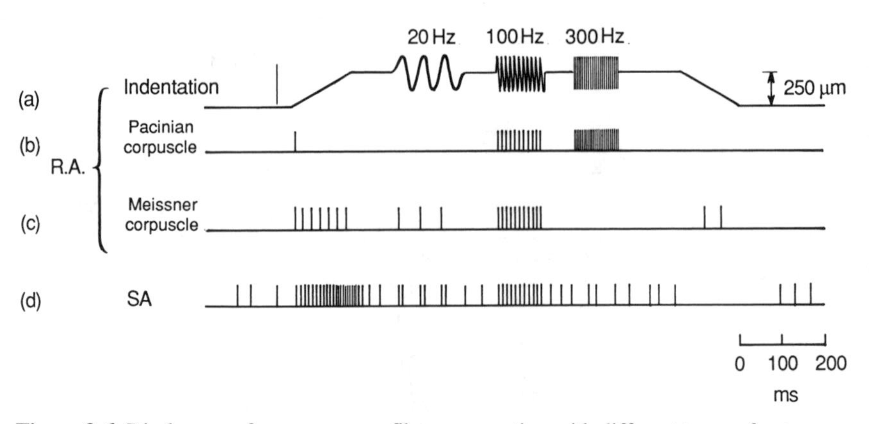

Figure 3.6 Discharges of sensory nerve fibres connecting with different types of cutaneous mechanoreceptor, rapidly adapting (RA) and slowly adapting (SA), in response to identical displacements of the skin, indicated in (a). The Pacinian corpuscle generates a single impulse at the start of the ramp and is excited by sinusoidal stimuli at 100 and 300 Hz, whereas the Meissner corpuscle fires repetitively during the ramp stretch, is excited by the sinusoids at 20 Hz and 100 Hz, but is not excited by 300 Hz. It generates impulses again when the indentation is finally removed. The slowly adapting mechanoreceptor generates a train of impulses in the absence of indentation. The intensity of discharge rises during the indentation and, although the discharge shows slow adaptation, the frequency of firing remains elevated throughout the indentation. This receptor responds to sinusoids at 20 Hz and 100 Hz but not at 300 Hz. As the indentation is removed, there is a silent period in the discharge train. Eventually, the resting discharge pattern reappears.

These three patterns of response are so dissimilar from each other that the response from the nerve can by itself be used to characterize the receptor to which it is attached.

Mechanoreceptors

As described earlier, there are two groups of mechanoreceptors, rapidly adapting and slowly adapting. The two principal types of rapidly adapting receptors are the Pacinian corpuscle, whose electrophysiology we have already considered, and Meissner corpuscles. The slowly adapting receptors are Merkel cells.

Each type of receptor shows a characteristic response pattern to different types of deformation; this is shown in Figure 3.6. The Pacinian corpuscle responds to a sine wave deformation at 300 Hz by generating in its sensory nerve fibre an action potential train at 300 Hz, whereas neither of the other receptors shows any response to an oscillatory deformation at this frequency. The slowly adapting receptor generates action potentials in its afferent nerve fibre even in the absence of deformation, showing the receptor to be prestressed. The action potentials increase in frequency with the onset of deformation and then the frequency falls to a lower level; it nevertheless remains above the resting frequency for as long as the deformation is maintained. It can therefore signal absolute values of deformation, not merely changes in deformation. When the steady deformation is finally removed, the frequency of firing of afferent impulses is transiently abolished. This behaviour, with the off response in the opposite sense to the on response, is more typical of receptors than the rapidly adapting receptors in skin, both types of which show responses both on application of deformation and on its removal.

Glabrous skin

The glabrous skin (the smooth hairless, fine-ridged skin of the hands and feet) contains all the types of mechanoreceptor described above. The technique of microneurography has allowed a correlation to be deduced between the anatomical type of ending and the sensation which its activation elicits in the normal human. Records are made from a single sensory nerve fibre and, by examining the response when an indentation is applied (Figure 3.6a), the identity of the receptor innervated by the nerve fibre is ascertained. The electrode used for recording action potentials in the nerve fibre is next used to stimulate, and the subject is asked to report on the sensation experienced. The results are that stimulation of a Pacinian corpuscle gives the sensation of vibration, a Meissner corpuscle gives the sensation of touch, and slowly adapting receptors give the sensation of pressure.

Fast and slow information from receptors

The foregoing account illustrates some general features of sensory systems. Different groups of receptors provide different types of information. One group is highly sensitive, responds with minimal delay, is rapidly conducted to the central nervous system, and a single action potential in the afferent fibre from a single receptor is sufficient to be perceived. This type of receptor is giving the organism early notice that something has changed. These receptors are rapidly adapting and so give little information about the time course of the change. The rapidly adapting receptors in skin respond to the onset and to the removal of the deformation and thus provide no information even on the sense of the deformation; they merely signal that something has changed, thus alerting the organism that something may be amiss.

The second group of receptors provides much more detail concerning the degree, time course etc. of the stimulus. Many impulses are required before awarenesss is reached and this obviously introduces much longer delays. This second system of receptors provides the alerted organism with detailed information on the nature of the change, to which a response may be required.

We will encounter other examples of a similar nature, with high sensitivity rapidly adapting receptors transmitting information quickly in order to prime the brain, while receptors of lower sensitivity, slowly adapting and projecting to the brain at a more leisurely pace, provide fine discriminatory information. It takes time for the free-wheeling brain to be aroused and to prepare itself for analysis and decision making. The second system of receptors has time in hand. This time is usefully spent in improving and refining the quality of information transmitted to the brain.

Thermoreceptors

Cutaneous temperature sense is served by separate receptors for hot and cold; these are free nerve endings. Sensory testing in human subjects indicates the existence of 'cold' and 'warm' spots, corresponding to the siting of cold and hot receptors in the skin. The firing frequency of the sensory fibre innervating a cold receptor increases as the temperature of the receptor falls, and *vice versa* for warm receptors. The afferent nerve fibres from cold receptors are small myelinated fibres (in the Aδ group) and those from receptors of heat are C fibres.

Nociceptors

These are also free nerve endings, connected to afferent nerve fibres with conduction velocities indicating them to be group Aδ and C fibres. This subject is taken up in Chapter 18.

FURTHER READING

Barlow, H.B. and Mollon, J.D. eds. (1982) *The Senses. Cambridge Texts in the Physiological Sciences* Vol 3. Cambridge, Cambridge University Press.

The neurone, neuromuscular transmission and synaptic transmission

4

4.1 MORPHOLOGY OF NERVE CELLS

There are many millions of nerve cells (also called **neurones**) in the nervous system. Every neurone has a cell body that houses the nucleus and the various other intracellular organelles needed for metabolic activity, protein synthesis and so forth. The cell body is the metabolic factory of the neurone.

There are differences in detail in the layout of different types of nerve cell. One type of nerve cell which we have already met is the motoneurone (see Figure 1.4); this is the type of cell which we now consider in more detail. The cell body gives rise to extensions, shown in Figure 4.1. These extensions are of two types. First there are the **dendrites**, so-called because they resemble the branches of a tree; their function is to collect information from other neurones. Second there is the nerve fibre or **axon**, which carries information away from the cell body. The axon arises from a small hillock, called the **axon hillock**, on one pole of the cell body. The first 0.5 mm of axon is unmyelinated; this is called the **initial unmyelinated segment**. Except for this segment, the axon is myelinated.

Excitation of one neurone by another occurs via contact points, called **synapses**, sited between the terminal of the axon projecting to the neurone and the dendrite receiving the input. Information flows in a particular direction through a neurone. Input is from other neurones through synapses on the dendrites. This input is transmitted from the dendrites to the cell body. When the cell body is sufficiently depolarized, an action potential is generated and this propagates thence along the axon to its far end. There the impulses influence another cell, either another nerve cell or an effector cell such as a muscle fibre. The neurone thus shows a 'dynamic polarization' of information flow, a point first established by Cajal (1954).

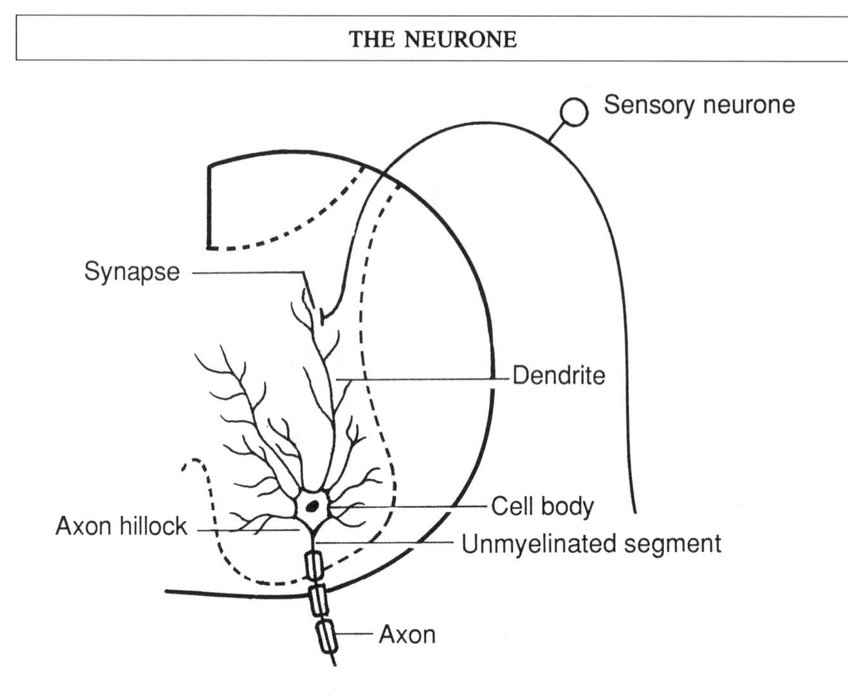

Figure 4.1 Transverse section through the spinal cord, to show a sensory nerve fibre entering the cord and making synaptic contact with a dendrite of a motoneurone.

As is also shown in Figure 4.1, sensory neurones are different from motoneurones in that the soma lies on a side branch of the main axon, in the dorsal root ganglion. There are no synapses on a soma of this kind. The centrally projecting branch of the axon extends to the spinal cord and makes synaptic contact with its target neurones.

Axonal transport

Many axons are very long. The cell body is the chemical factory for the whole of the neurone and transport mechanisms have evolved that carry vital chemicals to and fro along the axon. Axons contain microtubules along which transport of chemicals takes place in both directions. There are also vesicles which act as 'ferry boats'. There are two or more separate **orthograde** transport systems. (In this context, 'orthograde' means 'directed away from the cell body'.) As originally shown by Cowan *et al.* (1971), a fast transport system carries small quantities of synaptic vesicles and mitochondria directly to the axon terminals at rates in excess of 100 mm/day; this system provides the special materials needed for synthesis of neurotransmitter. A slower system transports particulate and dissolved proteins at a rate of 1–5 mm/day; these are distributed to the whole length of the axon. There seems to be just a single **retrograde** transport system which is fast ('retrograde' here means 'directed towards the cell body'). These transport mechanisms convey specific chemicals and this provides a technique for studying orthograde ([^{3}H]leucine) and retrograde (horseradish peroxidase) transport.

The microtubules along which transport takes place are part of the structural skeleton of the neurone, the so-called **cytoskeleton**. In normal young people, the tubular system is arranged in an orderly system of parallel tubes running from the soma into and along the length of the axon. In the cerebral cortex of ageing people, the cytoskeleton becomes less orderly and parts of the tubule system form tangles. In presenile dementia (Alzheimer's disease) this tendency to form tangles is very pronounced; the derangement of neuronal function which occurs as a result of this may be the basis underlying the clinical manifestations of the disease.

4.2 JUNCTIONAL TRANSMISSION

The nerve axon terminates close to the tissue that it innervates. When an action potential travels along a nerve fibre and reaches the axon terminal, a specific chemical transmitter is released. This chemical diffuses to the membrane of the cell to be influenced. On this membrane are specialized sites, called **chemical receptors**, which recognize the transmitter and initiate changes in the postjunctional cell.

The junctions are given different names, according to the type of cell which is influenced by the chemical transmitter. When this cell is another neurone, the term 'synapse' was coined by Sir Charles Sherrington. 'Synapse' is now used loosely to describe any neuro-effector junction. For the junction between a nerve fibre and a skeletal muscle fibre, the term **end-plate** is used.

Neuromuscular transmission

The neuromuscular junction is the junction between the motor nerve fibre and the skeletal muscle cell. In the adult human there is, as a rule, a single end-plate for each muscle fibre (the exception to this rule is the intrafusal muscle fibre, which will be considered in Chapter 7). As the motor nerve axon approaches its termination, it gives rise to many branches, each of which innervates a muscle fibre. These terminal branches are thin and devoid of myelin; conduction velocity is consequently low in this region. Each branch ends blindly, there being no cytoplasmic continuity between the nerve and muscle fibres (Figure 4.2b). The nerve ending has the form of a plate and this is the origin of the name 'end-plate'. The gap between the two cells is in continuity with the extracellular space. This **synaptic gap** is about 30 nm

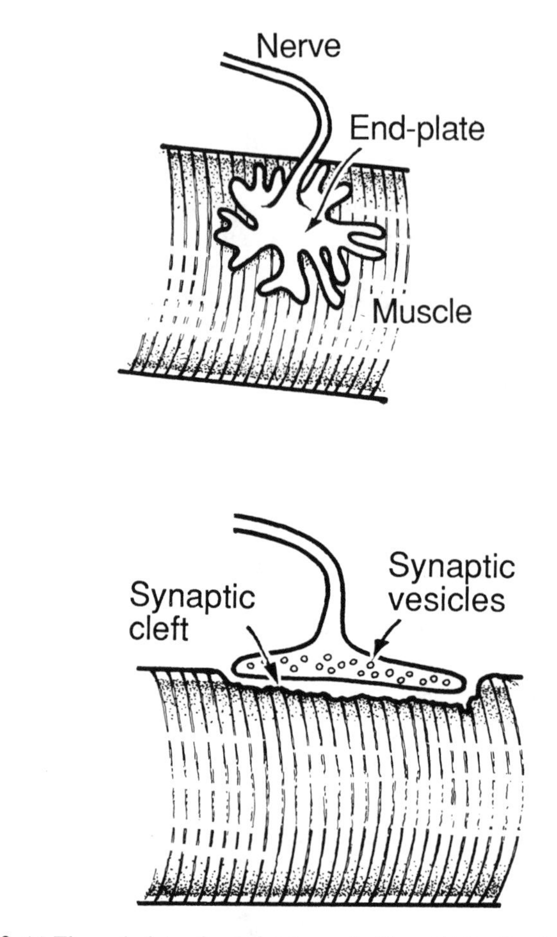

Figure 4.2 (a) The end-plate of a skeletal muscle fibre. (b) Section through the end-plate. The synaptic vesicles, the width of the synaptic cleft and the thickness of the plasma membranes are much smaller than the diagram shows.

across. The specific transmitter chemical is **acetylcholine**, evidence for which will be presented shortly.

In the presynaptic terminal, the transmitter is stored in small packets enclosed in envelopes of plasma membrane. These so-called **synaptic vesicles** (Figure 4.2b) are synthesized in the cell body and are carried the length of the axon to the terminals by one of the axoplasmic transport systems. The membrane of the presynaptic nerve terminal has voltage-sensitive sodium channels to support the action potential, just like the length of nerve fibre which feeds into it. This terminal region also possesses voltage-sensitive calcium channels. Invasion by the action potential of the presynaptic nerve terminal causes these calcium channels to open, with a resultant influx of calcium ions from extra- to intracellular fluid. The rise in internal calcium concentration induces the membrane of synpatic vesicles to fuse with the plasma membrane facing the synaptic cleft. At the site of fusion, the membrane breaks down and transmitter is released from the vesicle into the synaptic cleft, within which it diffuses. Some of the neurotransmitter reaching the muscle membrane which is folded and relatively thick in the end-plate region. On this thick membrane are protein macromolecules that recognize the acetylcholine molecules. These macromolecules are the receptors; their configuration is such that the acetylcholine fits them exactly and causes the opening of ionic gates which allow all small ions to flow down their electrochemical gradients, with a resultant depolarization of the muscle membrane. This is called the **end-plate potential**.

The depolarization spreads passively to neighbouring 'action potential' membrane and initiates an action potential in the muscle cell. The action potential is propagated along the whole length of the muscle fibre in both directions away from the end-plate. The action potential in the muscle fibre excites the myofibrils to contract.

The end-plate potential amounts to a depolarization of about 50 mV; this is much more than is needed to bring the nearby membrane to threshold for initiating an action potential in the muscle fibre. Since the end-plate potential is so large, the safety factor for neuromuscular transmission for normal human muscle is very high. Every action potential reaching the motor nerve terminal causes an action potential in the muscle fibre which it innervates.

Even in the absence of a nerve action potential, individual vesicles in the presynaptic nerve terminal fuse with the plasma membranes from time to time and release their packets of transmitter. Each packet is called a **quantum**. Every time one is released, a **miniature end-plate potential** results. This is 0.5–1 mV in magnitude, far too small to reach the threshold for generation of an action potential. The end-plate potential produced by an action potential in the motor nerve is due to synchronous release of about 200 quanta.

Agents used to block neuromuscular transmission, for instance to produce relaxation during a surgical operation, are of two distinct types. One reduces the number of quanta released by the motor nerve action potential, the other reduces the effectiveness of each quantum on the postsynaptic membrane.

Evidence that acetylcholine is the transmitter at the mammalian nerve–muscle junction

Loewi showed that when the vagus nerve to an isolated perfused frog heart was stimulated, the blood from that heart introduced into a second heart would slow the second heart. Acetylcholine released into the body fluids is rapidly broken down by a chemical reaction catalysed by an enzyme called

cholinesterase. In Loewi's experiment, it was necessary to add to the perfusing fluid a pharmacological agent which inhibited cholinesterase, a so-called **anticholinesterase** in order for the slowing effect to occur. Acetylcholine itself added to the perfusate had this same slowing action but again this depended on the presence of the anticholinesterase. This led to a series of experiments which established acetylcholine as the transmitter released by vagal nerve endings.

Sir Henry Dale and co-workers followed similar procedures to establish that transmission at the nerve–muscle junction is also cholinergic. They established three lines of evidence.

(a) stimulation of the motor nerve caused the release of acetylcholine into the blood perfusing the muscle. As in the heart, this required the presence of an anticholinesterase to inhibit degradation of acetylcholine.
(b) injection of acetylcholine into the artery supplying a muscle caused the muscle to contract.
(c) experiments with pharmacological agents showed that alteration in sensitivity of the end-plate to activity in the motor nerve was accompanied by a parallel alteration in the response to acetylcholine applied directly.

An example of such a pharmacological experiment is to inject succinylcholine into an anaesthetized animal. This chemical, similar in structure to acetylcholine but more stable, attaches to the postjunctional receptors and depolarizes the membrane by opening channels in the same manner as acetylcholine. Since succinylcholine persists at the neuromuscular junction, the depolarization is long lasting. This results in a brief period of repetitive excitation, manifested by transient muscle fasciculation. This is followed by block of neuromuscular transmission; the end-plate is depolarized and the muscle membrane accommodates so that no further action potentials are initiated in the muscle fibres. Stimulation of the motor nerve is now ineffective in producing muscular contraction. The end-plate is depolarized and released acetylcholine cannot depolarize it further. Equally, application of acetylcholine itself to the neuromuscular junction no longer initiates a contraction. This is called **depolarization block**.

Anaesthetists use succinylcholine to induce relaxation in patients and to allow a tracheal incubation to be performed without interference from reflex activity in the skeletal musculature. Succinylcholine is a short-acting relaxant and in order to procure the prolonged relaxation needed for subsequent surgical procedures, a drug such as curare, described later, is used. Such a relaxant allows the surgeon to cut without initiating reflex movements of the patient which would interfere with the operation. Patients must be ventilated artificially since the muscles involved in respiration are paralysed along with other skeletal muscles.

Histochemical observations add further information about neuromuscular transmission. The motor nerve terminal contains an abundance of the enzyme choline acetylase, which catalyses the synthesis of acetylcholine from precursors; this is of importance in maintaining a supply of transmitter in the presynaptic terminals. The postsynaptic membrane contains a high concentration of the enzyme cholinesterase, which breaks acetylcholine down to inactive metabolites. The acetylcholine is released from the presynaptic terminals, attaches to the acetylcholine receptors on the postjunctional membrane, contributing its part to the end-plate potential, and then detaches itself from the receptor. The acetylcholine is then rapidly inactivated by cholinesterase. This mechanism prevents the acetylcholine from re-attaching to other receptors to produce an inappropriately prolonged end-plate potential. Cholinesterase is congenitally absent from a strain of goats. When such a goat jumps, the muscles involved in the movement go into a prolonged contraction due to the end-plate potentials being grossly prolonged. When the goat lands, its limbs are consequently locked and it falls, unable to control the skeletal musculature at the speeds necessary for efficient movement.

The attachment of acetylcholine to its receptors on the postjunctional membrane is competitively blocked by the plant alkaloid, curare. Injection of curare produces blockade of neuromuscular transmission because the curare competes with released acetylcholine for the receptors; this is a quite different mechanism of blockade from that of succinylcholine described earlier. With curare the muscle end-plate is polarized. Curare was used by the South American Indians as the active agent of poisoned arrows; they prepared it from a species of plant called *Strychnos*. When the curare-containing point of the arrow penetrated the skin of the animal or human target, the curare rapidly proved fatal due to paralysis of the muscles of respiration. Curare is administered, along with artificial respiration, to produce prolonged muscular paralysis during surgical operations; curare is a long-acting neuromuscular blocker. When the operation is finished, the administration by the anaesthetist of an anticholinesterase such as physostigmine reduces the hydrolysis of acetylcholine released at the neuromuscular junction and thus reverses the neuromuscular blockade.

Transmission at other junctions

In outline the process of synaptic transmission is similar at other junctions, such as nerve to smooth muscle, nerve to gland, and nerve to nerve. The structure of the synaptic cleft differs, however. At the neuromuscular junction, the synaptic cleft is narrow and is insulated from the surrounding interstitial fluid. This limits the leakage of acetylcholine. The neuromuscular junction is also rich in the enzyme cholinesterase, which promotes the breakdown of acetylcholine to inactive component molecules. One of these molecules is choline, which is reabsorbed by the presynaptic terminal for recycling.

At adrenergic junctions in the autonomic nervous system (see Chapter 15) the cleft is often wide and connects relatively freely with the surrounding interstitial fluid. The chemical transmitter can diffuse and activate specific receptors on other cells. The transmitter may even enter the blood stream and be carried to other cells elsewhere in the body; in this case the transmitter is also a hormone.

4.3 SYNAPTIC TRANSMISSION (NEURO-NEURONAL TRANSMISSION)

The **synapse** is the junction between the axonal terminal of one neurone and the region of membrane of the neurone that it innervates. Unlike the neuromuscular junction, where each muscle fibre is innervated by a single end-plate, each neurone is innervated by many synapses, typically 15 000. A few synapses are shown in Figure 4.3a. In other respects, synaptic transmission has similarities with neuromuscular transmission. Structurally, there is **no physical continuity** between the cytoplasm of the presynaptic and postsynaptic cells. A chemical **neurotransmitter** mediates the transmission.

Mechanisms in synaptic transmission

Every step in the process of synaptic transmission is capable of being influenced by such factors as blood-borne hormones. The present account is of the principal mechanisms that have been identified. No synapse has all these features; the components provide a repertoire of features and an indivdual synapse exhibits a selection.

Vesicles are synthesized in the cell body and transferred by axoplasmic transport to the presynaptic terminal (Figure 4.3b). The nerve terminal is rich in mitochondria, which act as the energy factories for resynthesizing active transmitter. Neurotransmitter is stored in the vesicles. An action potential arriving along the presynaptic nerve causes the opening of voltage-gated calcium channels.

Figure 4.3 (a) The cell body of a motoneurone in the spinal cord. (b) The principal mechanisms involved in synaptic transmission. (c) An excitatory synapse and an inhibitory synapse in the cerebral cortex. Points of difference are: the shape of synaptic vesicles, round for excitatory and oval for inhibitory; density of vesicles, dense for excitatory, sparse for inhibitory; thickness of postsynaptic membrane, thickened for excitatory but not for inhibitory.

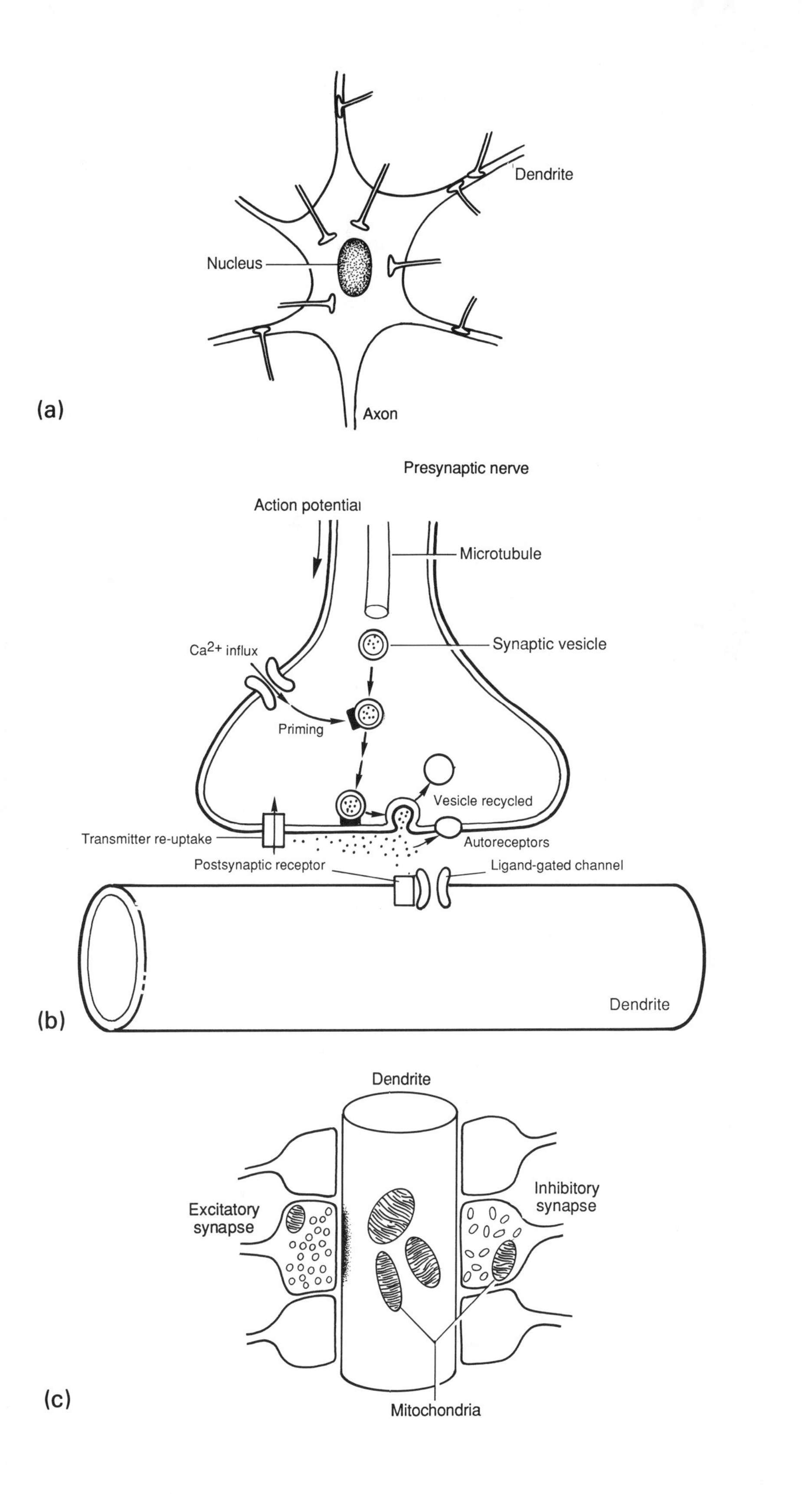
Dendrite
Nucleus
Axon
(a)
Presynaptic nerve
Action potential
Microtubule
Synaptic vesicle
Ca2+ influx
Priming
Vesicle recycled
Transmitter re-uptake
Autoreceptors
Postsynaptic receptor
Ligand-gated channel
Dendrite
(b)
Dendrite
Excitatory synapse
Inhibitory synapse
Mitochondria
(c)

Calcium ions flow into the axoplasm from the extracellular fluid and this leads to priming of vesicles, by attachment of a protein called synapsin I, so that they adhere to the synaptic membrane. The vesicles fuse with the membrane and release their contained neurotransmitter chemical into the synaptic cleft. The release of neurotransmitter is regulated by neuromodulator chemicals that attach to receptors on the presynaptic membrane; modulation may also occur as a result of binding of the neurotransmitter chemical to receptors on the presynaptic membrane. Once they have discharged their packet of neurotransmitter, vesicles are retrieved from the membrane, recharged by a new supply of neurotransmitter and used again. This process is not 100% efficient, and stores of vesicles must be topped up by transport of new vesicles from the cell body.

Neurotransmitter released into the synaptic cleft diffuses across and attaches to receptors on the postsynaptic membrane. The influence of neurotransmitter is limited by diffusion away from the synaptic cleft. Additionally, for some synapses, neurotransmitter is inactivated by hydrolysis. Yet another mechanism is the re-uptake of transmitter by the presynaptic terminal for recycling. The attachment to postsynaptic receptors causes a conductance change in membrane channels. Ions flow through these channels causing a change in the membrane potential of the postsynaptic cell. This potential is called a **postsynaptic potential**.

Excitatory and inhibitory synapses

There are two types of synapse: **excitatory** synapses that bring the postsynaptic cell nearer to threshold for firing action potentials, and **inhibitory** synapses that have the opposite effect of reducing the likelihood of an action potential in the postsynaptic cell. Interaction between these two types of synapse provides the basis for interaction of inputs to a single neurone. In humans, at the neuromuscular junction in skeletal muscle, all the end-plates are excitatory; there is no parallel to the inhibitory synapse in the central nervous system.

Structural correlates of excitatory and inhibitory synapses

In the higher centres, including the cerebral cortex, there are structural differences, at the electron microscopic level, between excitatory and inhibitory synapses (Szenthagothai, 1978). Figure 4.3c shows excitatory and inhibitory synapses. The differences between the two types of synapse lie in the shapes of the synaptic vesicles and in the thickness of the postsynaptic membrane. Considering the postsynaptic membrane first, we subdivide synapses into those with symmetrical and those with asymmetrical synaptic membranes (Gray, 1961). This refers to the relative density, as viewed under the electron microscope, of dense material adjacent to synaptic membrane. If this is of approximately equal density on pre- and postsynaptic sides, the synapse is labelled symmetrical. When the postsynaptic membrane is more dense than the presynaptic membrane, the synpase is asymmetrical.

In synapses with an asymmetrical synaptic membrane, the synaptic vesicles are round and densely packed, whereas in those with a symmetrical membrane the vesicles are oval and sparse (Uchizono, 1967). It seems likely that the synapses with an asymmetrical membrane and round vesicles are excitatory, whereas those with symmetrical membrane and oval vesicles are inhibitory. The difference in shape is probably an artefact of the fixation procedure through which brain tissue must be put to prepare it for electron microscopy; Gray (1969) gives a good review of the situation and concludes that the functional significance of these structural differences is entirely unknown.

The situation is shown in outline in Table 4.1.

Table 4.1 Comparison of excitatory and inhibitory synapses

	Presumed excitatory	*Presumed inhibitory*
Synaptic vesicles	Round	Flattened or oval
Symmetry	Asymmetrical	Symmetrical

This schema applies to only parts of the brain such as the cerebral cortex and cerebellum, not necessarily to lower centres such as the basal ganglia.

The resting potential of a neurone

This is recorded from a recording micropipette inserted into the cell body (Figure 4.4) At the start of the record, the resting potential shows two obvious differences from the recording made from a nerve axon. First, the membrane potential is less negative in the cell body. It is hazardous to give a value for the membrane potential because it is continually fluctuating. However, a typical value is about −65 mV, compared with −80 mV in the nerve axon. This depolarization is due to continual excitatory synaptic input. Second, the baseline is constantly varying, due to the asynchronous activation of different synapses.

The excitatory postsynaptic potential

The **excitatory postsynaptic potential** is the potential recorded in the postsynaptic cell when an action potential reaching the end of the presynaptic axon activates an excitatory synapse. The excitatory postsynaptic potential is a depolarizing potential; it consists of an upstroke lasting around 0.1 ms, and a much slower exponential relaxation back to the resting potential, lasting

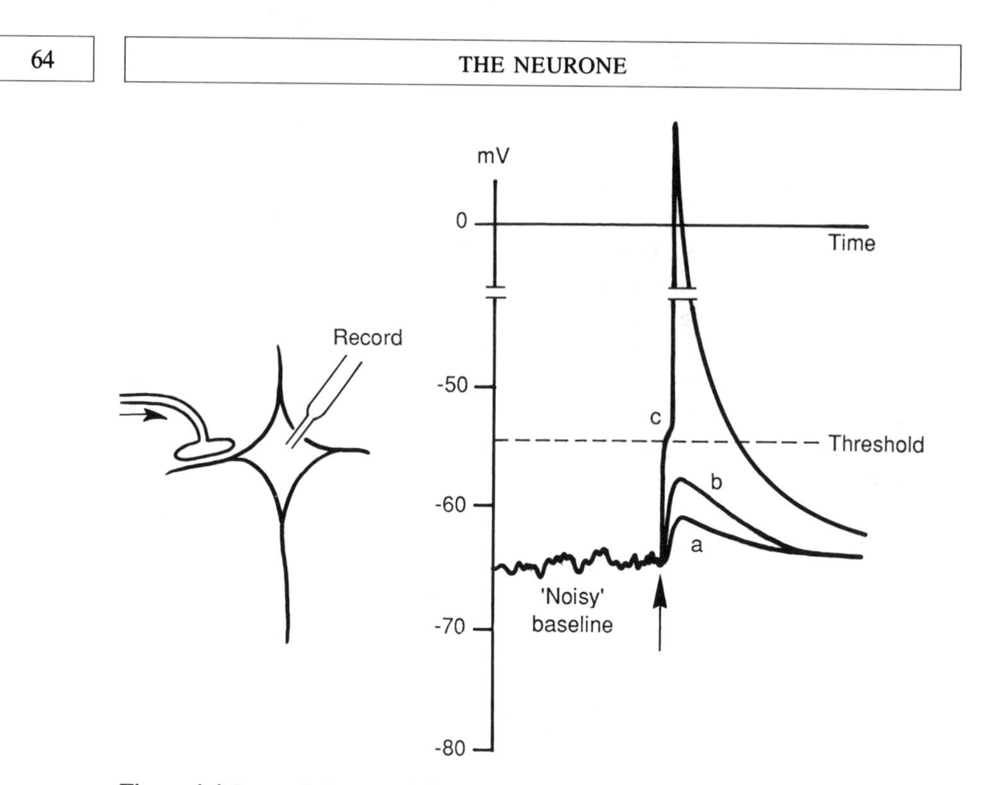

Figure 4.4 Intracellular record from a motoneurone. The arrow indicates the instant of synaptic excitation as a result of stimuli applied to a pathway sending excitatory nerve fibres to the neurone. The stimulus was weak in trace a, intermediate in strength in trace b and strong in trace c. In c, the synaptic excitation reached threshold for initiation of an action potential shown in the diagram with its top truncated.

about 3 ms. The magnitude of the response depends on the strength of the stimulus. A weak stimulus activates a small number of excitatory boutons, and the excitatory postsynaptic potential is small (Figure 4.4a). If more boutons are excited the excitatory postsynaptic potential is larger (Figure 4.4b). If yet more synapses are activated the depolarization reaches threshold and an action potential is fired (Figure 4.4c). As with the receptor potential, the excitatory postsynaptic potential is a graded potential; the magnitude of depolarization signals the intensity of excitation. The maximum postsynaptic potential caused by activity in a single excitatory synapse is less than 1 mV, quite inadequate to bring the neurone to threshold. Activation of 20 excitatory synaptic knobs is typically required to fire off an action potential in the postsynaptic neurone. As the strength of the excitatory stimulus is increased, more and more presynaptic nerve fibres discharge. The gradation in synaptic response is due to differences in the number of synaptic knobs that are activated.

The ionic mechanism of the excitatory postsynaptic potential is similar to that of the receptor potential at the sensory receptor. It is produced by the opening of membrane channels that allow the passage of all small inorganic ions. The ions present in significant amounts are sodium, potassium, and chloride. The excitatory synaptic effect is a short-circuit of the membrane. The equilibrium potential for the excitatory postsynaptic potential is close to zero (it is, in fact, about $-10\,mV$). The generator potential and the excitatory postsynaptic potential thus share a common ionic mechanism different from the action potential, which is an opening of membrane channels specifically for sodium ions.

Initiation of the nerve impulse

A neurone integrates synaptic potentials by summing these local events at a site specialized for impulse initiation (Eccles, 1955). This site is the **unmyelinated initial segment** of the axon, already mentioned. At a critical level of about 10 mV of depolarization, the initial segment generates an action potential. This then propagates along the axon away from the soma as the conducted nerve impulse. This situation is reminiscent of the generation of nerve impulses in the receptor, described in Section 3.1.

The cell body is itself electrically excitable, but its threshold for generation of an impulse is at a depolarization of around 25 mV. In normal circumstances this depolarization depends on passive spread of current from the impulse generated in the initial segment. The sequence of events (Figure 4.5a) is thus synaptic depolarization of the dendrites and soma, passive spread to the initial segment, an impulse generated in the initial segment, forward propagation of an action potential along the axon, and an action potential sweeping backwards over the cell body and proximal dendrites.

The reason for the high threshold of the soma compared with the initial segment is that much of the membrane of the soma is occupied by synapses and is thus not electrically excitable. Synaptic membrane and electrically excitable membrane presumably form a mosaic. When the soma is depolarized, membrane current is divided between the two types of membrane. The threshold for initiation of an impulse is therefore higher than in the initial segment, where a much higher proportion of the membrane is electrically excitable.

The inhibitory postsynaptic potential

Many synapses in the central nervous system are ihibitory: in the cerebral cortex about 10% of synapses are inhibitory, 90% excitatory. Activity in inhibitory synapses renders the postsynaptic cell less excitable.

The synaptic mechanism at this and other central inhibitory synapses is predominantly an opening of very narrow membrane channels. The only ions present in significant concentration that can pass through these channels are

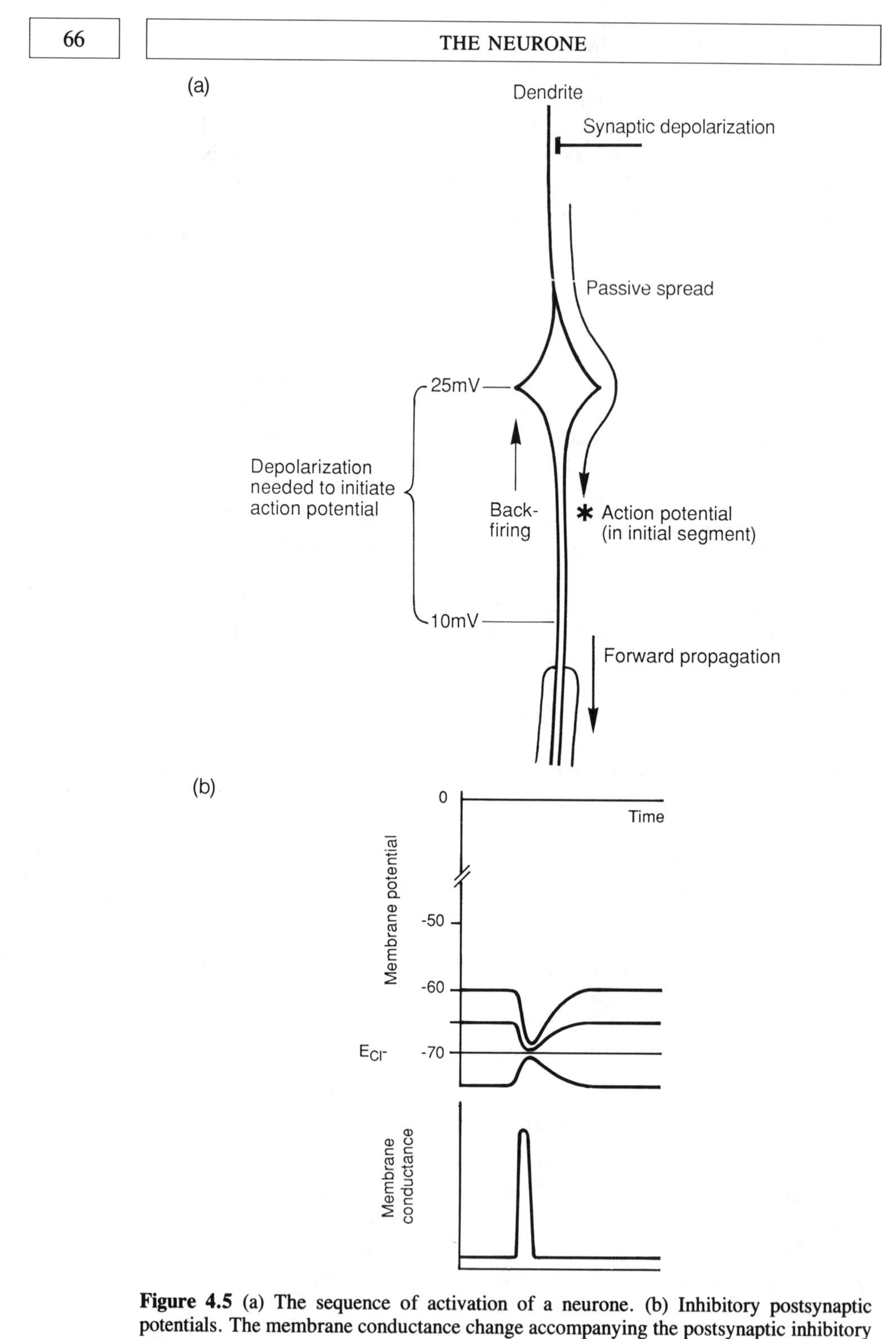

Figure 4.5 (a) The sequence of activation of a neurone. (b) Inhibitory postsynaptic potentials. The membrane conductance change accompanying the postsynaptic inhibitory effect.

chloride ions. Potassium ions, with an atomic weight of 39 are larger than chloride ions, which have an atomic weight of 35.5. Although the atomic weight of sodium (23) is less than that of chlorine, sodium ions are hydrated and are physically larger than chloride ions. For spinal motoneurones, the equilibrium potential for chloride is typically −70 mV. When chloride channels are opened by the inhibitory transmitter, the membrane potential moves towards this equilibrium potential for chloride ions. If the membrane is slightly depolarized, at perhaps −60 mV, the inhibitory effect is recorded as a hyperpolarization of almost 10 mV (Figure 4.5b). If the membrane potential is initially −65 mV, then the inhibitory effect appears as a smaller hyperpolarization of rather less than 5 mV. If the membrane potential is initially −70 mV, then the inhibitory effect yields no change in the membrane resting potential, and if the membrane potential is initially −80 mV, then the effect of activating the inhibitory synapse is a depolarization of the postsynaptic membrane. Measurement of the membrane conductance will show, in each of the four cases, a similar transient increase, signalling the opening of chloride channels (Figure 4.5b). (The quantitative effect of opening chloride channels is best understood by reference to the 'batteries and conductances' representation of the neural membrane, Figure 2.13.)

Features of junctional transmission

By contrast with the propagation of the action potential, which can occur in either direction along a nerve or muscle fibre, junctional transmission is unidirectional. Another point of contrast is that junctional transmission is much more liable to fatigue. Because of this, and the fact that junctional transmission is chemically mediated, transmission can be blocked or enhanced by means of chemicals similar in structure to the transmitter chemical. The junctions between excitable cells provide the sites at which many therapeutic agents act. Since different junctions use different chemicals for transmission, selective blockade or facilitation of a specific type of junction is possible, e.g. in blockade of sympathetic junctions for lowering blood pressure.

4.4 THE TWO TYPES OF NEURONAL MEMBRANE

At this stage we compare the behaviour of the two types of neuronal membrane which we have considered in previous sections: action potential membrane and synaptic membrane.

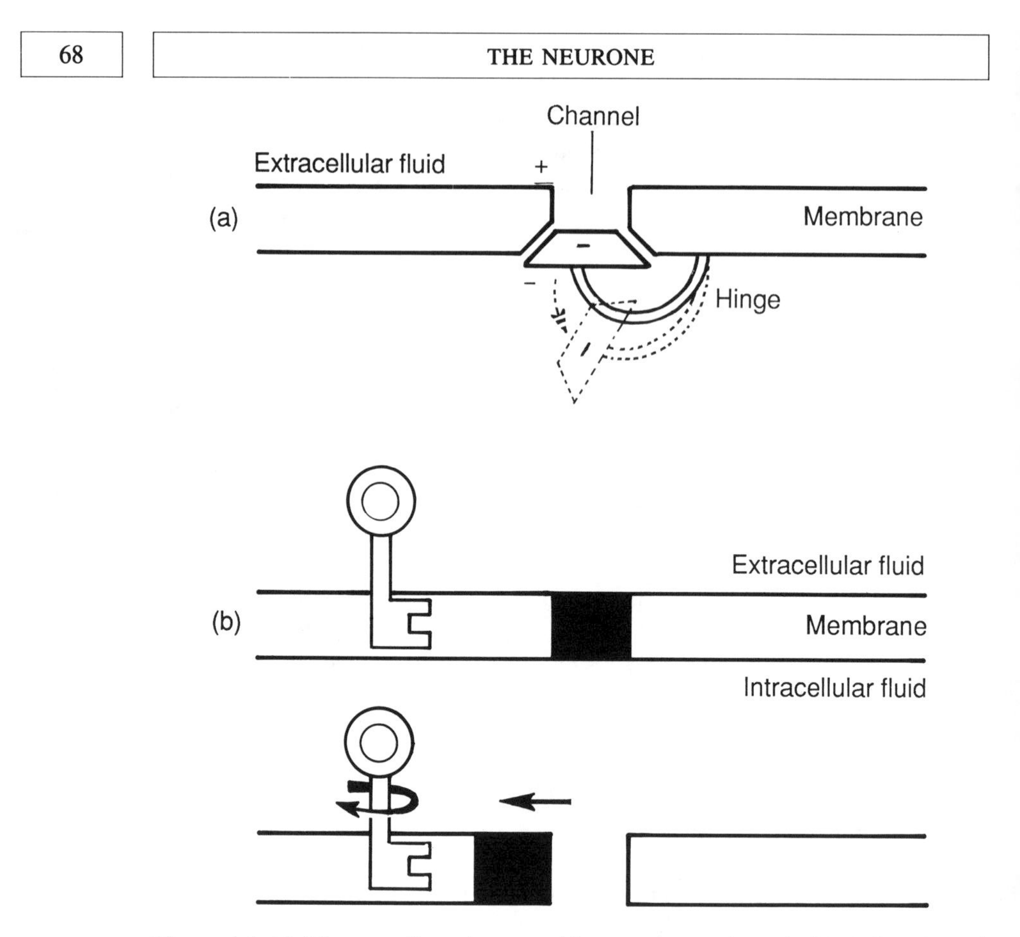

Figure 4.6 (a) Diagram of a voltage-sensitive membrane channel. (b) A ligand-gated membrane channel.

Action potential membrane (voltage-gated channels)

This is membrane which has conductance properties which are voltage dependent; the specific factor which changes the membrane conductance is a change in the membrane potential. Depolarization of the membrane causes an increase in the sodium conductance. As a result there is a net entrance of sodium ions which adds to the depolarization; the opening of the channels leads to a change in membrane potential which causes the opening of more sodium channels.

Figure 4.6a shows one way in which a sodium channel may be constructed. The channel is of the appropriate shape to allow sodium ions to pass; no other ions have the correct configuration to pass. At rest the channel is blocked by a plug on a hinge. The internal negativity repels the charge on the gate and pushes it firmly closed. A decrease in the membrane potential reduces this force and the gate tends to open. Sodium entry causes the internal potential to rise further and more gates are opened. This is positive feedback. This voltage-dependent sodium conductance is specifically blocked by a snake venom called **tetrodotoxin**.

Membrane housing channels with these properties covers the axon and also exists over the soma. The density of voltage dependent sodium channels is highest at the node of Ranvier of myelinated nerve fibres.

Synaptic membrane (ligand-gated channels)

The other type of membrane is synaptic membrane, in which membrane channels open as a consequence of a specific neurotransmitter acting on the postjunctional receptors. Figure 4.6b symbolizes the behaviour of such a channel. This membrane incorporates channels whose conductance is almost always increased by transmitters, though channels do exist whose conductance is decreased by transmitter (Section 4.6). The first postjunctional membrane to be studied was that at the skeletal neuromuscular junction: here the membrane conductance is not affected by the membrane potential (Katz, 1966). This is also true at most postsynaptic sites but for one important type of channel, the n-methyl-D-aspartate operated channel (considered in Chapter 17) the channel conductance is voltage-dependent.

The channels in the postsynaptic membrane can only be opened by a molecule of the right shape. Physiologically, the molecule to open the lock is the specific neurotransmitter chemical. Pharmacologically, the opening can be opened by other keys which have the right configuration to fit (so-called 'agonists') and blocked by a molecule that is almost but not quite the right shape (so-called 'antagonists').

In areas of membrane which are covered with synaptic membrane alone, the membrane conductance is not voltage dependent and the membrane will not generate action potentials. This is the case at the end-plate of the skeletal neuromuscular junction; an action potential travelling along a muscle fibre does not invade the end-plate region; the latter therefore acts as a shunt and reduces the voltage of the action potential recorded here (Katz, 1966, p 122).

Synaptic membrane completely covers much of the membranes of dendrites. It is also intermingled, probably as a mosaic with action potential membrane, over the soma and the regions of the basal dendrites closest to the soma. The correlation of this layout with the high threshold for generating an action potential at the soma has already been described.

4.5 APPENDIX A: NEUROMUSCULAR TRANSMISSON AS AN EXCEPTIONAL SYNAPTIC MECHANISM

The command formulated by the central nervous system is carried unmodified to the skeletal muscle. This requires high safety factors of conduction and junctional transmission. Modification of the message is not a feature. The neuromuscular junction is a structure with specialized features to ensure this high safety factor and as such is rather atypical as a junction. The specializations include:

(a) the release of a large amount of transmitter
(b) the narrowness of the synaptic cleft and the sealing of the cleft surround to minimize leakage of acetylcholine to other receptors
(c) the presence of densely packed receptors in the postsynaptic membrane. The infolding of this membrane is to accommodate more receptors.
(d) The presence of acetylcholine esterase to reduce rapidly the concentration of acetylcholine so that the end-plate potential declines rapidly. This is to prevent repetitive firing. Without this mechanism, acetylcholine in the middle of the end-plate region would influence many receptors as it diffuses away.

At junctions in the autonomic nervous system and central nervous system, there are interactions and modifications, as we shall see in later chapters. The efficacy of transmission there is modified by

(a) changes in amount of transmitter release
(b) modification of sensitivity of pre- and postsynaptic membranes by other chemicals, called neuromodulators
(c) the release of more than one transmitter chemical to influence different postsynaptic receptors.

For these reasons, the skeletal neuromuscular junction has some features that are the exception rather than the rule in neuro-effector junctions.

4.6 APPENDIX B: CLOSING OF MEMBRANE CHANNELS AS A SYNAPTIC MECHANISM

Until now, all the synaptic mechanisms which we have encountered have involved the opening of membrane channels by transmitter chemicals. Examples of transmitter chemicals achieving their effects by closing membrane channels are common. In the nervous system an example is afforded by certain of the sympathetic ganglia, such as the superior cervical ganglion whose function is described in Chapter 19. This ganglion is a synaptic relay, the transmitter chemical being acetylcholine. A volley of impulses in the preganglionic nerve causes a rapid synaptic depolarization of the neurones in the ganglion by opening of membrane channels that allow all small ions to cross, similar to the end-plate potential in muscle or the excitatory postsynaptic potential in the motoneurone. After this there occurs a second wave of depolarization that follows a slower time course than the first 'fast' synaptic potential; the second slow depolarization is brought about by closure of potassium channels. A reduction in potassium conductance results in the membrane potential being further from the potassium electrode potential. The result is depolarization. The time course is slow because the net membrane current that charges the membrane capacitance flows through a high resistance. The time taken to charge a capacitance depends on the magnitude of the capacitance multiplied by the magnitude of the resistance through which charging is taking place.

FURTHER READING

The synapse. In Shepherd, G.M. (1988) *Neurobiology* (2nd edn). Oxford, Oxford University Press.

Neuronal interactions 5

5.1 SYNAPTIC INTEGRATION

The concluding section of Chapter 4 was about synaptic potentials, a theme to which we now return. We noted that there are both excitatory and inhibitory synapses. Although these two types of synapse are structurally similar, they use different transmitter chemicals which open channels of different configuration, allowing different ions to pass. The existence of the two types of synapse provides a mechanism for controlling the excitability of the neurone in either the direction of increase or of decrease.

Consider two inputs to a motoneurone, the first excitatory and the second inhibitory (Figure 5.1a). Suppose that the input E alone, the excitatory input, elicits an excitatory postsynaptic potential sufficiently intense to fire an action potential. The inhibitory input I alone elicits an inhibitory postsynaptic potential (Figure 5.1b, E and I, respectively). Activation of both synapses simultaneously yields a postsynaptic potential that is small and fails to reach threshold for initiation of a nerve impulse.

Activation of the excitatory synapse alone opens channels which change the membrane potential towards the equilibrium potential for excitatory postsynaptic potentials; this is close to 0 mV. Activation of the inhibitory synapses alone opens channels that change the membrane potential towards the chloride equilibrium potential. Activation of excitatory and inhibitory mechanisms together opens both types of channel. The excitatory postsynaptic potential may then arise from a more hyperpolarized baseline and this in itself reduces the peak of depolarization to which the excitatory postsynaptic potential climbs. The crucial effect, however, is the concomitant opening of chloride channels, which reduces the excursion of the excitatory postsynaptic potential; this effect is shown in Figure 5.1b by the heights of the double-ended arrows. The arrow elicited by E alone is longer than that elicited by E occurring immediately after I. As a result of this reduction in excursion, the depolarization is insufficient to reach threshold; no action potential is initiated.

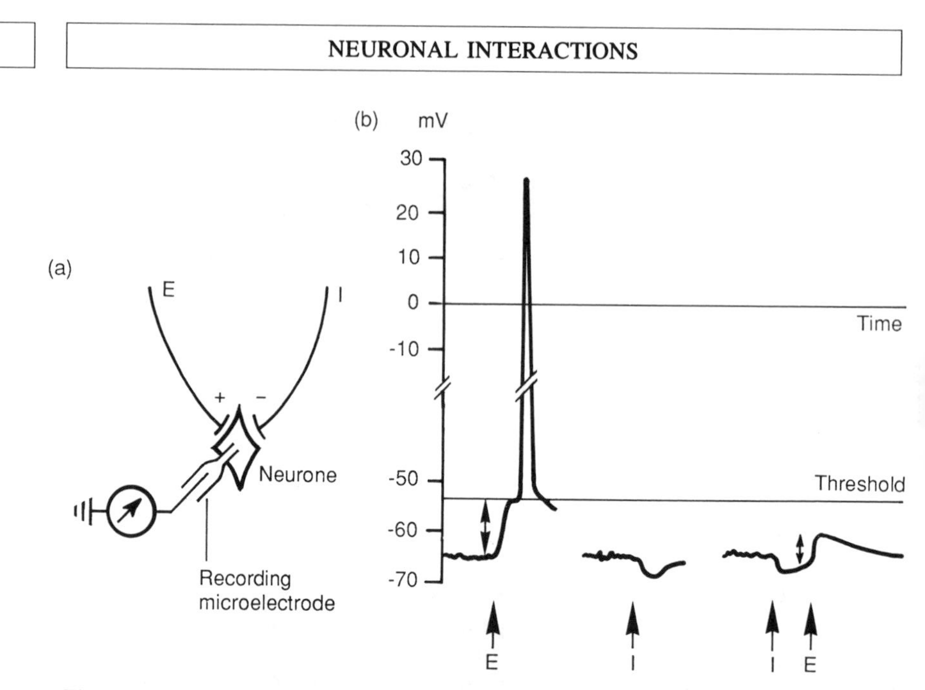

Figure 5.1 (a) Diagram of a neurone subject to excitatory (E) and inhibitory (I) synaptic input. A microelectrode records the membrane potential of the neurone. (b) Diagram of the neuronal membrane potential as it is influenced by input E alone, input I alone, and both inputs together. The excursions of the excitatory postsynaptic potentials are shown by arrows. The reduction in effectiveness of the excitatory postsynaptic potential when it is preceded by I is principally due to a reduction in the excursion of the EPSP consequent on the opening of chloride channels by the input I. The slight hyperpolarization of the membrane proudced by I is unimportant; if the ambient resting potential of the neurone had been −70 mV, input I would have caused no change in membrane potential but, by opening chloride channels, would have been just as efficacious in moderating the excursion of the EPSP.

This may be understood by appeal to the analogy of a tug of war. The excitatory synaptic input is tugging in the direction of zero membrane potential and the inhibitory input is tugging towards the chloride equilibrium potential of −70 mV. With the excitatory input alone, this effect easily tugs to the threshold potential and a nerve impulse is initiated. When the inhibitory input is added, it pulls towards −70 mV and prevents the excitatory input from pulling as far as the threshold.

This interaction of synaptic potentials produced by activity of different inputs is the basis of the ability of the central nervous system to integrate information coming from many sources.

The role of inhibition

The central nervous system consists of about 10 000 000 000 neurones each linked to many others. The profuse interconnections amongst neurones, particularly those of the cerebral cortex, means that activity in one neurone influences that of neighbouring cells which feed back to the original cell directly, or after a few more relays. If the interconnections were all excitatory, the system would quickly degenerate into uncontrolled positive-feedback oscillations. So an abundance of inhibition is necessary. When this inhibition is disturbed, large numbers of neurones discharge in synchrony, causing the pathological condition of epilepsy.

In man, as in other mammals, motoneurones supplying skeletal muscle are all excitatory. Muscular contraction is inhibited by stopping the firing of motoneurones supplying that muscle.

Some specific examples of the role of inhibitory pathways in the spinal cord will now be considered. The pathways involving postsynaptic inhibition involve a type of neurone that we have not encountered to date, the **interneurone**. An interneurone is a neurone with a relatively short axon which projects within the grey matter to form synaptic connections with other neurones close to its cell body. Interneurones are found in all regions of central grey matter. In the grey matter of the spinal cord, the axonal ramifications are confined to segments close to that in which the cell body lies. In the cord, interneurones outnumber motoneurones by at least 30-fold.

5.2 THE ORTHODROMIC INHIBITORY PATHWAY

Function of the pathway

Contraction of the muscles producing any movement is accompanied by relaxation of the antagonist muscles. This relaxation is mediated by the cord pathway which we now consider. The reason for the name 'orthodromic' will become clear in Section 5.3.

The neural circuit

In the tendon jerk reflex, a tap to the tendon initiates a burst of impulses passing along the Group Ia afferent nerve fibres, one of which is shown diagrammatically in Figure 5.2a. The axon projects directly to the motoneurone on the left without involving an interneurone. The motoneurones so activated innervate the extrafusal fibres of the muscle that was originally stretched and the action potentials in the motor nerve fibres cause the muscle to contract. This is the classical tendon jerk reflex, considered in more detail in Section 7.2. The sensory nerve fibre branches within the dorsal horn of the spinal cord; a side branch (collateral) projects to an inhibitory interneurone in the spinal cord that inhibits motoneurones innervating antagonist muscles.

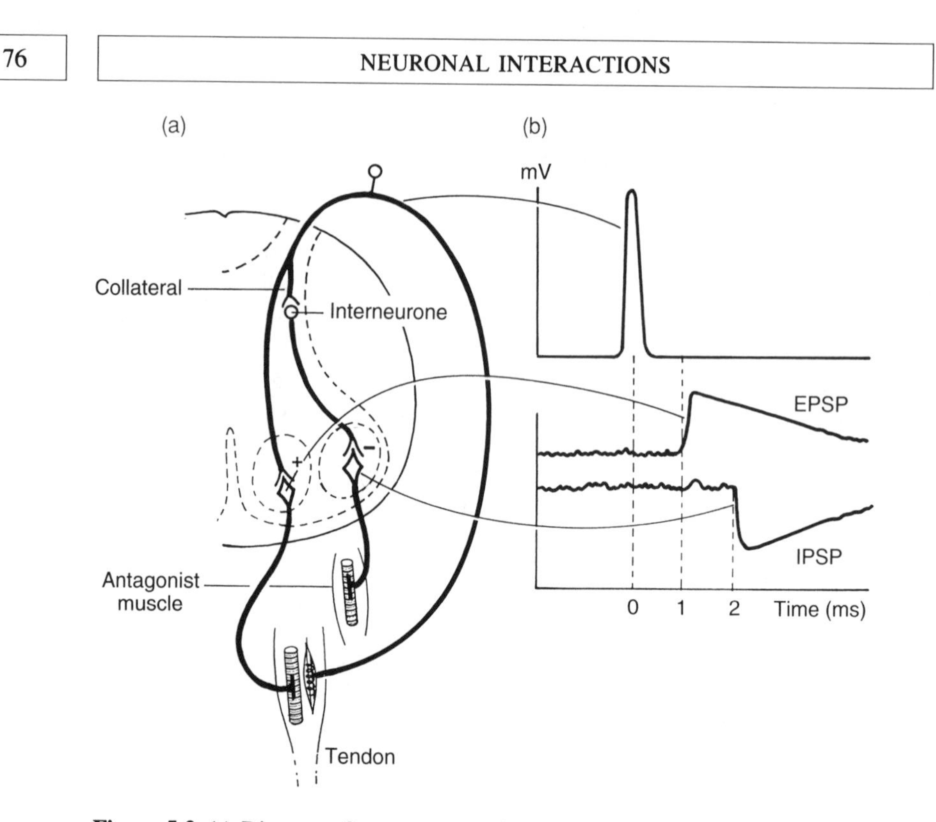

Figure 5.2 (a) Diagram of transverse section of the spinal cord showing the pathways for the tendon-jerk reflex and for orthodromic inhibition of motoneurones innervating antagonist muscles. (b) Records of the afferent nerve volley and intracellular records from the motoneurones.

The inhibitory pathway contains one synaptic relay more than the excitatory pathway. This is shown by an intracellular recording made from a motoneurone in each of the two motoneurone pools (Figure 5.2b). The time delay between the afferent volley recorded in the dorsal root and the excitatory postsynaptic potential in the excited motoneurone is around 1 ms. A further 1 ms elapses before the inhibitory postsynaptic potential in the motoneurone supplying fibres of the antagonist muscle. This is the extra time contributed by the second synapse. This inhibitory effect is produced by a **disynaptic** reflex.

The need for an interneurone on the inhibitory pathway is dictated by a phenomenon of neurophysiology that, **for a given neurone, either all the terminals of the neurone are excitatory or all are inhibitory**. This allows us to divide all neurones into two sets, **excitatory** and **inhibitory**.

The group Ia afferents from the muscle spindle receptors form excitatory synapses. Accordingly, impulses in these same afferent fibres cannot themselves mediate inhibition of motoneurones to antagonist muscles.

An inhibitory interneurone is therefore interposed in the pathway. The primary afferent excites this interneurone, which in turn inhibits the motoneurone innervating antagonist muscles.

Another important principle in spinal cord physiology is that **inhibition in the spinal cord is mediated by neurones with short axons**. There is no recorded instance of a central inhibitory action being directly produced by primary afferent nerve fibres. This is also true of fibres projecting from motor cortex to cord, which mediate inhibition by interneurones. The same is probably true of other long descending tracts. The interpolation of the interneurone in the inhibitory pathway is necessary to change the chemical transmitter from an excitatory type to an inhibitory type. This presumed necessity accounts for an antatomical arrangement that in other respects introduces two disadvantages, namely the hazard of an additional synaptic relay and the consequent additional delay of at least 0.8 ms (Eccles, 1964, p. 209).

In higher centres this rule is not observed. An example is to be found in the Purkinje cells of the cerebellar cortex, which project to the cerebellar nuclei and are inhibitory (Chapter 12).

To return to the spinal cord, all inhibitory neurones are interneurones, but the converse is not true; there is an abundance of excitatory interneurones. Indeed the excitatory interneurones outnumber the inhibitory interneurones.

5.3 THE RECURRENT INHIBITORY PATHWAY

Function of the pathway

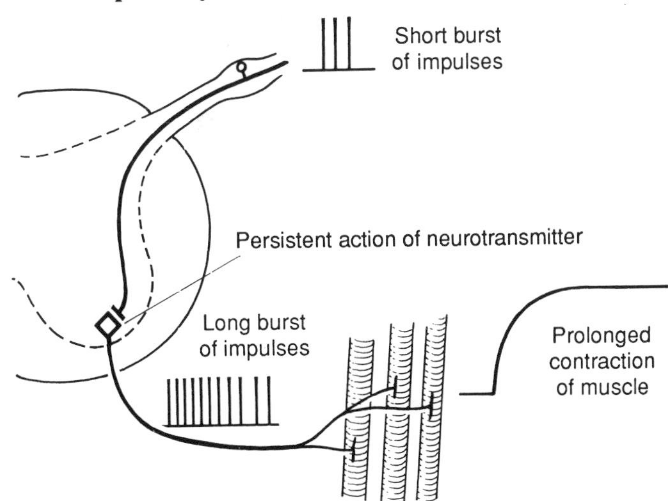

Figure 5.3 The effect on muscle contraction of persistence of neurotransmitter action at excitatory synapses, of which one is shown. A short burst of afferent impulses elicits a prolonged contraction of the muscle.

This is a negative feedback mechanism which improves the spatial resolution of neuronal action. An informal description of the function of this pathway is that it provides a mechanism to ensure that muscles do not contract for too long. In the production of a brief movement, a motoneurone is synaptically depolarized. This is caused by the release of a neurotransmitter chemical from the presynaptic nerve terminals. This chemical may remain in the synaptic gap and its action may persist for some time. By itself, this would result in a long-lasting train of action potentials and therefore a prolonged contraction (Figure 5.3). To avoid this undesirable feature, several mechanisms have evolved for limiting the duration of the effects of synaptic excitation; one of these involves the recurrent inhibitory pathway.

The neural circuit

The circuit for the inhibition following excitation is shown in Figure 5.4 (Eccles, 1955). The motoneurone A has its axon travelling out of the grey matter but, before it leaves, a branch arises from one side. This branch arises from the first node of Ranvier, stays within the ventral grey matter, and passes medially and dorsally. It is called a 'recurrent' branch because it runs back in the grey matter. This recurrent axon forms synapses on interneurones sited in the ventro-medial region of the ventral horn. These interneurones are known as **Renshaw cells** after the investigator who first described them. Their axons project to the motoneurones (both the one initially discharged, shown as motoneurone A, and its neighbours, represented by motoneurone B) and end in inhibitory synapses. This pathway thus mediates feedback inhibition. In accordance with the anatomical layout, this circuit is known as the **recurrent inhibitory pathway**.

Electrophysiological evidence for the pathway

Intracellular recordings from a motoneurone show that an action potential is followed by a profound hyperpolarization (Figure 5.5a). This hyperpolarization consists of a sequence of hyperpolarizing waves which add together to give the prolonged hyperpolarizing envelope. This hyperpolarization depresses the excitability of the neurone and prevents repetitive firing of the motoneurone. In the Renshaw cells a single impulse in a motoneurone results in a long train of discharges. This is due to a prolonged action of neurotransmitter. Each action potential in a Renshaw cell causes an inhibitory postsynaptic potential in the motoneurone, as indicated in Figure 5.5b. This is the origin of the ripple present during the hyperpolarization of the motoneurone. The hyperpolarization and ripple also occur in other motoneurones that are in the same

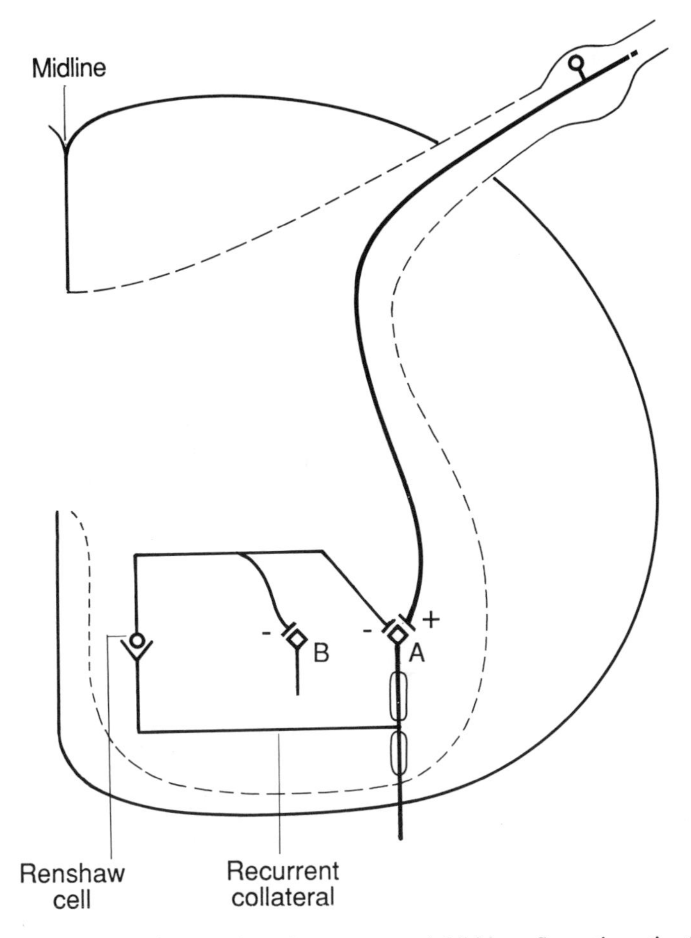

Figure 5.4 The neural circuit subserving recurrent inhibition. Synaptic activation of motoneurone A results in feedback inhibition, both to the motoneurone which discharged an impulse and to neighbouring motoneurones, represented by motoneurone B, which did not themselves discharge.

motoneurone pool but which did not themselves fire an impulse (Figure 5.5c).

Whereas in physiological circumstances the activation of the motoneurone is usually synaptic, the Renshaw circuit can be activated artificially by an electric stimulus to the motor nerve. This causes impulses to travel in both directions, orthodromic and antidromic. The antidromic impulse propagates back to the motoneurone and also along the collateral axonal branch to activate the Renshaw circuit. For this reason the recurrent inhibitory pathway is sometimes called the **antidromic inhibitory pathway**. This is

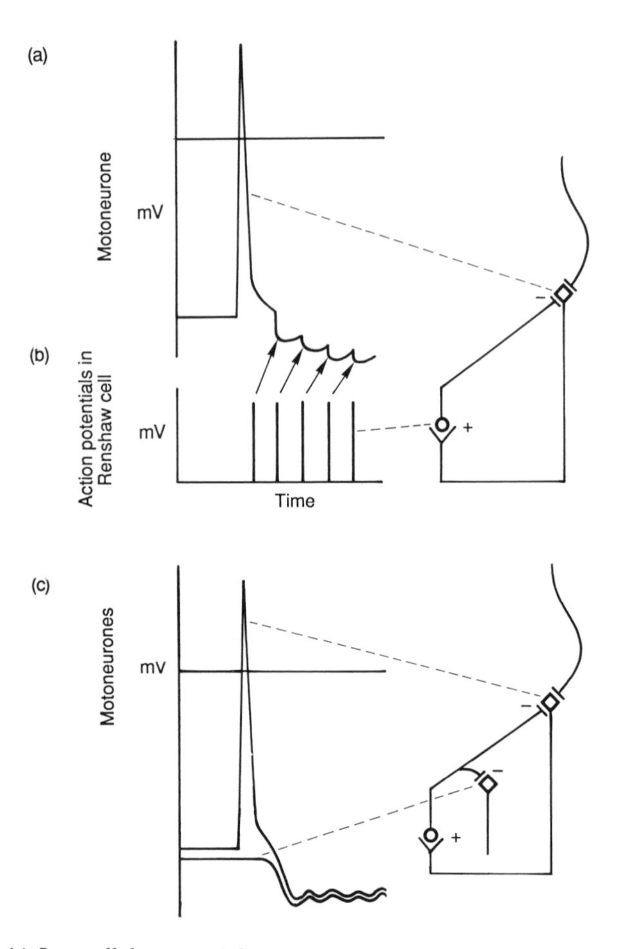

Figure 5.5 (a) Intracellular record from a motoneurone that is synaptically excited to discharge an action potential, together with (b) the train of action potentials occurring in the Renshaw cell. These show the correspondence between the ripples on the membrane potential record from the motoneurone and the action train in the Renshaw cell. (c) Record (a) shown once again together with the intracellular record from a neighbouring motoneurone which was not excited by the afferent volley. This cell nevertheless shows the Renshaw inhibition.

in contrast to the orthodromic inhibitory pathway, already described, which cannot be activated antidromically.

These are two well-documented inhibitory pathways in the spinal cord; there are many others.

5.4 PRESYNAPTIC INHIBITION

So far, the only way that has been considered for one nerve cell to influence another is by means of synapses from the presynaptic terminal to the postsynaptic cell (see, for instance, the synaptic connections in Figure 5.2). The

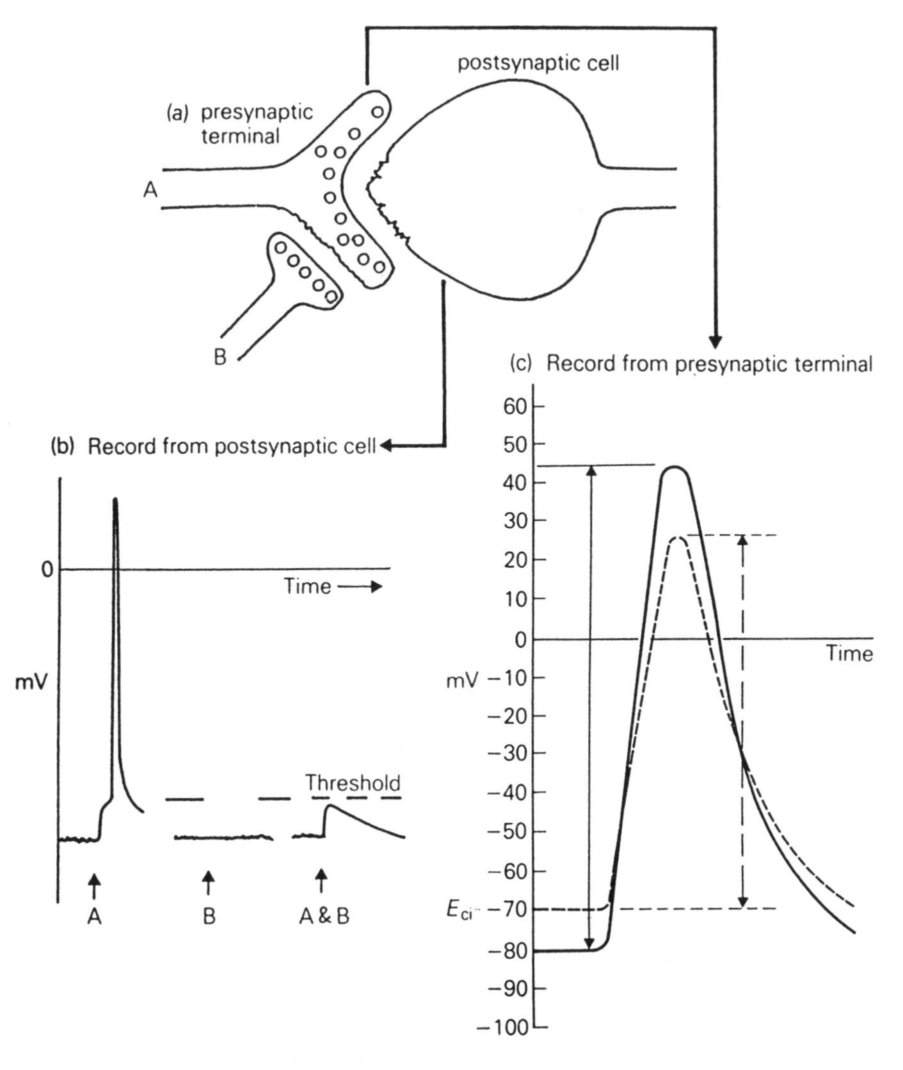

Figure 5.6 Presynaptic inhibition. (a) Diagram of synapses with synaptic vesicles in the presynaptic terminal (A) and in the axon (B) which ends on that terminal. (b) Intracellular record from the postsynaptic cell. (c) Intracellular record from the presynaptic terminal. The full graph is the action potential recorded when axon A alone carries an impulse. The dashed graph is that recorded when both axon A and axon B carry impulses. The vertical indicators show the voltage excursions of the action potentials in these two situations.

synaptic potentials can be recorded from the postsynaptic cell, hence our nomenclature, excitatory **postsynaptic potential** and inhibitory postsynaptic potential. Another mechanism of neural interaction is **presynaptic inhibition** (Dudel and Kuffler, 1961). A normal synaptic terminal (A) itself receives synaptic input from another nerve fibre (B), as shown in Figure 5.6a. Synaptic vesicles in the ending of axon A contain excitatory transmitter which, when released into the cleft between A and the postsynaptic cell, attaches to receptors on the folded postsynaptic membrane of the postsynaptic cell. Synaptic vesicles in ending B contain transmitter which, when released, attaches to the postsynaptic membrane of the presynaptic terminal A.

Let us suppose that the depolarization of the postsynaptic cell produced by an impulse in input A is sufficient to generate an action potential, as shown in Figure 5.6b, arrow A. An impulse arriving along input B causes depolarization of the presynaptic terminal but has no direct effect on the postsynaptic cell (Figure 5.5b, arrow B). There is no effect on the membrane potential of the postsynaptic cell and no change in the membrane conductance. The excitability of the nerve cell body tested by direct electrical stimulation is unaltered. However, if impulses occur in the two inputs together, the depolarization of the postsynaptic cell is less than if input A were active in isolation. As a result, threshold may not be reached; no action potential is fired. Hence the name 'presynaptic inhibition'. As with postsynaptic inhibition, presynaptic inhibition is mediated by interneurones.

Ionic mechanism

The ionic mechanism of this presynaptic effect is an increase in membrane permeability to chloride ions of the presynaptic terminal. (As already indicated, synaptic inhibition in the mammalian central nervous system is mediated by the opening of chloride channels.) The resting potential of the presynaptic terminal is around -80 mV, as in any nerve axon. The equilibrium potential for chloride is typically -70 mV. Consequently, the increased chloride permeability results in a small depolarization of the nerve terminals. More importantly, the influence on the presynaptic terminal membrane potential of the increased sodium conductance responsible for the action potential is partially counteracted by the increased chloride conductance, with the result that the peak voltage of the action potential is reduced. If an action potential propagates along pathway A when the terminal of axon A has an increased chloride permeability, the excursion of the action potential in the presynaptic terminal is reduced, as shown in Figure 5.6c. The amount of transmitter released from the presynaptic terminal depends on the excursion of the action potential invading the terminal; a smaller excursion results in the release of less transmitter, and hence a smaller excitatory postsynaptic potential in the cell body. In summary, presynaptic inhibition is due to a reduction of release of the excitatory neurotransmitter.

The phenomenon of presynaptic inhibition requires us to modify our view of the action potential being all-or-none. In nerve terminals, the amplitude of the action potential can be modulated. Trains of nerve impulses do not pass through relays in the nervous system unmodified. It is unusual for every impulse in a presynaptic nerve fibre to lead invariably to an impulse in the postsynaptic cell. Presynaptic and postsynaptic inhibition both play roles in modulating the transmission at synaptic relays.

The significance of pre- as opposed to postsynaptic inhibition

Presynaptic inhibition depresses transmission from presynaptic axon terminal to postsynaptic cell without altering the general excitability of the postsynaptic cell, allowing the nervous system to block one particular path specifically, leaving all others unmodified. Figure 5.7 shows that inhibition mediated by axon 1 reduces transmission along pathway 2, but leaves the transmissivity of pathway 4 unaltered. Postsynaptic inhibition, subserved by axon 3, offsets the excitation produced by either 2 or 4, since it alters the general excitability of the postsynaptic cell.

This provides the central nervous system with two inhibitory command mechanisms. *Via* presynaptic inhibition, it can target individual excitatory pathways to the motoneurone; *via* postsynaptic inhibition, it can depress the transmissivity of all excitatory pathways to the motoneurone.

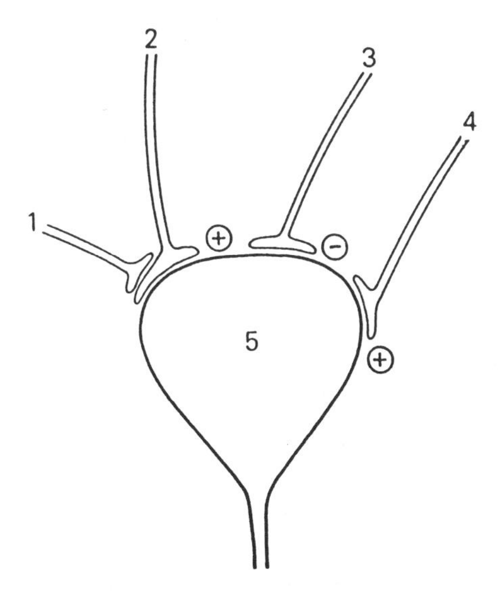

Figure 5.7 Pre- and postsynaptic inhibition compared (see text).

5.5 SURROUND (LATERAL) INHIBITION

Surround or lateral inhibition is found in many relay nuclei in the central nervous system, and is shown diagrammatically in Figure 5.8. The presynaptic axon, in addition to exciting synaptically its target postsynaptic neurone, gives rise to collateral branches which, *via* inhibitory interneurones (not shown in the diagram) or by means of presynaptic inhibition, reduce transmission along off-beam paths. The convention used in this and some later diagrams in the book is as follows. A solid arrow indicates that an increase in activity of the

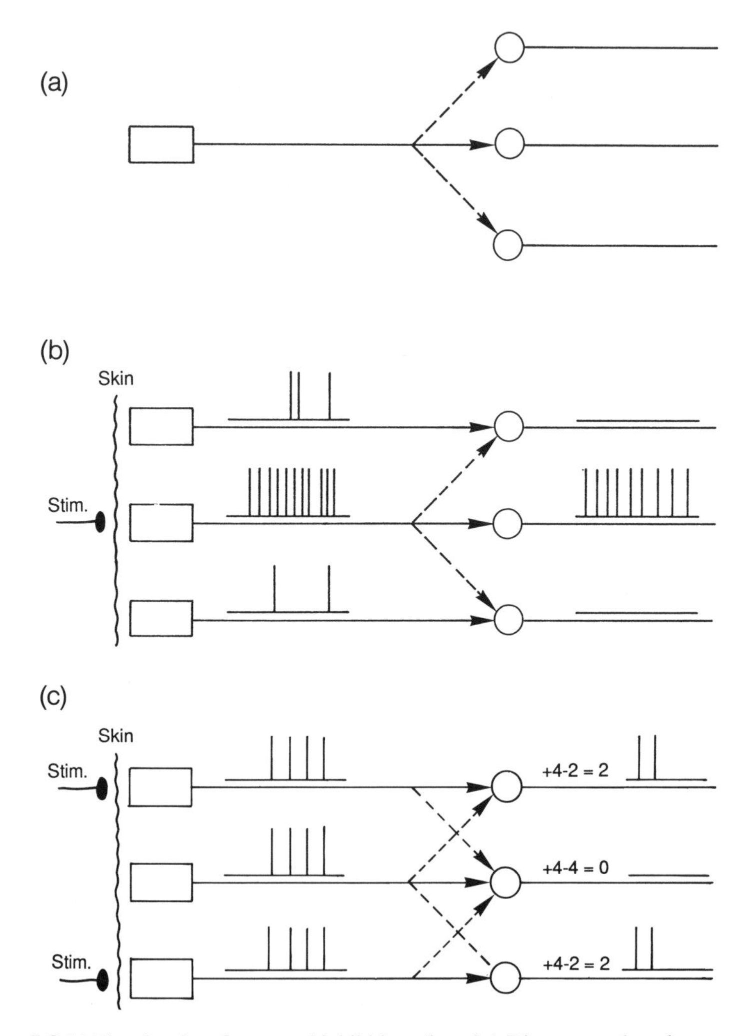

Figure 5.8 (a) The circuity of surround inhibition. (b and c) Diagram to show how surround inhibition sharpens the spatial distribution of activity in neuronal pathways.

receptor causes an increase in activity of the target neurone. A broken arrow indicates that an increase in activity of the receptor causes a decrease in activity of the target. The effect of this arrangement of excitatory and inhibitory effects is to limit and sharpen the profile of central neuronal activity set up by the stimulus, as will now be explained.

Consider three receptive fields of three skin receptors (Figure 5.8b). The stimulus is likely to influence the central receptor strongly and the neighbouring receptors weakly. This spread of stimulus to neighbouring receptors results from the physical properties of the skin: if a hot object touches one point on the skin, heat will be conducted through the skin to surrounding regions. If a point of light exists in the visual field, the optics of the eye can never be perfect and some scattering of light occurs; it is theoretically impossible to design a lens system which focuses a point source without any diffraction of the light. Consequently, a well-defined stimulus elicits activity in a group of receptors and not in just one. There is blurring of the edge of the stimulus profile. Surround inhibition results in suppression of the weak activity in the edge region with little alteration of the activity in the central region.

The intense stimulation of the central receptor results in intense activation of the postsynaptic neurone and inhibition of the neighbouring neurones. The weak stimulation of the receptors on either side results in weak activation of the appropriate postsynaptic neurones, but this excitation is cancelled by the lateral inhibition from the strongly excited path. Of the postsynaptic cells, only the on-beam cell fires impulses, the off-beam ones being silenced by lateral inhibition. The signal is spatially sharpened as it traverses the relay.

If two stimuli are applied to neighbouring sites, for example two points to the skin, or if the eye looks at two stars that are close together, the brain has the problem of deciding whether the information it receives indicates the existence of one distributed stimulus or two separated ones. Clearly, the tendency for stimulation of receptors around the one at which the stimulus is applied mitigates against discrimination of two stimuli. Figure 5.8c demonstrates how lateral inhibition operates here. Each of the two stimuli strongly excites the nearest receptor; this initiates a train of four impulses in each of the sensory nerve fibres. In the figure, it is assumed that there is sufficient spread of energy to excite the receptor lying in the middle and to initiate four impulses in its sensory nerve fibre. The activities of the first-order nerve fibres appears to have lost the information that there are in fact two separate stimuli rather than just a single diffuse stimulus. In order to calculate the distribution of activity in the second-order neurones, it is assumed that each first-order neurone elicits four units of excitation for its own on-beam second-order neurone and two units of inhibition for the neighbouring off-beam second-order neurone. The upper and lower second-order neurones each receive four units of on-line excitation and, from one side, two units of inhibition; they each therefore launch two impulses. The central second-order neurone receives two units of inhibition from each side, a total of four units of inhibition. The on-line excitation is thereby completely extinguished; no impulses pass along the middle second-order neurone. The spatial distribution of activity in the neurones now exhibits two lines of excitation separated by a quiescent line. The central nervous system is informed that the sensory excitation is by two spatially separate stimuli. By means of lateral inhibition, blurring is removed and two-point discrimination is retrieved.

This processing of information as it is handed across synaptic relays carries its own penalty; our brains are presented with predigested information from which certain features have been extracted. The brain is therefore capable of making mistakes in interpretation of this information. This is illustrated by a certain class of visual illusions, such as that shown in Figure 5.9, which consists of columns and rows of white stripes on a black background, like a map of intersecting roads. In the cross-road regions, we perceive a hazy dark ball; this is particularly prominent for crossroads which are far from the point of fixation. The right-hand part of this figure is labelled to give insight into the mechanism of this illusion. A retinal ganglion cell corresponding to a site on a road far from cross-roads (point **a** in Figure 5.9) is intensely activated by on-beam excitation of illuminated receptors. This ganglion cell also receives lateral inhibition from neighbouring receptors on two sides, as indicated by the two interrupted arrows. It thus passes back

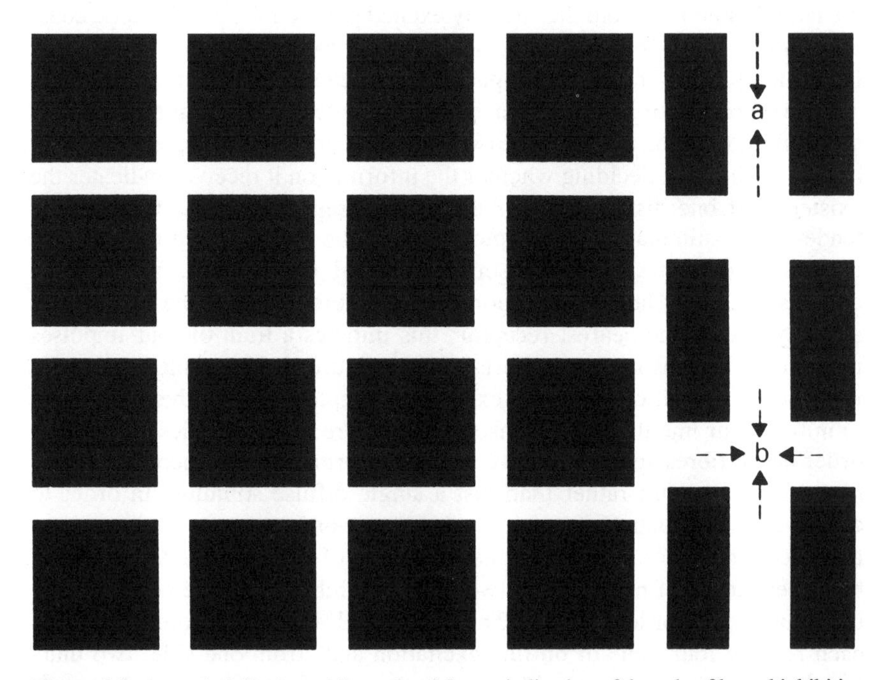

Figure 5.9 An optical illusion with, on the right, an indication of the role of lateral inhibition in producing the illusion.

excitation corresponding to the sum of intense excitation and two portions of inhibition. For a site in the centre of the cross roads, at **b**, the on-line excitation is exactly the same as for point **a**, but here lateral inhibition is acting from four sides, as shown by the four interrupted arrows. The excitation passed back is thus the sum of intense excitation and four portions of inhibition. The excitation passed back is less at cross-roads regions than at regions along the roads. Our perception is a correct representation of the spatial pattern of excitation projecting to the visual cortex.

Another illusion, shown in Figure 5.10, presents the brain with information which it thinks it can interpret and so allows for the surround inhibition. The upper and lower parts of the figure are similar in that in the middle there is a line of sudden change in brightness. Most people perceive both parts of

Figure 5.10 An optical illusion based on graded changes in brightness. The nature of the illusion may be appreciated by placing a finger to cover the edges separating the tones of grey.

the figure as consisting of a light panel on the left and a dark panel on the right. The actual distribution of brightness is different in Figure 5.10 (a) and (b), as indicated in Figure 5.11 (a) and (b). For Figure 5.10a, the left panel is uniformly light and the right uniformly dark; our perception of this distribution is correct. For the lower part, the difference in density of shading is confined to the central region, as shown in the input graph of Figure 5.11b. As one moves away from this central region, the shading changes gradually towards an intermediate value and at the edges of the panels, the darkness of shading is the same on the two sides. The truth of this is readily established by placing a strip of paper vertically across the figure to obliterate the visually confusing gradation.

To understanding the illusion, let us consider first the spatial distribution of intensity of excitation passing along the visual pathway after lateral inhibition has modified the image. At **A**, a point in the bright field close to the edge, the on-line neurones are strongly excited by the illumination. This on-line

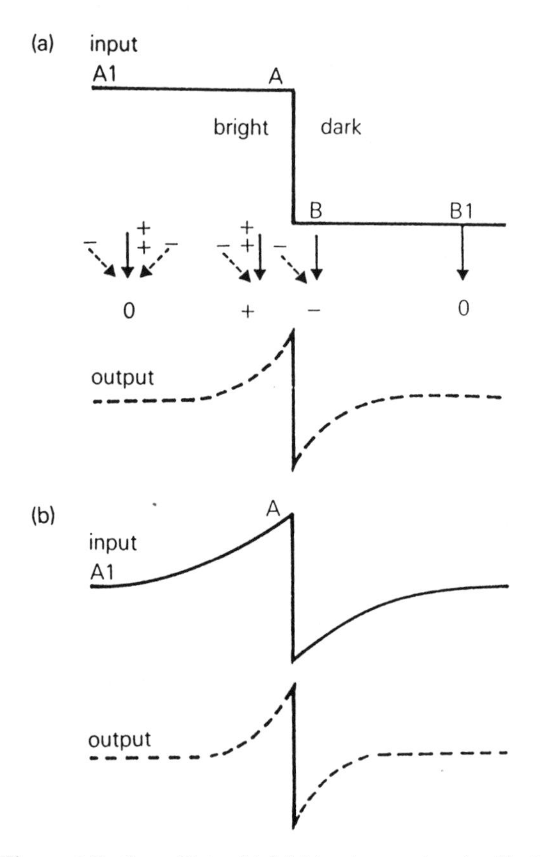

Figure 5.11 The contribution of lateral inhibition in creating the illusion of the previous figure. See text for explanation.

excitation is represented by two plus signs. Lateral inhibition operates from the bright field to the left of **A**; the amount of inhibition is given a value of one minus sign. Since to the right of **A** the field is dark, there is no lateral inhibition from the right. The two pluses and the single minus sum to pass on an intensity of excitation of one unit along the projection pathway. At **A1**, where the intensity of illumination is uniformly high, there is the same level of on-line excitation, but lateral inhibition operates from both sides. Hence, the intensity of excitation travelling along the projection pathway is zero, the sum of two pluses and two minuses. At **B**, just within the darker region, there is no on-line excitation because of the low level of illumination. From the left, there is lateral inhibition from the nearby bright field. The sum of these influences is -1 unit of intensity. This means that the spontaneous activity of neurones projecting along the pathway is inhibited to below its normal value.

At **B1**, where the intensity of illumination is uniformly low, the projection pathway receives neither excitation nor inhibition. The intensity of excitation projected along the pathway is zero, as at **A1**. The neurones exhibit a normal level of spontaneous activity, the level that would occur in the absence of any visual input.

From a consideration of these particular points, we build up the graph of output along the projection pathway shown in Figure 5.8a. This is shown as the output graph of Figure 5.11a in which the level of excitation (y-axis) is plotted as a function of position (x-axis). The brain is always provided with information that has been subjected to lateral inhibition, so this spatial pattern of excitation is correctly interpreted by the brain as being generated by two areas of uniform illumination and this is what we perceive.

For the lower part of Figure 5.10, lateral inhibition operates as usual. Figure 5.11b shows the spatial pattern of input. For point **A1**, the on-line pathway again provides excitation which is almost as intense as the corresponding point in Figure 5.11a. Again, lateral inhibition operates only from the left and again the result is approximately one unit of excitation passed along the projection pathway. At **A1**, where the illumination is intermediate, the projecting pathway receives virtually no excitation since on-line excitation and bilateral off-line inhibition cancel each other. This allows the construction of the output distribution of excitation (Figure 5.11b). The only difference between the output distributions shown in Figure 5.11a and b is that the excitation in b declines slightly more rapidly than that in a. The central nervous system is insensitive to gradual changes. Faced with the spatial distribution of excitation, as in Figure 5.10b, the central nervous system allows for the operation of lateral inhibition and interprets it in the light of previous experience. Our perception of this distribution of intensity is therefore as two fields of relatively constant illumination, since the true spatial distribution of the stimulus is one to which we are very unaccustomed.

To recapitulate briefly, both distributions of light intensity include a step change from light to dark in the middle of the figure. As one moves away from this step, the intensity is constant, as shown in the upper part of the figure, whereas in the lower part it gradually changes. This change is so gradual that it tends to be obliterated as the signal is transmitted through the relays between the retina and the cerebral cortex. The left half of the lower part of the figure is perceived as being almost uniformly bright and the right as being uniformly dark whereas, in fact, the outer parts on both left and right are of the same intensity. The brain perceives an edge in the field and the most likely interpretation is that the left is light and the right is dark. When this unusual edge is obliterated, the true situation of the figures becomes obvious.

5.6 CHEMICAL NEUROTRANSMITTERS IN THE SPINAL CORD

Sir Henry Dale, in 1935, proposed a principle, known as **Dale's** principle, which states that:

> *In the spinal cord, any one neurone releases the same transmitter chemical at all its axonal terminations.*

This principle has required modification in recent years because, in the autonomic nervous system and in the central nervous system, some neurones release two or more distinct transmitter chemicals, a phenomenon known as **cotransmission**. This is considered further in Chapter 19. Such complications do not, as far as is known, apply to motoneurones of the somatic nervous system. Referring back to Figure 5.4 and invoking Dale's principle, we know that the motoneurone releases acetylcholine at the motor end-plate and this same transmitter must therefore be responsible for transmission from the collateral to the Renshaw cell.

The probable transmitters at some of the synapses that we have considered are shown in Figure 5.12. At the synapse between an afferent nerve and a motoneurone in the anterior horn of the spinal cord, the excitatory transmitter is probably L-glutamate. For postsynaptic inhibition in the spinal cord, the transmitter chemical released by the inhibitory interneurone is probably glycine. Different chemical transmitters are inhibited by different poisons. For instance, strychnine competes with glycine for attachment to the receptors on the postsynaptic membrane and blocks inhibitory transmission in the spinal cord. The result is a release of excitation from inhibitory control. The motoneurones, subjected to unrestrained excitation, fire repeatedly causing the skeletal musculature to contract tonically, the phenomenon of **tetanus**. For presynaptic inhibition in the spinal cord, the neurotransmitter is γ-aminobutyric acid (GABA).

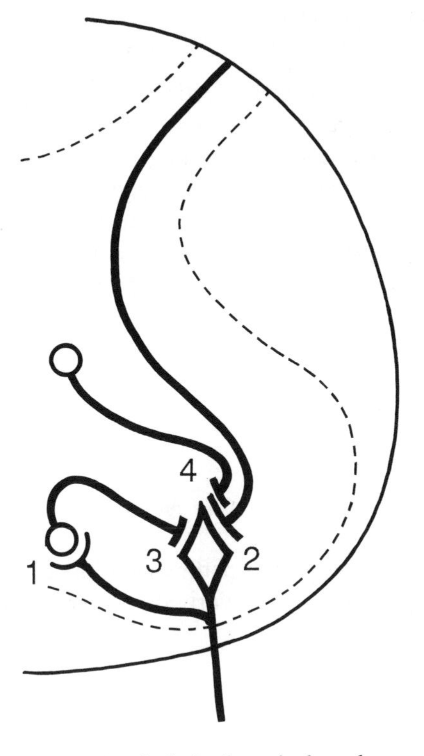

Figure 5.12 Neurotransmitter chemicals in the spinal cord.

	Type of synapse	Neurotransmitter
Synapse 1	Collateral branch of the motor nerve axon	Acetylcholine
Synapse 2	Excitatory synapse on motoneurone	Glutamate
Synapse 3	Inhibitory synapse on motoneurone	Glycine
Synapse 4	Presynaptic inhibition	GABA

Glutamate, glycine and GABA are amino acids.

5.7 APPENDIX

Three principles are observed by the central nervous system.

1. For a given neurone, either all the terminals of the neurone are excitatory or all are inhibitory.
2. Inhibition in the central nervous system is mediated by neurones with short axons. A corollary to principle 2 is that neurones can be partitioned into excitatory neurones and inhibitory neurones.

3. In the spinal cord, any one neurone releases the same transmitter chemical at all its axonal terminations. (This is known as 'Dale's principle'.)

These 'principles' are over-simplifications which are valuable at the present level of exposition. The reader will find exceptions in studying neurophysiology at a more advanced level. Dale's principle is stated above in a form that is much more authoritative than Sir Henry's original formulation (Dale, 1935). He merely proposed that, if a substance can be established as the neuro-transmitter at one synapse, this 'would furnish a hint as to the nature of' the transmitter chemical at other synapses made by the same neurone. Sir Henry proceeds 'The possibility has at least some value as a stimulus to further experiment'.

FURTHER READING

Eccles, J.C. (1968) *The Physiology of Nerve Cells*. Baltimore, The Johns Hopkins Press.

Eccles, J.C. (1969) The inhibitory pathways of the central nervous system. *The Sherrington Lectures IX*. Liverpool, Liverpool University Press.

6 Interruption of peripheral nerve and of the spinal cord: spinal pathways

6.1 INTRODUCTION

This account of the effects of damage to neurones applies to the human adult. The effects of damage to a neurone depend on which part of the neurone is involved. If the cell body is killed, its axon and dendrites all die. Neurones are incapable of mitosis; once a neurone is destroyed, it is lost and is never replaced by division of other neurones or from precursor cells.

The situation is different if the cell body is not itself directly damaged. Let us consider the common clinical situation of a nerve axon being cut across in an accident. The results are different for peripheral and central axons; we consider peripheral axons first. Such axons may regenerate, as the next section describes.

6.2 SECTION OF A PERIPHERAL NERVE

When a mixed peripheral nerve is completely anatomically divided, as in a stab wound, the damage results in a burst of impulses in each of the cut nerve fibres. Consider the consequences, for instance, of cutting the ulnar nerve at the elbow. Some pain is felt at the site of the cut due to the cutting of the skin. However the impulses in the afferent nerve fibres projecting to the central nervous system cause the subject to experience intense pain not at the site of the injury but in the region of the forearm receiving sensory innervation by the ulnar nerve. This is the medial aspect of the forearm and hand. A related everyday sensation is elicited when we inadvertently hit the skin over the ulnar nerve (sometimes colloquially called 'hitting the funny bone'). This initiates nerve impulses in the ulnar nerve, which

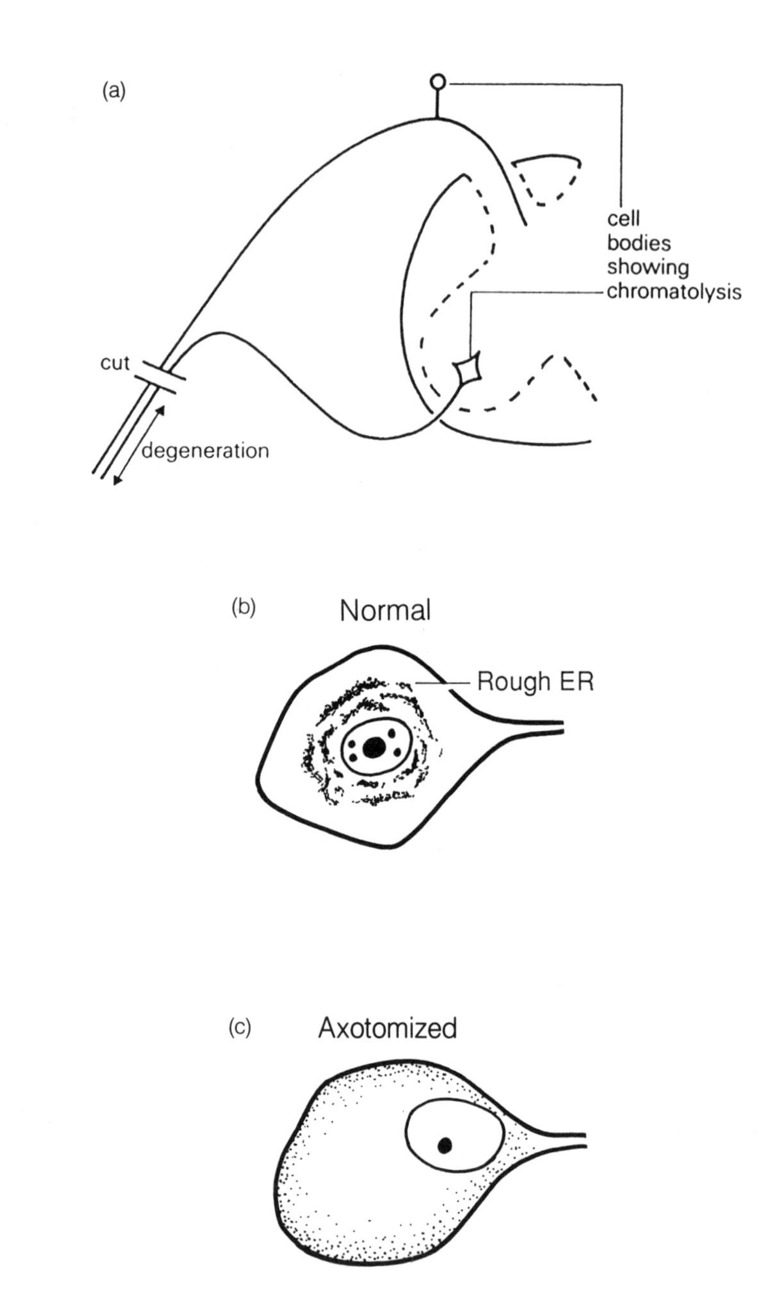

Figure 6.1 (a) Diagram to show the effects of a transection of a mixed peripheral nerve. (b) A motoneurone before and (c) one week after axotomy.

we interpret as pain and tingling in the medial aspect of the forearm and hand.

The pain caused by cutting a nerve is initially intense, but as the frequency of nerve impulses falls, due to accommodation of the nerve fibres, the pain wears off. The region previously supplied by the divided nerve is now anaesthetized and the muscles supplied by the motor fibres in the divided nerve are paralysed. In the example of the ulnar nerve, the denervated muscles are the flexor carpi ulnaris and the medial half of the flexor ditigorum longus.

Whereas the lengths of axon remaining atttached to the cell bodies remain alive, the lengths of the sensory and motor nerve fibres from the site of injury to their peripheral sites of termination are disconnected from their cell bodies and therefore die (Figure 6.1a). The axons degenerate. The surrounding myelin, although it is produced during nerve fibre formation by the Schwann cells, also degenerates as a response to the degeneration of the axon. The Schwann cells themselves, which surround the axon in the peripheral nerve, remain alive forming a cylinder around an empty **endoneurial tube**.

Histological changes in the cell body

The cell body giving rise to an axon which has been severed also shows histological changes as it prepares for the metabolic activity needed for the growth of the new axon (Lieberman, 1971). The cell bodies in question lie in the dorsal root ganglion and in the anterior horn (Figure 6.1a). The cell body swells; it may double in size. The nucleus also swells and moves from its usual central siting to an eccentric position (Figure 6.1c). The rough endoplasmic reticulum, which is responsible for protein synthesis and which normally lies around the nucleus, breaks down and moves to the periphery of the cell, probably to station itself near to the place where materials are needed to feed into the regenerating axon. This rough endoplasmic reticulum stains with basic dyes. Its breakdown is called **chromatolysis** and is readily observed microscopically within 24 h of injury. It is maximal at between 1 and 3 weeks and subsides as regeneration of the axon is completed.

After axotomy and accompanying chromatolysis, there is an early increase in nuclear RNA synthesis. This is followed shortly by an increase in cytoplasmic RNA content and increase in rate of cytoplasmic protein synthesis. A more extended discussion is contained in Appendix A.

Regeneration of peripheral nerve

Regeneration of nerve fibres occurs if the ends of the severed nerve are accurately apposed. A peripheral nerve consists of many bundles, or **fascicles** of nerve fibres, each surrounded by its own connective tissue sheath. The best recovery from a complete transection of a peripheral nerve is achieved when the neurosurgeon correctly apposes each individual fascicle within the compound nerve. Sepsis must be avoided since this interferes with regenerative processes. The living ends of the cut axons sprout and grow along the endoneurial tubes, eventually reaching receptors and muscle end plates. As sprouting of the nerve endings takes place, these endings initially form a lump, called a **neuroma**, at the cut end of the nerve. The neuroma consists of a tangle of nerve fibres, connective tissue and blood vessels.

The rate of growth of the sprouting nerve ends is about 1.5 mm/day (or about 1.5 in./month). The tips of the advancing nerve fibres show mechanical sensitivity: tapping over the regenerating end produces sensation. During the weeks following severance of a nerve, at a time before axons have re-established functional connections with receptors or end plates, the mechano-sensitivity of regenerating nerve tips affords the clinician a valuable method of determining whether or not a cut nerve is regenerating. If regeneration is occurring, tapping the skin overlying the regenerating nerve tips causes the subject to experience a sensation referred to the region of innervation of the divided nerve. In regenerating nerve fibres, the conduction velocity is low, about one fifth of normal a month after the accident and increasing to half the normal value by 3 or 6 months.

As functional connections are being re-established with the cutaneous receptors, cutaneous sensitivity begins to return. Initially, this has characteristics which distinguish it from normal cutaneous sensitivity. The threshold for sensation is initially high. When a stimulus of any type is strong enough to be perceived, the sensation perceived is unpleasant, badly localized, diffuse, and irradiates away from the site of sensory stimulation. The subject describes it as pain. This type of sensation, which was called 'protopathic' sensation by Head, is akin to the irritability exhibited by the lowliest worm in response to a potentially damaging stimulus. As regeneration proceeds, trivial stimuli can cause pain and, in some cases, pain occurs spontaneously. The pain can be intense and a source of great distress. The origin of this erroneous interpretation is considered in Chapter 18. The appreciation and accurate localization of light touch, together with tactile discrimination, do not recover for many months after the return of pain sensation. This type of sensation was called 'epicritic' sensation by Head; it is an attribute of animals with highly developed nervous systems. Epicritic sensation may never recover entirely after nerve section and regeneration. When the appreciation of light touch recovers, the uncomfortable painful sensitivity disappears (Walton, 1985, p. 497).

If a nerve is crushed but not divided, the regenerating axons sprout directly into the original endoneurial tubes. Restoration of function after regeneration may be nearly perfect. Following complete division of a nerve, a regenerating nerve sprout will almost never find its way into the original endoneurial tube. It will almost always enter a different tube and find its way to a different region of the body. Many fibres do not make contact with the same type of peripheral structure as originally; a motor axon may follow the endoneurial tube previously occupied by a sensory fibre and *vice versa*. It is not known whether a nerve tip, finding itself in an unfamiliar place, can explore and find an appropriate target. It is, however, well established that the restoration of innervation after a complete division of a nerve is much less complete than after a nerve crush.

Receptive fields after regeneration

Normally, a single afferent nerve fibre connects with receptors which innervate contiguous regions of the skin. After division of a cutaneous nerve, a single afferent nerve fibre often has two or more split receptive fields, because of branching of the regenerating nerve fibre in the neuroma, with different divisions projecting to different regions of skin. The central nervous system cannot discriminate between the siting of stimulation of these two regions of skin and, as a consequence, two-point discrimination is impaired.

Another effect of regeneration is on the pattern of endings of afferent nerve fibres in the grey matter of the dorsal horn in the spinal cord. In a normal subject, neighbouring points on the skin map to neighbouring points in the spinal cord: a map of the skin can be drawn on a histological section of the cord indicating the pattern of projection. This is called a **somatotopic** mapping of the skin surface to the dorsal grey matter. After transection and regeneration of a peripheral nerve, the map is completely disorganized, with neighbouring points on the skin projecting to widely separated points in the cord. This implies jumbling of input to the cord; it is not known how the nervous system analyses the jumbled information in this situation but the effect may account for the lack of complete recovery of epicritic sensation which was noted earlier.

Restoration of function

Restoration of function may be rated as ‘clinically perfect’ when as few as 1% of the original connections are functional. Mechanisms such as hypertrophy of re-innervated skeletal muscle fibres and enlargement of re-innervated motor units compensate functionally for the deficit of innervation.

6.3 LESIONS OF THE CENTRAL NERVOUS SYSTEM

By contrast with the regeneration that can occur after section of a peripheral nerve, once a nerve axon in the central nervous system of an adult person is cut, the portion severed from the cell body does not regenerate. This seems to be due to the existence of inhibitory chemicals in the brain and spinal cord.

After cutting central axons, the cell body undergoes chromatolysis as if in anticipation of protein synthesis, but then either proceeds to degenerate completely or remains severely atrophied. One of the key questions in experimental neurology is why damaged neurones are prevented from regenerating their central axons. It is possible for injected embryonic nerve cells to grow and become functional in carefully controlled animal experiments, but this has not yet been convincingly achieved in people. There are sporadic reports of successful neuronal grafting in patients in whom regions of the brain have been destroyed in stroke. There is no firm evidence that neurones have grown from the graft tissue and have established functional connections with host neurones (Dunnett and Richards, 1990).

Despite this inability of central neurones to regenerate, the central nervous system recovers much of its function after some types of injury. For instance, a patient who has suffered a severe subarachnoid haemorrhage is in deep coma, with all functions of the brain profoundly impaired. The patient is deeply unconscious, entirely unresponsive to stimuli and, except for breathing, appears dead. If the extravasated blood is suctioned off and the bleeding vessel is clipped, the patient may be sitting up in bed taking an active part in intelligent conversation and responding normally to neurological testing within 3 days. Large numbers of neurones are affected by the initial insult, but they are not killed. At the time of injury, they stop conducting impulses but are capable of recovering.

Effects of pressure

The brain is enclosed in a protective rigid bony case. This protection carries its own penalties, one of which is that any fluid accumulating or mass growing inside the skull must be balanced by the displacement of an equal amount of fluid. The fluid that is most easily displaced is blood. This jeopardizes the vascular perfusion of the brain; hypoxic damage to neural tissue ensues. Hypoxia of the cerebral capillaries damages the endothelial wall, making it unable to prevent the transmural passage of plasma proteins. Hence excess extracellular fluid forms, which is itself space-occupying. This is a self-perpetuating cycle of damage (Figure 6.2) and, if untreated, is quickly fatal. This sequence may be reversible if the condition is treated in time.

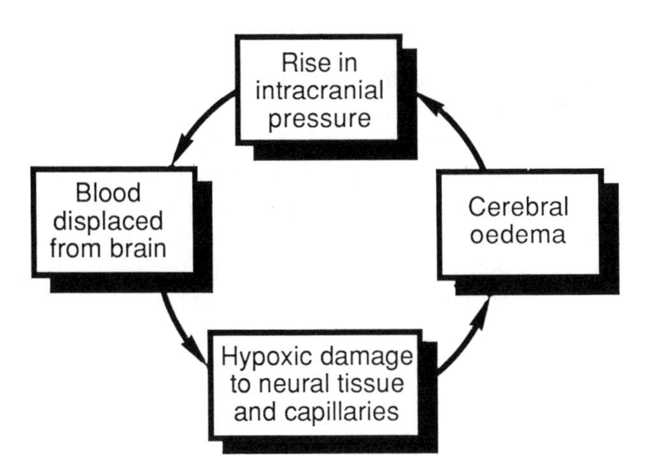

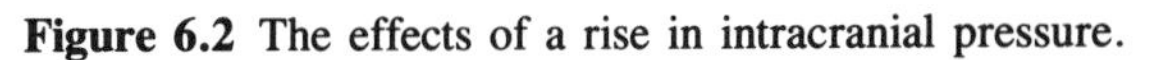
Figure 6.2 The effects of a rise in intracranial pressure.

Recovery of neural function

There are several factors, but **not** regeneration, which contribute to the recovery of function in neural tissues. As with all damage to tissues, mechanisms to repair the damage are recruited: monocytes from the blood migrate into damaged tissue and chemical repair mechanisms are called into play. Following damage to the central nervous system, several factors specific to the brain are also operative. One important factor that has already been mentioned is the removal of space-occupying fluid or tissue within the cranial cavity.

Resolution of oedema

Injury to the brain damages the walls of capillaries: this in turn results in the formation of excess interstitial fluid (oedema). As the tissue damage caused by the original damage is repaired by the cellular and chemical responses of the body, this excess fluid is reabsorbed. This 'resolution of oedema' is accompanied by restoration of adequate perfusion and a recovery of neural function.

Opening of collateral circulation

If brain damage is due to interruption of a cerebral blood vessel, the region supplied by the vessel stops functioning. Neighbouring patent cerebral vessels may provide a circulation to the edges of the ischaemic region by the formation of new blood vesssels. This is called a collateral circulation.

Improvement by substitution of function

Although different parts of the brain perform different functions, neural tissue, particularly that of the cerebral cortex, possesses the propensity to change its function, a phenomenon known as **plasticity**. When a region of cortex which performs a particular function is destroyed, other regions of cortex take over some of its functions. This is pronounced in very young children. The cortical centres controlling speech usually lie in the left cerebral hemisphere (Chapter 16). If the left cerebral hemisphere is damaged early in life, the speech centres develop in the healthy right hemisphere instead of on the left; the function of speech is transferred. The child often becomes left-handed, a further indication that a function previously performed by the left hemisphere has been taken over by the right hemisphere.

Neural shock (diaschisis)

This phenomenon is only found in the central nervous system. In evolution, as a new higher centre develops in the brain, the older lower centres are not replaced; they continue to function under the control of the higher centre. If the higher centre is destroyed, the lower centres over which it previously exerted control temporarily cease to function. The removal of normal control mechanisms leaves the lower centres at a loss as to how to behave. Initially, the lower centres stop functioning altogether. Later, they start autonomous activity of their own: this may not be to the advantage of the organism as a whole. The best example of the phenomenon of temporary cessation of function is that of spinal shock, a consequence of complete transection of the spinal cord.

6.4 SPINAL TRANSECTION

This happens not infrequently in serious car accidents and with gun-shot wounds. Flying glass or metal penetrates the spine and cuts right across the spinal cord. There is a phase, starting at the time of the injury and lasting 3 days or more in man, when all functions of the cord are profoundly depressed at all levels below the lesion. This cessation of function is despite the fact that the cord itself below the lesion is undamaged, with all its circuitry intact. This phase is called **spinal shock**, a subject to which we return shortly.

The levels of transection compatible with survival

The possibility of survival after a transection of the spinal cord depends on the level of the cut. The phrenic nerve, which carries the motor nerve fibres to the diaphragm, arises from the cord from the 3rd, 4th and 5th cervical segments. A section above this level results in disconnection of all motoneurones innervating respiratory muscles from the respiratory centres in the hind and midbrain. In this situation, breathing ceases and, unless the patient is artificially ventilated within a few minutes of the accident, he dies from respiratory failure.

When transection occurs below the origin of the phrenic nerve but above the thoracic region, which contains the motoneurones innervating the intercostal muscles, respiration after the cut is entirely diaphragmatic. The thorax is sucked in during inspiration instead of being expanded by the contraction of the internal intercostal muscles, as is normal. The sucking in of the thorax with diaphragmatic contraction is called **paradoxical breathing**. When the spinal cord is cut below the thoracic region of the cord, respiration is normal.

Sensory and motor effects of a spinal transection

There are sensory and motor effects shown schematically in Figure 6.3 and summarized in Table 6.1. The long tracts running up and down which connect the spinal cord with higher centres are cut. Their function is lost irreversibly. Interruption of the ascending afferent tracts results in a complete anaesthesia (lack of sensation) at all segmental levels below the transection; interruption of the descending efferent tracts to motoneurones results in a total muscular paralysis (loss of all volitional movement) at the segmental levels below the transection. The functions of the remaining intact length of cord below the lesion are also lost at the time of the transection, but later the cord recovers many of its functions.

Table 6.1 Spinal transection: effects below the segmental level of the interruption

	Immediate effects (including spinal shock)	*Delayed effects (chronic spinal man)*
Sensory effects	Anaesthesia	Anaesthesia
Motor effects	Paralysis	Paralysis
	Flaccidity	Spasticity
	Areflexia	Hyper-reflexia, clonus
Autonomic and allied effects	Areflexia	Hyper-reflexia (mass reflex)
	Cutaneous vasodilatation	
	No sweating	Profuse sweating in some subjects
	No shivering	No shivering
	Low blood pressure	Labile blood pressure
Bladder	Retention with overflow	Automatic bladder

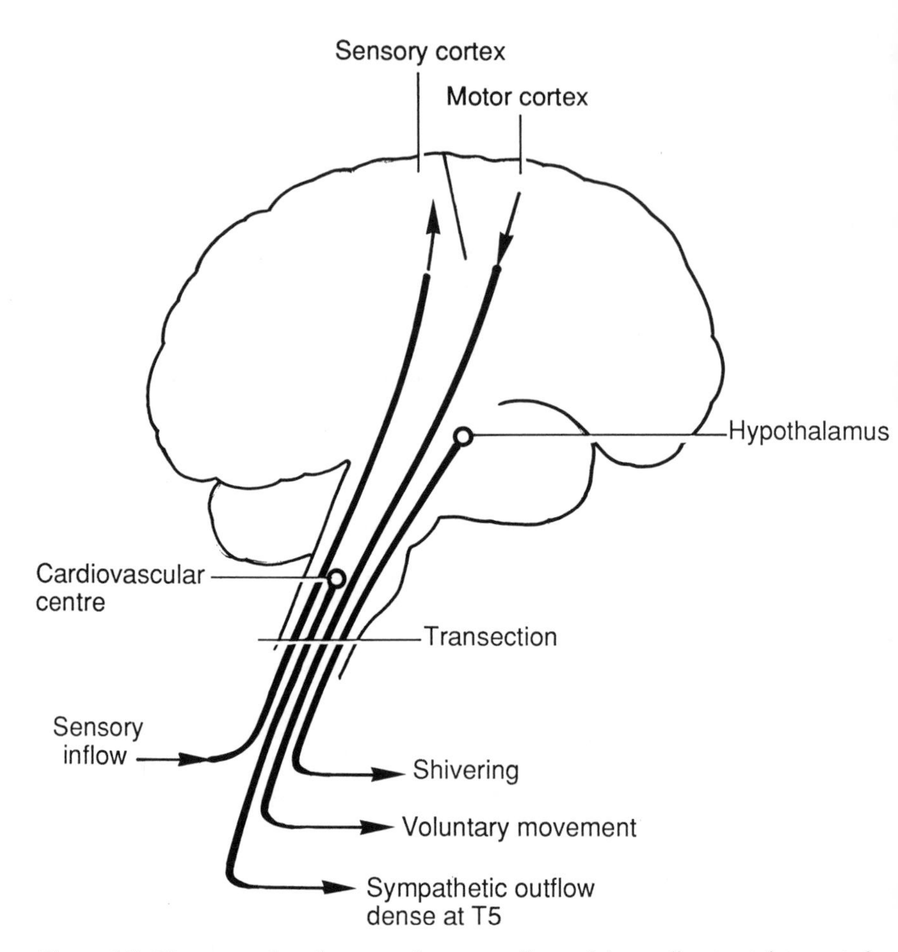

Figure 6.3 Diagram to show important long ascending and descending tracts interrupted by spinal transection.

The clinical picture

A spinal transection without other serious injury does not result in loss of consciousness; consciousness depends on the interaction of the brain stem reticular formation and the cerebral cortex (Chapter 11), connections which are spared in a spinal transection. The clinical picture is of a fully conscious patient who cannot feel his body and can make no body movements. His blood pressure is low. Being conscious, he is totally aware of his complete neural detachment from his body and he is, quite understandably, scared.

Lack of functioning of the spinal cord

In the phase of spinal shock, the skeletal muscles are completely relaxed, showing no tonus; this condition is called **flaccidity**. The subject is said to suffer from a **flaccid paralysis**. It is impossible to elicit any spinal reflexes; this is called **areflexia** meaning 'no reflexes'.

Autonomic and associated effects

In the phase of spinal shock, there is a complete suppression of all autonomic reflexes below the level of the lesion. In a normal person, the blood vessels in the skin are slightly constricted because of continuous activity in the sympathetic nerves which supply them. In the absence of this neural tone, the skin vessels are dilated; the skin is warm and pink. Sweating is also dependent on sympathetic innervation; in the absence of neural activity, there is no sweating and the skin is dry.

In a normal person, shivering is brought about by descending commands from the temperature regulating centres in the hypothalamus. After a spinal transection, these descending pathways are interrupted, and shivering is irrecoverably lost.

Blood pressure

In a normal person, blood pressure is regulated by an elaborate system of autonomic reflexes. The integrating centres are the cardiovascular centres which lie in the medulla. These centres project down via the spinal cord and project to the blood vessels as the sympathetic nerve supply. Compensatory mechanisms to stabilize blood pressure rely both on changes in the tone of the smooth muscle in the walls of the blood vessels supplying the viscera and on the release of adrenaline from the adrenal glands. The major sympathetic outflow to the viscera is from the cord at around the fifth thoracic segmental level; if the level of the spinal transection is above this, the blood pressure is low due to interruption of the controlling pathway. With transections below this level, the reflex maintenance of blood pressure is less disturbed.

Bladder reflexes

From the point of view of clinical care of the patient, the loss of bladder reflexes is very important. The bladder reflex is a spinal reflex. In a normal adult, as the bladder fills, the stretch receptors in the wall are activated. Micturition is inhibited by descending inhibition from the cortex until it is socially acceptable to void urine, when the spinal reflex is released from inhibition and micturition is initiated.

After section of the spinal cord, in the phase of shock, the micturition reflex is lost with other spinal reflexes. The bladder fills and when the pressure reaches a level such that the back pressure exerted by the sphincters is exceeded, a little urine dribbles away. The condition is called **retention with overflow**.

A bladder which does not empty completely is liable to infection. An important part of management is to introduce an indwelling catheter (under aseptic precautions) so that the bladder empties passively. It is also important to guard against bed-sores; the patient's skin is anaesthetic and, due to the state of flaccid paralysis, the patient lies motionless in one position.

The origin of spinal shock

The cessation of spinal function immediately after a spinal transection is due to removal of the normal physiological bombardment of the neurones in the cord from higher centres. It is not due to other factors such as the release of toxins from the neural tissue destroyed at the site of the lesion. This can be ascertained from observations on the effects of a second transection at a lower level some time after the first. The first transection causes the usual features of spinal shock but, after the second transection, there is no phase of loss of activity in the cord below the lesion. Release of toxins and allied phenomena are just as pronounced for the second as for the first transection. The lack of spinal shock following the second transection rules out these factors as important in the generation of spinal shock. The conclusion is that the removal of descending neural influences is indeed the origin of spinal shock.

The duration of shock and recovery from shock

As one ascends the phylogenetic scale, the duration of the shock increases. In the frog spinal shock lasts a few seconds, in the cat it lasts for about 1 hour whilst in man, shock persists for at least 3 days and often for as long as 2 weeks after injury. As spinal shock subsides, spinal reflex circuitry starts to function again. The effects to be described are summarized in the right-hand column of Table 6.1. This colum is headed 'chronic' which, in the medical sense, means 'of long standing'

The micturition reflex depends on the integrity of the lumbosacral region of the cord. If spinal transection occurs above this region, and if the lumbosacral cord escapes direct damage, the micturition reflex, as with other spinal reflexes, is re-established in the weeks following the transection. At this stage, the catheter can be withdrawn and the patient passes urine automatically once his bladder reaches a certain volume. When the sensory bombardment of the cord from stretch receptors in the bladder wall reaches a threshold, emptying is initiated. This is called an **automatic bladder**. Emptying of the bladder is complete.

Most patients void urine without any control, in the same way as a young baby. However, some learn to bridge the neurological deficit and to control their micturition by indirect means. As the bladder fills, input to the cord from stretch receptors increases. This input produces a reflex rise in blood pressure by spinal mechanisms. Blood pressure regulation is deficient in the patient with spinal damage, and the rise in blood pressure due to stretch receptor bombardment of the spinal cord is therefore greater than in normal people. The patient appreciates this rise in blood pressure as a flushing of his face. He goes to the toilet, there scratches his thigh to increase the sensory bombardment of the sacral region of the spinal cord, and facilitates the micturition reflex, allowing the bladder to empty.

Sequence of recovery of somatic reflexes as spinal shock wears off

Whereas the subject never regains any voluntary control of the skeletal musculature, autonomous activity of the cord gradually reappears and reflex activity recovers. Reflexes come back one by one, not all at once. The first reflexes to reappear are the flexor reflexes, the sequence of reappearance being ankle, knee, and hip. Then contraction of extensor muscles reappears, usually about 6 months after the transection. The extensor reflexes finally tend to become exaggerated, leading to spastic paralysis. In most patients, the final stage is of predominantly extensor activity with extensor spasms. Extensor tendon reflexes are hyperactive. Exaggeration of spinal reflexes is called **hyperreflexia**. Stretching the tendon causes the muscle to go into cyclic contractions; this is called **clonus**.

In some cases, trivial stimulation of the groin or the soles of the feet leads to an exaggerated spinal reflex response, with flexion of the legs, defaecation, micturition, and in the male, erection. There is profound sweating, an autonomic component of over-active reflexes. This complex of reflexes is called the **mass reflex**; it is distressing apart from being socially embarrassing.

Autonomic reflexes

When reflex activity of the cord is re-established after spinal shock, primitive control of autonomic function by spinal mechanisms is also re-established. There is some control of blood pressure, although this remains much more labile than normal. The blood pressure rises with filling of the bladder as already described. On changing posture from lying to standing (with crutches), pooling of blood in the legs, leads to a reduction of venous return. Autonomic compensating mechanisms are inadequate and the blood pressure falls. This is called **orthostatic hypotension** and the subject may faint.

Some autonomic reflexes may become exaggerated in a fashion analogous to the exaggeration of the stretch reflex. In some patients with chronic spinal injury an exaggeration of sweating is a major disturbance which is initiated by cutaneous stimulation. This contrasts with the situation in a normal person, in whom sweating is the response to a rise in core temperature.

Responses depending on connections of the cord with the cerebral cortex

Some reflexes found in a normal person never return or, if they do so, they assume a more primitive form. Because the normal reflex depends on the participation of supraspinal mechanisms, these reflexes are not spinal reflexes.

The plantar reflex

The most well-known example is the **Babinski** sign, also called the **plantar reflex** (Table 6.2). The motoneurones responsible for the muscular contraction lie in the second sacral segment. The normal form of the reflex depends on the integrity of the corticospinal tract which descends from the motor area of the cerebral cortex to the cord. If the lateral aspect of the sole of the foot is firmly stroked with a blunt object, the normal adult responds with plantar flexion of all the toes. This is part of the reflex armamentarium which maintains a standing position. The normal response is called **Babinski negative**. The muscles responsible for this movement are called flexors by anatomists; the normal Babinski response is said to be a 'flexor plantar response' although the movement is physiologically an extensor movement.

Table 6.2 The plantar reflex (Babinski sign) in response to a firm stroke to the sole of the foot

Normal	*Pathological (effect of interruption of the corticospinal tract)*
Plantar flexion of foot	Dorsiflexion of foot
The toes go down	Toes go up (there may be flexion of the knee and hip joints)
Babinski negative	Babinski positive
A 'flexor plantar response'	An 'extensor plantar response'
Physiological extension	Physiological flexion
Postural anti-gravity reflex	Flexor withdrawal reflex

After any interference with the corticospinal tract, the response to this stimulus is dorsiflexion of the great toe with fanning of the other toes. When the reflex is exaggerated, the postural reflex of the normal person is also replaced with the more primitive withdrawal reflex. This abnormal response is called **Babinski positive**. Following the anatomists' names for muscles, it is called an 'extensor plantar response', whereas it is a physiological flexion.

In the phase of spinal shock, the Babinski response is lost together with all other reflexes. When it reappears, it assumes the abnormal form. In some subjects, the Babinski sign is one of the first spinal reflexes to reappear as spinal shock subsides, and the fact that the Babinski sign can be elicited is an early indication that spinal shock is wearing off. This is not a consistent finding and in other subjects this response may fail to appear for some months.

The Babinski sign is positive in babies before the pyramidal tracts become functional. The pyramidal tracts carry the nerve fibres projecting down from the motor region of the cerebral hemisphere to the spinal cord; these axons acquire myelin and usually become functional at an age between 6 months and 1 year, at which time the adult form of the Babinski response appears. Using this sign, you can elicit reflexes in your own or your friends' infants to determine whether their pyramidal tracts have started to function.

Certain other reflexes depend on the integrity of this projection and are permanently abolished by spinal transection.

Abdominal reflexes

If the skin of the belly wall is lightly scratched in a person whose corticospinal pathways are functioning normally, the muscles of the belly wall contract, reflexly pulling the umbilicus towards the stimulus. This reflex is abolished bilaterally if the corticospinal tracts on both sides are interrupted, as in spinal transection. If the corticospinal projection is interrupted on one side of the cord, the ipsilateral abdominal reflexes are abolished.

Cremaster reflex

In the normal male, a scratch to the inner aspect of the thigh initiates a reflex contraction of the cremaster muscle, a reflex absent when the corticospinal projection is interrupted.

Placing reactions

These are more readily demonstrated in the cat or dog than in humans. If a blindfolded animal is held up in the air and brought to the edge of a table so that the backs of the paws touch the undersurface of the edge of the table, the paws are immediately brought up and placed on the table top, providing the body with support. These reactions are abolished by section of the corticospinal projection.

6.5 CENTRAL NERVOUS LESIONS: RELEASE OF LOWER CENTRES

As we have seen, one of the features of long-standing spinal transection is an overactivity of reflexes in the spinal cord. This illustrates a general feature of the results of injuries to the central nervous system. When the influence of a central nervous structure is lost, the contribution it made previously to central nervous functioning is obviously lost. For instance, when the corticospinal projection is interrupted as a component of a spinal transection, voluntary movement can no longer be intitiated. In addition, in later states after transection of the cord, there is an exaggeration of the spinal stretch reflex resulting in spasticity. The higher centres relinquish their control over the lower centres; the lower centres become autonomous, oveactive and exaggerated. This is a release phenomenon. These features were first recognized about a century ago by Hughlings Jackson, a London neurologist. He referred to the effects of damage to a brain centre as 'negative' (loss of function) and 'positive' (the release phenomena).

Phylogenetically, the oldest part of the central nervous system is the spinal cord. Even the lowliest worm has a nerve cord with a minimum of specialization of the upper end into primitive ganglia. As evolution proceeds, higher centres evolve to perform more complicated functions. The older components do not disappear: the higher centres probably produce many of their effects by commanding the discharge of the older components. The autonomy of the old components is kept under control by descending influences.

A related concept is that of upper and lower motoneurone lesions. Any interference with the corticospinal projection from the motor cortex to the anterior horn cell results in a condition known as an **upper motoneurone lesion**. A section of the cord interrupts this projection and results in an upper motoneurone lesion at all segmental levels below the lesion. Spinal reflex effects are uncontrolled and are usually exaggerated.

The name **upper motoneurone** derives from the fact that the cells of the primary motor area of the cerebral cortex, described in Chapter 15, are upper motoneurones because they project to the anterior horn cells of the spinal cord, forming monosynaptic contacts with the anterior horn cells. These anterior horn cells are the **lower motoneurones**. A patient with an upper motoneurone lesion shows overactivity of lower motoneurones, with exaggerated tendon jerks. This is a release phenomenon. The lower motoneurone is released from higher control; primitive reflexes become prominent. By contrast, with a lower motoneurone lesion, the muscles involved are flaccid and, with time, the muscle fibres atrophy.

6.6 CENTRAL AFFERENT PATHWAYS

The general body senses (or **somaesthetic senses**) are so-called to separate them from the special senses (Chapter 9). There is some separation of the routes taken up the spinal cord by the central projections of nerve fibres carrying information from different groupings of sensory modalities.

Table 6.3 shows these modalities grouped according to this separation, and also contains further information built up in the ensuing sections.

Table 6.3 Somaesthetic senses

Modalities	*Central pathway*	*Cross/uncrossed*
1. Position and vibration	Dorsal columns	Uncrossed
2. Pain and temperature	Lateral spinothalamic tract	Crossed
3. Touch and pressure	Anterior spinothalamic tracts	Both crossed and uncrossed

For all groups of modalities, there are two synapses en route from the periphery to the somatosensory area of the contralateral cerebral hemisphere, as shown in Figure 6.4. The second synapse is in the specific relay nuclei

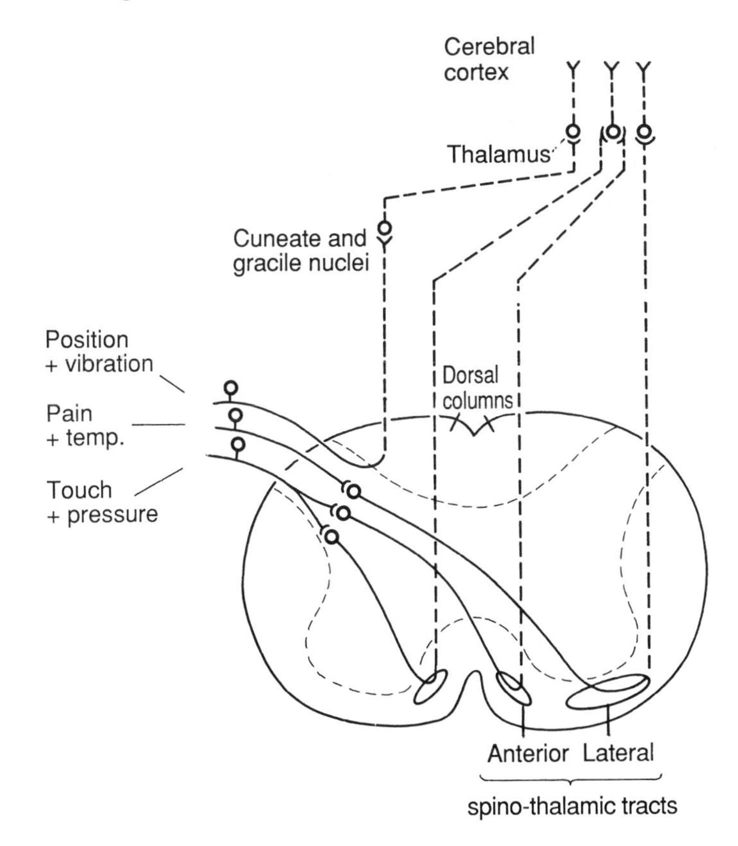

Figure 6.4 Diagram of the principal central specific afferent pathways to the cerebral cortex. 1: Position and vibration, 2: Pain and temperature, 3: Touch and pressure.

of the thalamus for all modalities. The site of the first relay is different for the different groups of modality.

Position and vibration

This group of modalities subserves fine tactile discrimination (including two-point discrimination), joint position sense (kinaesthetic sense) and the sense of vibration. The receptors responsible for the ability of a person to identify an object held in the hand connect with afferent nerve fibres which project along this route. These modalities ascend via fibres in the dorsal columns which are entirely uncrossed. Since these columns consist of axonal branches of dorsal root ganglion cells, we call them **first-order nerve fibres**. The nerve fibres end synaptically on the cells of the dorsal column nuclei (gracile and cuneate nuclei) which lie on the dorsal aspect of the medulla oblongata. From here, the second-order neurones cross the midline (decussate) in the medulla whence they ascend in a bundle of nerve fibres called the **medial lemniscus**. This name is derived from the fact that, in transverse section, the bundle is shaped like a lens. The medial lemniscus in turn projects to the thalamus. Thus information from one side of the body projects to the contralateral thalamus.

This ascending branch of the primary afferent neurone is only one branch of the nerve fibre. Some branches go to cells in the cord; others project via the spino-cerebellar tracts.

Pain and temperature

These modalities travel along first-order fibres which, after entering the cord and travelling up or down a few segments, form synapses on neurones in the posterior horn of the cord. From these cells, the second-order nerve fibres traverse the midline to enter the lateral spinothalamic tract, which is therefore entirely crossed. This nerve tract ends in the specific relay nuclei of the thalamus.

Touch and pressure

This group of modalities is for crude sensation. It allows the subject to detect whether or not a stimulus is being applied, but does not transmit the information necessary for precise localization of the stimulus or for making judgements about the nature of the stimulus. As with pain and temperature, the incoming fibres form synapses with cells in the dorsal horn of the spinal cord. Some of the second-order neurones pass to the ipsilateral and some to the contralateral anterior spinothalamic tracts. Thus this modality of sensation is carried up on both sides of the spinal cord.

The second-order neurones which are uncrossed in the cord decussate in the medulla to form synapses on the specific relay nucleus of the thalamus. The second-order neurones which cross the midline in the spinal cord travel straight up to the thalamus. By the stage at which the neurones reach the thalamus, they have all crossed the midline; the specific relay nuclei of the thalamus have strict contralateral connections with the periphery.

6.7 DESCENDING PATHWAYS

There are many higher centres involved in motor performance which project down to the spinal cord. These will be described in later chapters. The primary

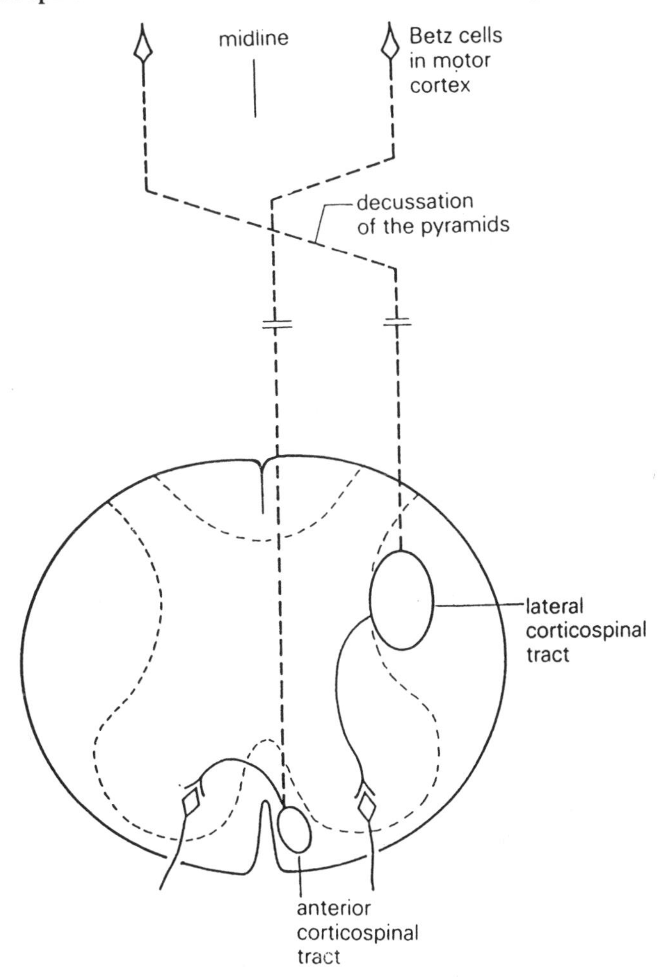

Figure 6.5 Diagram of the corticospinal tracts.

motor area of the cerebral cortex, to be described in Chapter 15, is one example, which projects via the corticospinal projection.

The corticospinal projection

The tracts descending from the motor area of the cerebral hemisphere to the cord are the lateral and anterior corticospinal tracts. The lateral tract, which derives from the pyramidal decussation, is by far the more important. The anterior is uncrossed and rather insignificant. There are few fibres in this tract and, in most individuals, it does not extend beyond the cervical region. The arrangement is shown diagrammatically in Figure 6.5.

6.8 SPINAL HEMISECTION (BROWN-SEQUARD SYNDROME)

We have just described the main roads for traffic up to and down from the cortex. This scheme of central pathways was largely based on a study of defects in sensation and in motor activity in patients with localized lesions in the spinal cord. One group of observations has been derived from subjects in whom the spinal cord is cut half way across, for instance in a road accident or following a war injury involving flying glass or metal. It is much rarer for the cord to be cleanly cut across with one side left intact (hemisection) than for the cord to be completely transected. Information from these subjects provides a clear picture of the layout of crossings of neuronal trajectories in the central nervous system.

At all levels below the lesion, section of the dorsal column results in ipsilateral loss of sense of position of joints and hence posture. Perception of vibration due to a vibrating tuning fork held against the skin is lost ipsilaterally. There is loss of tactile (two-point) discrimination. Interruption of the lateral spinothalamic tract, which is crossed, results in contralateral loss of pain and temperature at all levels below the lesion. This loss of certain types of sensation in one area and of other types in another area is called a **dissociated sensory loss**. Section of the anterior spinal thalamic tracts interferes with crude touch and pressure sense. Since this group of sensations is carried up both ipsilaterally and contralaterally, hemisection leads to impairment on both sides of the body, without there being complete loss anywhere.

This is the situation at levels several segments below the lesion but not within three or four segments of the level of the lesion. The branches of the primary afferents destined to project along the lateral spinothalamic tracts often ascend one or two segments before forming synapses on the neurones of the dorsal horn. It is the axons of these cells which then decussate and turn upwards as components of the spinothalamic tracts. Hence, the upper level of this crossed sensory loss is a few segments below the level of the lesion. Conversely, pain and temperature fibres entering ipsilaterally in the two or three segments below the lesion may be caught before they cross, giving a band of ipsilateral loss of pain and temperature (Figure 6.6).

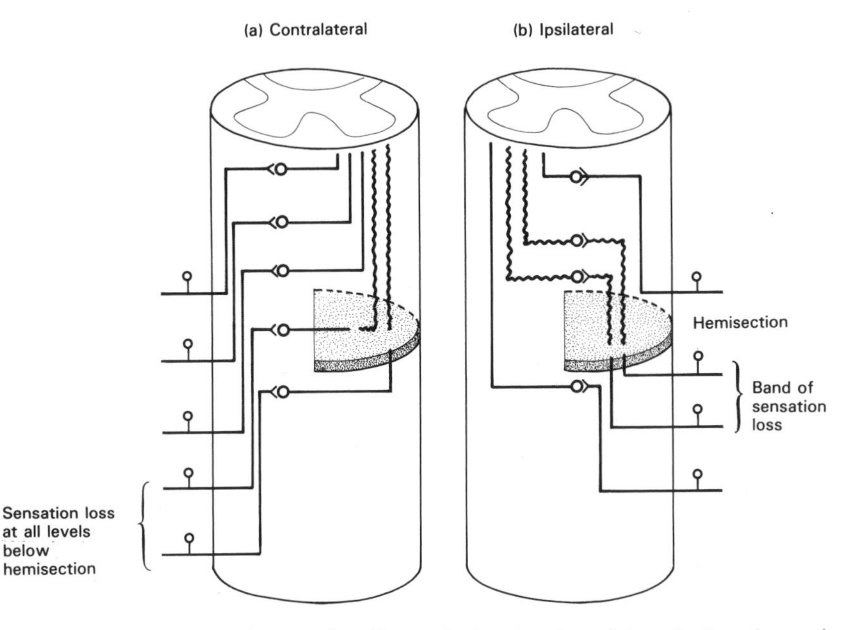

Figure 6.6 Diagram to illustrate the effects of a hemisection of the spinal cord on pain and temperature sensation. Interruption of projection pathways is indicated by the wavy lines. (a) Dorsal roots with incoming nerve fibres from the side of the body contralateral to the hemisection. There is loss of pain and temperature sensation on the side of the body opposite the hemisection, the distribution of this loss being all levels of the body downwards from a level about two spinal segments below the site of the lesion. (b) Dorsal roots with incoming nerve fibres from the side of the body ipsilateral to the lesion. There is a band of skin with no sensation of pain and temperature on the side of the lesion; this involves about two spinal segments just below the level of the hemisection.

On the motor side, interruption of the corticospinal projection results in an upper motoneurone lesion. Since the principal component of this projection is the lateral corticospinal tract that has arisen from the decussation of the medullary pyramids, the main motor effect of hemisection of the cord is an ipsilateral upper motoneurone lesion. The upper level of this is the segmental level of the lesion, and the upper motoneurone lesion extends down to all levels below the lesion. At the level of the lesion there may be a segment or two with motoneurones destroyed and hence a lower motoneurone lesion in that localized region, i.e. flaccidity, areflexia, and muscle wasting.

The tidy picture which has been described is of a loss of function, the distribution of which correlates with the site of lesion. For the sensation of pain, the subject is more complicated in that destruction of neural tissue may result in an exaggeration of pain rather than its abolition. This is described in Chapter 18.

6.9 APPENDIX A: THE REACTION OF THE NEURONE TO AXOTOMY

The displacement of the nucleus of an axotomized cell

The significance of nuclear displacement is not understood. In axotomized cockroach motoneurones, the nuclear migration is usually towards the axon hillock, suggesting that the nucleus may move towards the part of the cell body from which structural proteins must be synthesized and transported to the regenerating tip. In mammalian motoneurones, the nucleus tends to migrate to the pole opposite the axon hillock. It may be that nuclear eccentricity is simply a secondary phenomenon due to changes in bulk axoplasmic flow characteristics after axotomy.

Chromatolysis

Chromatolysis as part of the response of the cell body to axotomy poses an apparent paradox. There is disorganization of the rough endoplasmic reticulum, implying a reduction in the synthetic role of membrane-attached ribosomes in chromatolytic neurones at a time when net cytoplasmic protein synthesis is increased. Ribosomes are the intracellular organelles containing the ribonucleic acid (RNA) which acts as a template for synthesis of proteins. Protein synthesis is controlled by two types of ribosome, by ribosomes attached to the rough endoplasmic reticulum, and by free ribosomes. There is excellent evidence that free ribosomes are concerned with synthesis of cell structural proteins while membrane bound ribosomes are associated with the synthesis of proteins specific to the specialized functions of the cell. In the case of neurones, these proteins with special functions include peptide neurotransmitter chemicals and synaptic vesicles. The synthesis of these proteins concerned with neurotransmission is suspended in regenerating neurones. Lieberman (1971) speculated that chromatolysis corresponds to the cessation of synthesis of proteins needed for neurotransmission together with an increase in free ribosomes, invisible by light microscopy. This suggests that protein synthesis may be switched from production of synaptic vesicles etc., which are not needed by the neurone at this stage, to synthesis of the structural proteins of the cell membrane and cytoskeleton, which are essential for the rebuilding of the new axon.

Proprioception

7

In order for the central nervous system to control movement, information is needed about the relative positions of the different parts of the body. Receptors that signal this information are called **proprioceptive receptors** or **proprioceptors**. There are proprioceptors in the joint capsules around joints (these are called **joint receptors**), in tendons, and within skeletal muscles. Each group has its own characteristics, so each will be described in turn. The receptors themselves are all mechanoreceptors. Each group of receptors provides the central nervous system with proprioceptive information and initiates specific spinal reflexes.

7.1 JOINT AND TENDON RECEPTORS

Joint receptors lie in joint capsules. When a joint moves, parts of the capsule are stretched and the mechanoreceptors in these parts increase their activity. Other parts of the capsule are slackened and the receptors there decrease their firing frequency. The patterns of activity in the afferent nerve fibres from these receptors signal to the central nervous system the position of a joint and movement of that joint. Movement of a joint in particular increases the general level of bombardment of the central nervous system. This is perhaps an alerting system to signal that a change has occurred, since change may call for action on the part of the organism.

Figure 7.1 shows in diagrammatic form the effects of bending a joint on the discharge from the afferent nerve fibre connected with a stretch receptor which is stretched by this movement. The joint is moved as shown in Figure 7.1a. The joint is initially extended. It is then flexed at a constant rate until it reaches its new position, at which time movement abruptly ceases. This type of movement is called a 'ramp' movement. The time course of the movement is shown in Figure 7.1b.

Figure 7.1c shows the frequency of action potentials in the afferent nerve fibre at the different phases of the movement. Initially, before the

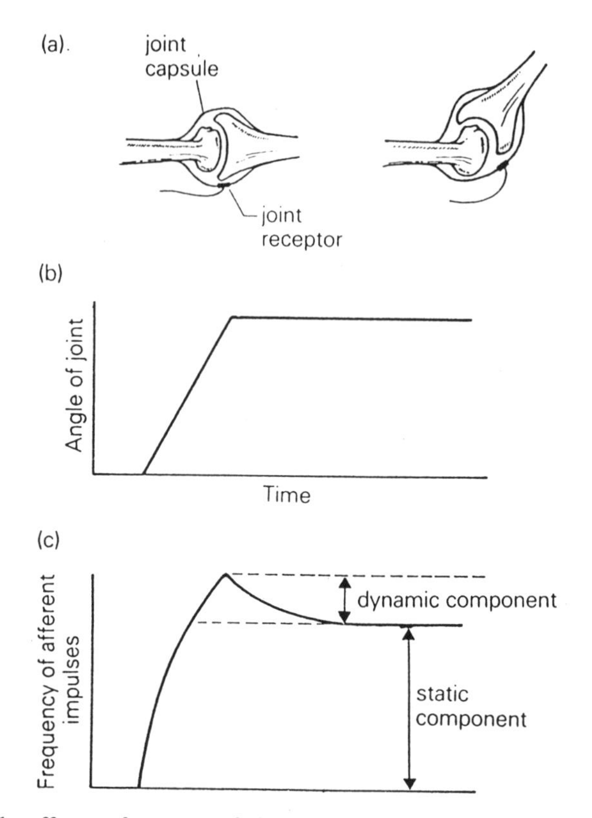

Figure 7.1 The effect on frequency of discharge of the sensory nerve fibre from a joint receptor of a ramp flexion of the joint.

movements, the nerve does not discharge because the receptor is scarcely stretched. As the joint is flexed and the receptor is stretched, the discharge rate of the afferent nerve rapidly rises. This frequency reaches a peak at the end of the movement. With the joint in its new position, the high initial frequency of afferent firing falls from its maximum and gradually settles to a lower rate. The relative prominence of the two components varies widely among different proprioceptors. This has led muscle physiologists to introduce a specialist terminology. The frequency curve is said to have two components, the transient **dynamic** component and the steady **static** component. The significance of this time course of firing is discussed in a later section.

Tendon receptors

The distribution of the terminal ramifications of a tendon afferent is shown diagrammatically in Figure 7.2. Because of their siting in tendons (which

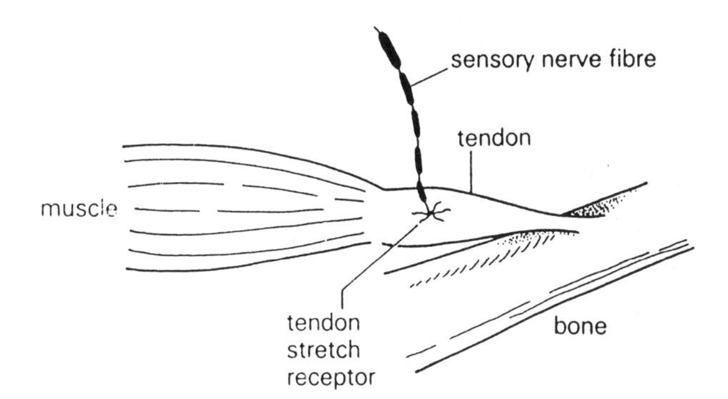

Figure 7.2 The siting of a tendon organ.

joint skeletal muscle to bone) these mechanoreceptors sense the tension in the skeletal muscle. Their sensitivity is low and firing starts only when the tension in the tendon is very high, approaching that developed when the whole muscle contracts maximally. One function of these receptors is probably a protective reflex to prevent its muscle from contracting so strongly that the tendon is torn. The reflex is an inhibition of the motoneurones supplying the extrafusal muscle fibres.

This reflex is the basis of the **clasp-knife** reflex. Patients with certain lesions of the central nervous system, for instance after a spinal transection of long standing, show excess of extensor tonus in the limbs. If an observer tries to flex the subject's limb, at first there is strong resistance. If the observer applies a strong force, the limb extensors suddenly give way, in much the same fashion as a clasp-knive. This relaxation of extensor tonus is due to excitation of tendon organs to a level that initiates reflex inhibition of the extensor motoneurones. The tendon receptors are not high-threshold in every situation. If the tendon is stretched by muscle contraction instead of by movement of the joint, then contraction of only a few specific motor units suffices to excite a particular receptor; other motor units are without effect on it. This is because the effective motor units pull on that part of the tendon in which the stretch receptor lies. Hence diferent types of stretching produce different patterns of afferent discharge. The central nervous system is informed of different aspects of stretch and of muscular contraction by different spatio-temporal patterns of activity in the afferents from the tendons. The sensory nerve fibres connected to tendon organs are Group Ib fibres.

7.2 THE RECEPTORS OF SKELETAL MUSCLE

These are also proprioceptors, but they are more sophisticated in that their response to stretch is controlled by the central nervous system. The receptors in skeletal muscle are part of an elaborate structure called the muscle spindle, so-called because of its shape. The spindle is up to 10 mm long, and there are around 100 spindles in a muscle the size of the gastrocnemius. Each spindle contains several specialized muscle fibres, called **intrafusal** (Latin for 'inside the spindle'), to distinguish them from the standard muscle fibres whose function it is to exert tension and perform work; these are called **extrafusal** fibres.

The layout of a muscle spindle is shown in Figure 7.3. The intrafusal fibres have both sensory and motor innervation. The sensory endings are of two types, **annulo-spiral** and **flower spray**, because of their histological appearance in stained sections. The annulo-spiral endings are also called **primary** endings; they are connected to Group Ia sensory nerve fibres, whereas the flower spray endings are referred to as **secondary** endings and are connected to Group II sensory nerve fibres. Each intrafusal muscle fibre consists of a central or equatorial sensory region which is non-contractile; the sensory ending winds around this. The sensory region is flanked on either side by contractile polar regions, each consisting of a thin striated muscle fibre innervated by one or more motor end-plates. The motor nerve fibres innervating the intrafusal muscle fibres are called **fusimotor** fibres. These are thinner and more slowly conducting than the A α nerve fibres that innervate the extrafusal muscle fibres. Most of the fusimotor fibres are in the A γ group (see Section 2.5). Each skeletal muscle receives almost as many γ as α motor nerve fibres.

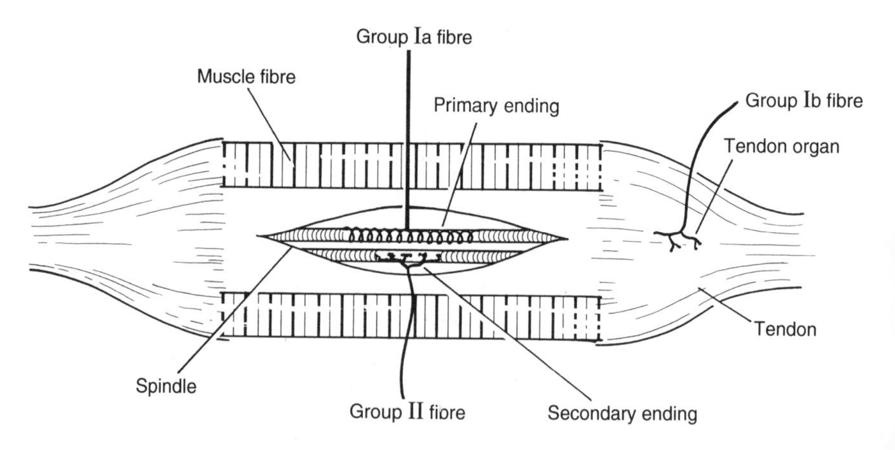

Figure 7.3 Diagram of a muscle spindle with an extrafusal muscle fibre on either side. Within the spindle, the two types of spindle stretch receptor are shown, with their sensory innervation. The innervation of the tendon organ is also indicated. The spindle receptors are in parallel with the muscle fibres whereas the tendon receptors are in series.

Figure 7.3 shows the mechanical relationships of the spindle and tendon receptors to the large muscle fibres which generate tension when activated. The spindle receptors lie in parallel with the large muscle fibres, whereas the tendon organs lie in series with them. When these large muscle fibres contract, tension is removed from the spindle receptors but applied to the tendon receptors. The significance of this difference will emerge shortly.

The tendon-jerk reflex

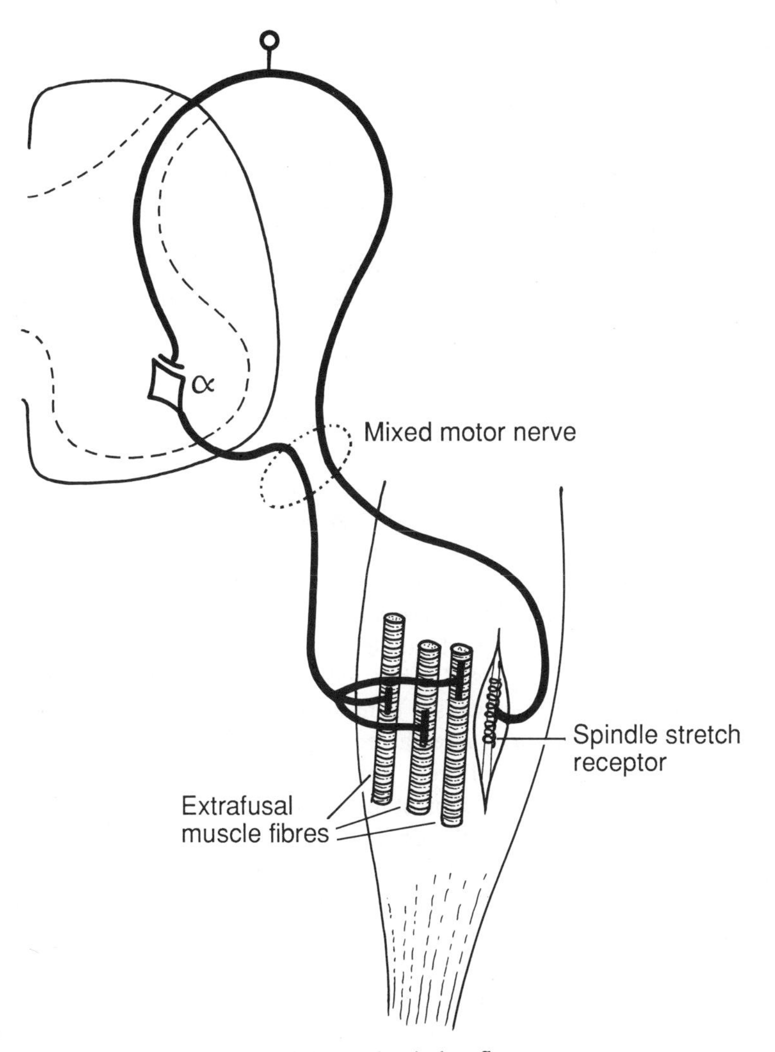

Figure 7.4 The reflex pathway of the tendon-jerk reflex.

The sensory component of the intrafusal fibre fires impulses when it is stretched. If a tendon is tapped with a patellar hammer, the tap mechanically stretches the whole muscle. This stretching of the muscle excites the primary sensory endings in the spindle receptors (Figure 7.4). These receptors generate action potentials which travel along the afferent nerve fibres to the spinal cord, where they make a direct excitatory synaptic connection with the motoneurones supplying the muscle whose tendon was tapped. Action potentials travelling down the motor nerve fibres from the α motoneurones cause contraction of the extrafusal muscle fibres and the muscle twitches. Since the reflex pathway of this tendon jerk reflex involves only one synapse in the cord, it is called **monosynaptic**.

The stretch reflex

As with the tendon-jerk reflex, the stretch reflex is initiated by receptors in the muscle spindle. The stretch reflex is the tendency for a muscle to resist stretch. It can be elicited by slowly stretching a muscle, this causing a reflex increase in tonus in the muscle. This increased tonus is sustained for as long as the muscle is stretched. The stretch stimulates both types of spindle receptor. In the spinal cord, at least two synapses are traversed in the reflex pathway. The stretch reflex is thus different from the tendon-jerk reflex which is a twitch response to rapid stretch, is initiated only by the primary endings and involves a monosynaptic pathway.

Gamma activation

The spindle receptors can be stimulated by another means. Since they sense stretching of the equatorial region of the intrafusal muscle fibres, the receptors are also excited when the polar regions shorten. Activity in γ motoneurones (the motoneurones from which Group A γ nerve fibres arise) has as a final result the contraction of extrafusal muscle fibres. Figure 7.5 will help to explain this; it is a repeat of Figure 7.4 but with the γ efferent route included. Activity in γ motoneurones causes contraction of intrafusal muscle fibres. This stretches the equatorial stretch receptor, yielding a burst of impulses in the muscle afferent nerve fibres. This in turn causes reflex excitation of the α motoneurones and contraction of the extrafusal muscle fibres. The central nervous system calls up muscular contraction by co-activation of both α and γ motoneurones; evidence for this will be presented later in the chapter. In some situations, α activation is predominant and in others, γ activation predominates.

How could one determine whether a muscular contraction is being produced predominantly by α or by γ activation? Let us consider the effect of interruption of the dorsal root (at the site marked with an arrow in Figure 7.5). This can be achieved by injecting the dorsal root with a local anaesthetic

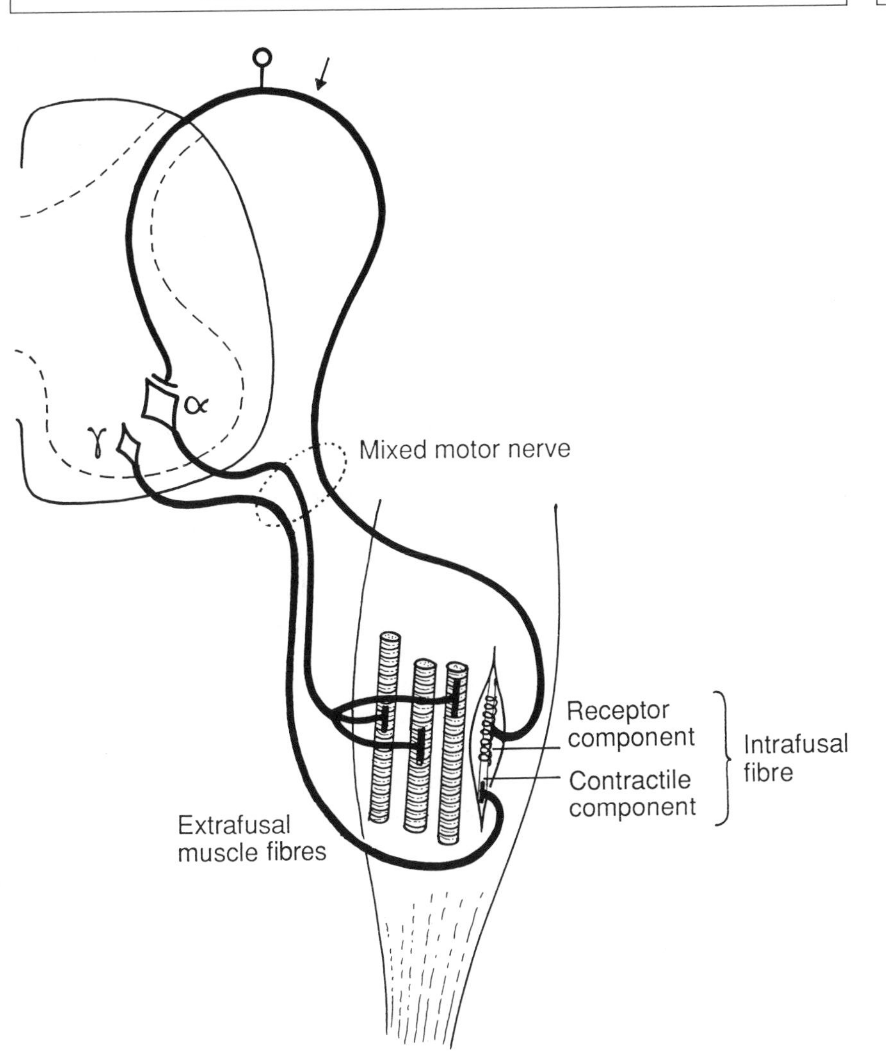

Figure 7.5 The circuit shown in Figure 7.4 with the addition of the γ-innervation of the intrafusal muscle fibre.

to block nerve conduction. For a contraction commanded directly by activation of the α motoneurones, the pathway is from the α motoneurones via the ventral nerve root and to the muscle as shown in Figure 7.5. Although conduction through the dorsal root is blocked, the ventral root is normal and so the motor command reaches its destination and the muscle contracts. For a contraction commanded by activation of γ motoneurones, the sequence of events following this activation is: nerve impulses along the fusimotor fibres which run in the ventral root, contraction of the intrafusal muscle fibres (which contributes negligible tension to the muscle as a whole), stretch of the spindle receptors, nerve impulses in the spindle afferent nerve fibres which travel up through the dorsal roots, synaptic activation of the α motoneurones, action potentials in the α motor nerve fibres travelling in the ventral root, and finally contraction of the extrafusal muscle fibres to give muscle tension. Since this pathway traverses the dorsal root, the muscular contraction is blocked if conduction in the dorsal root is blocked.

The foregoing lengthy description serves to emphasize that, for the pathway involved with γ activation, information must traverse the ventral roots twice: impulses are conducted firstly by the γ nerve fibres and then by the α fibres. In summary, a block of nervous conduction in the dorsal root abolishes contraction due to activation of γ motoneurones but does not interrupt the contraction produced by α activation.

The effects of α and γ routes can also be separated by differential nerve block. Injection of a small amount of local anaesthetic into the mixed motor nerve blocks fine nerve fibres more readily than thick ones. The thinnest of the nerve fibres concerned in motor control are the γ efferent nerve fibres. When these alone are blocked, movement due to activation of γ motoneurones is interrupted whereas movement due to α motoneuronal activity is intact. In some cases of pathological muscular rigidity, partial blockade of the mixed motor nerve supplying a rigid muscle can alleviate the rigidity without grossly disturbing voluntary movement. This proves that the rigidity was caused by overactivity of γ motoneurones and that voluntary movement involves direct activation of α motoneurones. Such a partial bock allows proprioceptive information from the muscle spindles to be conducted since the Group Ia afferent nerve fibres are very thick.

A third method of differentiating between direct and indirect pathways is by the use of chlorpromazine. Given systemically, this drug preferentially blocks the activity of γ motoneurones and hence abolishes activity initiated through the γ pathway. Activity initiated through the α pathway is left virtually unchanged.

The effects of fusimotor and skeletomotor activity on spindle stretch receptors

If the intrafusal muscle fibres contract, the stretch on the spindle receptors is increased, whereas if the extrafusal fibres contract and shorten, the stretch on the receptors is reduced (Figure 7.3). It is important to note the requirement for the extrafusal fibres to shorten; in an isometric contraction, in which the muscle contracts against a load that is so heavy that the muscle does not shorten, the stretch is not taken off the spindle receptors. If the load is lighter and the contraction of extrafusal fibres is accompanied by shortening of the muscle, it is possible to have a situation in which the stretch of spindle receptors by contraction of intrafusal fibres is exactly offset by the shortening of the muscle produced by contraction of extrafusal fibres. In this situation the firing of the spindle afferents is uninfluenced by this balanced contraction.

Why have two options, α and γ activation, for calling up muscular contraction?

The obvious advantage of the α route is that it is shorter and so introduces less delay. The extra delay introduced by the γ loop is about 30 ms for finger movements in an adult. For very quick movements such as the movements of the fingers in playing the violin or typing, speed of action is at a premium and the organism uses primarily the α route to command the musculature. The advantages of the γ route are subtle and less easily explained.

7.3 ACTIVATION OF γ MOTONEURONES

Suppose that you want to lift a bucket of water from the floor to a table; you do not know how much water the bucket contains. If the central nervous system were to command the movement by activation of α motoneurones alone, a given amount of activation of the α motoneurones would raise the bucket very high if the bucket were empty and scarcely at all if the bucket were full. Coactivation of the γ motoneurones overcomes this problem.

The component of the command signal sent by the central nervous system to γ motoneurones causes contraction of the intrafusal fibres and hence stretches the spindle receptors. The spindles therefore increase their discharge rate and cause a reflex excitation of α motoneurones and hence contraction of the extrafusal fibres. If the load that the muscle is lifting is light, the muscle shortens and this removes the stretch of the spindle receptors. The discharge of the spindle receptors thus declines back towards it resting level, the reflex excitation of α motoneurones declines, and the shortening of the muscle is achieved with an appropriately small muscular effort. If the load that the muscle is lifting is heavy, the muscle scarcely shortens; the stretch on the spindle receptors is maintained and the reflex excitation of α motoneurones continues. Thus the amount of muscular effort is appropriately much greater with the heavy than with the light load.

This description illustrates the general result that a command to the γ motoneurones produces a degree of excitation of extrafusal fibres that matches the load. The higher centres need not concern themselves with adapting the descending excitation to the amount of force needed for the muscle to do its job; this is all sorted out by the spinal cord and muscle spindles. The system does not entirely compensate for difference in load. Obviously, increased activity of extrafusal fibres produced in this way depends on an increase in stretch receptor activity. If the extrafusal fibres completely unloaded the spindle stretch receptor, then the spindle drive to the α motoneurones would be completely removed. The degree to which this mechanism compensates for differences in load depends on the gain of the receptors, that is, the amount of increase in stretch receptor activity for a given stretch of the receptor. The higher this gain, the greater the degree of compensation.

Muscle fatigue is also automatically compensated for by increased excitation of the α motoneurone pool for muscular effort commanded via the γ pathway. By commanding muscular effort through the γ system the central nervous system makes the movement relatively independent of changes in the loading of the muscle and of muscle fatigue. The shortening of the muscle is an accurate reflection of the command signal. This is at the expense of the time lag introduced by the γ reflex circuitry.

7.4 VOLUNTARY MOVEMENT

There is evidence that α and γ routes are both used in movement. Section of the dorsal roots interrupts the afferent fibres and hence abolishes any component of muscular contraction initiated by activation of γ motoneurones. In animals or humans, if the dorsal roots of a limb are cut, the limb is flaccid and shows very little voluntary movement. However, if the incentive to move is sufficiently strong, such as getting out of the way of a bus threatening to run the individual over, voluntary movement in the deafferented limb does occur. This proves that the individual does have volitional access to the α motoneurone pool, otherwise the limb could not be moved at all. The movements that are performed after deafferentation are much clumsier than normal. This is due to loss of the normal control mechanisms depending on feedback of muscle length and tension from proprioceptors. This condition of clumsiness is called **ataxia**.

Experiments have been performed to discover whether the human uses α, γ or both routes in performing a normal voluntary movement. To this end, spindle afferent discharge and electromyographic (EMG) activity have been recorded simultaneously during voluntary movements. The result is shown in Figure 7.6. The EMG activity starts to increase first and spindle activity then increases. The EMG activity signals action potentials in extrafusal muscle fibres. This activity is stimulated by action potentials in the motor nerve fibres originating in the α motoneurones in the cord. If this activity of α motoneurones had been produced entirely reflexly, by activation of γ motoneurones, then the activation of γ motoneurones would produce its effects by causing shortening of the intrafusal fibres. This would in turn have caused an increase in activity in the spindle afferent nerve fibre and then a reflex activation of α motoneurones. The activation of spindle receptors would have been signalled by an increase in the spindle afferent activity and this would have occurred before the onset of EMG activity. The figure shows that this did not occur.

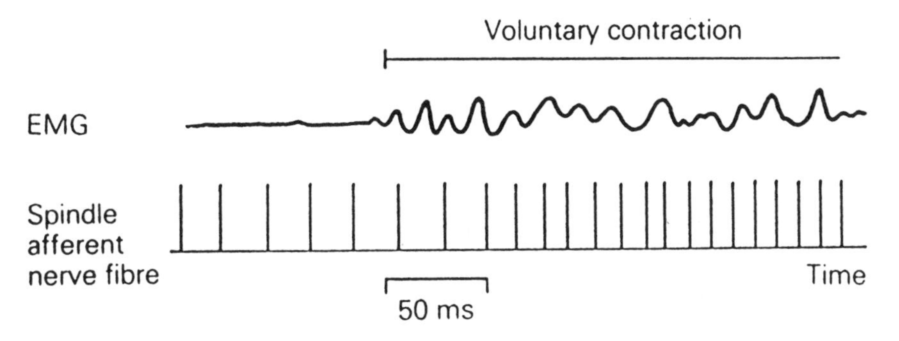

Figure 7.6 The electromyogram (EMG) recorded from skin electrodes overlying a muscle, together with the accompanying discharge of a sensory nerve fibre from a spindle stretch receptor. During the trace, the subject made a voluntary movement involving the muscle from which records were derived.

Hence the voluntary movement must have been commanded, at least in part, by direct activation of α motoneurones. However, if the command had involved the activation only of α motoneurones with no co-activation of γ motoneurones, then the unloading of the muscle spindle by the shortening of the muscle when the extrafusal muscle fibres contracted would have produced a silent period in the activity of the spindle afferent nerve fibre. The figure shows no slowing of discharge. Instead there is a delayed increase in firing rate. This proves that γ motoneurones were co-activated with α motoneurones, activation of both types of motoneurone being inseparable components of the motor command.

Hennemen size principle

The amount of excitatory synaptic current in a neurone depends on the number of excitatory synapses that are activated. For a given amount of synaptic current to the soma, the magnitude of the depolarization is roughly inversely proportional to the size of the soma. This is because the change in membrane potential depends on charging the membrane's capacitance by the synaptic current. A larger cell has a larger surface area and a proportionally larger membrane capacitance so that a given charge produces a proportionally smaller voltage change.

During a voluntary contraction of progressively greater power, the first α motoneurones in the cord to fire action potentials are the smallest (Hennemen *et al.*, 1965), presumably because there is an approximate equality of synaptic drive to all motoneurones in the motoneurone pool and hence a greatest depolarization of the smallest motoneurones. As the power of the contraction increases the small motoneurones increase their firing rate and larger motoneurones are recruited. This is known as the 'Hennemen size principle' (Hennemen *et al.*, 1965).

M and H responses

It is possible to study aspects of reflex activity by applying electrical stimuli to a mixed motor nerve percutaneously in man and to record the electromyogram (EMG), which is a compound action potential generated by the extrafusal muscle fibres (Figure 7.7). The central nervous system interprets the volley of nerve impulses as something unpleasant happening at the receptors connected with the nerve and not at the site of stimulation. The pain is

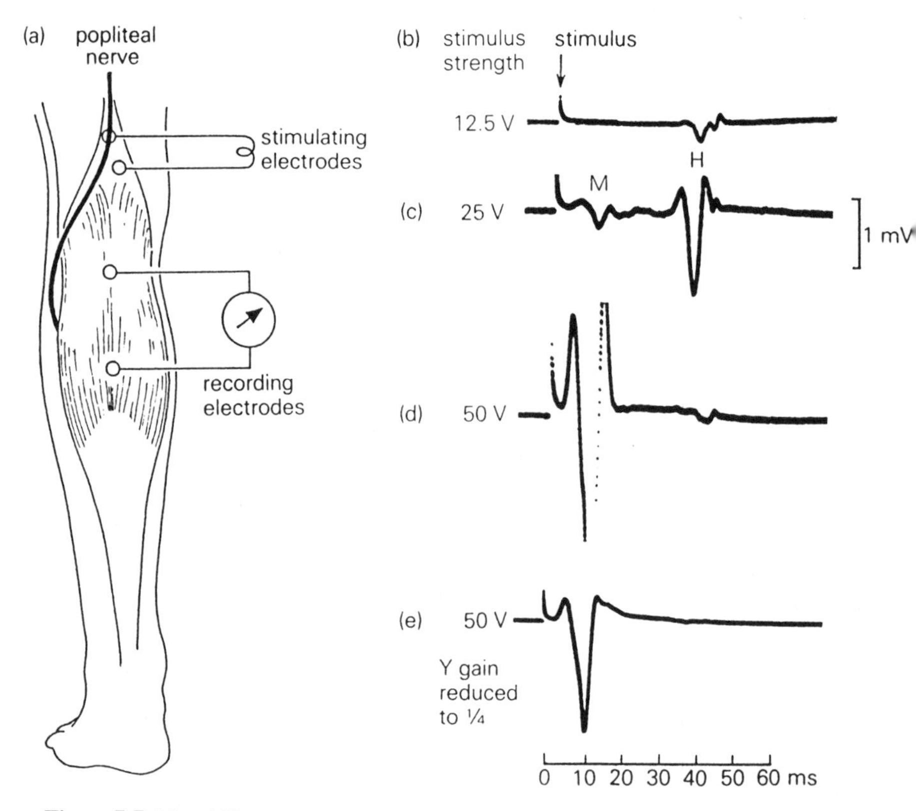

Figure 7.7 M and H responses. (a) The experimental set-up. The electric shock is applied to the skin overlying the mixed motor nerve innervating the gastrocnemius muscle. Electrical records are derived from a pair of skin electrodes overlying the gastrocnemius. (b) Stimulus strength 12.5 V, the weakest stimulus which evoked a response. (c) Stimulus strength 25V. (d) and (e) Stimulus strength 50 V. In (d), the M response was large and its peaks were lost. In (e) the amplification was reduced to a quarter so that the full excursion could be recorded.

perceived as coming from the region of cutaneous distribution of the nerve (see Chapter 6).

The waveform of the EMG depends on the strength of the stimulus. Since the largest nerve fibres have the lowest threshold for electrical stimulation, a weak shock stimulates just the Group Ia afferents; these innervate the primary spindle endings. These afferent fibres conduct impulses to the cord and initiate a monosynaptic reflex. The motor nerve volley causes a twitch in a small group of extrafusal muscle fibres. The compound action potential is picked up by the EMG electrodes. This response is labelled the H response, after Hoffman who first described it. The delay between the stimulus and the response is proportional to the length of the reflex pathway. For the gastrocnemius muscle of the calf, it is around 45 ms, most of which is occupied by nerve impulses travelling along the reflex pathway, as explained in Chapter 8.

At a stimulus strength of twice threshold, α efferent nerve fibres are also stimulated. This leads to the motor (or M) component, shown in Figure 7.7c; its delay is shorter because the nerve impulses pass directly towards the

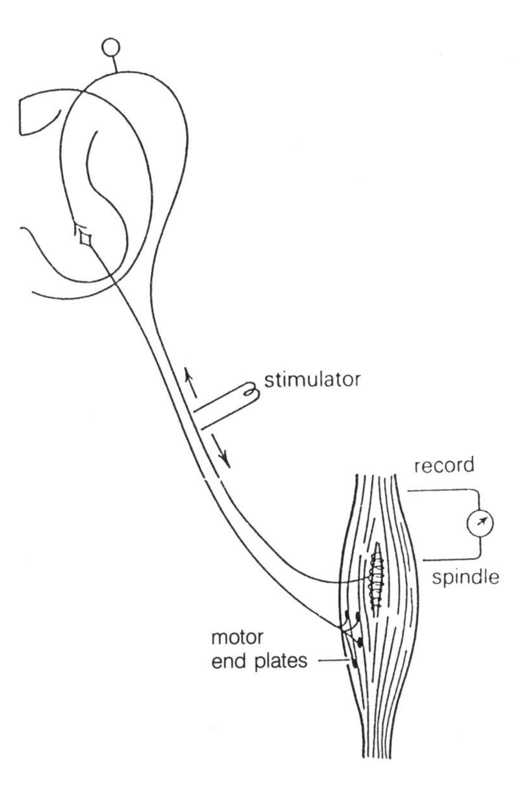

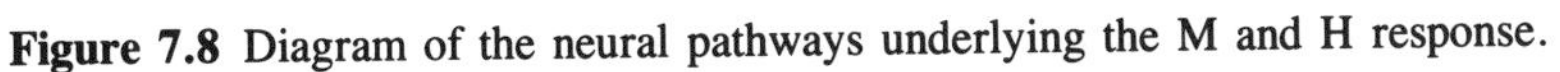
Figure 7.8 Diagram of the neural pathways underlying the M and H response.

muscle and do not traverse the cord; they therefore have a shorter distance to travel. The H response is larger than with the threshold stimulus because more afferent fibres are brought to threshold by the bigger stimulus. At still higher stimulus strength (Figure 7.7 d and e), there is a bigger M response and a much smaller H response.

The diminution of the H response with high stimulus strengths can be explained with the help of Figure 7.8. At this high stimulus strength, action potentials were initiated in most of the nerve fibres in the mixed motor nerve. In each nerve fibre, an action potential elicited by electrical stimulation passes in both directions away from the stimulator. The action potentials in the sensory fibres travelling towards the cord are those that initiate activation of the α motoneurones which then launch impulses along the motor nerve fibres towards the muscle. These same motor nerve fibres have been activated artificially by the electrical stimulus and so are conducting an impulse antidromically (backwards) towards the soma. The two impulses travelling in opposite directions meet and mutually extinguish each other. This is because the nerve membrane behind each propagating action potential is refractory. The overall result is that the reflex component of the response in the muscle is abolished and only the motor response survives.

Supraspinal control of spinal reflexes

Spinal reflexes, although they occur in subjects in whom connections with higher centres have been severed, are normally under the influence of descending influences which may either potentiate or depress them, as Figure 7.9

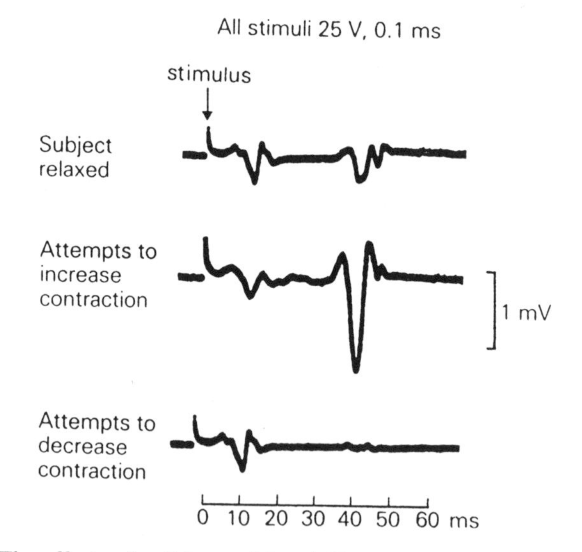

Figure 7.9 The effects of volition on M and H responses.

demonstrates. The recording shown in Figure 7.9 was obtained when an electric shock was applied to the nerve supplying the gastrocnemius muscle of a normal human subject at the level of the knee-joint. The electromyogram was recorded from an electrode on the skin over the gastrocnemius. The upper trace was obtained when the stimulus was applied to a relaxed subject; the stimulus was applied at the time zero on the x-axis. The subject was then asked to clench his fists in an attempt to increase the contraction; the response to nerve stimulation changed to that shown in the middle trace. There is no change in the M response; the number of motor nerve fibres excited under the stimulating electrode was unchanged. The H wave is potentiated, showing that the spinal motoneurones were diffusely facilitated with the result that more were brought to threshold by the afferent volley. The subject was increasing the central excitatory state of the spinal motoneurones. He then was asked voluntarily to inhibit the reflex contraction and the lower trace was obtained. Here the inverse result was obtained; the M response is still unaltered whereas the H response is depressed due to a general lowering of the level of excitation in the spinal motoneurones.

7.5 WHICH TYPES OF PROPRIOCEPTIVE ACTIVITY CAN BE CONSCIOUSLY PERCEIVED?

It is common experience that we can readily detect the position of our joints, so proprioceptors clearly project to conscious levels. From which types of proprioceptors does the cerebral cortex receive this information?

Stimulation of joint receptors evokes electrical activity in large regions of the cerebral cortex, including the primary sensory region and surrounding areas. Stimulation of the sensory nerve fibres from muscle spindle receptors evokes electrical activity confined to the primary sensory regions of the cerebral cortex. There is evidence that the relative importance of the two types of input in the conscious appreciation of position is different for different joints, as the following examples illustrate.

Finger and toe joints

In World War I, observations were made on soldiers whose hands or feet had been blown off. After healing of the wound, in some cases it was possible for the doctor attending the subject to pull on the exposed end of the tendon which had originally been inserted into the fingers or toes. This stretching of these particular muscles did not evoke any sensation in the subject.

A related observation is that if a finger or toe is anaesthetized by injection of a local anaesthetic into the digital nerves, the subject cannot tell the position of the interphalangeal joints. The long muscles which operate these joints lie in the forearm and are therefore unaffected by the anaesthetic. This suggests that finger joint position is signalled primarily from joint receptors and not from muscle spindle receptors. This is not surprising, since the muscles in question operate over several joints, so information of muscle length alone would not specify the positions of the different joints in the finger.

Elbow joint

For a muscle operating at a single joint, the spindle receptors would provide the brain with more precise information about the position of the joint; in this circumstance, spindle receptor activity does indeed contribute to the conscious appreciation of position, as the next observation shows.

A way of selectively activating the primary spindle receptors in normal people is to press a mechanical vibrator on the skin over the belly of the muscle. The experiment consists of sitting the blindfolded subject at a table, with both elbows resting on the table and the two hands held pointing vertically upwards. Vibration is applied to the skin over the tendon of the biceps muscle on one side and the subject is asked to match the angle of the elbows on the two sides. The result is that the subject extends the elbow on the unvibrated side relative to the vibrated side. This shows that the discharge initiated by primary spindle receptors in this muscle are perceived as indicating that the vibrated muscle is being stretched (Matthews, 1972, p. 497). There is a mismatch between muscle spindle input and joint receptor input; the brain believes the spindle input and ignores the joint receptor input.

The overall conclusion to be drawn from this section is that activity originating from muscle spindle receptors is perceived in situations when this contributes unambiguously to the perception of position. In situations where such information could be interpreted ambiguously, it does not reach consciousness; instead the brain relies on the unambiguous information from the receptors in the capsules of the relevant joints.

7.6 THE DIFFERENT TYPES OF SPINDLE RECEPTOR

Early histologists knew that the spindle receptors are of two histological types, annulo-spiral (primary) and flower spray (secondary) endings. The primary endings form complete spirals around the intrafusal fibre whereas the axons of the secondary endings form incomplete spirals, which histologists fancifully likened to a spray of flowers.

The primary ending

In response to a ramp stretch, these primary receptors show a prominent dynamic component; Figure 7.1c will remind you of what this means. The primary ending is connected with a Group Ia afferent nerve fibre, the fastest-conducting nerve fibre in the body. The function of the dynamic receptor is to initiate strong, rapid responses, such as the tendon jerk reflex. Consequently the dynamic component is of high frequency and immediate. Since it is necessary for the information to travel quickly, it is propitious that these receptors are connected with the fastest conducting nerve fibres. The dynamic component of the response is more pronounced the quicker the displacement which elicits it. This is again appropriate to mechanisms aimed at responding to rapid disturbances.

The secondary ending

This receptor is a static receptor, that is, it signals the degree of displacement and not the rate of change of displacement. It connects with a Group II afferent nerve fibre which conducts rather more slowly than the Group Ia fibres. The information which it signals is used for slower more accurate responses to change in position, for instance carefully following a slowly moving target with a finger. Since time is not at such a premium, receptors which signal primarily static information are usually connected to nerve fibres which conduct more slowly than those of receptors whose responses is mainly dynamic.

The anatomical relationship of receptors in skeletal muscle to the muscle fibres

The spindle receptors, both primary and secondary, lie in parallel with the extrafusal muscle fibres (Figure 7.3). When the muscle contracts and shortens, tension is taken off the spindle receptors and this reduces the firing rate of their sensory nerve fibres. The tendon receptors lie in series with the muscle fibres. When the muscle fibres contract and increase the tension in the tendon, the tendon, together with the tendon receptors, is stretched and the tendon organ's nerve fibre increases its discharge rate. We note that shortening is the factor which reduces the sensory discharge from the spindle whereas increase in tension is the factor which increases tendon receptor activity. When a muscle contracts, it usually increases its tension and shortens. The muscle can, however, contract isotonically, in which case the muscle shortens with no increase in tension; it can also contract isometrically, in which case the tension increases but there is no shortening. The muscle afferents carry information to inform the central nervous system about how much tension and how much shortening have occurred as a result of a motor command which it has issued.

7.7 APPENDIX: THE MAMMALIAN MUSCLE SPINDLE IN MORE DETAIL

In addition to there being more that one type of spindle receptor,

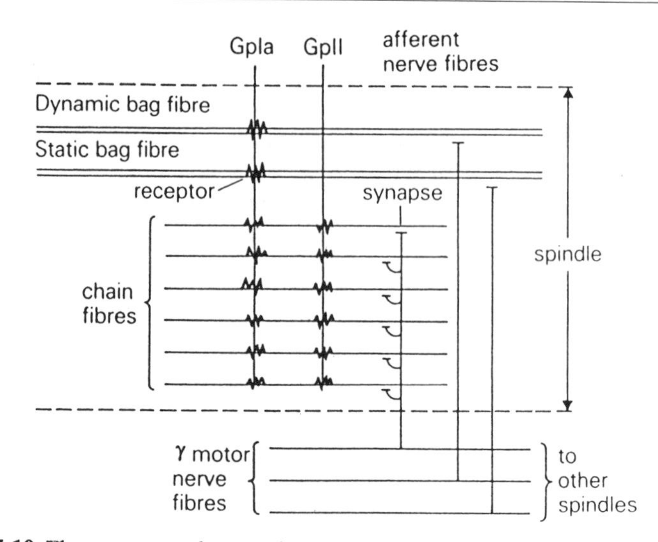

Figure 7.10 The sensory and motor innervation of muscle spindles.

there is more than one type of intrafusal muscle fibre. A mammalian muscle spindle is shown diagrammatically in Figure 7.10. Each muscle spindle contains several intrafusal muscle fibres. In the equatorial regions of each intrafusual fibre, the majority of the nuclei lie in the middle of the fibre, not just beneath the cell membrane as is the case in extrafusal fibres. In this equatorial region, the nuclei lie in a bag or in chains: this allows the intrafusal fibres to be classified into 'bag' and 'chain' fibres. Different spindles differ in the number and types of intrafusal fibres. A typical muscle spindle consists of two nuclear bag fibres and about six nuclear chain fibres. There are even differences between the two nuclear bag fibres found in a typical spindle. Such differences are highlighted when the motor nerve fibres innervating the two are separately stimulated. This has different effects on the afferent discharges from the sensory innervation. Activation of one of the fibres increases the dynamic component of the response to stretch; this is therefore called the dynamic bag fibre. Activation of the other increases the static component; this is the static bag fibre. The two bag fibres extend the whole length of the spindle and are rather thicker than the chain fibres, which extend less than the full length of the spindle.

An afferent nerve fibre from a spindle innervates only one spindle; this is in contrast with the motor innervation of the spindle, as we shall see. A Group Ia afferent nerve fibre from a spindle is connected with primary endings on many of the intrafusal fibres. There are always endings on both types of bag fibre and also endings on most or all of the chain fibres. A Group II afferent nerve fibre is connected with secondary endings on the chain fibres. As a consequence of this arrangement, a single afferent spindle nerve fibre is influenced by, and therefore integrates information from, several intrafusal muscle fibres.

A Group Ia afferent fibre exhibits both dynamic and static components in its discharge. The sensitivities of the dynamic and static responses can be separately changed by selective activation of either the dynamic or static muscle fibres to which the afferent nerve fibre is connected. The Group II afferent nerve fibres show a predominantly static response to stretch. Again the sensitivity of the response of Group II fibres to stretch is modified by activation of different intrafusal fibres.

A single fusimotor fibre sends branches to several muscle spindles in the same muscle. A fibre that innervates an intrasfusal muscle fibre of a particular type in one spindle is likely to innervate fibres of the same type in all the spindles to which it sends branches. Some fusimotor fibres innervate only dynamic bag fibres, others innervate only static bag fibres, and yet others innervate only chain fibres. Other fusimotor fibres are less selective in that branches from a single nerve fibre innervate both static bag and chain fibres.

The effects of fusimotor activation depend on which type of intrafusal muscle fibre is being activated. If the fusimotor fibre innervating a chain fibre is stimulated at a frequency up to 50 Hz, every impulse in the motor nerve fibre elicits an action potential in the Group II afferent nerve fibre. The motor nerve is said to drive the afferent nerve one to one. During activation of this type, mechanical stretching of the spindle does not alter the discharge rate of the afferent nerve fibre. The γ activation has rendered the afferent discharge insensitive to stretch.

Activation of the motor innervation of the static bag fibre has quite different effects. Although repetitive stimulation of the nerve fibre increases the discharge rate of the afferent nerve fibre connected with the receptor on that static bag fibre, there is no relationship between the timing of impulses in the motor and sensory fibres; there is no driving. Moreover, far from abolishing the sensitivity of the receptor to stretch, stimulation of the motor nerve fibre actually increases the sensitivity of the receptor on the static bag fibre to stretch. Activation of the motor innervation of dynamic bag fibres similarly increases the sensitivity of dynamic receptors to stretch.

In summary, the central nervous system can control the responsiveness of the muscle spindle afferent discharge provoked by stretching a muscle fibre. It can increase the dynamic component by sending impulses along γ efferent fibres to dynamic bag fibres; it can increase the static component via the motor innervation of static bag fibres. Conversely, it can render the afferent discharge entirely insensitive to stretch applied to the muscle via the motor innervation of the chain fibres. This latter is a mechanism available to the central nervous system for switching off a proprioceptive spindle reflex when the central nervous system pre-empts reflex movement during certain types of voluntary movement commanded by higher centres, as described in a later chapter.

Spinal mechanisms 8

8.1 PERCEPTION AND SENSATION

Our perception of a stimulus depends on many factors other than the raw input along sensory nerve fibres. We can discriminate between a 10p piece or a 50 p piece when blindfolded far more easily if we are allowed to manipulate the coin than if we are constrained to hold our hand still and a colleague moves the coin across the skin of the hand. When we manipulate, we bring to bear motor information in our sensory discriminating task. Another example, showing that the discharge pattern in sensory fibres is only part of our perceptive armoury, is that we cannot tickle ourselves. We can provide ourselves with the same sensory input as that provided when a friend tickles us but the perception is different. The reader will readily think of other situations in which sensory stimulation by a partner is far more evocative than self-stimulation.

The arrangement of neuronal connections in the spinal cord, with many synapses impinging on a single neurone, provides the basis for interactions of information passing along tracts from sensory nerve fibres, from nerve fibres from central nuclei, and from motor nerve fibres. Neural discharges are seldom, if ever, transmitted through the nervous system unmolested; they are modulated by influences from many sources. This section is about the modulation of sensory information as it passes from the incoming sensory nerve fibres and ascends to the higher centres of the brain.

Cortical and primary afferent modulation of cutaneous sensory information in the cord

In the spinal cord, there is only one synaptic junction along the specific afferent pathways (Section 6.6); for the spinothalamic projection, this relay is within the segmental levels of the spinal grey matter; for the dorsal column projection it is in the dorsal column nuclei. Information is modulated at this first synaptic junction by two groups of influences. First, there is spatial

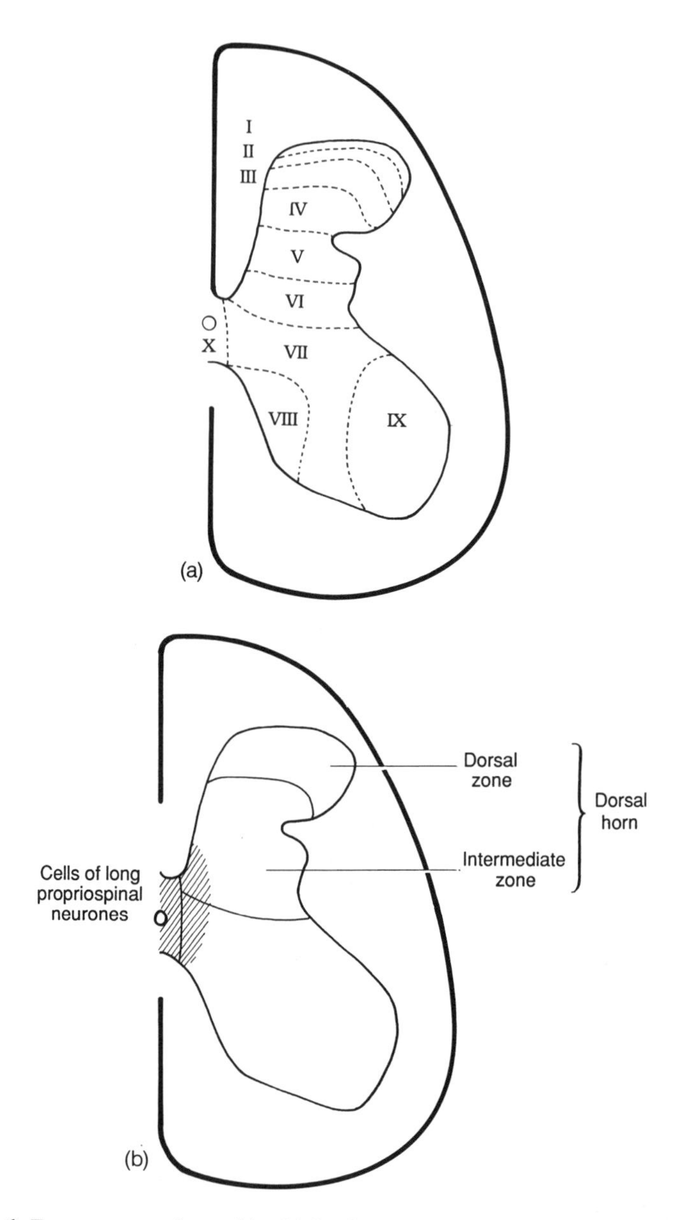

Figure 8.1 Transverse sections of the 5th lumbar segment of the spinal cord. (a) Rexed's division of the spinal cord into laminae. (b) Functional subdivisions of grey matter. The region labelled 'dorsal zone', Rexed laminae I to III, is primarily concerned with the transmission of nociceptive information and the intermediate zone, Rexed laminae IV to VI, with low threshold mechanoreceptor information. The cells of origin of long propriospinal neurones lie in the shaded area, the medial parts of Rexed laminae V and VI plus the central grey matter, Rexed lamina X.

interaction amongst the various afferent nerve fibres bringing information in from the periphery. Second, there is descending control from the cerebral cortex and other higher centres. The relay cells of the spinothalamic and dorsal column projection pathways will be considered in turn.

The cells of origin of the spinothalamic tract

The grey matter of the spinal cord has been subdivided into laminae by Rexed (Figure 8.1a). As shown in Figure 8.1b, afferent nerve fibres subserving nociception, with a low conduction velocity, project to cells in the dorsal zone (laminae I, II, and III) whereas those serving receptors responsive to low-threshold mechanical stimuli and hair movement, with a higher conduction velocity, project to cells located in the intermediate zone of the grey matter (laminae IV, V, and VI). The grey matter of the cord thus shows some semblance of functional organiztion, propitious for interaction of inputs from neighbouring receptors with a common function.

The propriospinal system

This system of interneurones originates, travels, terminates, and exerts all its influences within the spinal cord itself. There exist long ascending and descending systems of propriospinal neurones connecting the parts of the cord providing segmental innervation of the neck, forelimbs, and hindlimbs. These coordinate analysis of information and formulation of integrated patterns of motor command for these regions of the body. The cells of origin of these long propriospinal tracts are aggregated in the intermediate zone (laminae VI to VII) of the central core of grey matter and in lamina VIII. These long propriospinal connections are principally with the intermediate region of the dorsal horn, which is concerned principally with the transmission of low-threshold mechanoreceptor input.

One function of the long propriospinal neurones is to transmit motor commands from the cerebral cortex; axons of the corticospinal projection terminate on these cells. Another is the transmission of slow nociceptive information up to the higher centres, to be described in the chapter on pain, Section 18.2. Evidence for this comes from the observation that, in an animal, hemisections of either half of the cord, when separated by four or more segments (Figure 8.2), permit the reappearance of standard reactions of the animal to noxious cutaneous stimuli applied below the level of the lower lesion (Basbaum, 1973). This shows that information was reaching suprasegmental levels by a route not unilaterally confined. Nociceptive input from either side of the body can follow a winding course to and fro across the midline as it ascends.

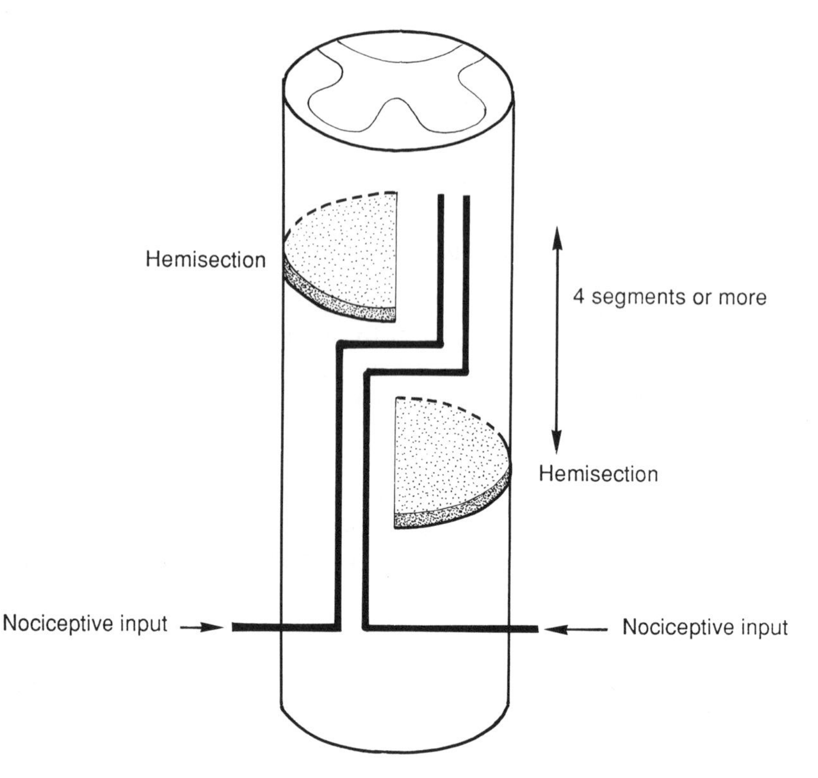

Figure 8.2 Two hemisections of the spinal cord separated by at least four segments are shown. After such an operation, nocifensive behaviour persists, showing that nociceptive transmission passes up the cord crossing to and fro across the midline, as indicated by the pathways in this figure.

The dorsal column nuclei (gracile and cuneate)

These are two nuclei which lie anatomically on the dorsal aspect of the medulla oblongata. Physiologically, these nuclei constitute a rostral extension of the spinal dorsal horn. Within the nuclei, there is a distinct somatotopic organization; that is to say, the body surface maps to different regions of the nuclei. Cells receiving primary afferent input from the hindlimbs are located in the gracile nucleus situated medially, and those from the forelimbs lie in the cuneate nucleus, situated lateral to the gracile.

The cells of the dorsal column nuclei are of two types. One is of relay cells of the dorsal column – thalamic system (Figure 8.3). These cells relay information principally from low-threshold hair and cutaneous mechanical receptors. This information is largely unprocessed, with little convergence and modality mixing. The other type comprises interneurones intrinsic to the dorsal column nuclei. These cells do not receive input directly from

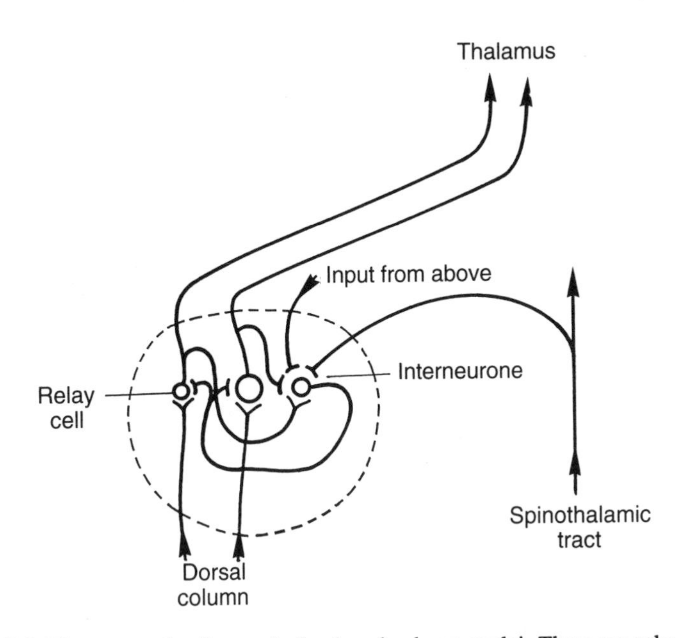

Figure 8.3 The neuronal cell types in the dorsal column nuclei. There are relay cells and interneurones.

the primary afferent axons; their input is from second-order relay neurones of the dorsal column and the spinothalamic tracts. These cells convey information that is modality mixed from high-threshold mechanical and thermal stimuli (via collateral branches of axons travelling up in the spinothalamic tracts, Figure 8.3) in addition to low-threshold mechanical input from second-order neurones of the dorsal columns. These cells have large receptive fields, indicating convergence on the cells of influences from many primary afferent fibres. These cells also receive input from higher centres.

In the relay cells of the dorsal column nuclei, lateral inhibition, or spatial interaction of channels carrying information from different cutaneous sites, operates. The mechanism for this is presynaptic inhibition. Lateral inhibition is highly selective. For instance, a primary afferent fibre from a Pacinian corpuscle projects to its on-line relay cell and laterally inhibits the relay cells for neighbouring Pacinian corpuscles, but it does not project to relay cells for the other types of mechanoreceptor. The input from Pacinian corpuscles is thus segregated and the process of lateral inhibition, to sharpen spatial discrimination, operates among sensory channels of the same nature. This specificity is of physiological advantage, as illustrated by considering the localization of a vibratory stimulus at 300 Hz.

Such a stimulus activates only Pacinian corpuscles, as shown in Figure 3.6. When several sensory fibres are activated as a result of a vibratory stimulus touching a region of skin, spatial discrimination depends on circuits activated only by fibres from these receptors, since other afferent fibres have nothing to add.

Another feature of obvious physiological value is that lateral inhibition operates most effectively between sensory fibres from neighbouring receptors; the intensity of interaction falls off rapidly with increasing distances between receptors. For discriminating whether two separate sites, rather than one diffuse area, are being stimulated, lateral inhibition operating over short separations between receptors sharpens the image but, if it occurred over a larger distance, two separate sites of stimulation would provide input which would tend to provide mutual cancellation. This would be counter-productive for two-point discrimination.

Lateral inhibition seems to be a prominent feature of sensory fibre projections from rapidly adapting receptors, but is less important for those from slowly adapting receptors. This is related to the fact that rapidly adapting receptors signal information about the occurrence of change and localizing the site at which change is occurring. Slowly adapting receptors are more important in signalling the time course of changes and related features.

All these features of function in which modality information remains segregated require a high degree of spatial specificity of projection pathways, with terminal branches of a primary sensory fibre from one type of receptor targeting relay cells activated by exactly the same type of receptor. There is almost no mixing of inputs from different types of receptor to a given second-order neurone.

Extrinsic influences on the dorsal column nuclei

Afferent projection systems other than the dorsal columns themselves influence transmission through the relay between first- and second-order neurones in the dorsal column nuclei. These influences are probably exerted via the interneurones in the dorsal column nuclei. One phenomenon of great importance is that synaptic activation of the relay cells in the dorsal column nuclei by impulses travelling up the primary afferent fibres is profoundly depressed by simultaneous noxious stimulation of the skin. This demonstrates the existence of extrinsic influences mediated by fibres of the ascending nociceptive pathways (the spinothalamic tracts and central grey interneuronal pathways, described in Chapter 18). The greatest depression of transmission is for afferent projections serving rapidly adapting mechanoreceptors. The survival value of this mechanism is that central transmission of nociceptive information and responses to it, which are essential for the subject to react to potentially damaging or lethal situations, are improved by attentuation of extraneous information which, in an emergency, the organism can do without.

Auditory and visual stimuli also influence transmission through the dorsal column nuclei. This influence may take the form of facilitation or suppression, depending on the experimental circumstances. This may be connected with the need for different sensory systems to cooperate in interpreting signals from auditory and visual scanning of an object which is being manipulated.

Corticofugal modulation

This means modulation as a result of nerve impulse along nerve fibres leaving the cerebral cortex. In monkeys, there exists a projection from the cerebral cortex to the relay cells of the dorsal column nuclei. Activity in this projection causes facilitation or depression of transmission according to the experimental circumstances. This allows an individual to 'scan' the sensorium; we can voluntarily pay particular attention to one part of the body: for example, we concentrate on a hand when it is being used to feel an object in the dark.

8.2 SPINAL REFLEXES

The simplest spinal reflex is the tendon-jerk reflex, described in Chapter 7. As a reminder, when the tendon of a muscle is tapped with a rubber hammer, the tap stretches the muscle and stimulates stretch receptors which lie in the belly of the muscle in structures called muscle spindles. The afferent discharge which this initiates passes to the spinal cord. The primary afferent axons project to form synapses directly with the motoneurones in the motoneurone pool which innervates the muscle whose tendon was tapped. The synapses are excitatory and so a volley of nerve impulses is conducted along the motor fibres to cause a muscle twitch. This is the tendon-jerk reflex.

Reflex time

For any reflex, the time between the stimulus and the onset of electromyographic (EMG) activity is called the **reflex time** (Appendix). Figure 8.4 shows the reflex pathway for the knee-jerk reflex and the timing of activity for different sites along the pathway. Zero time is the instant when the patellar tendon is tapped. It takes about 1 ms for the receptor to sense the stretch and generate activity in the afferent nerve fibre. It takes 6 ms for the nerve impulse to be conducted to the spinal cord, 1 ms for it to traverse the cord, 10 ms for it to be conducted back to the muscle. Delay in the nerve terminal and

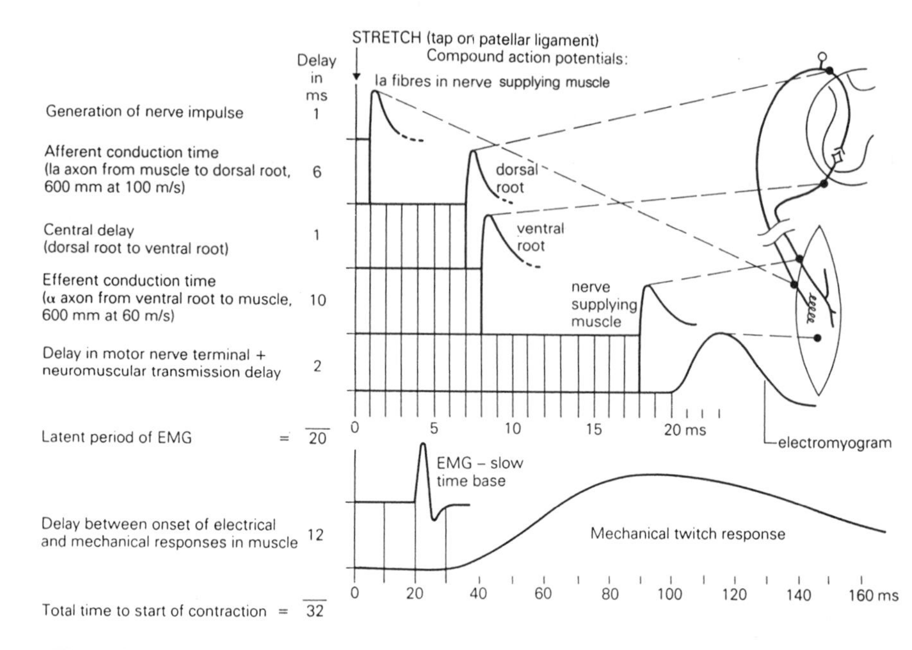

Figure 8.4 The knee-jerk reflex of an adult person: the timing of activity around the reflex arc.

neuromuscular transmission accounts for 2 ms. This gives a total reflex time, from stimulus to onset of electromyogram, of about 20 ms.

The two lower traces are on a slower time scale and show that there is a further delay of around 12 ms between the action potential in the muscle and the onset of contraction. So between the stimulus and the start of the mechanical response there is a total delay of around 32 ms. Most of the reflex time is taken up by conduction along the afferent and efferent nerves: only 1 ms or so is taken up in conduction through the spinal cord including the synapse. The time taken for conduction through the cord is called the **central delay**. It is measured as the time interval between the arrival of the afferent nerve volley entering the spinal cord through the dorsal root and the time when the efferent nerve volley leaves through the ventral root. The central reflex time depends mainly on the number of synapses on the reflex pathway in the cord. Each synapse introduces a delay of just under 1 ms. The tendon jerk is the only monosynaptic reflex in the body; all other spinal reflexes involve at least two synapses in the cord.

The number of synapses involved for common reflexes, some of which are described later in the chapter, are shown in Table 8.1.

Table 8.1 Number of synapses on the cord pathways for different reflexes

Tendon-jerk reflex	1
Inhibition of antagonists	2
Flexor withdrawal	3 or 4
Crossed extensor	>10
Scratch reflex	>20

Types of reflex

There are two main categories of spinal reflex. The first category is of **postural** reflexes; these allow an animal to maintain its posture. The stimuli which initiate these reflexes are the stretching of muscle, which elicits a stretch reflex (considered in Chapter 7) and pressure on the skin of the foot, which elicits an extensor thrust. The second category is of **protective** reflexes. These are reflexes that protect the animal from injury. The types of stimulus which elicit these reflexes include irritation and pain. The 'flexor withdrawal reflex' causes the withdrawal of a limb away from a painful stimulus. Another example is the scratch reflex; if the skin of the trunk is irritated, for instance by a flea, the appropriate reflex is not withdrawal, because the flea moves too. Instead, there is a rhythmic scratching movement of a limb, so that the offending insect is scratched away. A third example is 'abdominal guarding'. Obstruction of the bowel leads to bowel distension and severe abdominal pain. This elicits a reflex contraction of the abdominal muscles, which protects the abdominal contents against mechanical disturbance. Protective reflexes are also sometimes called **nociceptive** reflexes, since they are in general a response to potentially harmful stimuli.

Some characteristics of reflexes

The reflex response is specific to the stimulus that evokes it. For instance, a painful stimulus to the sole of the foot, such as standing on a sharp stone, will evoke a reflex withdrawal involving the flexor muscles of the leg. Firm pressure applied to the same site elicits a plantar thrust, which is the opposite movement from the flexor withdrawal reflex. It is a postural reflex, a supporting reaction involved in standing and walking.

A reflex response seldom, if ever, consists of a simple twitch in a single muscle. The spinal cord initiates a co-ordinated muscular response to produce a movement. The contraction of any one muscle is appropriate to this movement. The time-course of contraction, in the case of complicated reflexes such as the scratch reflex, bears no obvious relation to that of the input stimulus.

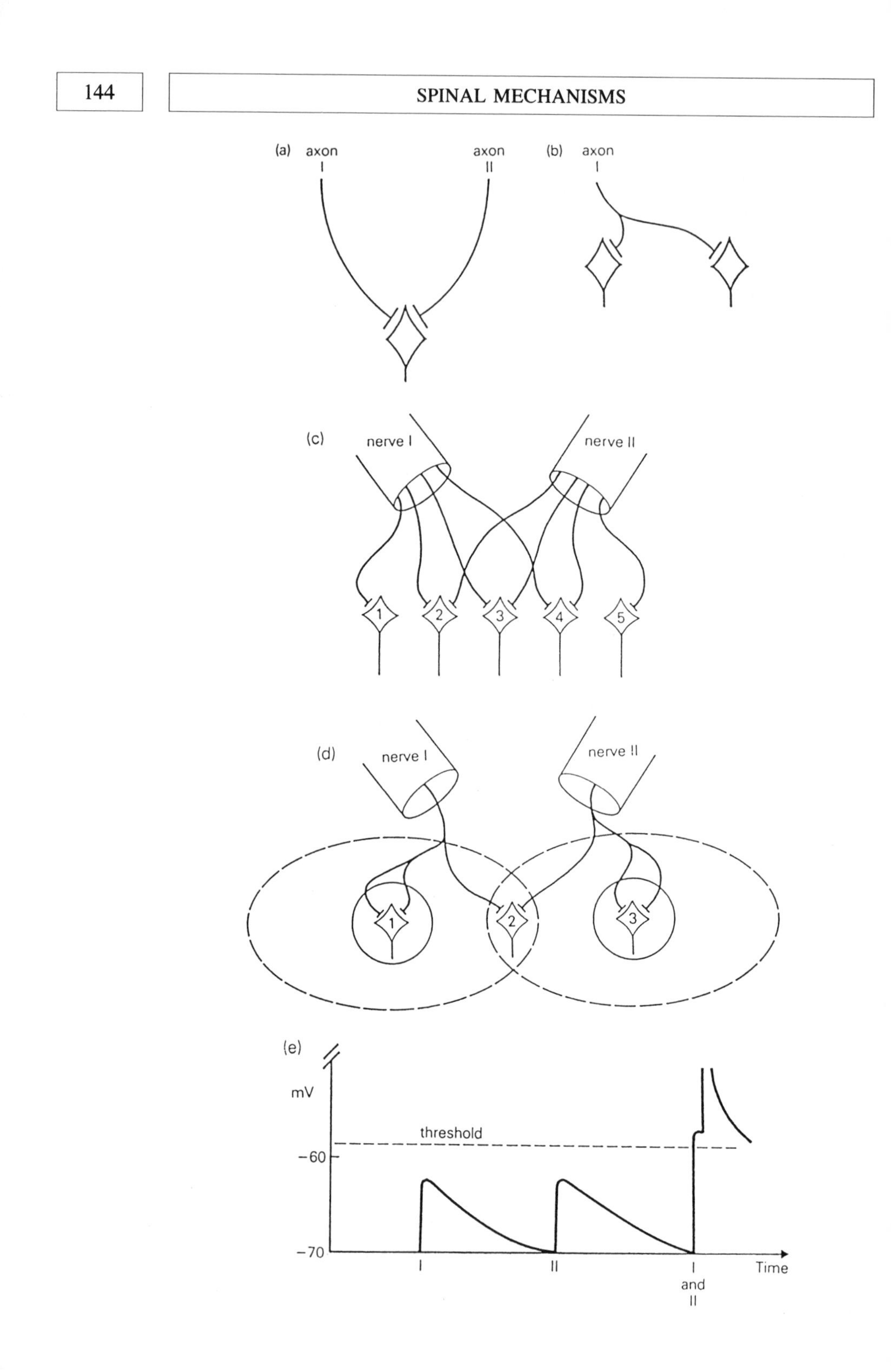
(a)
axon I
axon II
(b)
axon I
(c)
nerve I
nerve II
1
2
3
4
5
(d)
nerve I
nerve II
1
2
3
(e)
mV
threshold
−60
−70
I
II
I and II
Time

Convergence and divergence

The terms convergence and divergence are used to describe anatomical arrangements which can be deduced from the physiological behaviour of reflexes, as we shall see. The neurones of the central nervous system integrate information from a variety of sources. Of necessity axonal branches from many cells converge on a given neurone. This phenomenon of **convergence** is shown diagrammatically in Figure 8.5a. Conversely, in elaborating an appropriate response, the cord mechanisms may command activity at several segmental levels. The information entering the cord must be spread. One mechanism for this is the division of nerve fibres. Different branches from a single fibre form synapses on many cells. This **divergence** is shown in Figure 8.5b.

The existence of such anatomical arrangements can be deduced from the strength of reflex contractions of muscles when stimuli are given alone or together. When strong stimuli are applied to neighbouring regions of the skin, the contraction produced by the two stimuli together is less than the sum of the contractions produced by each stimulus separately. This **occlusion** can only mean that some motor units are fired by both inputs, and indicates convergence of inputs on some motoneurones. The way in which convergence operates is shown schematically in Figure 8.5c. This schema assumes that excitation of one of the synapses shown in the figure is sufficient to initiate an impulse in the postsynaptic neurone. The two inputs I and II are shown as each containing four nerve fibres. There is convergence of inputs from nerve I and nerve II onto postsynaptic neurones 2, 3, and 4. Since the stimuli are strong, all the fibres in each nerve are excited when the nerve is stimulated. Stimulus I thus initiates impulses in the postsynaptic neurones numbered 1 to 4; stimulus II activates neurones 2 to 5. Each stimulus alone activates four postsynaptic

Figure 8.5 (a) Convergence. Axons I and II both converge onto the single postsynaptic neurone. (b) Divergence. The single axon I divides into branches to influence each of the two postsynaptic neurones. (c) A schema showing how convergence accounts for the physiological property of occlusion. (d) Schema illustrating the neuronal circuitry underlying facilitation. The full circles show the regions where postsynaptic neurones are sufficiently strongly excited to fire action potentials. The larger dashed circles show the regions where neurones are excited but not necessarily brought to threshold. Neurones lying inside the dashed circles but outside the full circles are not brought to threshold when the stimuli are given singly; they are called the subliminal fringes. When both inputs are active, the region where the two subliminal fringes overlap contains neurones which are brought to threshold only if the two inputs are excited. (e) Intracellular electrical record from motoneurone 2 in (d), lying in the region of overlap between subliminal fringes of the inputs I and II. Each input separately elicits an excitatory postsynaptic potential insufficient to bring the motoneurone to threshold. When the stimuli are delivered together, they sum and threshold is reached so that an action potential (truncated in the diagram) is initiated.

neurones. The two stimuli together activate five neurones. This shows why convergence results in occlusion.

The amount of occlusion is a measure of the overlap of the motoneurones brought to threshold by the two inputs. When reflex responses of the muscle are recorded with stimuli of decreasing strength, the amount of occlusion decreases to zero and then, for some reflexes, reverses so that weak stimuli together elicit more reflex muscular contraction than the sum of the responses to the stimuli given separately. This phenomenon is called **facilitation**. Figure 8.5d is a schema to illustrate the anatomical substrate of these physiological relationships. In this schema, it is assumed that activation of one synapse is insufficient to bring a postsynaptic cell to threshold and that two synapses are needed to initiate an action potential. The figure shows nerves I and II each with a single nerve fibre activated by the weak stimuli. Each nerve fibre, as it approaches its termination, divides into branches to innervate two postsynaptic neurones. The weak input from a single stimulus is sufficient to bring to threshold only one postsynaptic neurone; for input I this is neurone 1. In addition to the small number of neurones which input I brings to threshold, a much larger number of surrounding neurones is excited, but not sufficiently to generate action potentials. These cells are indicated as the larger dashed circle on the left of Figure 8.5d. This group of subliminally excited neurones is called the **subliminal fringe**. Although they are excited, they do not contribute to the muscular contraction because they do not initiate impulses in their motor nerve fibres. Input II alone has similar effects on its target neuronal cell bodies.

Now let us consider what happens when the two weak inputs arrive simultaneously along adjacent channels. There is a zone of overlap of neurones in the two subliminal fringes – the region of overlap of the two large dashed circles in Figure 8.5d. Cells in this overlap region, of which cell 2 is an example, are now brought to threshold. The response to the two stimuli together consists of the cells brought to threshold by each input alone plus the cells in the region of overlap of the two subliminal fringes. Hence two weak stimuli together produce a larger reflex effect than the sum of the responses to the stimuli given separately.

Figure 8.5e is a schematic representation of the intracellular record from motoneurone 2 of Figure 8.5d. The excitatory postsynaptic potentials elicited by inputs I and II separately were insufficient to reach the threshold for initiation of an action potential, whereas the summed excitatory postsynaptic potential produced by the two inputs together (I and II) exceeded the threshold to generate an action potential.

Reciprocal innervation

In producing efficient co-ordinated movement, it is important that the contraction of muscles producing the movement should be accompanied by cessation of activity in muscles which oppose the movement (antagonist muscles). The orthodromic inhibitory pathway which we studied earlier is one neuronal circuit subserving reduction of activity in antagonists. A simpler and more generalized schema is shown in Figure 8.6a. Excitation of pathway I causes contraction of the extensor muscle and relaxation of the flexor, and hence an efficient extensor movement of the limb. Activity in pathway II produces the converse effect. The schema is grossly simplified to show only the final effect of the mechanism. It has been previously noted that inhibition is always mediated by an inhibitory interneurone: this has been omitted from Figure 8.6a for clarity. Sherrington called this phenomenon whereby contraction of a muscle is accompanied by reduction of activity in antagonist muscles, 'reciprocal innervation'.

Crossed extensor reflex

This reflex is a postural reflex to prevent overbalancing of the body. If a painful stimulus is applied to the skin of one foot, this foot is reflexly withdrawn. To prevent the individual from collapsing, a postural reflex comes into play with contraction of the extensor muscles in the leg contralateral to the stimulus. This is called the crossed extensor reflex. Because its spinal pathway involves more synpases than does that of the flexor withdrawal reflex, the reflex time for the crossed extensor reflex is longer than that for the flexor withdrawal reflex.

Competition of reflexes

A limb cannot flex and extend at the same time when stimuli commanding opposing reflex movements are delivered simultaneously. If stimuli that elicit opposing reflexes when given separately are delivered simultaneously, then only one reflex response is produced. Of the two reflexes, the one that is of more importance to the survival of the subject switches off the other. In general, protective reflexes have a greater survival value than do postural reflexes. The protective reflex is said to be **prepotent** to the postural reflex.

8.3 THE AXON REFLEX

This is not a spinal reflex, since it does not require participation of the spinal cord. It involves peripheral afferent nerve fibres. When the skin is slightly damaged, for instance by scratching the palmar aspect of the forearm firmly with the blunt end of a pencil, there is a series of reactions. One of these is a diffuse flare which appears a few seconds after the insult and involves the skin extending 1 cm or so on either side of the scratch. This is mediated by the so-called 'axon reflex'. The nerve fibres involved are unmyelinated

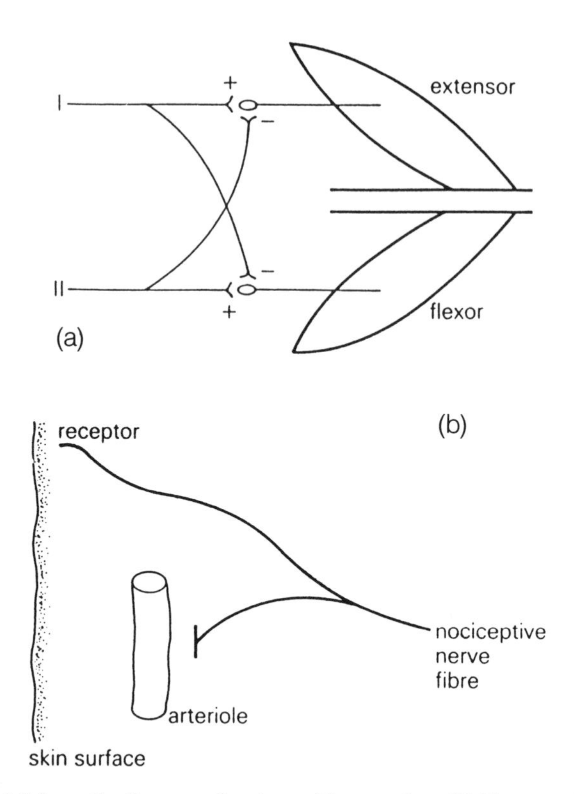

Figure 8.6 (a) Schematic diagram of reciprocal innervation. (b) The axon reflex (see text for explanation).

nociceptive afferent nerve fibres (described in Chapter 18); these comprise 50% of all the nerve fibres in a cutaneous nerve. The arrangement subserving the axon reflex is shown in Figure 8.6b.

Activation of the nerve ending triggers an impulse which travels centrally towards the spinal cord. A side-branch of the afferent nerve also carries impulses, which result in the release of Substance P at the nerve endings. Substance P causes arterioles to dilate; this is the origin of the flare.

If a cutaneous nerve is cut, there is immediate loss of sensation in the area of skin previously supplied by the nerve. The skin continues to exhibit an axon reflex, however, since conduction in the distal segment of nerve is not immediately abolished. The nerve fibres severed from their cell bodies do degenerate, and stop conducting impulses a day or two after the injury. At this stage, the axon reflex in the involved area of skin is lost. If the nerve regenerates, the axon reflex is re-established together with the other functions of the nerve.

8.4 THE SEQUENCE OF RECRUITMENT OF MOTOR UNITS

We now return to the subject of muscular contraction and its organization by the central nervous system. Skeletal muscle contains two types of twitch fibre, small and large, whose physiological properties are rather different. The small fibres are innervated by small α motoneurones, whereas the large muscle fibres are innervated by large α motoneurones. As described in Section 7.4, the small α motoneurones are the first to fire impulses in a voluntary or reflex movement. For the motor units of small muscle fibres innervated by the motoneurones, there are rather few muscle fibres in each motor unit (Figure 8.7a), so that

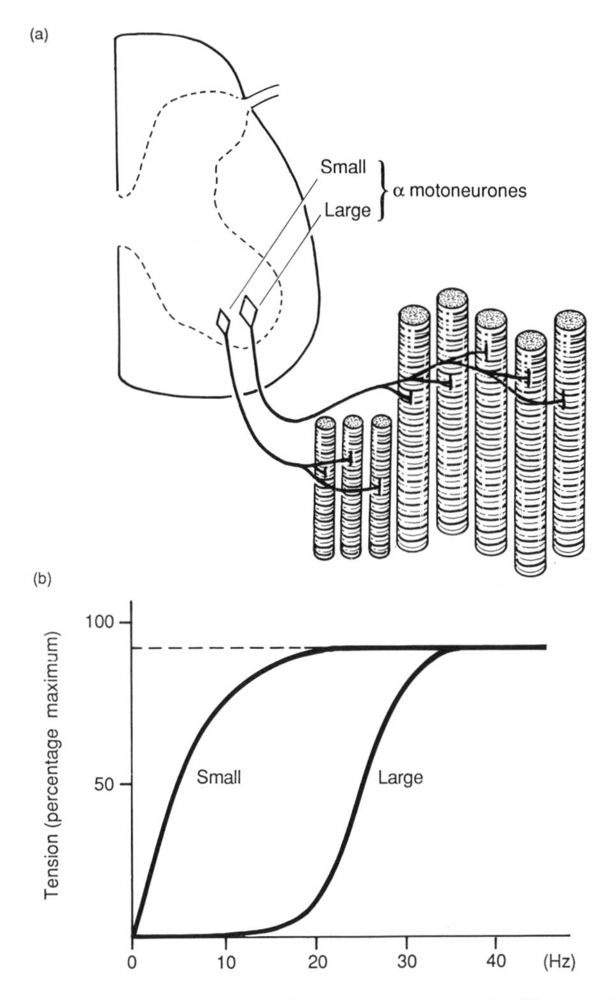

Figure 8.7 (a) Small α motoneurones innervate small muscle fibres, whereas large motoneurones innervate large muscle fibres. (b) The small muscle fibres require a lower frequency of stimulation to produce a tetanic contraction than do the large fibres.

the central nervous system can exert a fine degree of control. These muscle fibres are red and metabolize aerobically. They can therefore replace efficiently energy stores that are becoming depleted during a prolonged contraction; they are fatigue-resistant. Repetitive stimulation at a relatively low frequency produces a fused tetanic contraction (Figure 8.7b). The central nervous system can therefore produce smooth prolonged contractions without using high frequencies of impulses in nerve and muscle fibres. This is economic and efficient. The only factor missing is a very high contracting power.

The large α motoneurones are recruited when a very strong contraction is required; they innervate large muscle fibres. There are many muscle fibres in each motor unit and so each nerve impulse in the motoneurone gives a very strong twitch. The fibres are white and metabolize anaerobically. They are not able to replace depleted energy stores efficiently and are readily fatigued. A high frequency of stimulation is required to produce a fused tetanic contraction. These motor units are thus suited to provide short-lived contractions of immense power.

This is a beautiful example of the matching of electrophysiological phenomena with functional efficiency. The electrophysiological factor is the inverse relationship between sensitivity of a motoneurone and its size (the Hennemen size principle): when a motoneurone pool receives excitatory input, small motoneurones fire first. The functional factor is the relationship between muscle fibre size and biophysics: small muscle fibres are suited to finely controlled smooth prolonged contractions which are relatively weak. As excitation increases, signalling the need for a forceful contraction, the muscle fibres with the appropriate properties are called into play.

8.5 SPINAL REFLEXES ELICITED BY STIMULATION OF RECEPTORS IN SKELETAL MUSCLE

This section starts with gathering information from the relevant sections of previous chapters. Figure 8.8a shows the most direct intraspinal pathways for Group I afferents. The Group Ia afferents come from the primary endings within the muscle spindle; they project monosynaptically to the motoneurone running back to innervate the extrafusal fibres of the muscle in which the stretch receptor lies. This is the spinal mechanism of the tendon-jerk reflex. These Ia afferents also activate the pathway for reciprocal innervation, with inhibition of motoneurones innervating antagonist muscles.

The Group Ib afferents, from tendon organs, have as their most direct path inhibition of activity in motoneurones supplying the extrafusal fibres of the muscle to which the receptors belong. This provides a mechanism to prevent a muscle from contracting so strongly that it tears the tendon away from its insertion into bone.

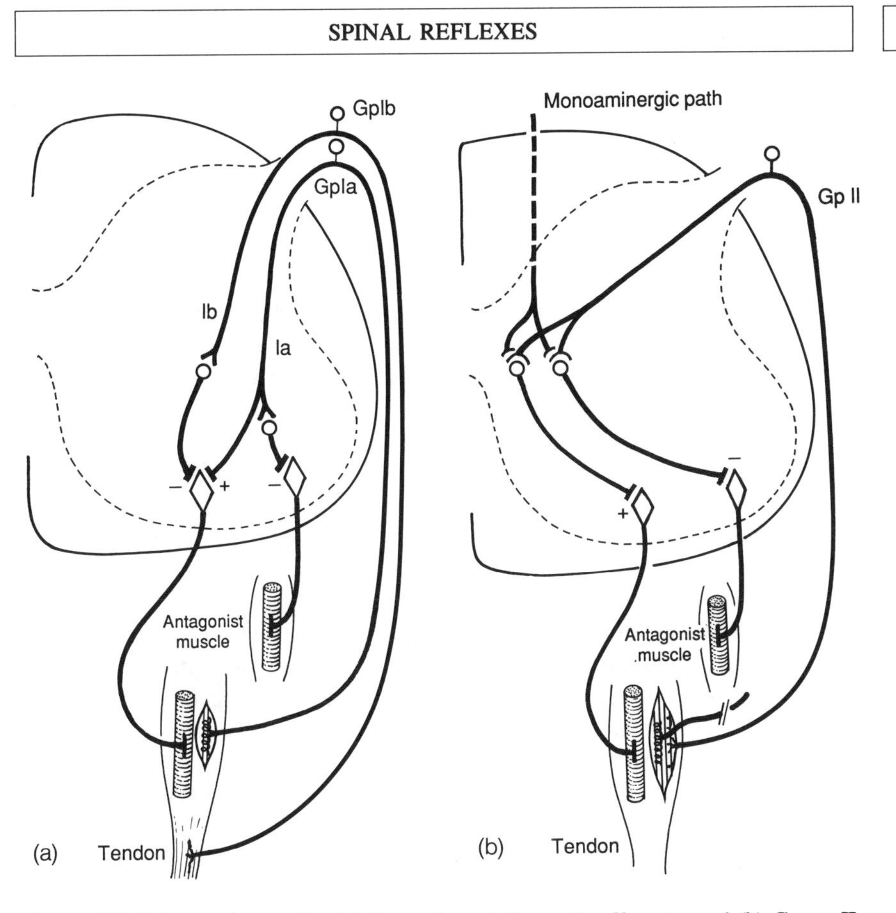

Figure 8.8 Spinal pathways for (a) Group Ia and Group Ib afferents, and (b) Group II afferents. The monoaminergic pathway is considered in a later section.

Spinal reflex activity activated by Group II afferent fibres

Figure 8.8b shows the pathways activated by input from the secondary receptors in the muscle spindle. All Group II pathways to motoneurones involve interneurones. The Group II afferents cause excitation of the motoneurones supplying the muscle in which the receptors lie: this is the basis of the stretch reflex. The afferents also project in an inhibitory fashion to motoneurones to antagonist muscles, again contributing to reciprocal innervation. Thus, in outline, the effects of activity in Group Ia and in Group II afferent fibres act cooperatively. In detail, the two paths differ.

The interneurones on the Group II pathway form aggregations in the medial part of the grey matter of the spinal cord, as shown in Figure 8.8b. Local information processing is performed in such aggregations. The feature of interest in the present context is that there is descending monoaminergic inhibition which is confined to pathways involving the Group II afferents, and which has no influence on pathways from the Group Ia or Group Ib afferents.

Raphe nuclei

The monoaminergic pathways originate in cells in the nucleus of the raphe and the locus coeruleus in the brain stem. The monoamines involved are noradrenaline and serotonin. The descending pathways inhibit synaptic transmission (Figure 8.8b) at the level of the interneurones which impinge on the interneurones – the so-called 'last order interneurones' (Noga *et al.*, 1992). Both excitation of the motoneurone to the muscle of origin and the inhibition of antagonist motoneurone are switched off by raphe stimulation. The equivalent Group 1a pathways, monosynaptic excitatory pathways, and disynaptic orthodromic inhibitory pathways are not influenced.

There is evidence that this projection from above is part of a mechanism for rhythmic limb movements. The raphe nuclei receive input from the mesencephalic locomotor region. During rhythmic limb movements the neurones of this locomotor centre are rhythmically active in time with the limb movements, as are cells in the raphe–locus coeruleus complex. The fact that the influence is inhibitory is explained by the rhythmic discharge from above periodically preventing the spontaneous activity of the last-order neurones and producing cyclic activity by this mechanism.

The descending inhibition is presynaptic, as shown in Figure 8.8b. Evidence for this is derived from experiments in which a stimulating micro-electrode is inserted into the cord to lie close to and stimulate directly the presynaptic terminals. This direct stimulation initiates action potentials which travel antidromically and are recorded from single Group II fibres in the peripheral nerve. During raphe stimulation, the threshold of the terminals is dramatically reduced, supporting presynaptic inhibition as the mechanism.

8.6 THE CONTRIBUTION OF THE SPINAL CORD TO THE ORGANIZATION OF COMPLEX MOVEMENTS

When a chicken's head is cut off, the decapitated animal may run or fly for half a minute or so. This is because the spinal cord contains circuitry which can organize locomotion without the need for the higher centres. Spinal rhythm generators project to motoneurones; muscle afferents feed back to these. Additionally, in the intact organism, the rhythm generators are influenced by nuclei in the brain stem.

Organization of spinal rhythm generators in mammals

In the first quarter of this century Graham Brown demonstrated that the spinal cord itself is capable of generating locomotor-like activity. He hypothesized that alternating activity between flexors and extensors might be explained by the existence in the cord of two so-called 'half-centres', one projecting to flexor muscles and the other to extensor muscles (Hultborn, 1992). A spinal network with the required properties was discovered. Jankowska *et al.* (1967), who experimented subsequently, found that, in animals in which the spinal cord was separated from higher centres by a spinal transection, the rhythm generator network became operative after the administration of DOPA. In the spinal cord, this is metabolized to yield dopamine and noradrenaline. It is not known why these monoamines have the effect of lighting up quiescent cord circuits which produce rhythmic activity. One can speculate that it is to do with the monoaminergic descending pathway from the mesencephalic nuclei, known to be involved with locomotor rhythms in the intact subject, described in the last section.

In animals with a transected spinal cord, the rhythm generator circuit revealed by DOPA is activated by high-threshold afferent stimulation. This stimulation elicits reflexes with latent periods much longer than those encountered without DOPA. The centres are tonically active and are connected to each other via mutual inhibitory projections. As the dose of DOPA is increased, the high threshold afferent stimulus gives a cyclic response, with flexor and extensor contractions alternating. With still higher doses these cyclic responses are spontaneous: they probably involve the same networks which are operative in normal locomotion. The cyclic discharges are referrred to as 'fictive locomotion' because there is no true walking.

Sensory feedback has been demonstrated to influence the rhythm generators. For instance, if the limb is passively moved during the cyclic activity of the motoneurone pools to extensor and flexor muscles, the sensory input to the cord may reset the centrally generated locomotor rhythm.

Reflex effects of Group Ib afferents

For extensor muscles, with intracellular recording in motoneurones, a stimulus to Group Ib afferents (from tendon organs) gives an inhibitory postsynaptic potential (Figure 8.8a). This provides a reflex pathway for autogenetic inhibition, that is, interruption of muscular contraction when the force of contraction becomes too great. Administration of DOPA results in a change to Group Ib causing depolarization of the same motoneurones. This shows that these afferents activate many circuits playing on the motoneurones and that DOPA changes the routing of information from the afferents to the motoneurones. The circuit revealed by DOPA is shown in block diagram form in Figure 8.9, which indicates the mutual inhibition between the

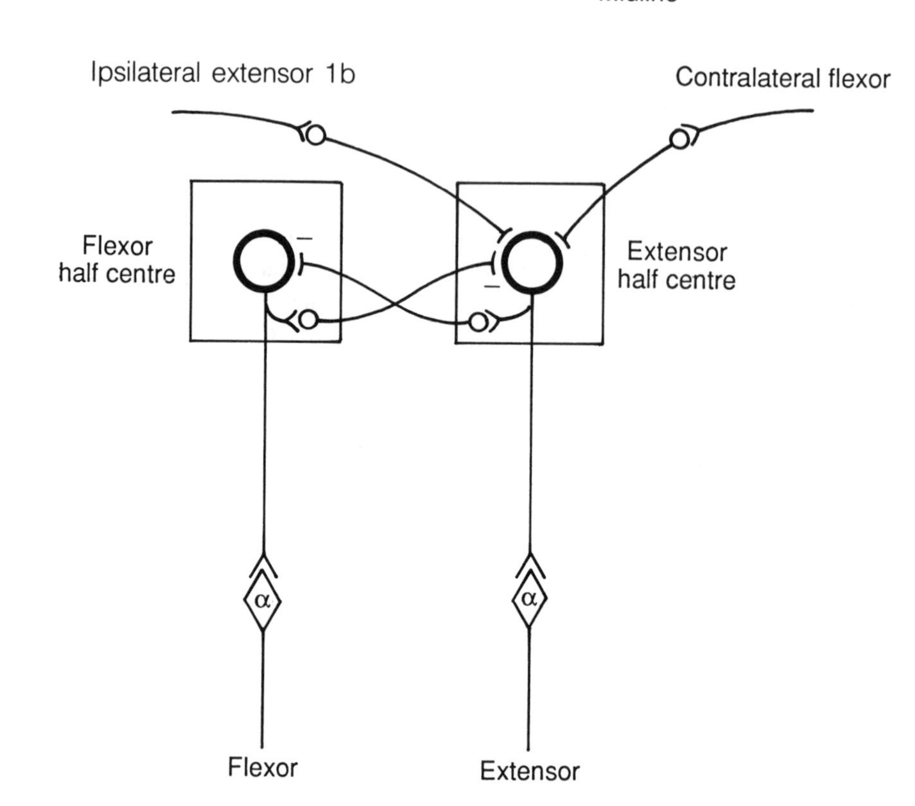

Figure 8.9 Half-centres in the spinal cord and some of the principal afferent inputs to them.

half centres. If one half centre is active, the other is held quiescent. If for instance during the flexor phase of the cycle, input is launched along the ipsilateral extensor Group Ib pathway, this excites the extensor half centre sufficiently strongly to spur it into activity and to switch off the activity in the flexor half centre. The Ib input therefore resets the cycle by interrupting the flexor activity and imposing an abrupt extensor activity (Hultborn, 1992). It is not yet known whether there is one single oscillator for a limb or whether there are separate oscillators, one for each joint in a limb.

Conclusion

We are only on the threshold of understanding the neural mechanisms involved in locomotion. One problem is that, for ethical reasons, much experimental work is necessarily performed on animals with experimental lesions of the spinal cord or on animals under anaesthesia: these procedures radically modify the behaviour of the neural networks of the cord. Another problem is that observations based on artificial stimulation of pathways can reveal neuronal connections but gives no idea of how the networks operate during normal locomotion. The difficulties involved in devising experiments which will show us how the cord works, and in analysing the activity in a structure which is so complex and has such profuse neuronal interconnections are almost insuperable with conventional techniques. However, new tools are continually appearing. One example of the type of technique which may provide a means of making inroads into the unknown is the '*cfos*' method of determining how many impulses a particular neurone has recently conducted in its natural functioning. This is described in outline in Chapter 18.

8.7 APPENDIX: THE DEFINITION OF REFLEX RESPONSE TIME

The reflex time is the time between the stimulus and the start of the reflex response. Since there may be some confusion about the definition of the response in this context, this appendix gives a brief description together with some historical notes.

Let us consider the tendon jerk reflex. The response is activity in the effector muscle. One could choose between the electrical activity in the muscle fibres, as recorded in the electromyogram, or the mechanical contraction as the response. Although the mechanical contraction is the feature which is of survival value to the subject, its use in timing the delay around the reflex arc is limited. There is a delay between the action potential in the skeletal muscle fibres and the onset of the mechanical response. The muscular contraction itself is long in duration by comparison with the delay introduced by the neural components of the arc. The duration of the contraction varies according to the type of muscle. For instance, the soleus muscle gives a slower contraction than the gastrocnemius. The contraction rises slowly and the identification of the exact time at which the contraction starts is subject to error. To quote from Fulton (1926, p. 273), 'mechanical records of the knee jerk have been made, but the method of recording was not of sufficient accuracy to admit of detailed analysis of the shape of the response'. Interest in neurophysiology usually centres on delays introduced by the neural components of the reflex rather than on the mechanics of the effector muscle, and for this purpose the electromyogram provides a much more useful reference waveform. Since the EMG rises sharply from the baseline and is of short duration, a decision about the timing of its onset is unequivocal.

Jolly (1911) measured the reflex time for the knee jerk reflex, calling it 'the latent period of the knee jerk'. He used a string galvanometer to record the action currents from wick electrodes placed on the skin over the quadriceps muscle. His photographic records were made on a glass photographic plate, shot, by means of a spring, vertically upwards behind a horizontal slit.

Reaction time

The reaction time is defined, consistent with the definition of reflex time, as the time between the stimulus and the onset of the EMG activity signalling a voluntary response. Modern research into the subject continues to use EMG records to signal the onset of activity in the effector muscle (Marsden *et al.*, 1978) in measuring voluntary reaction times. Clearly it is necessary to use the same indicator of effector activity when studying reflex time and voluntary reaction times if one wishes to use differences to calculate central delays.

Reaction times for intended movements triggered by stimuli involving displacements of the responding limb are faster than reaction times triggered by auditory or visual stimuli, but exactly how much faster is as yet not settled (Evarts and Vaughn, 1978). The uncertainty arises from the fact that intended limb movements triggered by limb perturbations are preceded by reflexes, and the experimenter has to sort out which is reflex and which is voluntary.

FURTHER READING

Creed, R.S., Denny-Brown, D., Eccles, J.C., Liddell, E.G.T. and Sherrington, C.S. (1932) *Reflex Activity of the Spinal Cord*. Oxford, Oxford University Press.

Sherrington, C.S. (1906) *The Integrative Action of the Nervous System*. New Haven, Yale University Press.

Sherrington, C.S. (1929). Some functional problems attaching to convergence. The Ferrier Lecture. *Proceedings of the Royal Society*, B, **105**, 332–362.

The special senses

9

9.1 INTRODUCTION

The special senses are housed in the head. They provide us with information about aspects of the environment distant from the body and about the orientation of the head in space. They comprise the nose for olfaction, the eyes for vision, the ears which have a dual role for hearing and for balance, and the tongue and palate for taste. They contrast with the somaesthetic senses (the body senses), which provide us with information about the body itself.

9.2 THE EYE

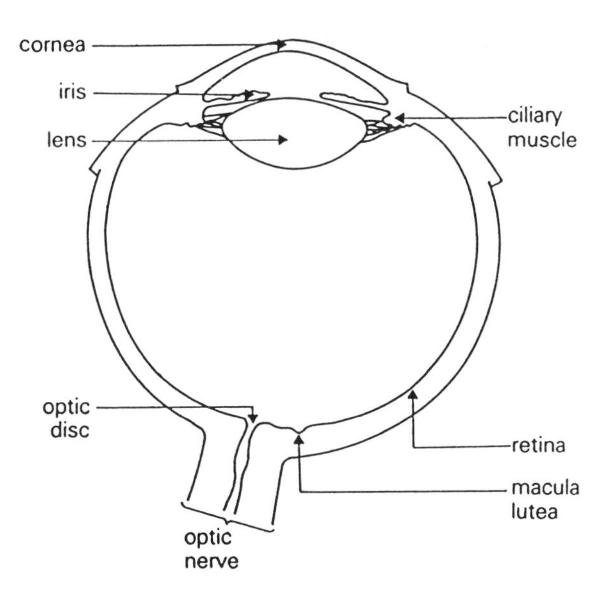

Figure 9.1 A section through the eye.

The eyeball is approximately spherical. It moves within the orbit under the influence of six small skeletal muscles inserted into the eyeball. They are called the **extrinsic ocular muscles**, because they lie outside the eyeball (this contrasts with the **intrinsic ocular muscles**, which are smooth muscles lying inside the eyeball; they are described later). The extrinsic ocular muscles are innervated by cranial nerves III (oculomotor), IV (trochlear), and VI (abducens). The motor nuclei for these lie in the brain stem, described in Chapter 11.

As shown in Figure 9.1, the eye contains a sheet of photosensitive receptors in the retina onto which is thrown a real image of the field of vision. This image formation is achieved by the refractory media of the eyeball through which light passes en route to the retina. The most powerful refractory surface is the anterior surface of the cornea; it separates media with refractive indices which differ greatly. The lens contributes much less to the refractory power of the eye since its refractive index is only slightly greater than that of the fluids on either side. The iris ensures that only the central part of the optical system transmits light. As in a camera, a small pupil aperture ensures good resolution of the eye as an optical instrument. A small pupil aperture also results in little light falling on the sensors. The pupil size is a compromise between these two conflicting requirements of good resolution and a sufficient amount of light falling on the receptors.

Photoreceptors

The photosensitive elements on which vision depends lie in the retina, which consists of the photoreceptors and nerve cells. There are two general requirements for an efficient visual system. First, it should operate over a wide range of intensities of illumination. In particular, the ability to see at low levels of illumination at night confers an obvious survival advantage; if two animals hostile to each other meet in the night, the one that can see will escape or hunt with considerable advantage over the one which is night blind. Second, the ability to resolve detail of the visual world is an advantage. As we will discover, the physiological adaptations for these two requirements are mutually competitive; the visual system must arrive at a compromise and this is achieved by having two systems, one operating for low intensities of illumination at night and the other for high intensities of light during the day.

To meet these requirements, the photoreceptors are of two types, rods and cones. Absorption of light energy by pigments in the receptors is an essential step in photoreception. The rods, which are very sensitive to light (**sensitivity** is the amount of stimulus energy which is just detectable by the visual system) contain the pigment rhodopsin. In the fully dark-adapted normal eye, measurements of the threshold for seeing dim flashes of light indicate that a rod is capable of responding to a single photon falling on it. Rods are responsible for night vision: cones are excited only by higher light intensities. The penalties which one pays for being able to see in very dim light are two-fold. First, rod vision does not allow the discrimination of colours. Second, the visual **acuity** (the ability to resolve two points of light) is low. These two features are dependent on cone vision. In daytime vision, rhodopsin is completely bleached and the rods do not contribute to vision.

Anatomy of the retina

Light entering the eye directly on the visual axis falls on a specialized part of the retina. An observer inspecting the retina through an ophthalmoscope sees this area as a yellow disc, the **macula lutea** (yellow spot). The yellow tint is due to a yellow pigment in the nervous layers. In the centre of the macula is a pit called the **fovea**; it measures about 1.5 mm across. Rods and cones are distributed over the retina in a characteristic manner (Figure 9.2a). Most of the 6 million cones lie within the fovea, although there are some cones sparsely distributed through the peripheral retina. There are few rods in the fovea and, within the central region of the fovea, there is a rod-free zone, about 0.5 mm across. The central fovea is therefore useless for night time vision. The 120 million rods are most densely packed immediately around the fovea; the part of the visual field to which this corresponds is where visual sensitivity in the dark-adapted eye is greatest. The reader may readily prove that sensitivity is greater off the visual axis than on the axis by looking at faint stars at night. If one looks directly at a faint star, it is invisible because the retinal image is on the fovea. As soon as one looks slightly to the side, the star is readily seen, because the retinal image falls outside the fovea and here rods are abundant.

The structure of the retina is such that the receptors are facing away from the incident light. Consequently, this light must pass through the nervous layers before reaching the light-sensitive elements. Such a situation is not ideal for obtaining a good optical image, as light must be considerably scattered on passing through the nervous layers. This is partially counteracted by the differentiation of the fovea. The nervous elements run obliquely away from the centre of the fovea; this reduces the thickness of nervous tissue through which light must travel.

Acuity of vision

The smaller the distance between two points of light which can be resolved by the visual system, the higher the acuity of vision. Visual acuity is thus a measure of the fineness of detail with which the visual system can transmit information about the visual world around. Although convergence along a neural pathway favours sensitivity, it prejudices acuity since two light sources close together in the visual field are likely to fall on receptors within the

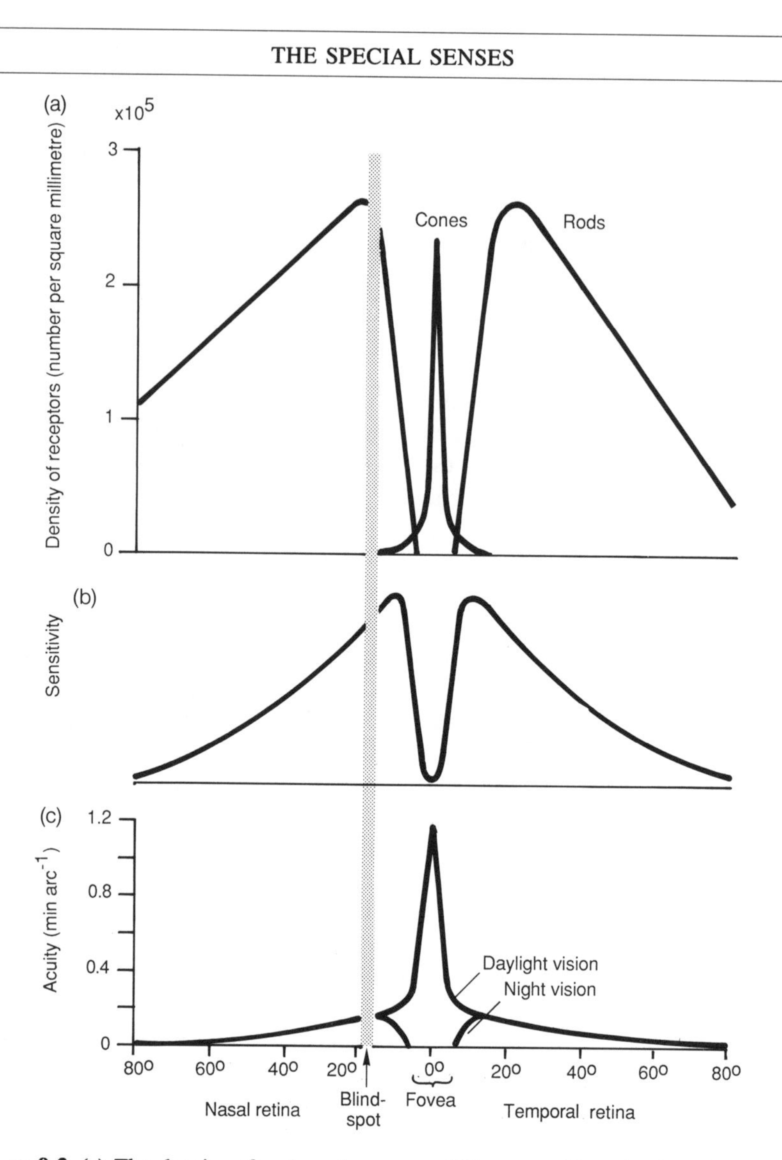

Figure 9.2 (a) The density of rods and cones at different positions in the retina along a line through the optic axis but not through the optic disc. (b) The threshold for seeing a point of white light as a function of position in the visual field. Sensitivity is the reciprocal of the ordinate scale. (c) The visual acuity, i.e. the reciprocal of the minimum discernible spatial separation of two points of light.

receptive field of a single ganglion cell. In the fovea, the cones are very thin and closely packed (Davson and Eggleton, 1962, p. 1247). In the central part of the fovea, each cone is connected with a bipolar cell; this in turn is connected with a retinal ganglion cell. Each cone has its own private line back to the optic nerve and thence to the central nervous system. These features all favour high visual acuity.

At low intensities of illumination, when only the rods are excited, the lower acuity curve in Figure 9.2c is obtained. The fovea shows zero acuity since there are no rods there. Further out in the visual field, there is a low level of acuity.

Convergence along the visual pathway

Both types of photoreceptor connect synaptically with nerve cells called bipolar cells (see Figure 9.5). These in turn relay to ganglion cells, which are also situated in the retina. The ganglion cells are the cells of origin for the nerve fibres travelling back in the optic nerve. Since there are 1 million optic nerve fibres carrying information from 6 million cones and 120 million rods, there is much convergence. The degree of convergence is greatest in the receptors of the periphery of the retina, where about 600 rods innervate a single bipolar cell.

Contrast sensitivity

Figure 9.3 A sinusoidal grating whose spatial frequency is increasing from left to right and whose contrast is decreasing from bottom to top.

In order for the human visual system to convey information to the brain, it must be able to signal differences in intensity of various parts of the visual field. This is investigated by asking a subject to look at a grating for which the intensity of illumination varies sinusoidally across the visual field. The contrast is adjusted until the subject can only just see that there is a grating rather than a uniformly illuminated plane. The contrast that can just be perceived depends on the spatial frequency of the grating. Figure 9.3 shows a sinusoidal grating whose spatial frequency is increasing from left to right and whose contrast is decreasing from bottom to top. The contrast threshold is lowest for the middle range of spatial frequencies: one can perceive the grating at a higher level on the figure than for low or high spatial frequencies.

Colour vision

In mammals, colour vision, which depends on the cones, is found only in a few primates. The colour of a light is determined by its spectral composition. In people with normal colour vision, the minimum number of pure colours from the spectrum which, by suitable mixing, can be made to match any colour (including white), is three. Different combinations of three primary spectral wavelengths can be used. The combination used by physiologists is red, green, and blue, since these correspond to the spectral sensitivities of cones, as we shall see.

The most generally accepted theory of colour vision is the Young-Helmholtz theory, which assumes the existence in the retina of three separate colour detecting mechanisms, all colour sensations being resolved from an unequal stimulation of these three, and white from universal stimulation. The sensation of black is due to the absence of stimulation, but a normal retina is required as, for example, at the blind spot nothing is seen rather than blackness.

Helmholtz's theory can be presented as a triangle, each point of which corresponds to a determined proportion of the three primary colours (Figure 9.4a). The coordinates at any point can be read by following the three lines from that point. A human can match every conceivable colour by a point in this triangle.

In a given individual, the different cones have many spectral sensitivities due to differences in the pigments that they contain. The spectral sensitivity curves of different cones tend to be grouped so that, very approximately, we can regard the cones as being of three types, each containing different photopigments (red, green, and blue). These spectral sensitivities, together with that for rods, are shown in Figure 9.4b. In this figure, the sensitivity has been scaled separately for each type of receptor so that the maximal sensitivity is 1.00; this hides the fact that the absolute sensitivity of rods is much higher than that of cones.

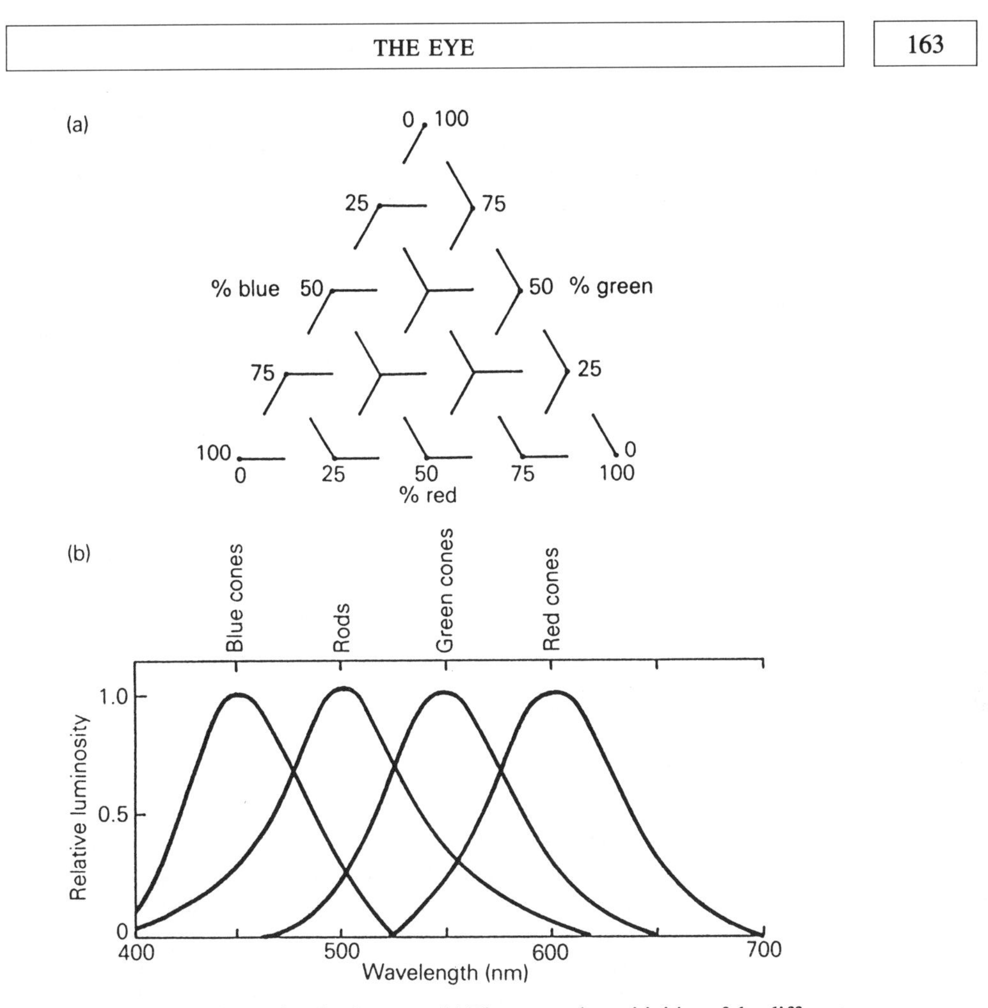

Figure 9.4 (a) The colour triangle. See text. (b) The spectral sensitivities of the different types of retinal receptors.

If a visual field is mainly of one colour but contains a different colour in a small region, the perception of the colour in the small region is profoundly influenced by the colour of the surround. Such observations are not easily explained by the classical theory, but considerations of such phenomena are beyond the scope of this book.

The different spectral sensitivities of rods and cones

In a daylight environment, a person uses cones for vision. As already stated the rhodopsin (the pigment in rods) is entirely bleached and the rods contribute nothing to daytime vision. When a person moves from a daylight environment to the dark, the rhodopsin must be chemically changed to its unbleached form before the rod function is restored. This is called **dark-adaptation**. It is a slow process; after exposure to bright sunlight, it takes hours for full sensitivity to be re-established. This slow time course of dark adaptation is sometimes a nuisance, for instance if one wants to work in a darkroom without waiting for an hour for one's eyes to dark adapt. There is a manoeuvre which can expedite things. Reference to Figure 9.4b shows that the spectral sensitivity for rods overlaps that of both blue and green cones, but the overlap with red cones is slight. If a subject in a daylight environment wears spectacles coloured so that only light with wavelengths in the far red part of the visible spectrum is allowed to pass, the glasses protect the rods and the subject becomes dark-adapted, much as in a dark room.

Colour blindness

In some individuals one or more photoreceptor types may be absent, or an abnormal photopigment may be present. Such defects reduce the subject's ability to distinguish colours leading to the condition known as colour-blindness.

The blind spot

All the axons of retinal ganglion cells course across the retina and converge to a region clearly visible ophthalmoscopically as the optic disc. This is the origin of the optic nerve and, being devoid of photoreceptors, it forms the blind spot. The reader can appreciate its existence by closing the left eye, looking at the cross in Figure 9.5b, and slowly bringing the page towards the right eye. The white circle, at first clearly visible, disappears as its image falls on the blind spot, and then reappears again when the image is thrown on a region of retina medial to the optic disc.

Now position the page again so that the white circle falls on the blind spot and is not seen. At this stage, open the left eye and close the right eye. The white circle can now be seen again. The blind spot for the visual field of one eye falls within the visual field of the other eye: with binocular vision there are no regions within our visual fields where seeing is defective.

Integration within the retina

Much processing of visual information takes place along the visual pathway from the photosensitive elements themselves to the ganglion cells. In addition to the direct pathway from receptor to bipolar cell to ganglion cell, there are two sets of neurone which spread information horizontally between adjacent lines of projection. These are the horizontal cells and the amacrine cells (Figure 9.5a). The horizontal cells lie close to the receptor cells. Their

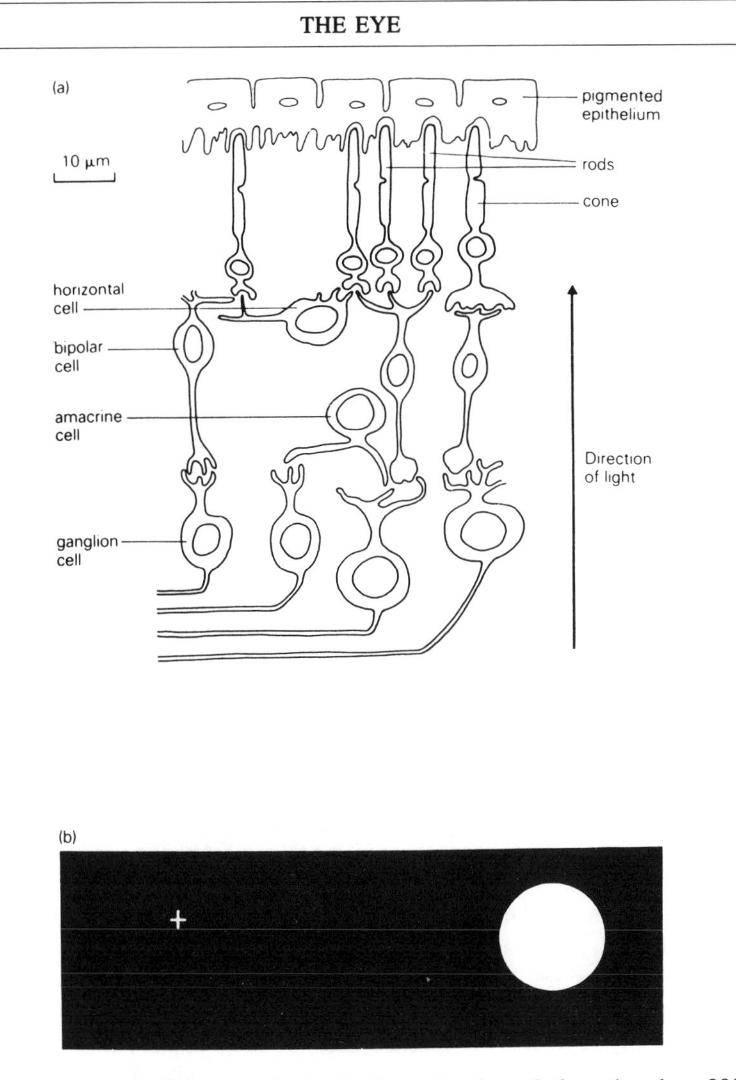

Figure 9.5 (a) Highly diagrammatic sketch of a section through the retina about 20° off the optic axis. (b) The blind spot.

dendrites receive input directly from many rods and cones. Their axons run horizontally and form synapses on other receptor cells and on their junctions with bipolar cells.

The **amacrine** cells (amacrine means 'without an axon') lie close to the layer of ganglion cells. Such neurones are phylogenetic relics from ancestral forms in which neurones had not yet become polarized. In amacrine cells, all conduction is along dendrites, which subserve two functions. As in more conventional neurones, the dendrites act as the input to the neurone, by receiving synaptic input from neurones that influence them. The dendrites also act as the effector element of the neurone and form synapses on the neurones which they influence (ganglion cells in this case). The dendrites from one amacrine cell make contact with many ganglion cells.

The photoreceptors themselves do not generate action potentials. This is also true of bipolar cells and horizontal cells; both these types of neurone conduct information from one part of the neurone to another by means of passive spread of current. The first neurones to generate action potentials are the amacrine cells, but even these operate mainly by passive conduction; they only generate occasional impulses and these ride on large non-regenerative potentials. The ganglion cells generate frequent action potentials and operate according to the usual rules of neurones with which the reader is by now familiar.

The transduction of light energy depends on the bleaching of visual pigment. This causes the closing of ion channels in the membrane of the receptor cells, resulting in hyperpolarization. This forms the raw data on which the circuitry of the retina operates to synthesize the information transmitted back along the optic nerve fibres.

The visual pathway

As shown in Figure 9.6, the optic nerve from either side runs back to join its fellow in the **optic chiasma**. Here, the optic nerve fibres innervating the medial half of each retina cross the midline to project back to the contralateral cerebral hemisphere. The optic nerve fibres from the lateral half of each eyeball project back to the ipsilateral hemisphere. Since the retinal image is inverted, the overall result is that an object on one side of the visual field projects back entirely to the cerebral hemisphere on the opposite side of the body. Behind the optic chiasma, the projection continues as the **optic tract** to the lateral geniculate body, which is the specific thalamic relay nucleus for vision. From here, the projection continues back to the primary visual cortex.

A branch of the optic nerve leaves the main nerve trunk before the latter reaches the lateral geniculate body. This is the **retino-tectal projection** (Figure 9.6), which projects to the midbrain nuclei that control the smooth muscle in the eyeball. The smooth muscle of the iris controls the diameter of the pupil. The smooth muscle of the ciliary body controls accommodation, i.e.the change in optical power necessary for focusing the eye. There is an interesting interaction of pupillary and ciliary reflexes. When one accommodates for near vision, this is normally accompanied by constriction of the pupil. This pupillary constriction, which does not depend on the change of light intensity, is called the **accommodation reflex**. This reflex depends on brain stem mechanisms, perhaps involving feedback from the spindle receptors in the extrinsic ocular muscles.

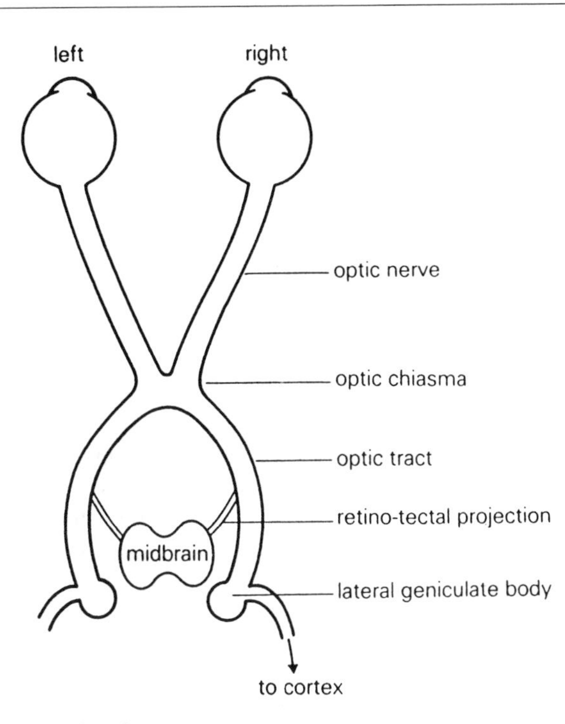

Figure 9.6 The visual pathways.

Pupillary reactions to light

Light shone in an eye causes pupillary constriction of the same eye via a midbrain reflex, to be considered further in Chapter 11. This is called the **direct light reflex**. Furthermore, light shone in one eye also causes the pupil of the other eye to constrict. The so-called **consensual response**.

Argyll Robertson pupil

In this condition, the subject can see, the pupil constricts during accommodation, but there is a loss of direct and consensual light reflexes. The abolition of the light reflexes is due to interruption of the retino-tectal projection. Information about light entering the eye no longer reaches the midbrain nuclei and the pathway for the light reflexes is blocked. Pupillary constriction accompanying accommodation depends on brain stem mechanisms but is independent of visual input and therefore survives interruption of the retino-tectal projection. The projection through the lateral geniculate body to the visual cortex is unharmed, accounting for the fact that the subject can see.

9.3 THE EAR

The ear consists of the outer ear, the middle ear, and the inner ear. The middle and inner ears lie in cavities in the temporal bone of the skull and are connected with the body surface by the outer ear (Figure 9.7).

The outer ear

The pinna or ear flap consists of cartilage covered by skin. It funnels sound energy into the column of air in the external auditory meatus.

The middle ear

The middle ear is an air-filled cavity lined with mucous membrane. It contains a chain of three interconnected small bones, called the auditory ossicles. The outer and middle ears are separated by the ear drum which, in a normal person, forms a complete seal. Attached to the inner aspect of the eardrum is the handle of the malleus, the first of the ossicles (malleus means 'mallet'). Sound, which consists of longitudinal vibrations of the molecules in the air, enters the external ear and causes the eardrum to

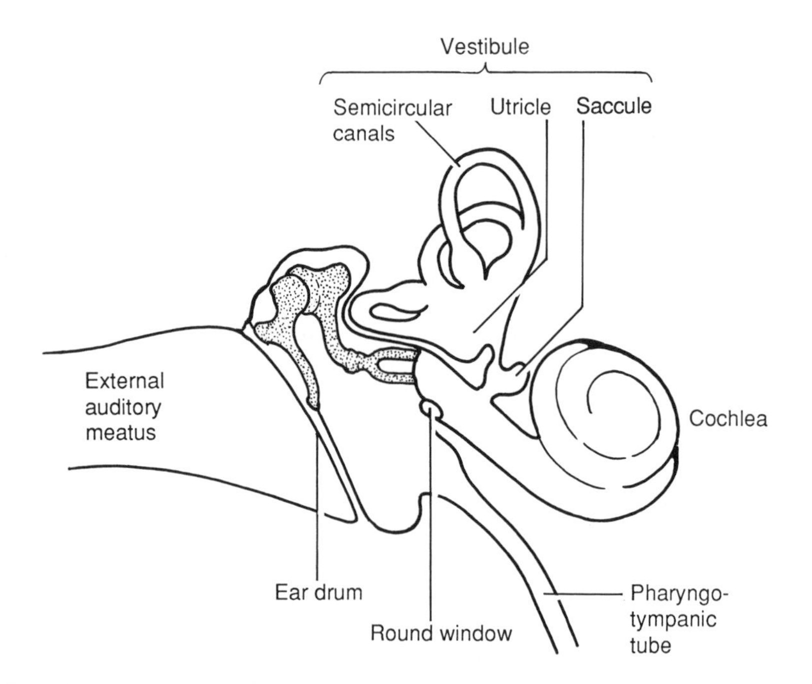

Figure 9.7 Diagram of the ear. The outer, middle, and inner ear all lie in bone. The middle ear contains the three ossicles; the maleus is attached to the ear drum, the incus is the middle ossicle and the stapes has a base plate which fits into the oval window.

vibrate. This vibration is transmitted to the malleus and thence to the second ossicle called the incus (incus means 'anvil') via a joint between the bones. From here, the vibration is transmitted via the third ossicle called the stapes (stapes means 'stirrup'), whose base fits into an oval window (called the foramen ovale) through which vibrations are transmitted to the inner ear.

The middle ear connects with the nasopharynx along the pharyngo-tympanic or Eustachian tube. This tube is closed most of the time but it is opened during swallowing by muscles inserted into its wall. When the tube is open; the pressures between the air in the middle ear and the atmosphere equilibrate to avoid the build-up of any standing pressure difference across the eardrum. During swallowing, when the tube is open, the subject hears a brief roaring noise that obscures normal hearing.

The ossicles are arranged in a lever system. When sound waves are being transmitted, the movements of the eardrum are relatively large in amplitude but are weak. The lever system of the ossicles is such that movement of the foot of the stapes is only about one-twentieth of that of the eardrum, but the force of the movement is increased by almost the same proportion.

The ossicles also control the amount of vibrational energy that is transmitted through the middle ear. Two small muscles are attached to the ossicles, the tensor tympani into the handle (the middle part) of the malleus and the stapedius into the neck of the stapes. Contraction of these muscles reduces the proportion of acoustic energy that travels from the eardrum to the oval window. The muscles contract reflexly in response to loud sound. This is a protective reflex to prevent too much movement of fluid in the inner ear. The muscles are also activated in other circumstances by the central nervous system; this is a mechanism for the control of sensory input.

The inner ear

The inner ear houses the receptors responsible for hearing and balance. The parts of the inner ear concerned with balance have evolved as an infolding of the lateral line organ found on the body surface of fish and amphibia; they have the character of a system of tubes with patches of sensory epithelium facing the inside of the tubes. In man, the inner ear lies within a geometrically complicated system of cavities, called the **bony labyrinth**, in the temporal bone. Within this lies the **membranous labyrinth**, which is filled with lymph.

The labyrinth consists of the cochlea for hearing and the vestibule for balance. The main cavity of the vestibule is called the utricle. This connects with the saccule via a stalk. The utricle and saccule have similar functions, as we shall see. The utricle is also connected with the three semicircular canals through three small apertures.

Each of these parts of the inner ear contains deformation receptors and the relationship of these to the surrounding structures determines the nature of the disturbance which excites the receptor. The receptors are called hair cells because of hair-like projections from the outer aspect of the receptor.

The cochlea

The cochlea is the portion of the inner ear that subserves hearing. It consists of a tube in the form of a spiral with 2¾ turns. It is similar in shape to a snail's shell, the Greek name for which is 'cochlea'. The diameter decreases towards the apex, where the bony labyrinth comes to a blind end. The tube is divided into three chambers by the basilar and Reissner's membranes (Figure 9.8).

The two outer chambers connect at the apex of the spiral. The lymph that they contain is called perilymph: this differs in composition from endolymph, which lies in the scala media. At the base of the spiral of the cochlea, the scala vestibuli is plugged by the foot of the stapes. A ligament connects the edge of the footplate of the stapes with the wall of the labyrinth, sealing the middle and inner ears from each other. Movements of the ossicles are transmitted mechanically to the perilymph of the scala vestibuli and thence to the scala tympani, the base of which connects with the round window.

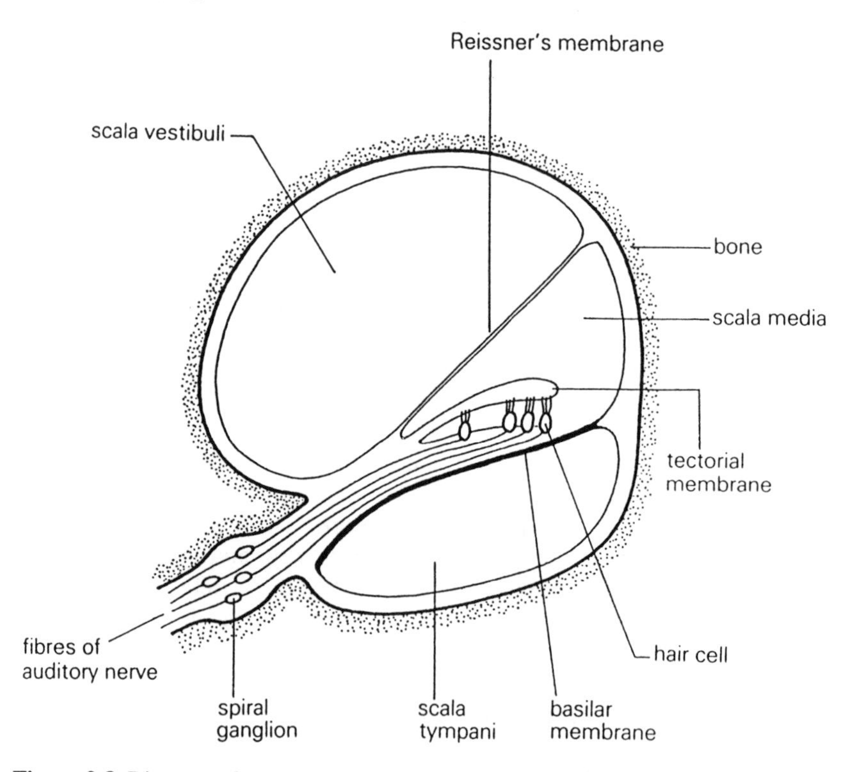

Figure 9.8 Diagram of a transverse section through one coil of the cochlea.

This separates the scala tympani from the middle ear and is covered with a membrane which bulges in and out under the forces transmitted through the cochlea from the auditory ossicles. This allows movement of the incompressible lymph in the cochlea to occur. It is this movement that excites the cochlear receptors.

The receptors are arranged in two rows, the internal row being a single row of receptors and the external row being three or four receptors deep. The base of each receptor cell is supported by the basilar membrane, and the hairs are embedded in the gelatinous tectorial membrane. Afferent nerve fibres cross the basilar membrane and, on its course, each nerve fibre contacts several hair cells. The nerve fibres have their cell bodies in the spiral ganglion in the bony labyrinth: this ganglion is analogous to a dorsal root ganglion.

Sound vibrations transmitted to the cochlea deflect the hairs on receptor cells and this initiates action potentials in the auditory nerve. Different frequencies of sound are distinguished by two mechanisms. First, the cochlear lymph and the membranes dividing the cochlea resonate at a frequency which depends on the magnitude of the inertia of the lymph and the elasticity of the membranes. A relatively short path through the scala vestibuli across the Reissner's and basilar membranes and back through the scala tympani resonates at a relatively high frequency. A longer path resonates at a lower frequency. The recognition process in the central nervous system takes into account the spatial pattern of impulses from the cochlea to deduce the pitch of a sound. Evidence for this comes from the phenomenon of **boiler-maker's deafness**. In former times, boiler makers worked long hours in an environment in which the reverberations from steel plates being rivetted together to form a boiler were at a sufficiently constant pitch and so loud that, over the course of years, these workers became deaf to this noise and to others at a fre-quency close to it. They were able to hear higher and lower notes. Some component of the receptor apparatus maximally excited by the environmental noise was damaged, whereas other components remained functional. When these workers were subsequently examined at post mortem, damage was found to be localized to one region of the cochlea. A particular frequency of sound maximally vibrates a particular part of the basilar membrane, and receptors there are excited. The highest frequencies activate receptors close to the base of the cochlea, and lower frequencies are detected further away. The frequency is encoded in the siting of the receptors which are stimulated.

For low frequencies in the auditory range, a second mechanism is more important. Each cycle of the tone elicits a burst of impulses in auditory nerve fibres. The frequency is encoded in the pattern of action potentials traversing the fibres of the auditory nerve.

The receptors are epithelial cells which release a chemical transmitter to initiate impulses in the terminals of the primary afferent nerve fibres, as

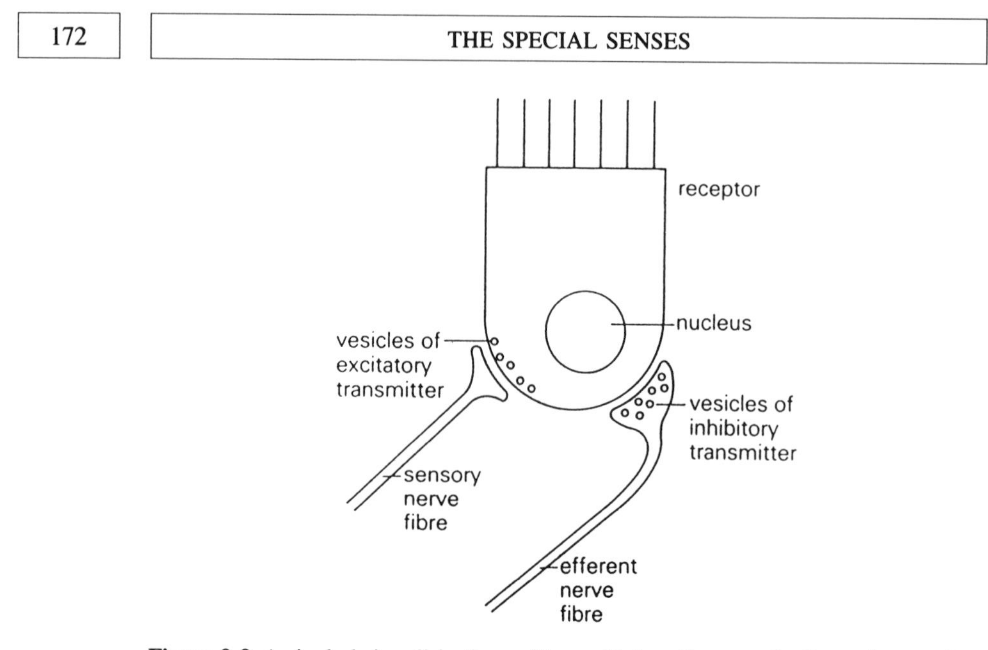

Figure 9.9 A single hair cell in the cochlea, with its afferent and efferent innervation.

shown in Figure 9.9. The receptor cells of the cochlea receive an efferent innervation via the olivo-cochlear bundle (Figure 9.9). Nerve impulses in these nerves hyperpolarize the receptors and inhibit them. By this mechanism, the central nervous system has direct inhibitory control over the sensitivity of the receptors. This situation is unusual in two respects. First, we have efferent neural control of receptors and second, the efferent nerve fibres, which are several centimetres in length, are inhibitory (Desmedt and Monaco, 1960).

Receptors in the vestibules and semicircular canals

These receptors occur in patches; in the semicircular canals these are called **cristae** (crista means crest) and in the vestibule they are called **maculae** (macula means spot). As for cochlear receptors, the hairs of these receptors project into an overlying jelly and so are bent if there is relative movement between the jelly and the wall of the labyrinth. Bending of the hairs causes a change in discharge rate of the afferent nerve fibre connecting with the receptor.

A hair cell has one long thick hair (the kinocilium) and about 70 short ones (stereocilia). The hairs are probably prestressed during development, so that even when no external bending force is applied to the free end of the hairs, the receptor generates action potentials at a constant rate in the afferent nerve fibre. Bending the hairs alters this rate of discharge, and the direction of bending is important. For a given amount of bending there is one direction which causes the greatest increase in discharge rate: this is indicated by

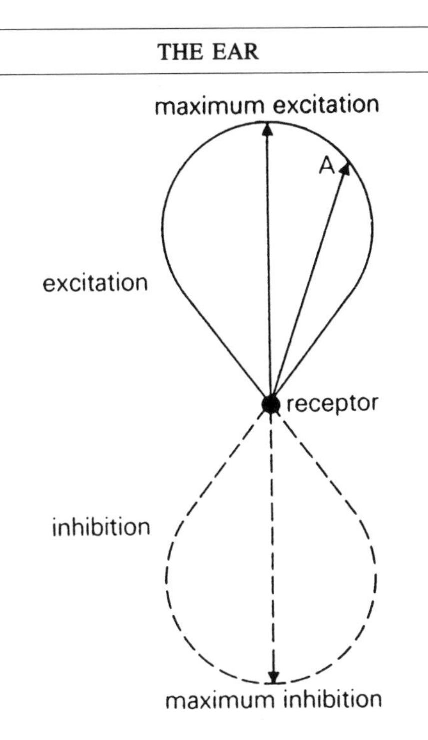

Figure 9.10 A polar plot of the responses of deflection of the hairs of a hair cell in different directions.

the vertical full arrow in Figure 9.10. Bending the hairs in the opposite direction causes an equivalent reduction in discharge rate, indicated by the broken arrow in Figure 9.10. Intermediate directions of deformation cause intermediate alterations. If the hairs are bent by a standard amount in the direction A, the discharge rate increases by an amount plotted as the length of the arrow. This procedure is repeated for all directions and the heads of the arrows are joined to give the double loop. The direction of movement which causes the maximum increase in discharge rate is called the orientation of the receptor.

The semicircular canals

Although the canals are called semicircular, each forms about two-thirds of a complete circle. At one end, each is expanded to form an **ampulla** in whose walls the receptors lie. The ampulla contains the **cupula**, a wedge of mucus into which the hairs project. Ink injected into the semicircular canal on one side of the cupula does not reach the fluid on the other side, showing that the cupula seals the canal.

The stimulus that activates the semicircular canals is angular movement around an axis through the centre of the semicircular canal and perpendicular to the plane of the canal. Because of the inertia of the fluid, angular acceleration of the semicircular canal tends to leave the fluid behind. (Inertia is the reluctance of a body, or in this case fluid, to change from a state of rest or of uniform movement in a straight line.) The movement of fluid relative to the membranous labyrinth displaces the cupula and bends the hairs of the hair cells.

This movement of the fluid relative to the membranous labyrinth due to inertia is opposed by viscous drag on the wall of the canal and the restraining force of the cupula, which tends to cause the fluid to move along with the wall. The relative magnitudes of these forces are such that the hair cells effectively signal the angular velocity of turning during short lasting rotations in everyday life. A less usual type of rotation is experienced if a subject sits on a rotating chair and is rotated for a considerable period at a constant angular velocity. Receptors in the semicircular canals initiate reflex eye movements described in Chapter 10. These eye movements continue for about 20 s, this being the time during which the semicircular canal receptors are activated. Such a prolonged rotation is non-physiological and, after about 20 s, the receptors cease to signal any rotation. With the eyes closed, the subject is no longer aware of the rotation. If the chair and subject are now brought to an abrupt halt, the inertia of the fluid results transiently in its continued movement in the direction of the previous rotation. The subject interprets this as indicating a renewal of rotation but in the opposite direction from that during the real rotation. If the eyes are opened at this stage, the subject sees that he is in fact stationary. Such conflicting information from the sense organs leads to the feeling of dizziness. The world seems to go round in the direction opposite to that of the previous rotation. The effect passes in about 20 s. During this time, the displacement of the hair cells falls to zero as the restoring forces return the cupula to its rest position. These restoring forces are opposed by the viscous drag, just as is the initial deflection at the onset of turning and it is this drag which prolongs the sensation of rotation.

All the receptors in one ampulla have the same orientation, along the axis of the semicircular canal at the site of the ampulla. Movement of the mucus plugging the ampulla causes similar changes in the discharge rate of the afferent nerve fibres from all the receptors. Movement in one direction causes an increase in the discharge whereas movement in the other direction causes a decrease in the discharge of the afferent nerve fibres.

The plane of the imposed angular acceleration of the head is deduced from information provided by the whole array of receptors in all the semicircular canals on both sides of the body. This information is not contained in the discharge patterns from the receptors of any one ampulla alone.

The utricle and saccule – balance

The utricle and saccule sense the direction of 'down-ness'. They contain beds of sensory hair cells and supporting cells called the maculae. These beds cover the lower surfaces and walls, but not the roof of the utricle and saccule. As in the semicircular canals, the endolymph over the hairs is of a jelly-like consistency but, unlike that in the semicircular canals, it is loaded with small hexagonal crystals of calcium carbonate called **otoliths** (literally 'ear stones'). The otoliths have a specific gravity of almost three. The jelly-like material loaded with otoliths is displaced by any linear acceleration, including gravity. This sets up stress forces which bend the hairs of the hair cells. The orientations of different hair cells in the vestibular maculae vary in a regular manner so that tilting in any direction is signalled, as shown in Figure 9.11. Hairs in different parts of each macula respond optimally to accelerations in different directions. This contrasts with the hair cells in an ampulla, all of which respond similarly to displacement in a single direction. The utricular and saccular receptors do not differentiate between the static tilt and the constant linear acceleration in the appropriate direction because the forces exerted are similar as sensed by the receptor.

The otoliths operate as tiny plumb lines. A plumb line is used for finding the vertical, for instance when building a wall. The direction of the string of a plumb line indicates the direction of the linear acceleration being applied to the weight. If the top of the plumb line is held stationary, as is always the case when it is being used to construct an upright wall, the forces acting on the weight are the gravitational force downwards counteracted by the tension exerted by the string. The string indicates the direction of 'up-ness'. Moving the head about is analogous to moving the hand that is holding the top of the plumb line string. The string no longer points to the centre of the earth. It indicates at each instant the direction of the force being applied to the weight through the string. One component of the force is to counterbalance the gravitational force; the other component accelerates the weight as it follows the movements imposed from the far end of the string.

The plumb line differs from the otolith system in that it is liable to swing like a pendulum when the top of the plumb line string is held stationary

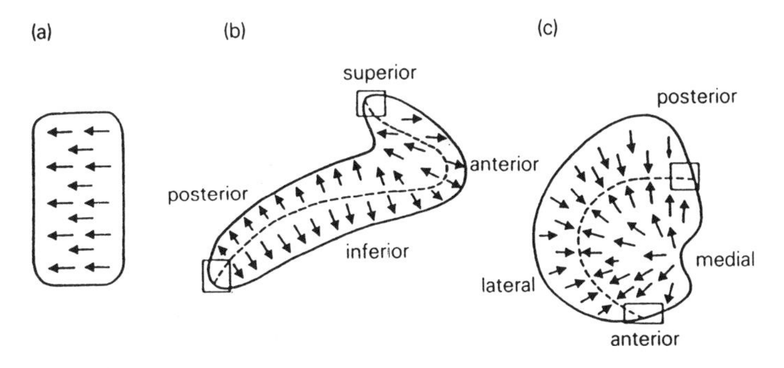

Figure 9.11 The directions of deflection which maximally excite hair cells in the crista of a semicircular canal (a) and in the maculae in the saccule (b) and the utricle (c).

after being moved around. The otolith system is heavily damped and consequently slightly sluggish in its response.

9.4 OLFACTION

Olfaction plays a relatively minor role in providing man with information about the world around him. In phylogenetic development, olfaction has been supplanted by vision, which is our major means of sensing the environment. The structure to which the olfactory nerves project is the olfactory bulb and, during phylogenetic development, this has become progressively less prominent.

Lower animals use olfaction as their primary source of information about the environment. Predators sniff out their prey, fish find their way back to breeding grounds via the sense of smell. Communication is also via olfaction. Many lower animals release specific chemical signals which are sensed by fellows around through the sense of olfaction. This is a method of providing sexual attraction and location of partners. It persists in an atavistic form in humans, with females using perfumes to lure their prey.

The olfactory receptors

These lie in the olfactory epithelium lining the nasal cavity. The outer aspect of the receptor cell projects beyond the surface of the epithelium, as shown in Figure 9.12. Up to 20 cilia project from this aspect. The outer aspect also contains vacuoles which actively take up colloidal gold applied to the surface. The particles of gold are so large that the only possible mechanism for this uptake is for the vacuole to approach the cell membrane and for the two membranes to fuse. This provides a channel between the surface fluid and the content of the vacuole. This mechanism of uptake is called **pinocytosis**; it provides a continual flow of chemicals from the fluid bathing the nasal mucosa right to the olfactory bulb. Whether the odorous chemicals conveyed in this way can be sensed in central neurones is unknown.

The sensitivity of olfaction in species that rely on this sense is such that the absorption of one or two molecules of certain chemicals by an olfactory receptor can excite it to threshold. An olfactory receptor does not respond specifically to one stimulus. Every receptor responds to a selection of olfactory stimuli and, to date, any classification of olfactory receptors or indeed of stimuli has eluded investigators. There are no primary olfactory stimuli, as there are primary colours in colour vision. There are no simple rules of summation, analogous to the specification of colours according to their siting in the colour triangle. Increasing the concentration of an olfactory stimulus can alter the nature of the smell that is perceived. The lack of

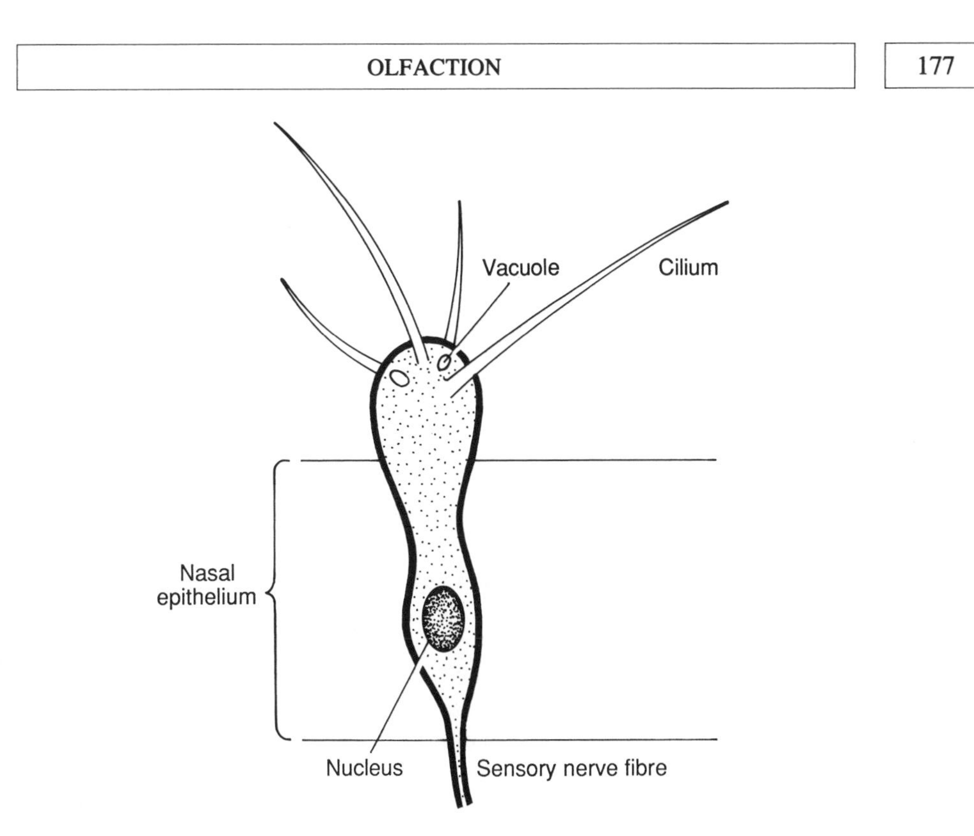

Figure 9.12 An olfactory receptor.

any useful system of classification of olfactory stimuli, together with the relative unimportance to humans of olfaction and a consequent lack of motivation to research the subject, is the reason for rather little being known about this aspect of sensory physiology.

The olfactory receptors live for only a few weeks and are constantly being replenished by developing cells which lie within the mucous membrane. Whereas for all other special sense receptors there is a chemical synapse between the receptor cell and the afferent nerve fibre connecting the receptor to the central nervous system, the olfactory receptor cell doubles as the cell body of a neurone and sends back an axon to the olfactory bulb. These axons are very fine. In the olfactory bulb they form synapses on large cells called mitral cells, which in turn project back via the olfactory tracts to phylogenetically ancient regions of the cortex; these regions form a ring from which newer regions of the cortex sprouted, as will be described in Section 13.4.

Together with the involution of olfaction as an important sense, the regions of the brain devoted to the analysis of olfactory stimuli in lower animals have taken over other functions in humans. Since olfaction was of such significance to our phylogenetic ancestors, providing them with all the information which they could handle relating to such basic drives as sexual activity, it is not surprising that the regions of the brain which previously analysed olfactory information evolved into centres primarily concerned with emotion and memory. Many readers will themselves have experienced that an olfactory stimulus reminding them of an incident or situation in their childhood brings back memories that are so vivid that they seem to transport one back in time: memories conjured up by other types of sensation are seldom as intense.

9.5 TASTE

Our perception of the taste of something that we eat is heavily influenced by smell. If smell is temporarily abolished, as a result of a head cold, food becomes 'tasteless'. The information provided by the taste buds on the tongue is relatively crude. The taste buds lie around the edge and across the back of the tongue; the central area of the tongue is almost insensitive to taste. The taste buds are located on papillae, each of which is a small hill of mucous membrane. Four varieties of taste stimulus are recognized: salt, sour, sweet, and bitter. Each of these varieties of taste sensation is best developed in one particular area of the tongue. As with smell, two stimuli applied simultaneously do not result in simple summation of perceived taste. A striking example with which we are all familiar is that the taste of sourness, evoked by acids, can be counteracted by adding sugar to an acid drink. Sugar does not change the hydrogen ion concentration of the drink but by adding stimulation of receptors for sweetness to that of receptors for acid cancels the perception of acid in the mouth.

Whereas the other special senses have their own nerves which project to the brain, the afferent nerve fibres connected with taste receptors run with the cranial nerves VII (facial) and IX (glossopharyngeal). The pathway to the cerebral cortex is analogous to the central afferent pathways of other senses, including a relay in the thalamus. Taste projects to a small region of infolded cortex at the posterior end of the lateral sulcus; this region is called the parietal operculum.

Reflexes of balance 10

10.1 STANDING AND WALKING

Standing

A standing human has a relatively high centre of gravity, which is at the level of the waist. The person is supported by the area of the soles of the feet in contact with the ground, and must maintain the centre of gravity vertically above this small area of support to avoid toppling. Many reflex mechanisms contribute to standing and enable the individual to return to an equilibrium position if someone bumps into him. Even more mechanisms come into play during walking.

The balance of a standing human is precarious. He sways to and fro, making postural corrections in anticipation of toppling over. His postural reflexes are continually being invoked to maintain the centre of gravity over the support. There is an instant during the sway when there is very little tonus in the muscles, with the body held in exact balance and the weight supported passively by the bones. Without continuous adjustment of muscular tone, however, the human will fall. It is impossible to balance in a standing position a human who has no muscle tonus and reflexes; one cannot balance a corpse, for instance.

As long as the centre of gravity is vertically above a point within the span of the base support, changes in muscle tone are sufficient to remedy the tendency for the body to fall without the need to move the support. If the centre of gravity moves beyond this, stepping or hopping reflexes are initiated to readjust the position of the base. If such adjustments cannot occur, such as when a human balances on a support of small area such as tightrope, then frantic attempts are made to avoid falling. The subject may make circular movements of the arms or sway about. When the centre of gravity does move outside the base support, these attempts are in vain and the subject falls.

Walking

Walking can be regarded as a deliberate act of throwing oneself forward and out of balance; the centre of gravity is moved in the direction of the proposed movement, and reflex mechanisms to correct for the lack of balance then come into play automatically. This is an example of the principle mentioned in Chapter 6, of higher centres commanding movement via in-built mechanisms in more primitive parts of the brain.

Overview of the receptors important in the maintenance of posture

In order to maintain an upright posture, the organism needs information concerning the orientation of the body in space (e.g. the direction of free fall) and the geometry of the body (e.g. whether the head is in line with the trunk). These two quite different types of information are supplied by the special senses and the somaesthetic (body) senses respectively. We consider these in turn.

For the orientation of the body in space, the **balance receptors** of the vestibule are the most important and Section 10.2 of this chapter is devoted to them. The **eyes** and **ears** also contribute. As the body moves in relation to the environment, the subject sees that the environment is apparently moving in the opposite direction. Auditory information is also used, although in man it provides much less precise information than the visual system about the position and movement of the body in space.

The **geometry of the body** is signalled by proprioceptors in muscles, tendons, capsules, and ligaments. This information is needed if the central nervous system is to synthesize smooth, efficient, and purposeful movements. For instance, the stretch receptors in the spindles of skeletal muscles signal the state of stretch in the spindles. This is turn depends on two factors: the length of the muscle, and the activity of the intrafusal muscle fibre controlled by the γ motoneurones of the spinal cord. Since the central nervous system has access to both the afferent discharge from the receptors and the intensity and distribution of efferent neuronal discharge, it has the information necessary to compute muscle length and state of contraction.

The muscle stretch receptor system is the afferent limb of both the tendon jerk reflex and the stretch reflex. The stretch reflex is particularly active in the antigravity muscles, in which continued tonus is necessary when we are standing. For instance, when one sways forwards, the posterior crural muscles (muscles at the back of the calf), of which the most important are the gastrocnemius-soleus, are stretched. This initiates reflex contractions to cause plantar flexion at the ankle joint, which pulls the body back into the upright position. As one sways backwards, the stretch is taken off these posterior muscles, relieving the tension on muscle stretch receptors, removing drive to the stretch reflex and relaxing the muscles. At the same time, the anterior crural muscles (tibialis anterior etc.) are stretched. The reflex contraction so initiated in these anterior muscles causes dorsiflexion at the ankle which pulls the body upright again.

A third factor of relevance to balance is the direction of support. **Cutaneous pressure receptors** in the region of the skin supporting the body provide a limited amount of information about the direction of the support. Pressure on the skin of the soles of the feet (and to a lesser extent on the palms of the hand) evokes a reflex extensor thrust.

Table 10.1 is a list of the receptors which initiate postural reflexes.

Table 10.1 Receptors important in the maintenance of posture

Information yielded	*Sensors*	*Example*
Orientation of the head in space	Special senses	Balance receptors of the vestibule; eyes, ears
Geometry of the body	Somaesthetic proprioceptors	Stretch receptors in muscles, tendons, capsules and joints
Direction of thrust	Somaesthetic proprioceptors	Cutaneous pressure receptors in soles of feet

10.2 RIGHTING REACTIONS

Even if deprived of visual and auditory information, the normal human can remain standing. Movements that return a person to a standing posture whenever that posture is disturbed are called **righting reactions**. They comprise **vestibular righting reflexes** to bring the head back to its upright position and **neck righting reflexes**, which bring the trunk back into line with the head. We will consider these two in turn.

Vestibular reflexes

The vestibules contain two groups of receptors which respond to different types of stimulus and initiate different reflexes. The receptors in the otolith organs respond to linear acceleration; they signal the direction of down-ness by afferent nerve impulses indicating the direction in which the otoliths are pressing or hanging. If one is standing still, this is the direction of the gravitational field. Their reflex effects are vestibular righting reflexes. The receptors in the semicircular canals respond to angular movement (rotation) of the head. They initiate reflex movements of the eyes.

Reflexes with semicircular canals as receptors (vestibulo-ocular reflexes)

The prime function of the sensors in the semicircular canals is to stabilize the visual image on the retina. The effector muscles are the extrinsic ocular muscles. Under their influence, the eyes rotate in the orbits when the head moves so that the eyes continue to look in the same direction. For instance, at the start of a rotation on a revolving chair, a subject's eyes move in the orbits in the opposite direction from the rotation of the head.

There is a limit to the extent to which the eyeballs can move. If they were to rotate indefinitely, they would break off their stalks. Once the eyes are maximally deviated, they rapidly flash back to a central position. The alternation of slow tracking movements and fast flick-back movements is called **nystagmus**. The slow component depends on the integrity of the brain stem, whereas the fast component depends on the cerebral cortex.

The reader can perform a simple experiment to demonstrate the efficacy of vestibulo-ocular reflexes. Hold this page at arm's length and look at it. Move the script from side to side at an increasing rate until the wording is no longer legible. Now hold the script still and rotate the head from side to side to achieve about the same relative movement between page and head. If semicircular canal function is normal, the writing will now be clearly legible. This is because the semicircular canals provide information about movements of the head and the central nervous system can compensate by reflex control of the extrinsic ocular muscles. One way of looking at these vestibulo-ocular reflexes is to regard them as increasing the apparent inertia of the direction of gaze.

The siting of the vestibule in the head rather than in the trunk is necessary for the receptors to be able to initiate movements of the eyes to compensate for movements of the head. For the purpose of maintaining uprightness of the whole body, the vestibules should ideally be sited in the trunk, as we shall see. The siting of the vestibules in the head is witness to the importance for the organism of maintaining a stable retinal image.

Reflexes with otolith organs as receptors

These reflexes serve to keep the centre of gravity over the support. Different effector muscles are called into play according to the disposition of the body and neck.

Neck muscles as effectors

If the head and body are tilted to one side, reflex contraction of the neck muscles brings the head back into the upright position. This reflex is abolished by removal of otoliths from the maculae of the vestibule, but does not depend on vision.

Limb muscles as effectors

The vestibules also initiate reflex contractions of the muscles of the limbs. These reflex effects may be explained by reference to a cat standing on a tilt table (Figure 10.1). When the support is horizontal, as in Figure 10.1b, there is moderate extensor tonus in fore and hind limbs sufficient to maintain the trunk with its axis in the horizontal plane. If, as in Figure 10.1a,

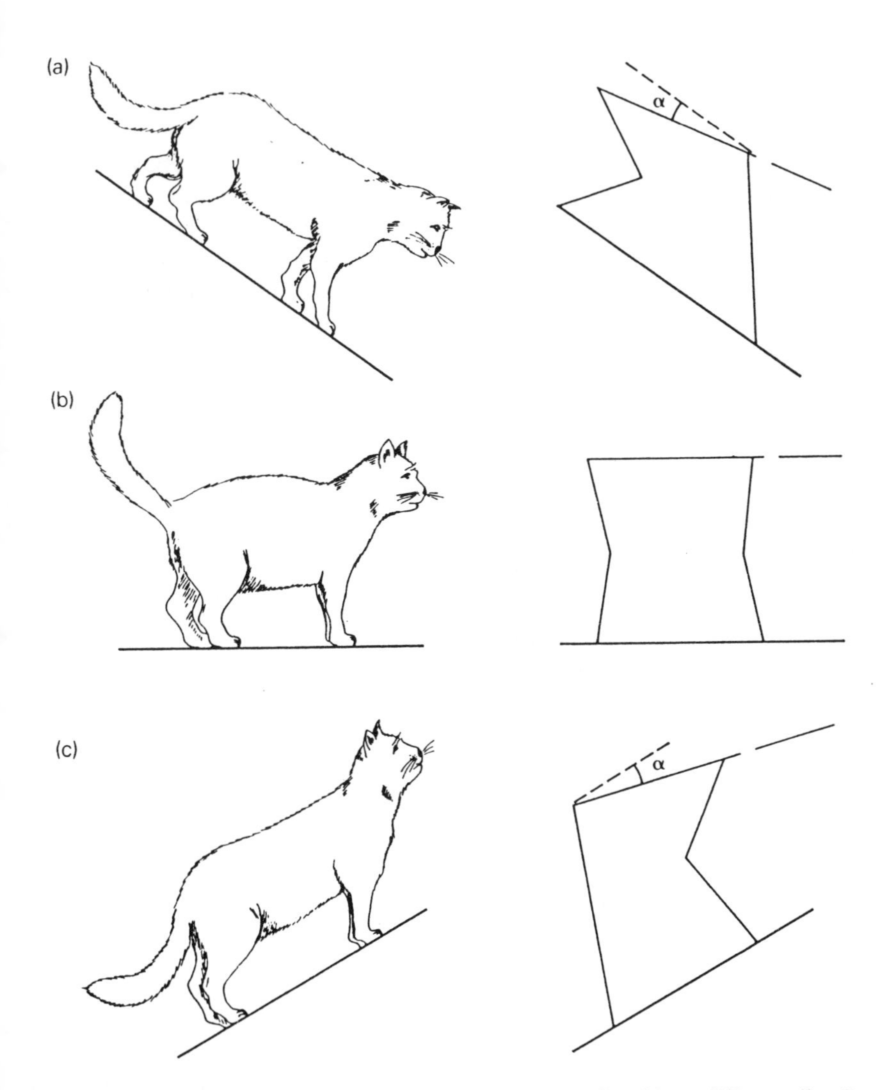

Figure 10.1 Vestibular reflexes. Left: the cat stands on a tilt table at different tilts. In all three cases, the head is in line with the trunk. Right: sketch of the state of flexion or extension of the limbs.

the table is tilted so that the support of the cat's front legs is lower than that of the hind legs, in the absence of reflex adjustments the axis of the trunk would be tilted by the same amount as the tilt table. This is indicated by the dashed line in Figure 10.1a. The animal would topple forwards. Toppling is prevented by reflex adjustments. The tilt is detected by the vestibular receptors which initiate reflex redistribution of tonus in the limb muscles so that the forelimbs extend and the hindlimbs flex; the consequence is that the tilt of the trunk is reduced. This reduction is by the angle α. When the tilt is in the opposite direction as in Figure 10.1c, the reverse occurs; the hindlimbs extend and the forelimbs flex. Again, the effect is to reduce the tilt of the trunk and hence to counteract the tendency to topple.

Similar adjustments occur if the ground support on one side suddenly drops. The whole body is thrown sideways and the vestibules initiate reflex extension of the limbs on the downhill side with flexion of those on the uphill side; this adjusts the animal's posture in a sense to bring the body upright and consequently to bring the vestibules back to the neutral position.

All of these reactions are encapsulated in the rule that **the vestibular reflexes initiate limb movements such that downhill limbs extend**. For a standing biped such as a human, this applies mainly to the legs.

Trunk muscles as effectors

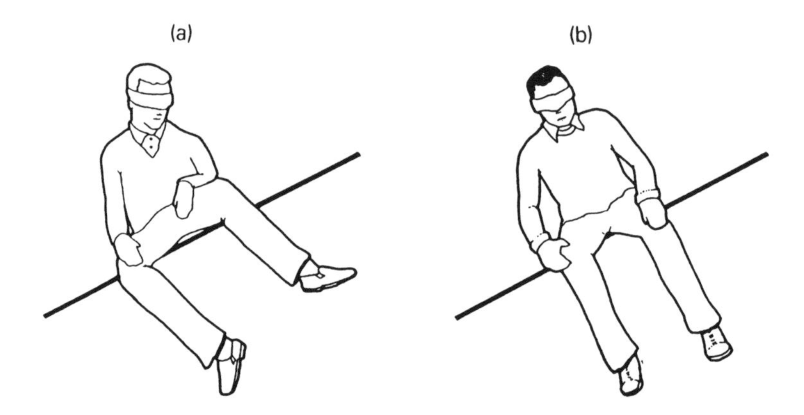

Figure 10.2 Subjects on a tilt table. (a): Normal individual. (b): Vestibule-defective.

The trunk musculature in the normal human acts as an important effector system to maintain the spinal column in the upright position, particularly in the sitting subject. The tilt table is used clinically to assess vestibular function. The normal person keeps his centre of gravity above his base by adjusting the tonus in his flank muscles to provide the lateral deviation of the trunk needed to counterbalance the tilt (Figure 10.2a). The vestibule-defective subject shown in Figure 10.2b does not show reflex adjustments in posture as the table is tilted. As a result, he is much more easily toppled over than the normal person (Purdon-Martin, 1967).

In summary, all the reflexes initiated by the otolith receptors tend to return the vestibule, and hence the head, to the upright position. Table 10.2 summarizes this section.

Table 10.2 Reflexes with vestibules as sensors

Sensor	*Effector muscle*	*Result*
Semicircular canals	Extrinsic ocular muscle	Stabilization of retinal reflex image
Otolith organ	Neck muscles	Stabilization of head in space
Otolith organ	Limb muscles	Downhill limbs extend
Otolith organ	Trunk muscles	Centre of gravity remains over base

10.3 NECK REFLEXES

Introduction

In order to elaborate motor commands to prevent the body from falling over, the central nervous system needs to compute the direction in which the organism must push against the support. Ideally, for this purpose, the detectors for uprightness (vestibules and eyes) should be sited within the trunk. A highly integrated system of reflexes has evolved to cope with the siting of the vestibules and eyes in the head, rendering them mobile on the trunk.

Referring to Figure 10.1, it can be seen that the head was in line with the trunk in all of the situations considered. If the head is tilted forwards on an upright trunk, the changes in the input from the vestibules (and, incidentally, the eyes) are far in excess of those that occur during a sway forward of the whole body. The postural reflexes initiated by these inputs therefore have to be countermanded when they are produced by movements of the head on the body rather than by movements of the whole body in space.

Figure 10.3 illustrates the point. In 10.3a, the subject is upright. In 10.3b, he tilts his head forward. The direction of down-ness, signalled by his otolith receptors, has changed; this is indicated by the relation of the plumbline to the head. However, scarcely any reflex adjustment in limb musculature is warranted. In 10.2c, where the whole trunk and head have tilted forwards, the same otolith input as in B is elicited, but this time extension

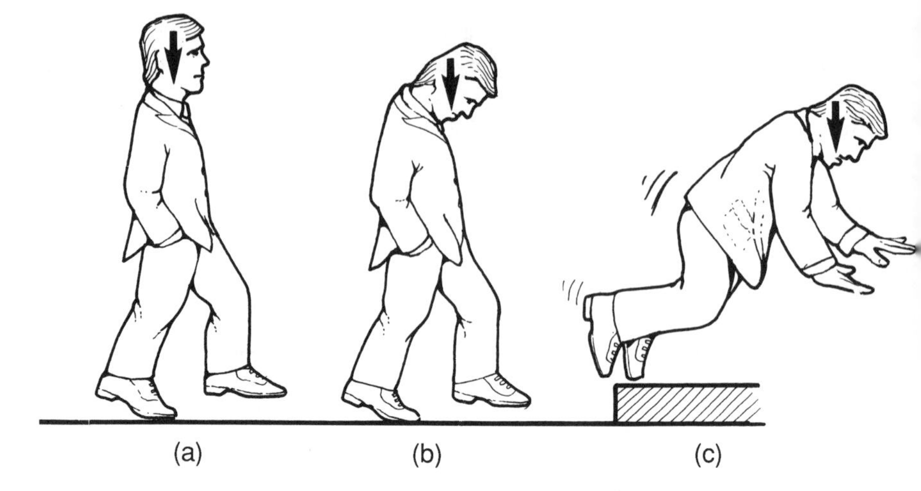

Figure 10.3 Vestibular input with different postures. (a) Head and trunk upright. The arrow shows the direction of 'down-ness' signalled by the otolith receptors. (b) Head tilted forwards over an upright musculature. (c) Head and trunk in line and both tilted forwards: vestibular reflexes are prominent with extension of the downhill limbs (arms) and flexion of the uphill limbs (legs).

of the arms and flexion of the legs occur to counteract and break the fall.

The crucial information needed by the central nervous system in this context is the position of the head with respect to the trunk, i.e. the degree of flexion, extension, lateral deviation, or of rotation at the cervical intervertebral joints. This information is provided by muscle spindle receptors in the small muscles very close to the capsules of the intervertebral joints of the uppermost cervical vertebrae and the base of the skull. The most important joints are the atlanto-occipital and atlanto-axial joints.

Afferent nerve fibres from these receptors are the origin of a powerful series of postural reflexes called the **neck reflexes**. To simplify, attention will be confined to the **tonic neck reflexes**, which occur when the head is held stationary at an angle to the body's axis. As with vestibular reflexes, different muscles contract in reflex response to different deviations of the neck from the neutral position. The reflex effects of deviations of the neck can be summarized thus: **the neck reflexes initiate limb movements to bring the trunk back into line with the head.**

Flexion and extension movements of the neck

To visualize neck reflexes free from vestibular reflexes, consider situations in which the head remains in the upright position whilst the trunk is displaced from the neutral position. For instance, consider a cat with the same

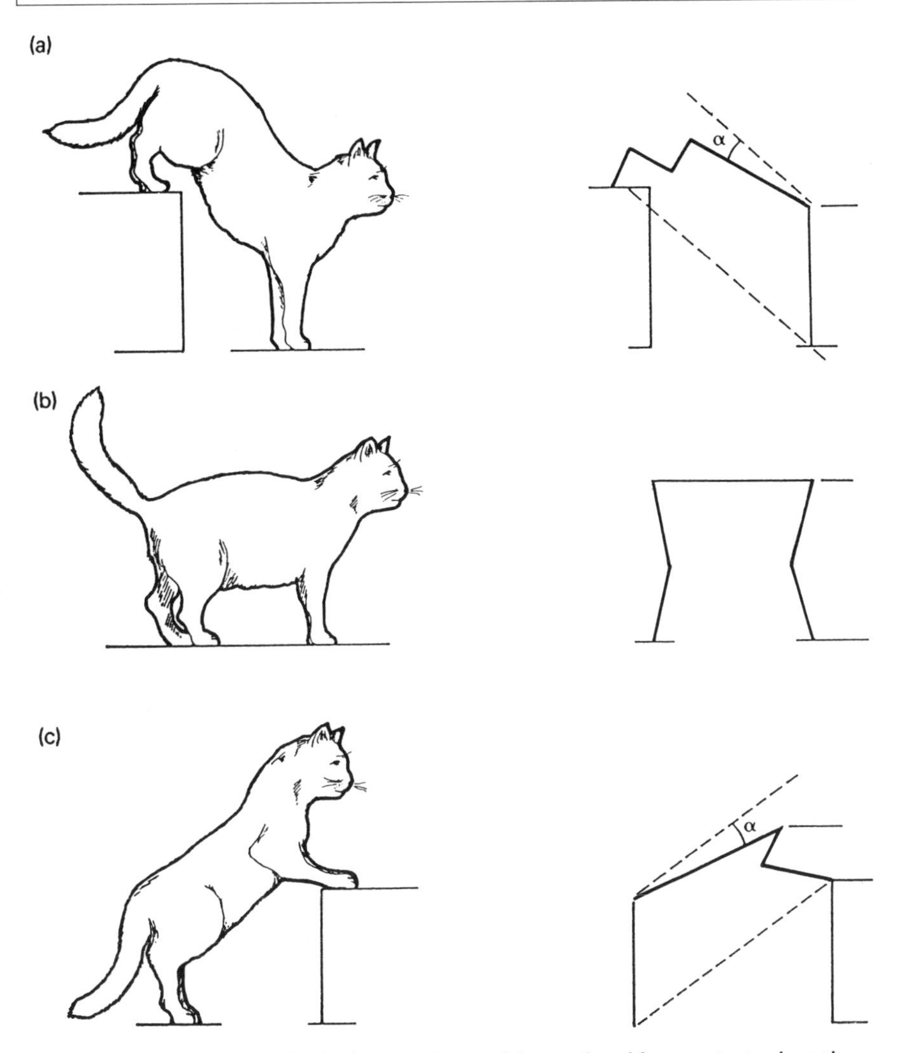

Figure 10.4 Neck reflexes. Left: three positions of the trunk, with a constant orientation of the head in space. Right: sketch of the disposition of the limbs.

three tilts as previously considered, but this time with the head held horizontal, in the normal position for a cat. The vestibular input to the central nervous system is constant. The deviation of the trunk from the normal neutral position is compensated by flexion or extension of the neck (Figure 10.4). Differences in the limb muscle tonus are then dictated by the neck reflexes uninfluenced by vestibular effects.

In Figure 10.4b the neck is in line with the trunk. In this intermediate position, the tone in fore and hind limbs is approximately equal, appropriate when the body's weight is shared equally by the four limbs. In Figure 10.4a, when front of the animal is tilted down, the neck is extended. The reflex effects are extension of the forelimbs with flexion of the hindlimbs. This combination brings the trunk back towards being aligned with the head. These effects are indicated by the stick diagram. If there were no reflex adjustments, the axis of the trunk would be parallel with the inclination of the support, indicated by the dashed lines. The reflex adaptations reduce the deviation by the angle α.

Conversely, in Figure 10.4c, when the front of the animal is tilted up, the neck is flexed. The reflex effects on limb muscles are of flexion of forelimbs with extension of hindlimbs. Again, this combination minimizes the deviation of head from being in line with the trunk. These are particular examples of limb movements that tend to bring the trunk back into line with the head. Comparison of Figures 10.1 and 10.4 shows that the reflex effects of changes in tilt of the tilt table on the tone in limb muscles are the same whether the reflex effects are initiated by vestibular or neck reflexes.

Rotation

The neck reflexes elicited by rotation of the head are best illustrated by considering an individual immersed in water. If a quadruped makes forceful extensor thrust movements with the limbs on one side of the body, this tends to rotate the trunk away from that side of the body. This movement is brought into play reflexly by input from the neck proprioceptors when the head is rotated on the trunk. The limbs on the side to which the head is rotated perform the extensor thrust movements whilst the limbs on the opposite side show flexor movements. As with flexion and extension at the neck, rotation results in movements that bring the trunk back into alignment with the head.

The neck reflexes are obtained in an exaggerated and abnormal form in infants born with hydrocephalus, a condition in which there is an increase in volume of cerebrospinal fluid within the skull. The cause is an obstruction to the circulation of the cerebrospinal fluid, for example, obstruction of the foramina of Luschka and Magendie in the roof of the fourth ventricle, or obstruction of the aqueduct of Sylvius. The normal circulation of cerebrospinal fluid involves passage of the fluid through these canals and, if they are blocked, the development of the higher brain centres, including the vestibular nuclei, is prevented. The brain of such an infant consists of a huge thin-walled bag of cerebrospinal fluid. The neck reflexes, mediated as they are by lower centres, are present and uncontaminated by vestibular reflexes. If the head of a hydrocephalic is rotated, the extensor tonus of fore and hind limbs increases dramatically on the side to which the chin is turned and is diminished on the other side (Figure 10.5). In hydrocephalus, only tonic movements are seen and it is in this respect that they are abnormal.

In summary, the neck reflexes are initiated by deviations of the neck from its normal netural position and their role is to initiate limb movements to bring the trunk back into line with the head and reinstate the normal position of the neck. With the head itself in the neutral position, the neck reflexes thus have the effect of returning the centre of gravity of the body to its neutral position of being vertically above the base support.

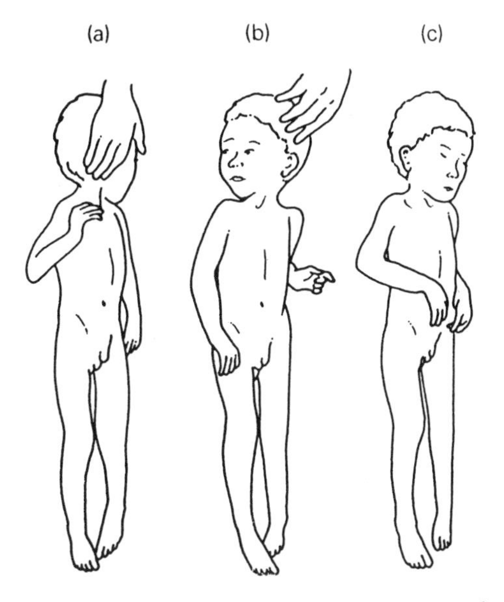

Figure 10.5 Neck reflexes in a hydrocephalic infant. (a) Head rotated to the left. (b) Head rotated to the right. (c) Head in the midline position.

10.4 SYNTHESIS OF VESTIBULAR AND NECK REFLEXES

Overview of vestibular and neck reflexes

The sensors for vestibular receptors lie in the head and, by detecting their own orientation in space, signal the orientation of the head in space. The vestibular reflex effector mechanisms restore the uprightness of the head in space; they are unrelated to the geometric relationships of the various parts of the body to each other. When downhill limbs extend, the result is to push the whole body, including the head and the otolith sensors back into the normal upright position. When this is achieved, the otolith organs no longer signal any deviation from stability and the reflex has achieved its objective.

Complementary considerations apply to the neck receptors and the neck reflexes. The sensors signal the geometric disposition of the upper intervertebral joints, that is, the relationship between the head and the trunk, and are uninfluenced by the orientaiton of the head in space. The effector mechanisms have, consistently, the object of restoring a normal, non-deviated status in the neck. Here, limb muscle activation is such as to restore the trunk to being in line with the head and is again unrelated to the orientation of the head in space, this latter being the province of the vestibular otolith receptors. We note that it is the trunk which is brought into line with the head and not the other way round; this is because in the overall design of postural mechanisms, the vestibules look after the uprightness of the head and the neck reflexes look after the alignment of the trunk with this upright head.

An understanding of these principles should help to avoid an error which students frequently make of stating that the neck reflexes bring the head back into line with the trunk. This is wrong and would not serve any stabilizing function since it is the trunk, not the head, which has to be re-positioned to preserve balance.

Interaction of vestibular and neck reflexes

When an individual flexes his head on a stationary trunk, both neck and vestibular reflexes are brought into play. The vestibular reflexes are those of 'nose down' and the neck reflexes those of neck flexion. The effects of vestibular and neck reflexes are shown separately in the stick diagram of Figure 10.6, copied from Figures 10.1a and 10.4c. The effects are copied into the first two rows of Table 10.3. The third row shows the net effect when both reflexes are operating. Since, in any limb, the effects of the two reflexes are opposed, the net effect is a cancellation, with little change in muscle tonus.

Table 10.3 The effects of flexing the head without moving the trunk.

	Hindlimb	*Forelimb*
Vestibule alone, nose down	Flexed	Extended
Neck reflexes alone, neck flexed	Extended	Flexed
Both reflexes together, body neutral, nose down, neck flexed	Intermediate	Intermediate

When the head is flexed on the stationary trunk, vestibular and neck reflexes tend to cancel each other. This is also true of other deviations of the head on the stationary trunk. The head can be voluntarily moved on the trunk without calling up inappropriate postural reflexes. If, however, the head is tilted because of a change in support, for instance by the ground giving way under one foot, then appropriate righting reflexes operate. The combination of otolith and neck reflexes with limb muscles as effectors adjust for imbalance of the whole body. The system works as if the receptors signalling position in space were in the trunk, not in the head.

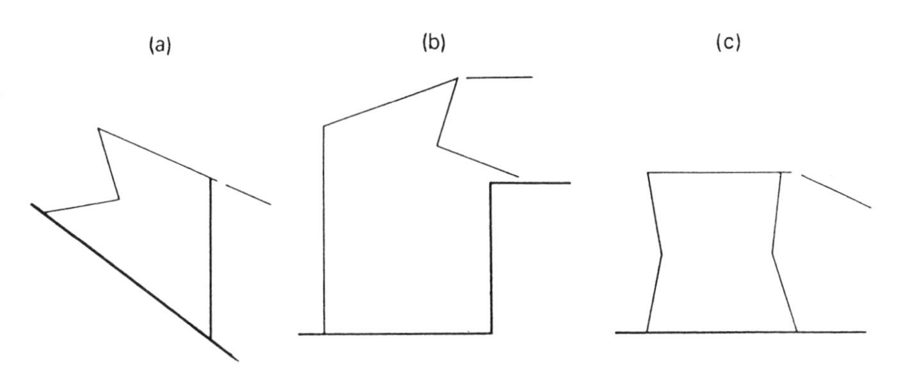

Figure 10.6 Stick diagrams to show the separate contributions of vestibular and neck reflexes to the distribution of muscular tonus when an animal flexes its neck on the trunk. See text.

Vestibular and neck reflexes initiate appropriate limb movements irrespective of whether the subject stabilizes his head in space or allows his head to follow the trunk. The latter may occur in order to allow deviation of the subject's visual axis so that he can see why the ground has suddenly collapsed.

Suppose a person is kneeling on a tilt table facing in the direction of the tilt. The table is tilted so that the front of the body goes down. If the head is kept in line with the trunk, muscle limb tone will alter from reflexes originating in the vestibular otolith receptors; the forelimbs extend and the hindlimbs flex (Figure 10.3c). If, however, the head is maintained in the upright position, this entails extension of the head on the trunk. The extension of the neck initiates neck reflexes which involve extension of the forelimbs and flexion of the hindlimbs (Figure 10.4a). The effect on the limb musculature is the same whether the head stays upright or remains in line with the body.

Linear acceleration

This section is about the adjustments that we make when our base support moves about. The weight of the body behaves as if it were concentrated at the point called the centre of gravity; this is indicated in the subject sketched in Figure 10.7. Due to the gravitational attraction between the earth and the mass of the subject, the body experiences a downward force, shown as the downwardly directed arrow in Figure 10.7a. This force is offset by an equal and opposite force from the support.

Now let us consider a person standing in a bus and looking in the direction in which the bus will move. The person is not holding on. With the bus stationary (or moving at a constant linear velocity), the disposition of forces is as shown in Figure 10.7a. The force exerted by the floor of the bus on the soles of the man's feet (the area of support) is called the **thrust force**.

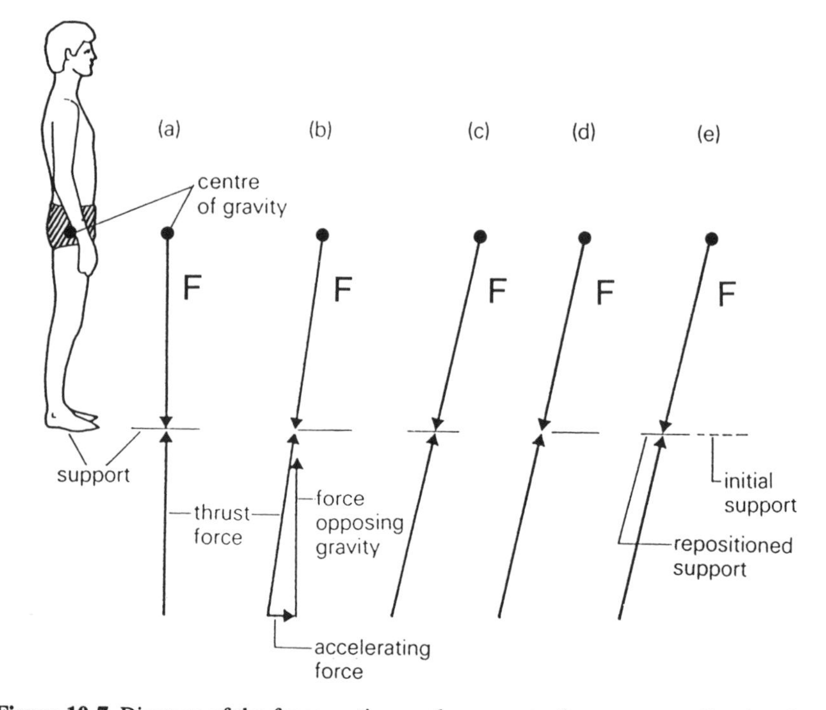

Figure 10.7 Diagram of the forces acting on the support of a person standing in a bus. In each part, the horizontal lines indicate the support area, which is the area of contact of the feet and the floor of the bus. F is the force vector representing the person's mass. (a) The bus is stationary; (b) the onset of an acceleration in the direction in which the person is facing; (c) equilibrium during a maintained acceleration; the person is now leaning forwards with the force F passing through the middle of the support area; (d) the onset of an acceleration which starts abruptly; the force F now passes behind the support area and equilibrium requires the step backwards shown in E; (e) the person has now stepped backwards and is leaning forwards. Force F once again passes through the centre of the base support.

It passes through the centre of the span of the support, as shown in Figure 10.7a and consequently the passenger is in equilibrium.

When the bus accelerates forwards, the upward force of the floor has added to it a horizontal component forwards, as shown in Figure 10.7b. With respect to the bus, the man's body is accelerated backwards and his centre of gravity is behind his base. The acceleration of the bus momentarily leaves the passenger behind. The vector sum of the force opposing gravity and the accelerating force is the thrust force (Figure 10.7b); this determines the direction of the thrust force vector. The position of the thrust force vector is determined by the fact that its projection must pass through the subject's centre of gravity. When the acceleration is small enough for the thrust force to fall within the contact area of the feet with the floor, as in Figure 10.7b, the passenger does not topple. He adapts to the acceleration by leaning forwards. If the acceleration is gradual, the subject leans progressively further forward, and this allows the standing posture to be maintained.

Suppose that, as a result of a gradually increasing acceleration, the passenger has adjusted his posture so that, as in Figure 10.7c, the axis of his trunk is in line with arrow F. His centre of gravity lies in front of the contact area of the feet. The thrust force passes through the contact area of the feet and the passenger is in equilibrium. He does not fall. If acceleration occurs suddenly and is so large that the arrow 'F' meets the floor behind the support area, the disposition is as shown in Figure 10.7d, with the support shown as the full horizontal line. The required thrust force vector falls outside the available area of support. If this is not corrected, the passenger must fall backwards. To avoid falling, the passenger steps back to reposition his support (Figure 10.7e), so that the thrust force vector passes through this support. The passenger also leans forwards. He has arrived at a similar final position as that in the case of the gradual acceleration, but instead of achieving this by gradually leaning forwards, he has taken a step to give the required purchase for the forwardly tilted thrust. Once the acceleration has reached a constant value, the subject remains in equilibrium by leaning forwards. When the acceleration ceases as the bus reaches its cruising speed, the reverse series of postural adaptive changes occurs.

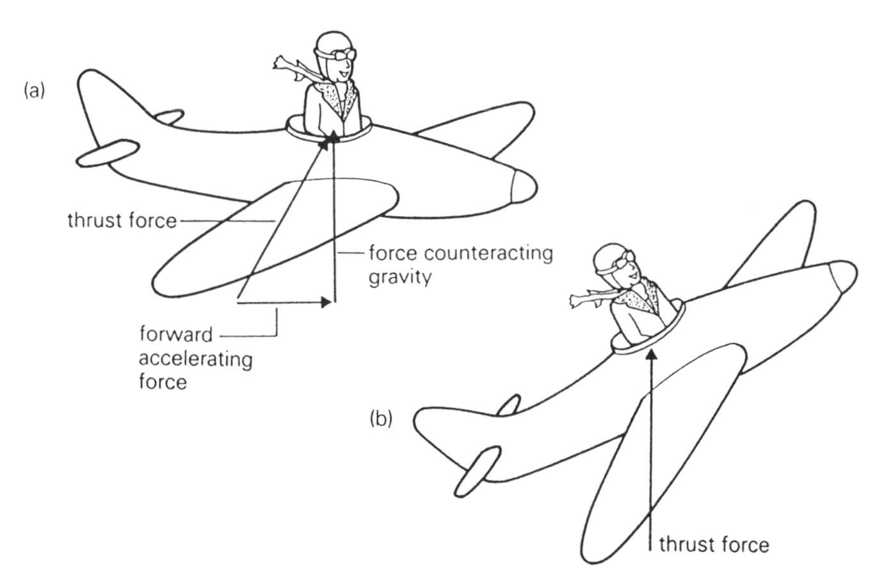

Figure 10.8 The forces acting on a pilot in a plane. (a) Plane horizontal, accelerating forwards. (b) Plane ascending at a constant speed.

We next consider the information relayed by sense of balance of a person in an aircraft. If the aircraft is accelerating on the tarmac, the accelerating force is much greater than in a bus. Passengers and pilot are usually seated and the forward acceleration is provided by the thrust of the back of the chair. The sensation of acceleration is due to the tilting of the thrust force, as shown in Figure 10.8a. Once the aircraft has accelerated to take-off speed and is airborne, it climbs at a constant speed until it reaches its cruising height. During this climb, the thrust force on the pilot is merely to counteract the force of gravity, since the speed is constant. The direction of the thrust force is inclined to the axis of the pilot's body, because the aeroplane is tilted with its nose upwards, as shown in Figure 10.8b. The directions of the thrust forces are identical in a and b. The magnitude of the thrust force is slightly larger in a than in b but directional changes alter the pattern of vestibular stimulation much more dramatically than changes in magnitude. The pilot perceives his vestibular input in the two situations shown in Figure 10.8 as almost identical.

This similarity in thrust forces in very different situations led to many accidents in the early days of aviation. When the pilot had landed his plane too late on an airfield and realized that he could not stop before he ran out of runway, he wished to take off again quickly into a climb, ready to go round for another attempt at landing. In order to take off, he would open the throttle wide and accelerate rapidly. With the aircraft not yet airborne, the rapid acceleration gave the pilot the impression that the aircraft was climbing in the air and that the climb was so steep that the aircraft was in danger of stalling. The fear of stalling prompted the pilot to push his stick forwards. This put the nose of the aircraft down into the tarmac, with fatal consequences. After a few accidents of this type, pilots learnt to distrust the information from their own senses and to rely instead on their instruments.

Conflicting information from the sense organs

The different sense organs usually provide the central nervous system with information that is consistent. If a man falls forwards, his vestibules and eyes both indicate the fact. In some situations this is not so, however; for instance, in an aeroplane which is performing a properly banked turn. When the plane is flying in a straight line at constant speed, the passenger is supported by the thrust force that counteracts gravity. In order to change the direction of movement of the body, an additional force must be applied to accelerate the passenger in the direction towards the centre of curvature of the flight path. The faster the speed, the greater the force to turn the direction of movement through a given angle. If the pilot were to turn the aircraft without banking, the inertia of the passengers would carry them straight ahead and they would be thrown to the side of the aircraft; the aeroplane company would soon be out of business. To avoid this, the pilot inclines (i.e. banks) the aircraft so that the passengers are tilted towards the direction of the turn, as shown in

Figure 10.9. The thrust force then remains directly upwards from the floor of the plane or from the seat in which the passenger is sitting. This provides the force to counteract gravity and to accelerate the passenger towards the centre of curvature of the flight path.

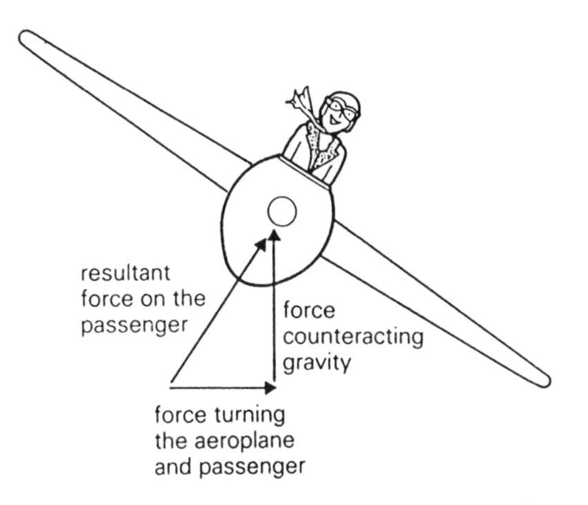

Figure 10.9 The forces acting on a passenger in an aeroplane during the performance of a properly banked turn.

As long as the passenger does not look out of the aircraft, the turn is made without the passenger being aware of it. A plumb line held in the hand will point to the floor of the aircraft, since the weight is acted upon by the same forces as is the passenger's body. However, if the passenger looks out of the window, his true orientation will be sensed by his visual system. It looks as if the ground below is tilted since the information from the vestibules indicates no tilt. Situations where different senses give apparently contradictory information tend to lead to travel sickness.

We have considered specific examples illustrating the general principle that information from the vestibules can be misleading. The pilot must learn to disregard his own vestibular information and rely instead on his instruments. If the aircraft is in fog so that visual information is not available, the vestibular signals are liable to be entirely misleading. One cannot tell whether one is flying straight or going round in banked circles. If the pilot relies on his own vestibular information, he can fly round in circles, climb or dive without being aware of it.

FURTHER READING

Roberts, T.D.M. (1978) *Neurophysiology of Postural Mechanisms* (2nd edn.). London, Butterworths.

Brain stem mechanisms 11

11.1 INTRODUCTION

The previous chapter was about the physiology of vestibular reflexes and neck reflexes in the context of the behaviour of a normal animal or human standing, walking, or being tilted on a tilt table. This chapter is about the physiology of higher brain centres, largely in the context of damage to the brain. A consideration of the interactions between the forebrain and the lower centres which result in consciousness leads on to a study of the functions of the brain stem. First, let us define the brain stem.

The brain stem

The brain stem is the stem joining the forebrain above with the spinal cord below. It consists of the mesencephalon, the pons, and the medulla. The brain stem is important in many vital functions of the body, including consciousness. It also contains the nuclei which mediate the pupillary reaction to light (Chapter 9) and the vestibulo-ocular reflexes (Chapter 10).

Consciousness

Consciousness is an important function of both the brain stem and the cerebral cortex. A normal level of consciousness, in which an individual is awake and responding to the environment in a fashion that we recognize as normal, depends on the harmonious interaction of the cerebral cortex and the reticular formation (a component of the brain stem, to be described shortly). Consciousness manifests itself by the activity of the cerebral cortex, but normal cortical activity depends on excitatory input from the reticular formation. Consciousness therefore requires the integrity of these two structures and of their interconnections. It is impossible to know whether any sort of consciousness can exist in a subject possessing an intact brain stem but no cortex.

As a brief digression, most of the methods used for the execution of convicted criminals leave the brain stem and cortex intact so that consciousness persists for a short time. The guillotine, used in the French revolution, may not result in as instantaneous a death as one might expect, since the neural substrate of consciousness is not immediately destroyed. This may account for contemporary reports that, when, after performing his duty, the executioner held up the head for the crowd to witness, the eyes in the severed head surveyed the crowd for a few seconds before asphyxia killed the brain. Although these eye movements may have been merely reflex, there is no reason to assume that consciousness was immediately lost. It is not recorded whether in any instance the severed head was commanded to move its eyes, so the question must remain open.

To return to contemporary causes of head injury, damage to the cerebral cortex and damage to the brain stem produce different groups of effects with disturbances of consciousness as a common feature.

11.2 BRAIN DAMAGE

Brain damage is a common consequence of accidents, and is a frequent cause of death. Accident is the most common cause of death in people under the age of 45; in Britain there are about 7000 deaths from accidents each year. Brain

Table 11.1 Widespread brain damage

	Widespread brain damage	
	No cortical function	*No brain stem function*
Causes	Anoxia Cardiac arrest Hypoglycaemia	Rotational injuries of head
	Disconnection of cortex from brain stem	Spacde occupying lesion. Supratentorial: pinching of brain stem against tentorium Infratentorial: coning
Effects	Subject unconscious; vocalization, if it occurs is incomprehensible	Subject unconscious; no vocalization
	Spontaneous respiration	No spontaneous respiration
	Good heart action indefinitely	Heart stops in a few days or weeks at most
	Reflex swallowing	No reflex swallowing
	Eyes rove and will follow	Eyes fixed
	Primitive reflex movements	
	No indication of intelligence	
	Vegetative survival is possible for years 'persistent vegetative state'	Death

damage is the only common cause of persisting disability in survivors of accidents; at least 15 000 individuals are permanently brain damaged annually and, of these, one half are never able to return to work (Jennett, 1980). Brain damage is therefore a subject of great importance to doctors and to society.

Widespread brain damage commonly takes one of two forms. The patient is left with either no cortical function or no brain stem function. The effects of these two types of brain damage are summarized in Table 11.1.

No cortical function

The cerebral cortex is the first to suffer irreversible damage if the whole body lacks oxygen for a few minutes. Lack of oxygen may occur when a person is exposed for a long period to a barometric pressure so low that there is insufficient oxygen in the inspired gas. Pathologically, it occurs as a consequence of cardiac inadequacy, such as when the heart stops beating for a few minutes or when cardiac action is impaired following an extensive coronary thrombosis. The extreme sensitivity of the cerebral cortex to hypoxia is a consequence of the fact that oxidative respiration is the only source of energy for brain tissue. The brain cannot metabolize anaerobically; 'oxygen debt' is not possible.

In another situation, the cortex, although intact, cannot express its function. This occurs when there is extensive damage of the cortical white matter which normally acts as the neural connection between the cortex and the rest of the central nervous system. All the lower centres are intact; the nuclei controlling the respiratory muscles (responsible for air flow) and the striated muscle of the larynx (responsible for operating the vocal cords) are intact. Spontaneous vocalization may occur, but the sounds are incomprehensible. Such a subject may exhibit an alternating sleep–wake rhythm.

With no cortical function or expression of cortical function, the subject exhibits no behaviour which we would interpret as indicating intelligence. Some individuals recover from coma, in that they have long periods of apparent wakefulness when their eyes are open and move. They show no adaptive response to the external environment; they fail to respond to commands by eye movements, although their eyes sometimes follow moving objects reflexly. Their responsiveness is limited to primitive reflex movements. Considerable EEG activity may be present. In this survival 'as a vegetative wreck', the vegetative physiological mechanisms persist; the subject breathes spontaneously and has good heart action. He will swallow appropriately prepared food and drink when these are placed in the mouth, since swallowing depends on reflex mechanisms which depend on the brain stem. Such people may survive for years if properly nursed. This condition is known as the 'persistent vegetative state' (Jennett and Plum, 1972).

Coma and stupor

Coma is pathological complete unconsciousness, a condition in which the patient cannot be aroused by any stimulus, not matter how strong and painful. Between normal consciousness and coma there exist many states of impairment of consciousness. *Confusion* is an intermediate condition in which the subject is conscious but there is lack of clarity of the mental processes. *Delirium* is confusion with excitement. In order to provide a numerical scale for recording the level of consciousness, Jennett and Teasdale (1977) introduced the *Glasgow coma scale*, reproduced in the Appendix. This allows different clinics to assess the level of consciousness on an agreed score system. It allows the clinician to follow the progress of a patient; if the score remains constant, the patient is stable, if the score falls the patient is deteriorating, and if it rises, the patient is improving.

No brain stem function

When the brain stem ceases to function, the heart stops beating, for reasons which are not yet understood, after a few days or weeks at most. The doctor managing a patient whose brain stem has ceased to function knows that the patient will soon die. The doctor is faced with a body perfused by a heart which will soon fail. Such a body may be a valuable source of organs for transplant; the doctor has the responsibility of deciding whether it is ethical to ask relatives to give permission for this and thereby give another patient with a good brain but a bad heart or kidney the chance of life. It is therefore important to be able to recognize whether the brain stem is functioning or not.

Causes of brain stem damage

Brain stem damage often accompanies mechanical injury to the head. This occurs despite the fact that the brain stem is well protected from direct injury: the skull affords protection to the whole brain and the massive forebrain provides additional protection to the brain stem lying deep to it. It is an indirect effect, explained below, that accounts for the brain stem's susceptibility to damage.

The skull provides the brain with a strong rigid protection, and serves its function excellently in everyday life. It protects the brain from direct injury that would otherwise result from blows to the head by sharp objects. The skull also provides a defence against damage to the cerebral tissues by infection or injury, such as scalding, spreading from the scalp. The skull cavity provides a contoured container matching the shape of the brain, within which the brain is supported by the cerebrospinal fluid (CSF). The CSF cushions the brain against mechanical shocks and bumps to the head. The sheets of connective tissue comprising the **falx cerebri** and the **tentorium cerebelli** provide a large area of contact with the surface of the brain, and this spreads the forces applied to the brain.

The housing of the brain in the skull, however, carries its own penalties when trauma is extreme. A brain freshly removed from a recently deceased human has the consistency of porridge. It has insufficient mechanical rigidity to support itself; if placed on the post-mortem table, it spreads out. Its weight in air is about 1.4 kg. *In vivo*, it is supported indirectly by the skull, and directly by the CSF, in which it is semi-buoyant. The specific gravity of the CSF is slightly less than that of the brain itself; consequently, the brain's effective weight *in vivo* is only 0.05 kg. The brain is almost floating in the CSF, analogous to a jelly fish supported by the sea in which it lives. Because of its large mass, the brain has high inertia. This means that if the head is suddenly accelerated, the brain tends to remain in the same position as before the acceleration, and collides with the inner aspect of the skull.

A high linear acceleration causes bruising of the forebrain when it is dashed against the inner aspect of the skull and against tough meningeal bands such as the falx cerebri and the tentorium cerebelli. If rotational acceleration is applied to the head, the movement of the forebrain can lag behind the movement of the skull much more than in the case of a linear acceleration. This is a consequence of the geometry of the almost spherical brain lying in a similarly shaped cavity. After the intial lag, the more slowly moving brain catches up with the skull and tends to overshoot. The forebrain swirls about relative to the skull, but the spinal cord is tethered by the nerve roots and cannot rotate. The brain stem connecting the swirling forebrain above and the spinal cord below is twanged. This indirect mechanism often causes damage to the brain stem (Ommaya and Gennarelli, 1974).

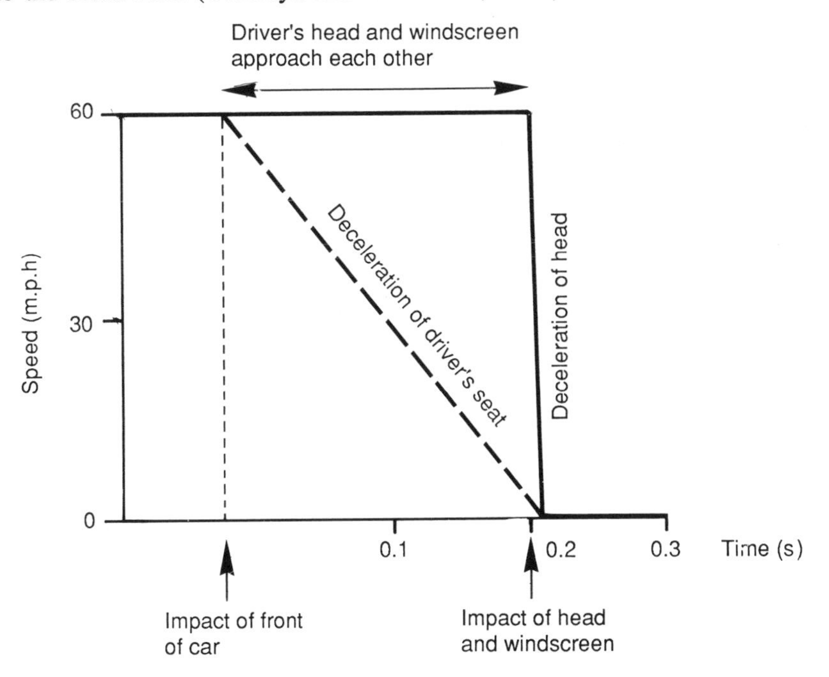

Figure 11.1 Graphs of speed as a function of time, to compare the deceleration of the car and of the head of a driver not wearing a seat belt.

A driver not wearing a seat belt may suffer extreme skull deceleration if his car crashes. Suppose that the driver and car are originally travelling at 60 miles/h (26 m/s) before the car crashes into a stationary barrier (Figure 11.1). In the deceleration resulting from the crash, the impact is broken by crumpling of the front of the car. As a result of this, the deceleration of the body of the car and of the driver's seat may take place over one fifth of a second, as shown in Figure 11.1. This corresponds to a deceleration of 132 m/s^2 or 13 **g**. During this one fifth of a second, if the passenger is not strapped into the car, he continues to move forwards at the original speed until his head hits the windscreen, which is by this stage stationary. Unlike the car, the driver has no front to crumple in order to break the impact. The head is brought to a standstill during a period of perhaps one fiftieth of a second. This corresponds to 132 **g**, a force far greater than the brain can endure without damage.

Other causes of brain stem damage: space-occupying lesions

The brain stem is also liable to damage when the intracranial pressure rises. It is the mechanical distortion of the brain, not the pressure as such which causes damage. A space-occupying lesion above the tentorium cerebelli pushes the brain across to one side so that brain stem becomes pinched against the tentorium cerebelli (see Chapter 17). Space-occupying lesions below the tentorium impact the hindbrain into the foramen magnum, impairing its functioning. This condition is called **coning**, because the space into which the hindbrain is forced is conical in shape.

Short-term survival of patients with brain stem death

Since the brain stem houses the respiratory centres, which are essential to life, it is appropriate to ask why patients with brain stem damage survive. Until the 1970s, a subject with a badly damaged brain stem died from respiratory failure. Since then tracheal intubation and artificial maintenance of respiration of head-injured patients has become a routine, so that as brain stem function declines and ceases, the subject may survive in the short term. The patient is apnoeic but his heart function may initially be good. The patient with brain stem death who survives for a few days on a ventilator is a product of our times.

Brain stem function

As we have seen, the brain stem is an alerting system for consciousness. Since the respiratory centres lie in the brain stem spontaneous respiration depends on its integrity. Spontaneous movements of skeletal musculature in the limbs and trunk also depend on an interaction between the brain stem and the cerebral cortex. If there is no brain stem function, there are no spontaneous movements. Reflex activity of the spinal cord can occur without participation of the brain stem and other higher centres, so spinal reflexes may be brisk even if the brain stem is not functioning.

11.3. BRAIN STEM REFLEXES

These are the reflexes which depend on nuclei in the brain stem. All the motor nuclei of the cranial nerves lie in the brain stem; all reflexes depending on motor cranial nerve thus depend on the integrity of the brain stem. We consider the most important of these reflexes in turn.

Pupillary reaction to light

The autonomic motor nucleus (the Edinger-Westphal nucleus) controlling the smooth muscles in the pupil lies in the brain stem. Two brain stem reflexes depending on this nucleus are tested in head-injured patients. The **direct light reflex** is a reflex pupillary constriction when light is shone in the eye. The **consensual light reflex** is a pupillary constriction when light is shone in the opposite eye.

Since a reflex depends on the integrity of the entire pathway plus its receptor and effector components, the pupillary reaction to light also depends on the normal functioning of the photoreceptors in the eye itself, on the optic (second cranial) nerves and on the oculomotor (third cranial) nerves which carry the autonomic motor nerve supply to the pupillary muscles. The reflexes also depend on the motor nuclei being in a normal physiological state and able to respond to changes in activity of afferent nerve fibres. In some cases of cerebral irritation, as in meningitis, the pupils may be fully constricted (the so-called 'pin-point' pupil). In such cases, the pupillary reflexes cannot be elicited despite the fact that the brain stem may be functioning.

Vestibulo-ocular reflexes

When the head is rotated, the semicircular canals initiate reflex activity in the extrinsic ocular muscles to stabilize the visual image on the retina, as described in Chapter 9. The stimulus is movement of endolymph relative to the walls of the semicircular canals. Such a movement of endolymph can be produced by two means, each of which leads to a reflex deviation of the direction of gaze if the reflex pathways are intact.

Cephalo-ocular reflexes (passive head rotation)

If the head is rotated from side to side, the eyes move reflexly in the orbits in a direction opposite to that of the head. In the conscious subject, this reflex maintains the visual image stationary on the retina. This response is known as 'the doll's head phenomenon' because some dolls are provided with high-inertia eyes which show a similar behaviour.

Caloric testing

In this test, the endolymph is made to move by the non-physiological manoeuvre of heating or cooling it. Hot or ice-cold water is slowly run in to the external auditory meatus. The lateral semicircular canal is nearest to the meatus. The movement of endolymph is due to convection currents generated by the difference in density of fluid at different temperatures. In a normal subject, the manoeuvre makes the subject feel dizzy. An observer notices nystagmus, as described below.

If cold water is run into the external meatus of a subject lying supine, the endolymph cools, becomes more dense, and flows downwards. This movement of endolymph is that which would occur if the head were being rotated away from the ear receiving the cold water. The reflex response to stabilize the visual image in this situation is a slow deviation of the eyes in the opposite direction from that of the simulated head movement. In this situation, this is achieved by deviation of the visual axis towards the cooled side. If the procedure were repeated with subject lying prone, then the reflex response would be a slow deviation of the eyes away from the side of infusion. The two eyes move in the same direction; this is called **conjugate** deviation of the eyes.

Nystagmus was described in Chapter 10. It consists of two components, the slow conjugate deviation, integrated by brain stem mechanisms, and a fast flick back, commanded by the cerebral cortex. A normal subject exhibits nystagmus in response to a caloric test, whereas a subject with cortical damage exhibits only conjugate deviation. In a subject with no brain stem function, the eyes do not move at all in response to caloric testing.

Before performing this test, it is necessary to ensure that there is clear access to the ear drum on either side, and that the ear drums are not perforated. Otherwise one risks running non-sterile water through a perforated ear drum into the middle ear and starting an infection there.

Table 11.2 Summary of brain stem mechanisms and reflexes

Brain stem mechanisms
Alerting system for consciousness (consciousness also depends on the cortex)
Spontaneous movements (also depend on the cortex)
Respiration
Brain stem reflexes
Pupillary reaction to light (direct and consensual light reflexes)
Vestibulo-ocular responses (passive head rotations, caloric testing)
Facial nerve reflex
Corneal reflex
Gag reflex
Brain stem evoked potentials

Facial nerve reflex (grimacing reflex)

Pressure on the supra-orbital notch produces a reflex grimacing. The afferent pathway is via the ophthalmic nerve, which is the first division of the trigeminal (V cranial) nerve. The motor nerve fibres arise from the facial (VII cranial) nerve nucleus, which lies in the lower part of the pons.

Corneal reflex

Touching the cornea with a wisp of cotton wool causes reflex blinking. Again, the afferent pathway is via the ophthalmic division of the trigeminal nerve and the efferent pathway is the facial nerve. Normal blinking is mediated by the facial nerve nucleus.

Gag reflex

Touching the back of the pharynx causes gagging, which is an attempt at vomiting. Here the motor nuclei lie in the hindbrain.

Brain stem evoked potentials

Evoked potentials are electrical deflections which can be recorded from the surface of the scalp at a fairly constant delay after a stimulus is applied to the integument or one of the special senses. The auditory pathway relays in the brain stem and this generates components of the auditory evoked potential. These components are depressed when the brain stem is damaged. Central afferent pathways from the cord pass straight through the brain stem without relay. Cortical potentials evoked by somatic stimuli may therefore be normal even when the brain stem is damaged. Similarly, visual evoked potentials do not depend on the integrity of the brain system.

Table 11.2 summarizes this section. We next proceed to consider the interactions amongst brain stem nuclei in the production of normal tone in limb muscles.

11.4 DECEREBRATE RIGIDITY

This is a situation in which excitatory neural centres are overactive. It occurs as a result of transections of the neuraxis at the midbrain level. Such injuries occur in humans as a result of direct trauma or by indirect mechanisms which we have already considered. In animals, the brain stem may be sectioned surgically under anaesthesia.

Transections at levels between C4 in the spinal cord and the respiratory centres in the medulla are incompatible with independent existence since they result in paralysis of respiration. If the neuraxis is transected at the level of the midbrain, which is above the bulk of the respiratory centres, spontaneous respiration occurs. The most striking feature of the condition is an extensor rigidity throughout the body, called **decerebrate rigidity**. This rigidity is much more intense than that occurring in a subject with a transected spinal cord: it is possible to balance a decerebrate animal on its extended limbs. The rigidity is due to overactivity of the spinal stretch reflex.

The degree and distribution of rigidity in the fore and hindlimbs depends on the exact level of the section. Standard experimental decerebration is performed between the superior and inferior colliculi, which lie on the dorsal aspect of the midbrain.

The mechanism of decerebrate rigidity

The most important structures in the genesis of decerebrate rigidity are the red nucleus above and the lateral vestibular nucleus and inhibitory reticular formation below (shown in boxes in Figure 11.2).

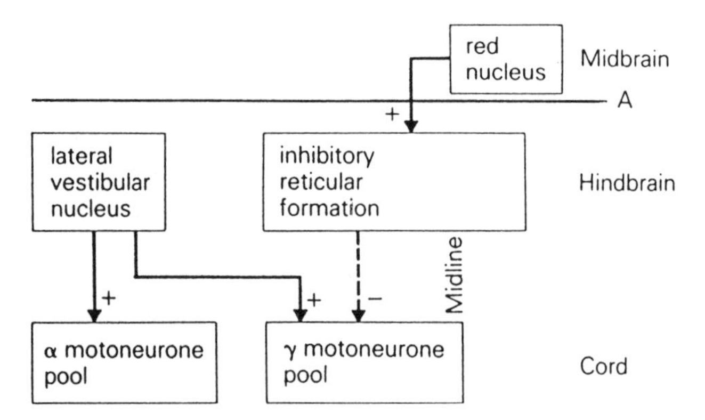

Figure 11.2 Nuclei important in the genesis of decerebrate rigidity; A indicates the level of a transection after which decerebrate rigidity occurs.

The role of the lateral vestibular nucleus in the maintenance of normal posture

A normal human who stands and walks needs continuous tonus in the skeletal muscles, particularly in the extensor muscles since these are the 'antigravity' muscles. In the intact organism, this is achieved by a tonic discharge projecting down to the cord to motoneurones which innervate the extensor muscles. An important contribution to this is made by the lateral vestibular nucleus. The input to this nucleus is from the vestibules, whose importance in initiating reflexes of posture has already been described. This nucleus bombards both α and γ motoneurones with excitation. It is on this background of excitation that humans can stand, walk and run.

All central nervous nuclei exhibit spontaneous electrical activity. Such activity in the lateral vestibular nucleus increases the general level of excitability of both α and γ motoneurones. This powerful excitation is normally balanced by an intensity of inhibition which results in a degree of tonus appropriate for standing and walking. This descending inhibition comes primarily from the inhibitory reticular formation, a region of the brain stem which will be described later in this chapter. Spontaneous activity in the red nucleus excites the inhibitory reticular formation; this in turn inhibits the motoneurones, primarily the γ motoneurones, in the cord. Transection of the midbrain upsets this balance. This input from the red nucleus to the inhibitory reticular formation is interrupted, with a consequent decrease in the inhibition descending to the cord. Ongoing excitation from the lateral vestibular nucleus is then not held at bay. Sections either below the lateral vestibular nucleus or above the red nucleus do not result in decerebrate rigidity.

Consistent with this scheme, destruction of the lateral vestibular nucleus on one side abolishes the rigidity in the ipsilateral musculature but leaves the rigidity in the contralateral musculature unimpaired. Evidence that decerebrate rigidity is primarily a γ rigidity is given in the following section.

Decerebrate rigidity: α or γ?

Figure 11.3a represents the normal animal, with a balance of flexor and extensor tonus in the limb musculature; the limbs are shown as slightly flexed. An animal with a mid-collicular decerebration has a moderately severe extensor rigidity in all four limbs (Figure 11.3b). If, as shown in Figure 11.3c, one forelimb is deafferented, the rigidity in that limb declines, identifying the rigidity as predominantly γ rigidity, as explained in Chapter 7.

Decerebrate rigidity has similarities with the rigidity resulting from lesions of the basal ganglia or of the motor area of the cerebral cortex, to be described in later chapters. These centres also tonically excite the inhibitory reticular formation.

Mesencephalic fits

In some cases, decerebrate rigidity waxes and wanes. This is due to damage of the respiratory centres; respiration becomes **periodic**, with periods of adequate ventilation and periods of hypoventilation in alternation. When oxygenation of the blood is adequate, the brain nuclei function well and the centres driving decerebrate rigidity are unleashed. In periods when oxygenation is inadequate, the activity of brain nuclei is depressed by anoxia and the central drive to decerebrate rigidity is curbed. The patient is said to have **decerebrate attacks** or **tonic mesencephalic fits**.

A painful stimulus applied during a phase of depression can provide sufficient extra excitatory input to revive activity in the flagging central nuclei;

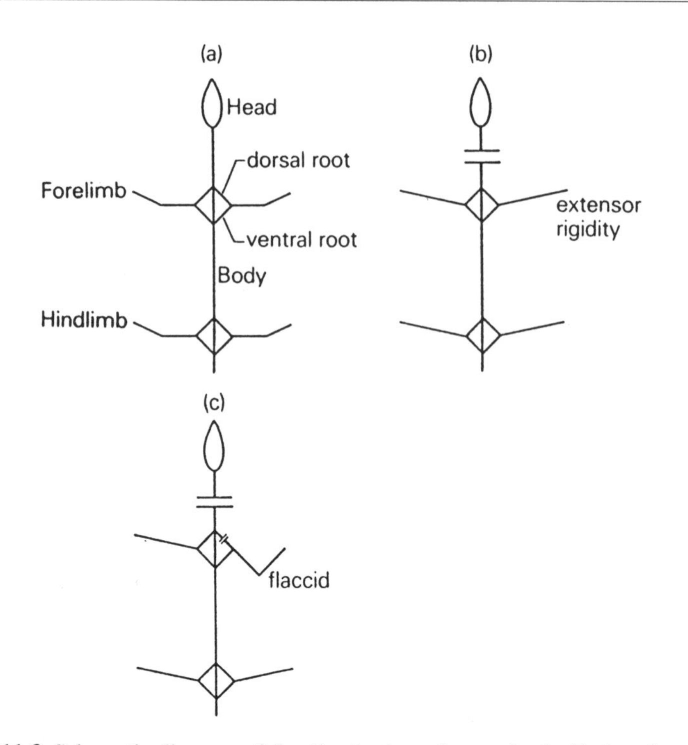

Figure 11.3 Schematic diagram of the distribution of tonus in the limbs of an animal. (a) Normal. (b) Decerebration. (c) Decerebration and section of the dorsal roots of the right forelimb.

this is evidenced by a renewal of extensor rigidity either confined to the limb stimulated (if the effect is confined to raising the central excitatory state of the segment of cord bombarded by the input) or generalized throughout the body (if the effect is to stimulate respiration).

Effect of transection of the midbrain on movements in limbs

High midbrain section leaves an animal with relatively normal tone in the limbs. Such an animal suspended on a treadmill will walk at exactly the speed required to keep the trunk over the support area. Input from contact of the feet with the moving ground provide the afferent information necessary for this adjustment. If the animal is similarly suspended but over a turn-table, when the turn-table turns, the animal walks as if walking around a circle in order to maintain stable posture; again input comes from cutaneous pressure receptors (Graham Brown).

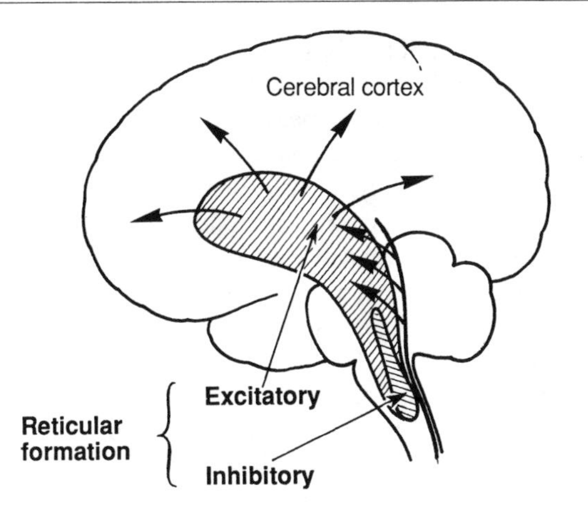

Figure 11.4 Medial aspect of the right half of the brain. The reticular formation is shaded.

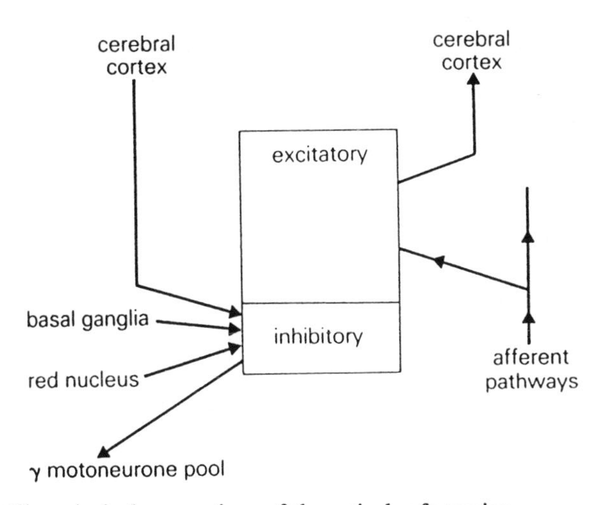

Figure 11.5 The principal connections of the reticular formation.

11.5 THE BRAIN STEM RETICULAR FORMATION

Unless lesions of the forebrain involve the whole of the cerebral cortex, as described at the outset of this chapter, they do not interfere with consciousness, whereas lesions compressing the brain stem readily cause clouding or loss of consciousness. Since the early days of surgery on the brain conducted under local anaesthesia, it has been known that the cerebral cortex could be displaced or cut, or parts of it removed, without the slightest interference with consciousness. By contrast, if the third ventricle is opened and a spatula touches its wall (which houses the upper part of the brain stem), there is an immediate profound disturbance of consciousness.

Drowsiness and impairment of consciousness in many patients who had encephalitis and subsequently died was observed to be associated with lesions in the central grey matter of the brain stem detected at post-mortem. Lesions of the forebrain were not potent in causing loss of consciousness, whereas lesions compressing the brain stem readily caused clouding of consciousness. Such clinical observations indicated that consciousness depends on the integrity of brain stem function.

The neural tissues of importance in consciousness lie in the core of the brain stem. The neurones form nuclear groups with indistinct boundaries and fibres not organized into well-circumscribed tracts. Because of this histological appearance, this system is called the **reticular system**. It extends from the medial portion of the thalamus through the midbrain and pons to the caudal medulla. It consists of excitatory and inhibitory components, as described below.

Figure 11.4 shows a diagram of the antero-posterior extent of the reticular formation. The principal connections of the reticular formation are shown in Figure 11.5. The reticular formation is divided anatomically and physiologically into two portions.

The excitatory reticular formation

This is so-called because localized electrical stimulation of this region causes widespread excitation of the brain. Anatomically, it is very extensive, from the thalamus rostrally to the hindbrain caudally. Its input is mainly ascending, from the spinal cord, and its output projects principally upwards to the cerebral cortex; it projects down to the cord to a lesser extent.

Generalized activation of the cortex occurs as a result of a volley of impulses along sensory pathways. Moruzzi and Magoun (1949) showed that the ascending reticular formation, rather than the pathway through the specific thalamic relay nuclei, was responsible for this generalized activation. The delay between a stimulus and generalized activation of the cortex is some 50 ms longer that that of the potential which is evoked from the primary somatosensory cortex. All the main sensory tracts send collaterals to the brain stem reticular formation and provide tonic excitatory input. Even if the classical afferent tracts are divided above the point where the collaterals arise, sensory stimulation of any modality causes arousal.

The inhibitory reticular formation

This is so-called because local electrical stimulation of this region causes widespread inhibition in the brain. Anatomically, the inhibitory reticular formation is localized, lying mainly in the caudal part of the medulla. It projects mainly downwards to the spinal cord, with little inhibitory effect projecting up to the cortex. It relays information from higher centres such as the cerebral cortex, the basal ganglia, and the red nucleus down to the spinal cord. The input from the red nucleus is the most powerful and hence the importance attached to it in the genesis of decerebrate rigidity. The input from the other two centres accounts for the observation that decerebrate rigidity has similarities with the rigidity produced by lesions of the basal ganglia or the precentral gyrus of the cerebral cortex. This is because, in the intact animal, these centres tonically excite the inhibitory reticular formation.

Chemical transmission and the reticular formation

Two groups of transmitter are important in the reticular formation and its projections: acetylcholine and the monoamines. These transmitters are phylogenetically very old and are used by structures such as the reticular formation, which are themselves phylogenetically old. As with their actions in the periphery, their actions in the central nervous system tend to be mutually antagonistic.

The excitatory reticular formation is almost certainly cholinergic (Lewis and Shute, 1963; Shute and Lewis, 1967). Its neurones and their axons contain the necessary enzymes for the synthesis and hydrolysis of acetylcholine. Activity in this system is accompanied by the release of acetylcholine into the cerebral cortex. This chemical can be collected by superfusing the cortex surface with a physiological saline solution. The release of acetylcholine from the cortex is reduced during sleep. One component of the reticular formation which provides a particularly rich diffuse cholinergic projection to the cerebral cortex is the basal forebrain nucleus, a collection of cholinergic neurones lying in the grey matter of the brain stem inferior to the lentiform nucleus. Alzheimer's disease, a neurological disease characterized by the occurrence of a dementia similar to senile dementia in people who are not yet old, is characterized by degeneration of the cholinergic cells of the basal forebrain nucleus.

All acetylcholine released into the cortex originates from axons projecting up from subcortical structures; there are no cholinergic neurones in the cortex itself (Fibinger, 1982). Acetylcholine probably does not itself act as a neurotransmitter at synapses in the cerebral cortex. It facilitates excitatory synaptic transmission at synapses that use other neurotransmitter chemicals, such as amino acids. Such action is known as **neuromodulation**. In the case of acetylcholine, this facilitatory effect is produced by reducing the conductance of potassium channels in the neuronal membrane (Krnjevic *et al.*, 1971). This increases the likelihood that an excitatory synaptic input will bring the neurone to the threshold of firing an impulse.

Sleep

Sleep is an active physiological process, not merely a failure of arousal, and is clearly separable from stupor and coma. It is periodic depression of the parts of the brain concerned with consciousness. Since consciousness depends on the activity of the reticular formation, it is not surprising that the physiological alterations in conscious level which we call sleep and wakefulness are regulated by components of the reticular formation.

There are two types of sleep, orthodox and paradoxical. These are also known as slow wave and rapid eye movement (REM) sleep respectively because, during orthodox sleep, the electroencephalogram shows prominent slow wave activity and, during paradoxical sleep, there are rapid eye movements.

In a normal human, the usual cycle is

Wake → Orthodox sleep → Paradoxical sleep
25 min ← 5 min

A night's sleep consists of alternating periods of orthodox and paradoxical sleep. Some authorities subdivide these two types of sleep further. Various features allow one to distinguish between the two types of sleep. In orthodox sleep, the subject is easily aroused by minor stimuli; orthodox sleep is therefore regarded as light sleep. A person in paradoxical sleep is aroused with considerable difficulty and is undisturbed by minor stimuli; from this point of view paradoxical sleep is deep sleep. Tonus in skeletal muscles is present in orthodox sleep, but absent in paradoxical sleep. However, in paradoxical sleep, the flaccid muscles show occasional twitches. The eyes show few movements in orthodox sleep and any such movements are slow; in paradoxical sleep there are frequent rapid eye movements. The eyeballs can be observed to be moving beneath the closed eyelids and this observation led to paradoxical sleep being called Rapid Eye Movement (REM) sleep.

A subject awakened from REM sleep routinely reports that he was in the middle of a dream, whereas one awakened from orthodox sleep reports no dreams. In orthodox sleep the heart rate and blood pressure are both stable whereas in REM sleep they are labile. The electroencephalogram (EEG) of a subject in orthodox sleep is markedly different from the pattern in the waking subject. The waking subject shows a very low voltage of irregular high frequency waves, signalling asynchronous activity in the neurones of the cerebrum. In orthodox sleep, the EEG is dominated by high voltage waves of low frequency, indicating that large number of neurones are discharging in synchrony. In REM sleep, the EEG is low voltage and high frequency, very like that of the waking subject. In REM sleep, although the subject is deeply asleep and aroused only with difficulty, the brain, although detached from the realities of the environment, is in a very active state.

The features of these two types of sleep are summarized in Table 11.3.

Table 11.3 Sleep

Type of sleep	*Orthodox*	*Paradoxical*
Ease of arousal	Easily aroused, sleep is light	Aroused with difficulty sleep is deep
Skeletal muscle	Tonus present	Muscles are flaccid with muscle twitches
Eyes	Infrequent slow movements	Rapid eye movements
Dreams	None	Dreaming
Heart rate blood pressure	Stable	Variable
Electro-encephalogram	Slow wave (3 Hz)	High frequency (12–15 Hz) very similar to EEG of waking subject

Paradoxical sleep accounts for 20–25% of the total night's sleep. Sufficient paradoxical sleep seems to be very important to a human's well-being. Specific deprivation of paradoxical sleep can be brought about by placing electrodes to the skin around the eyes and recording the electrical changes associated with the rapid eye movements. If the subject is aroused whenever paradoxical sleep occurs, he is very drowsy the next day. If deprivation continues for a number of nights, the subject soon becomes disorientated.

Random deprivation of an equal amount of sleep (by awakening at times unrelated to the rapid eye movement) is accompanied by very much milder degrees of sleepiness and disorientation. A subject who has been deprived of paradoxical sleep for some nights and is then allowed to sleep undisturbed makes up for his lack of paradoxical sleep by a much greater proportion of his night in paradoxical sleep. No-one knows why paradoxical sleep is so important to the brain

Sleep centres

Nuclei in the brain stem important in the wake–sleep cycle are known as **sleep centres**. Their siting can be inferred from observations on sleep patterns after selective lesioning of different groups of neurones. Such procedures reveal the existence of different centres for orthodox and paradoxical sleep.

The **locus coeruleus** (Latin, 'blue place') is important in sleep. It is bilaterally placed in the rostral pons and consists of a collection of pigmented neurones forming nuclei that are cerulean blue in colour and can be seen in the floor of the 4th ventricle adjacent to the pontine reticular formation.

In 1972 Jouvet showed that lesions of the locus coeruleus disturb paradoxical sleep but not orthodox sleep. Nerve fibres arising from the locus coeruleus terminate massively in all layers of the cerebral cortex and also in subcortical structures. The transmitter released by these axons is noradrenaline: virtually all the noradrenaline secreted into brain comes from this pair of nuclei.

Lesions of the midline raphe nuclei (sited in the mesencephalon, pons, and medulla) interrupt orthodox sleep. The total amount of sleep is reduced to about 10% of normal. Sleep that does occur has the characteristics of REM sleep. The neurones of the raphe nuclei are rich in serotonin (5-hydroxytryptamine). An increase in brain concentration of serotonin increases the amount of sleep. These observations suggest that the raphe system may be involved in the generation of orthodox sleep by releasing serotonin and opposing the actions of the reticular activating system.

The sleep centres are monoaminergic and the reticular activating system, as we saw in a previous section, is cholinergic. Both project diffusely to the cerebral cortex. There is a nice parallel here with neurotransmission in the autonomic nervous system. The phylogenetically ancient brain stem centres use phylogenetically ancient neurochemicals. In the cortex, as in the periphery, the two groups of chemicals tend to produce effects in different directions. In the brain, acetylcholine keeps the cortex alert while the monoamines send it to sleep. The nervous system, both peripherally and centrally, operates by a balance of cholinergic and monoaminergic mechanisms.

11.6 APPENDIX: THE GLASGOW COMA SCALE

Table 11.4 Glasgow 'coma' or responsiveness scale

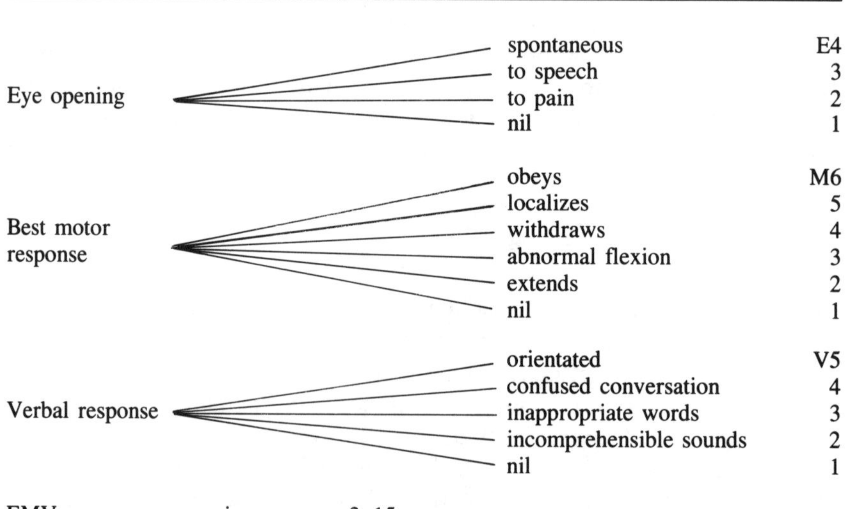

Eye opening	spontaneous	E4
	to speech	3
	to pain	2
	nil	1
Best motor response	obeys	M6
	localizes	5
	withdraws	4
	abnormal flexion	3
	extends	2
	nil	1
Verbal response	orientated	V5
	confused conversation	4
	inappropriate words	3
	incomprehensible sounds	2
	nil	1

EMV score or responsiveness sum 3–15

FURTHER READING

Jennett, B.J. and Teasdale, G. (1981) *Management of Head Injuries*. Philadelphia, F.A. Davis.

The basal ganglia and cerebellum

12

12.1 THE SUPRASPINAL CONTROL OF MOVEMENT

All the centres of the nervous system have motor activity as their final mode of expression. The control of motor activity is directly or indirectly the concern of every neurone. As a prelude to descriptions of each individual centre, we consider the general architecture of motor control.

The lower motoneurones are the final common path and directly command muscle contraction; they lie in the spinal cord or in homologous cranial nerve nuclei of the brain stem. The cerebral cortex influences the motoneurones by axons that project along the corticospinal tract. This passes down to the cord via the medullary pyramids. This has led to the concept of the **pyramidal system**, which is defined (Ruch and Patton, 1979, p.63) as those axons which originate in the cerebral cortex and pass to the spinal cord through the medullary pyramids. All other paths projecting from higher centres to the cord are known as **extrapyramidal pathways**. The medullary pyramids themselves contain many descending fibres from subcortical nuclei; these fibres are extrapyramidal although they run in the pyramidal tracts. The nomenclature is certainly confusing and some authorities regard the use of the terms pyramidal and extrapyramidal as obsolete.

In outline, the central nervous structures concerned with motor control may be classified into three levels:

I Precentral gyrus of the cerebral hemispheres
II Cerebellum
Basal ganglia
Vestibular nuclei
III Spinal cord

This classification is phylogenetic, the spinal cord being the first component to appear in evolution, the cerebellum, basal ganglia and vestibular nuclei appearing next, and the cerebral hemispheres appearing most recently. It will emerge in Chapter 15 that the phylogenetic sequence does not coincide with the sequence of activation of neural centres in the production of a voluntary movement.

During the voluntary movement of a limb, there is discharge from all these structures. In physiological terms, it is questionable to subdivide a system the whole of which is concerned with the production of efficient movement. However, lesions in different parts of the brain do produce different general types of disturbance of muscular activity. The **initiation** of voluntary movement requires the participation of the precentral gyrus. The **timing and degree** of contraction and relaxation of the different muscles involved, i.e. coordination, are organized by the cerebellum. **Postural adjustments** are largely under the control of the basal ganglia, the vestibular system, and the spinal cord. Postural adjustments include the vestibular and neck reflexes which were considered in Chapter 10. The foregoing is summarized in Table 12.1.

Table 12.1 Voluntary movement: the contributions made by the different parts of the nervous system

Precentral gyrus	Participates in the initiation of voluntary movement
Cerebellum	Important in co-ordination
Basal ganglia Vestibular system Spinal cord	Control of postural adjustments

Skilled voluntary movements can be classified into ramp and ballistic (Desmedt and Godaux, 1978).

Ramp movements

An example of a ramp movement is tracing carefully a ramp with a pencil. A more homely example is picking up a cup of tea. Such movements are slow and controlled. There is time for continuous modification of the motor output by feedback from receptors, particularly proprioceptor and visual receptor input. This allows a skilled movement to be performed smoothly even if the load being moved varies during the execution of the movement. There is evidence that such movements are primarily under the control of the basal ganglia, where ramp generators have been described.

Ballistic movements

'Ballistic' means 'throwing'. Ballistic movements are fast movements, such as hitting the key of a piano or typewriter, or striking the fingerboard of a violin. The duration of a ballistic movement is less than the reflex time around proprioceptive reflex arcs. Proprioceptive feedback is too slow for modification of the motor command. The whole act must be pre-programmed by the nervous system; once the command has been launched, nothing can stop it. Amateur typists experience this infuriating characteristic of ballistic movements frequently. They realize too late that they have commanded the wrong key to be struck and so, despite the fact that they know that they are about to type the wrong letter, they cannot interrupt the command. Ballistic movements are primarily under the control of the cerebellum. Since proprioceptive feedback is too slow to contribute to the control of ballistic movements and since reflex effects occurring after the end of a quick movement are undesirable, proprioceptive reflexes such as the stretch reflex are actually switched off by descending inhibition from higher centres during the performance of a ballistic movement. Most motor acts involve a highly complex sequenced mixture of the two types of movement.

Referring to the list of centres important in motor control, we have already considered the cord and the vestibular nuclei; we next proceed to the basal ganglia.

12.2 BASAL GANGLIA

As we have already seen, the basal ganglia are involved in ramp movements and in postural adjustments. Anatomically, the basal ganglia lie within the substance of the brain and cannot be visualized, as can for instance the cerebellum, in the whole brain removed from the body. The most prominent component of the basal ganglia is the **corpus striatum**, so called because it has a striated appearance. It is made up of the lentiform, the caudate, and the amygdaloid nuclei (Figure 12.1a). The **lentiform** nucleus is soalled because it is shaped like a lens. From the anterior pole arises the **caudate nucleus** which, as its name implies, has a long tail. This curves round to end in the **amygdaloid nucleus**. A section through the basal forebrain (Figure 12.1b) shows that the lentiform nucleus is subdivided into the **globus pallidus** (pale globe) medially and the **putamen** laterally. Lateral to the lentiform nucleus is the **claustrum**, the function of which is unknown.

Phylogenetically, the oldest part of the corpus striatum is the globus pallidus; it is called the **paleostriatum** (old striatum). The younger parts of the striatum are the putamen and the caudate nuclei; they are called the **neostriatum** and they function as a single entity. The amygdaloid nucleus, although anatomically part of the corpus striatum, is functionally part of the limbic system, described in Chapter 13.

There are two other basal ganglia, the **subthalamic nucleus** and the **substantia nigra**. The foregoing is summarized in Table 12.2. The physiological and functional grouping is not the same as the anatomical grouping. It is the functional grouping that is of interest to the neurophysiologist.

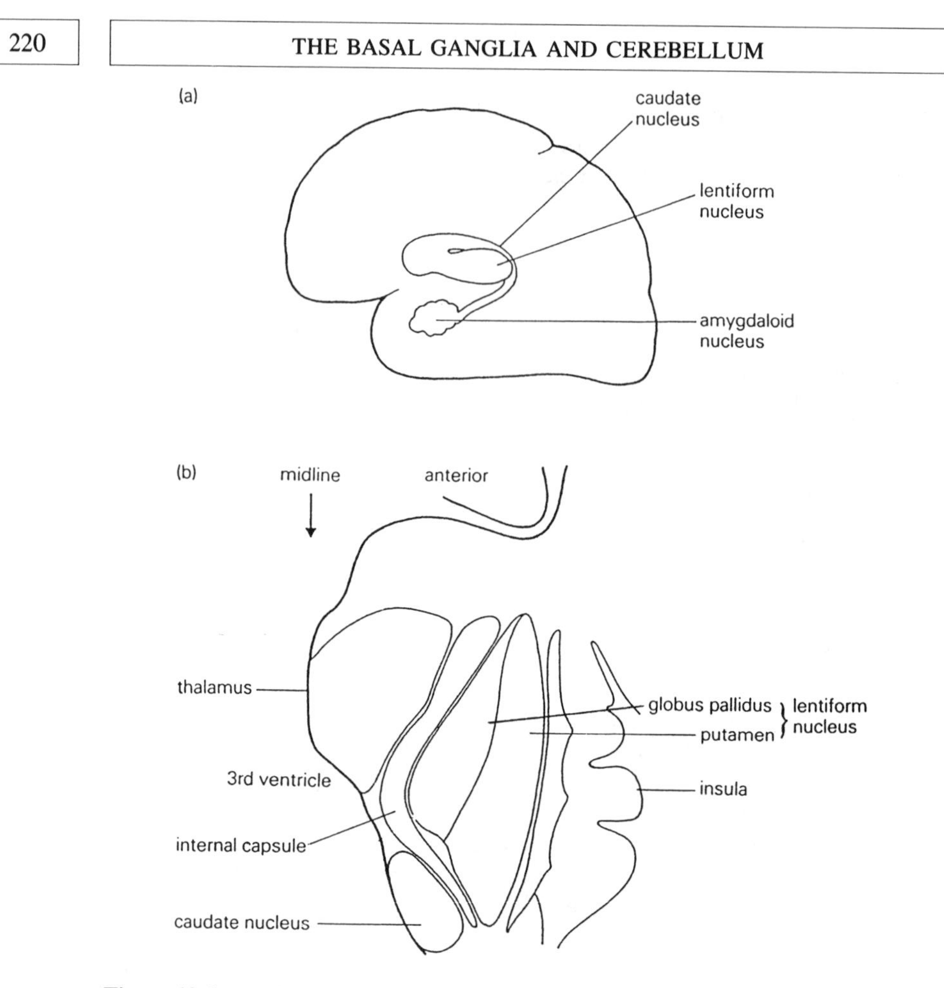

Figure 12.1 (a) The corpus striatum on the left projected on an outline of the brain. (b) A horizontal section through the lentiform nucleus.

Table 12.2 Basal ganglia

Corpus striatum	Lentiform nucleus	Globus pallidus	Paleostriatum
		Putamen	Neostriatum
	Caudate nucleus		Neostriatum
	Amygdaloid nucleus		Part of limbic system
	Claustrum		Function unknown
Subthalamic nucleus			
Substantia nigra			

Connections of the basal ganglia

Within the corpus striatum, the flow of information is from new structure to old: from neostriatum to globus pallidus (Figure 12.2).

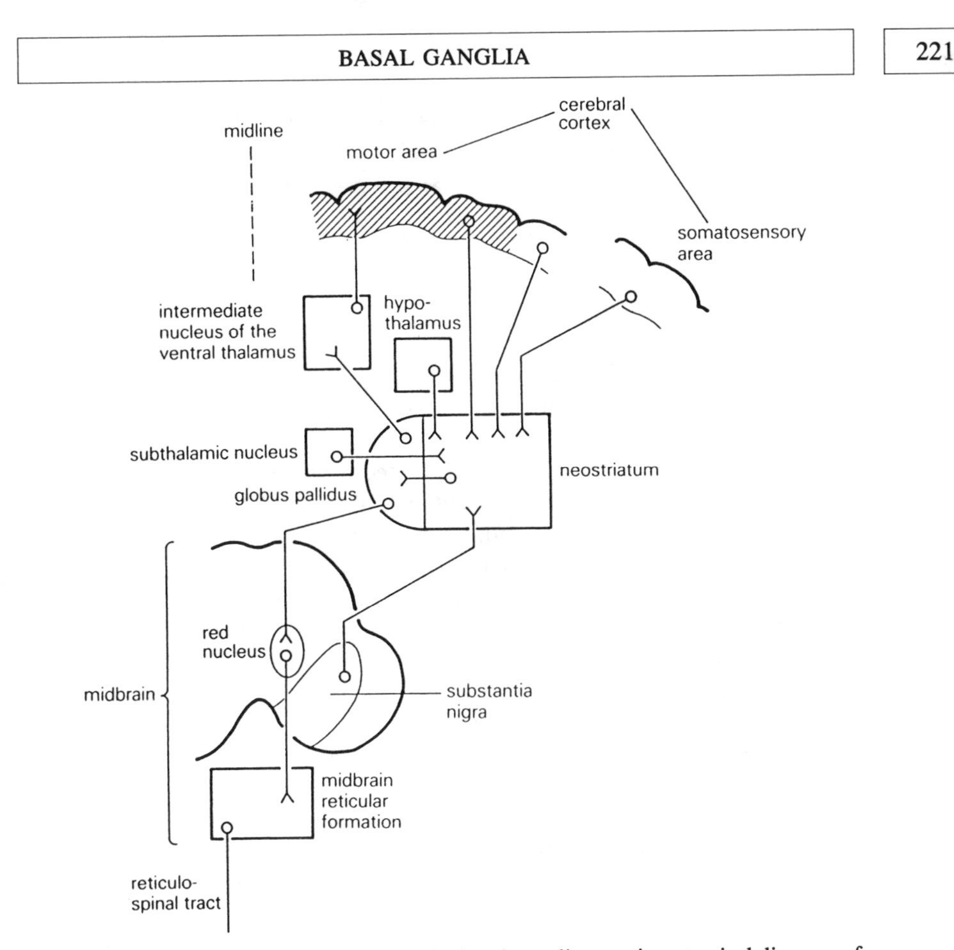

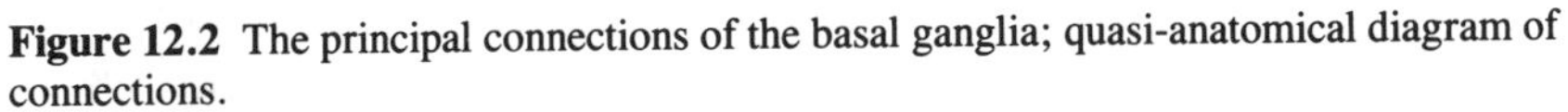
Figure 12.2 The principal connections of the basal ganglia; quasi-anatomical diagram of connections.

Afferent projections

The bulk of the input is to neostriatum. From the cerebral cortex, there is input from all areas: primary sensory and motor areas, secondary areas, and association cortex. This is a source of information about the external environment and about ongoing motor activity commanded by the neocortex. Information about the internal environment is provided by input from the hypothalamus. There is also a rich input to the neostriatum from both the substantia nigra and the subthalamic nucleus; these are considered later in this section.

Overall, the basal ganglia receive input from brain centres but relatively little direct input from the cord. They operate on information which has already been analysed and generated by the other brain centres, and they have little access to raw data from the periphery. In this respect, the basal ganglia contrast with the cerebral hemispheres and cerebellum, both of which receive extensive input from the cord, as described in later sections. Within the basal ganglia, there is some rudimentary somatotopic organization, although this is much less well spatially resolved than in the motor area of the cerebral cortex.

Efferent projections

The only neurones in the basal ganglia that are known to send axons to other parts of the nervous system are those of the internal part of the globus pallidus (Ruch and Patton, 1979). This projection is downwards and upwards. The downward projection is via the red nucleus and reticular formation to the spinal cord. The red nucleus in lower mammals consists of magnocellular and parvocellular parts. The magnocellular part gives rise to the rubrospinal tract, which is involved in motor control. In man, there is a regression of the magnocellular part of the nucleus and the rubrospinal tract is rudimentary (Watkins, 1972). The motor influence of the red nucleus on the cord is via the reticular formation (Snider, 1972), along the **rubro-reticulo-spinal** tract. The globus pallidus also projects directly to the reticular formation; the motor influence of the globus pallidus on the cord is via the reticular formation.

The lack of the rubrospinal tract in man is part of the move to replace the control of old centres on finely controlled muscle action by direct control from the precentral gyrus (Chapter 15). Such cortical control brings the output close to the input from the sensory innervation of the glabrous skin of the fingertips, which is so important in the sensorimotor feedback involved in stereognosis.

The upward projection is to the precentral gyrus via the thalamus. The thalamus thus acts as a relay nucleus to the cerebral cortex for the basal ganglia, just as it acts as a relay nucleus for the specific afferent projection to the cortex from the cord. The significance of the siting of the thalamic relay nucleus will emerge in an ensuing section.

To recapitulate briefly, the main connections of the basal ganglia are to and fro with cerebral cortex. These connections are with all parts of the cerebral cortex (Kemp and Powell, 1970), although the density of connections differ for different regions. There are two configurations. Firstly, there is a corticotopic projection, such that one region of the basal ganglia has reciprocal connections with one particular region of the cortex. Secondly, there are particularly rich projections from the posterior parietal cortex to the basal ganglia and from the basal ganglia via the thalamus to the motor cortex. These projections, shown in Figure 15.1, are a crucial part of the pathway in the transference of voluntary commands, which originate in the posterior parietal cortex, to the primary motor cortex (to be described in Chapter 15). Deficiency of this pathway accounts for the extreme difficulty experienced by Parkinsonian patients in the initiation of voluntary movements.

The functions of the basal ganglia

The details of the functions of the basal ganglia in mammals are poorly understood. In some groups of animals such as birds and reptiles, the cerebral cortex is rudimentary or absent. The basal ganglia are prominent and are the higher centres which control all movement.

Much of our knowledge concerning central nervous system function comes from clinical observations on humans with localized disease in the brain. The correlation between the site of a lesion in the basal ganglia and its clinical manifestations is obscure and variable from patient to patient. The account given in this book aims to be inherently consistent, and this involves deliberate simplification.

In the higher mammals and humans, neurones in the basal ganglia discharge action potentials throughout ramp movements, but they are silent during ballistic movements. The hallmark of lesions of the basal ganglia is involuntary movement. Many of the involuntary movements of basal ganglia disease look like voluntary ramp movements out of control. These observations provide one of the various lines of evidence pointing to the basal ganglia as the controlling centres for ramp movements. The nature of the involuntary movements depends on the site within the basal ganglia of the lesion. Different types of involuntary movement have been given different names. For instance, there are

a) **choreic movements** (dancing movements)
b) **athetosis** (slow writhing movements)
c) **ballismus** (flinging movements).

In many cases these involuntary movements are not present all the time. They are brought on by attempts by the subject to make voluntary movements. Voluntary movements in the affected limbs are also often paralysed. Of the many clinical conditions associated with disease of the basal ganglia, the most common is Parkinson's disease.

Parkinson's disease (also called paralysis agitans, paralysis with agitation)

Of all the nuclei in the basal ganglia, the one which always undergoes degeneration in Parkinson's disease is the substantia nigra, sited in the midbrain. Different projections from the substantia nigra play different roles in the various manifestations of the disease. There are three classically described components ot this disease; akinesia, involuntary movement, and rigidity (Table 12.3). Different patients exhibit these various components to differing degrees: one patient may show uncontrollable, disabling involuntary movements, whereas another may show no such movements but may be incapacitated by rigidity. There is some clinical evidence to suggest that differences in the distribution of lesions correspond with differences in clinical disorders.

Table 12.3 Parkinson's disease

Features	*Degeneration in projection from substantia nigra to*	*Helped by*
Akinesia (disinclination to voluntary movement)		
Involuntary movement (pill-rolling)	Ventral lateral nucleus of thalamus	Thalamotomy
Rigidity (lead-pipe, cogwheel)	Neostriatum	DOPA

Akinesia

This is a disinclination to move the affected part in the normal manner. The name 'paralysis agitans' derives from the presence of this akinesia, the *paralysis* component, along with involuntary movement, the *agitans* component. The signs of akinesia include a paucity of voluntary movement, a mask-like expression with an unblinking stare, a loss of swinging of arms while walking and a shuffling gait.

Paralysis is a misnomer because there is no true paralysis: the patient can be provoked to activate his previously immobile muscles. A severely incapacitated Parkinsonian patient may deftly catch a cricket ball which is thrown to him and escape with alacrity from the road when a bus approaches and threatens to knock him down. The specific disability of the Parkinsonian patient is the production of self-generated voluntary movements; without strong motivation the subject remains immobile and expressionless. The slowing of voluntary movement is much more pronounced when two movements are attempted either together (e.g. squeezing and flexing the hand simultaneously) or in immediate succession than when a single voluntary movement is required.

The akinesia of Parkinson's disease is explicable in terms of lack of excitation reaching the supplementary motor area of the cerebral cortex from the dense projection from the basal ganglia via the thalamus. The thalamocortical component of this projection is excitatory: lowering of activity therefore lowers the activity of the cortex. This projection is also on the pathway of voluntary movement, making a consistent picture. The patient cannot switch on voluntary movement; movement must be provoked from outside. Throw a tennis ball at the patient and he will catch it, but he then finds it totally impossible to initiate the throwing of the ball back again.

Some subjects learn to provide themselves with afferent stimuli to initiate voluntary movement which they find otherwise impossible. An example is the old Parkinsonian man who carries an umbrella. When he finds it impossible to start walking, he throws the umbrella on the floor ahead of him and this provides sufficient input for him to step over the umbrella; he can then carry on walking.

Involuntary movement

This term is used to describe movement not under voluntary control and occurring while the patient is otherwise at rest. Disturbances of voluntary movements, for instance in cerebellar disease, although they are not under voluntary control, are not included in medical parlance in the term 'involuntary movements'.

Characteristically, in Parkinson's disease, the involuntary movements consist of alternating abduction and adduction of the thumb, occurring at about five cycles per second; this is called **pill rolling**. This name was given to the condition by doctors in olden days when pharmacists rolled pills by hand. The tremor continues during voluntary movements, rendering them clumsy and poorly controlled. If the disease of the basal gangla is unilateral, the tremor is unilateral and is confined to the musculature contralateral to the lesion.

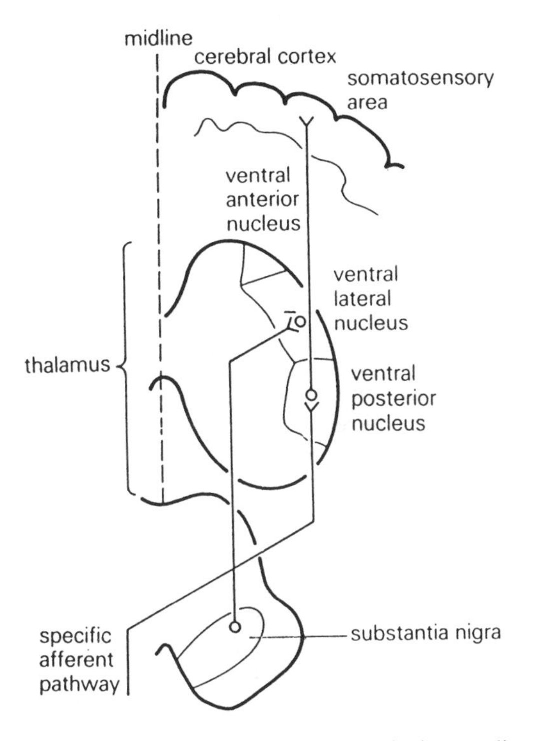

Figure 12.3 The projection from the substantia nigra to the intermediate nucleus of the thalamus. Interruption of this projection results in tremor.

Degeneration of the substantia nigra in Parkinsonian tremor

In healthy people, the substantia nigra projects in a predominantly inhibitory manner to the ventral lateral nucleus of the thalamus (Figure 12.3), holding its activity in check. The ventral lateral nucleus of the thalamus acting by itself is a **tremorogenic** zone; overactivity of this region results in tremor. In patients in whom tremor predominates, this nigrothalamic deficiency seems to be the important lesion. The tremorogenic zone is released from inhibition from the substantia nigra.

The pathway for the involuntary movements of Parkinson's disease passes via the precentral gyrus. In a subject with Parkinson's disease who subseqently suffers a destructive lesion of the precentral gyrus of the cerebral cortex, the involuntary movements are abolished.

Stereotaxic destruction of the ventral lateral nucleus of the thalamus (thalamotomy) results in a reduction or abolition of tremor. The operation is particularly effective when the tremor is unilateral, when thalamotomy is needed on one side only (the side opposite the tremor). The ventral lateral nucleus of the thalamus lies adjacent to the ventral posterior nucleus of the thalamus; this latter is the part of the thalamus which acts as the specific relay nucleus for the somatosensory projection. Great care has to be taken not to damage the specific relay nuclei of the thalamus. This operation frequently produces dramatic improvement although, in common with many neurosurgical operations, the subject is then deprived of two brain centres, in this case the substantia nigra, which is destroyed by the original disease, and the ventral lateral nucleus of the thalamus, destroyed by the operative procedure.

Rigidity

A patient with Parkinson's disease in whom rigidity is prominent shows excessive tonus in all the limb musculature, extensors and flexors alike. In this respect the rigidity is different from decerebrate rigidity, which is primarily an overactivity in anti-gravity muscles. In Parkinson's disease, when an observer attempts to move a limb passively in any direction, the movement is resisted. This is rather like the feeling when one tries to bend a lead pipe, which is the origin of the name **lead-pipe rigidity** for this type of rigidity.

In rigid subjects the stretch reflex is exaggerated. In a normal person, if an observer passively flexes and extends the wrist joint, asking the subject to let go, there is insignificant stretch reflex activity evident in the EMG recorded from muscles acting over the wrist. In a Parkinson's subject with rigidity in the wrist, the same manoeuvre results in a strong reflex contraction as a muscle is stretched. This is observed in both extensor and flexors, a corollary to the rigidity involving both flexor and extensor muscles.

In a patient with both rigidity and tremor, an attempt by the observer to bend a limb gives the feeling of a quick succession of interruptions of the rigidity rather than a smooth resistance to movement. This interrupted resistance is called **cog-wheel** rigidity.

In normal subjects, the substantia nigra sends an inhibitory projection to the neostriatum, via the nigrostriatal pathway (Figure 12.4). Degeneration of this pathway is a feature of Parkinson's disease; the striatum is released from inhibition and is free to work overtime to produce rigidity (Section 12.4). Once again, the clinical disability is primarily a release phenomenon.

Parkinsonian rigidity and chemical transmitters

The basal ganglia contain at least 40 different neurotransmitters and the list grows longer every month. We consider two neurotransmitters of importance, acetylcholine and dopamine. In the basal ganglia, acetylcholine is predominantly excitatory and dopamine predominantly inhibitory. A normal balance between dopamine and acetylcholine in the corpus striatum is needed for normal muscle tone.

The cholinergic innervation is probably from intrinsic cholinergic interneurones (Figure 12.4), although there is some evidence for the existence of a cholinergic input from the cerebral cortex. Dopamine is the immediate precursor in the biochemical synthesis of noradrenaline. In certain parts of the brain, notably the corpus striatum, catecholamine synthesis stops at dopamine and this amine is secreted as a synaptic transmitter. The pigmented cells of the substantia nigra contain dopamine. Their axon terminals release dopamine in the corpus striatum.

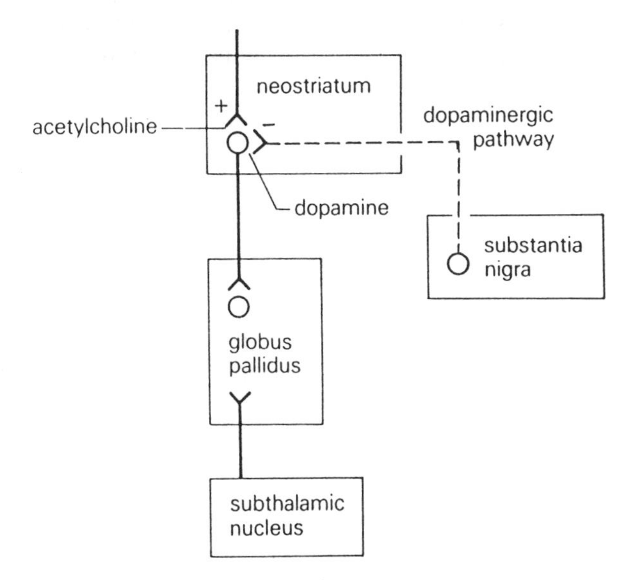

Figure 12.4 Simplified diagram of the role of dopamine and acetylcholine in the neostriatum.

In Parkinson's disease, the dopamine content of the presynaptic nigrostriatal terminals in the neostriatum is reduced by 50% or more. The Parkinsonian patient is helped by both boosting dopaminergic transmission and partial blockade of cholinergic transmission (by anticholinergic drugs). The former pharmacological manoeuvre is the more effective for most patients.

Administration of dopamine itself does not help the patient with Parkinson's disease because dopamine does not cross the blood–brain barrier. **Levodopa (L-dopa)**, which does cross this barrier, is the immediate precursor of dopamine and reduces rigidity. L-Dopa is the treatment of choice and specifically relieves rigidity; it does not improve tremor. Indeed, the effects of L-dopa are unpredictable and may vary with time. Parkinsonian patients treated with L-dopa for rigidity may develop involuntary choreiform movements not previously present. L-Dopa is contraindicated in patients with involuntary movements since these may be exacerbated by the treatment.

Hemiballismus

Hemisballismus is a condition of the basal ganlgia which is much rarer than Parkinson's disease. The correlation between the pathology and the disorder of function is, however, very clearly defined. The lesion of hemiballismus is in the subthalamic nucleus (Figure 12 .4); this causes an intermittent flinging movement, often associated with paralysis. In monkeys, destruction of 20% or more of the subthalamic nucleus results in hemiballismus; in humans lesions in or around the subthalamic nucleus produce the condition. This ballism can be partially or totally abolished by destroying the globus pallidus. Ballism is attributed to overactivity of the globus pallidus, due to removal of inhibition normally provided by the influence of the subthalamic nucleus on the globus pallidus.

Disorders of the basal ganglia as release phenomena

In retrospect, the reader will appreciate that most of the effects of malfunctioning of the basal ganglia are release phenomena. Involuntary movements and rigidity are an exaggeration of activity in centres released from control when centres that previously provided this control cease to function normally. Akinesia is the only feature that represents a primary deficit in function.

12.3 THE CEREBELLUM

Structure

In the whole brain removed from the body, the cerebellum is seen lying below the posterior pole of the cerebral hemispheres. The cerebellum consists of two hemispheres connected by the vermis (worm-like in appearance). The cerebellum is like a fist and, when spread it out, its appearance is as shown in Figure 12.5.

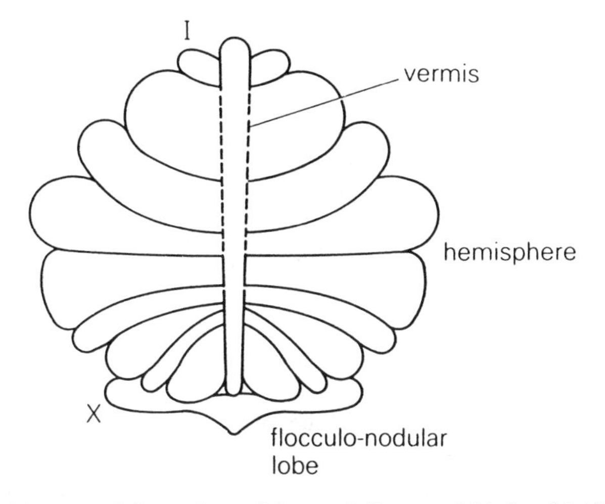

Figure 12.5 Diagram of the surface of the cerebellum, unfolded and laid out flat.

The cerebellum is anatomically self-contained, connecting with the rest of the brain at the level of the pons via tracts of fibres called the **cerebellar peduncles**. Each cerebellar hemisphere consists of a cortex, or surface layer, of grey matter which is highly convoluted and exhibits extensive transverse folds or folia, a medulla of white matter and, embedded in this white matter, four paired cerebellar nuclei.

The cortex of the hemispheres is further subdivided into a total of 10 lobes on either side, each with its own name. Modern anatomists have replaced these names with a system of numbering but lobe X is still known also by its former name of the **flocculo-nodular lobe**.

Functions and connections of the cerebellum

The cerebellum is concerned with balance and receives a profuse input from the vestibules. The cerebellum is also very important in coordinating the movements, particularly of fingers, where some of the finest of all movements in the body are executed. The cerebellum is also involved in the learning of skilled movements such as writing and playing the violin. It is therefore not surprising that the principal input from the spinal cord originates in proprioceptors, particularly those in the distal limb musculature and joints. The cerebellum was called the 'head ganglion of the proprioceptive system' by Sherrington.

The vestibular part of the cerebellum was the first to appear phylogenetically. Next to appear were the parts interacting with the spinal cord. The parts interacting primarily with the cerebral hemispheres developed most recently. This connection is important in the integration of motor commands from the higher motor centres.

Localization

The integrative role of the cerebellum is reflected in an extensive overlap of the projections. The regions to which the different sense organs project overlap each other. A local stimulus to the skin gives an evoked potential over a large area of the ipsilateral cerebellar hemisphere. Stimuli to two separate points of skin evoke activity over similar regions of the cerebellum. On the output side also, the cerebellum broadcasts far and wide. Strong stimulation of the cerebellar nuclei results in gross movements of the limbs. Attempts to draw maps showing the regions of the cerebellum connected with particular inputs and outputs are futile.

Despite this, there is some preference for different regions of the cerebellum to be targetted by a particular type of input and to project to particular structures. There is, for instance, some correspondence between the segmental level of input from the spinal cord and the region of densest projection to the cerebellum. The following account is of these principal connections.

Evolution

The oldest part of the cerebellum is the flocculo-nodular lobe. Since its connections are largely with the vestibular system, it is called the **vestibulocerebellum**. Next to develop, in evolutionary terms, was the vermis and contiguous parts of the cerebellar cortex. These regions receive extensive input from the spinal cord and are called the **spinocerebellum**. Most recently on the phylogenetic scale, the lateral parts of the cerebellar hemispheres developed pari passu with the development of the cerebral hemisphere, to which they are intimately related. The projections from the cerebral cortex to the cerebellum relay in pontine nuclei and from this is derived the name **pontocerebellum** to describe this part of the cerebellum. In order to emphasize the evolutionary sequence, the pontocerebellum is alternatively called the **neocerebellum** and the spino- and vestibulocerebellum are together called the **paleo-** or **archicerebellum**.

Connections

The cerebellar hemispheres are connected ipsilaterally to the spinal cord and contralaterally with the basal ganglia and cerebral hemispheres. We consider the connections of these different components in turn.

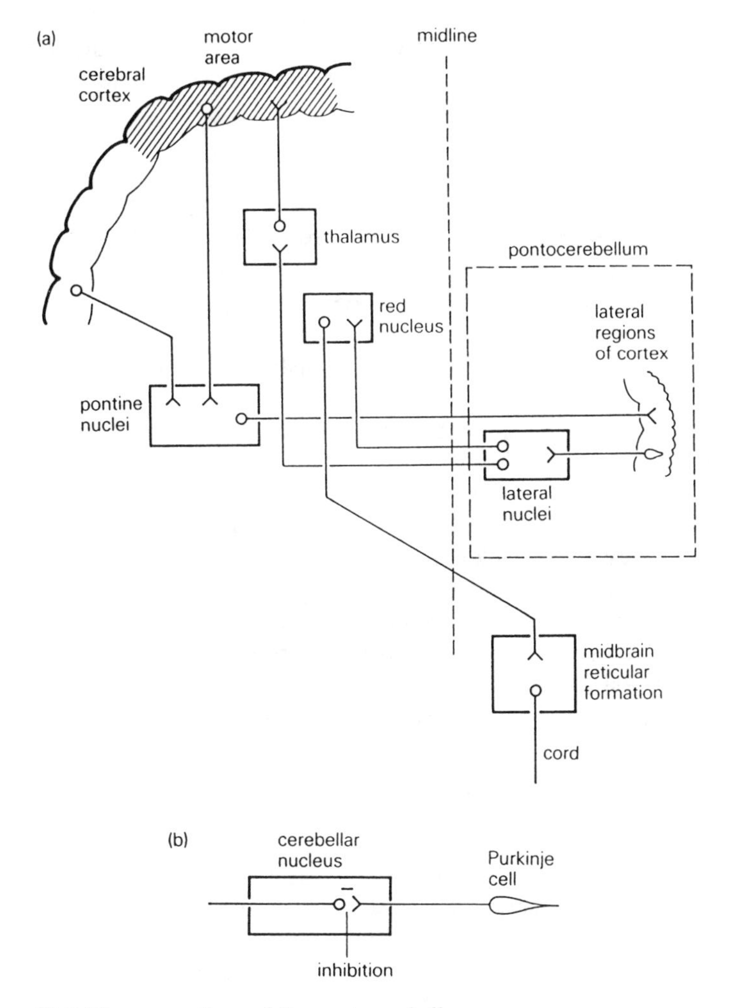

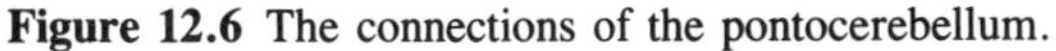

Figure 12.6 The connections of the pontocerebellum.

Connections of the pontocerebellum

Input

The cerebral cortex projects to all parts of the cerebellar cortex except the flocculo-nodular lobe, although by far the most dense projection is to the pontocerebellum. The principal input to the cerebellum is via pontine nuclei (Figure 12.6) although some cortico-cerebellar input is via the olive. This is a prominent nucleus in the hindbrain, the connections from which are exclusively with the cerebellum. It acts as a relay nucleus for some of the input to all three parts of the cerebellum, relaying from the cerebral cortex, the basal ganglia, and the cord. The olive on one side feeds to the cerebellar hemisphere of the opposite side. Although the connections are profuse, the functions of the olive are unknown.

Output

The Purkinje cells are the only output cells of the cerebellar cortex. The cerebellar cortex produces its effects by axons from Purkinje cells whose endings are purely inhibitory. The cerebellar Purkinje cells are spontaneously active, firing typically at 50 Hz, and they exert a tonic inhibitory effect on the cells of the cerebellar nuclei (Figure 12.6b). These cells in the cerebellar nuclei are themselves spontaneously active and their effects are excitatory.

The details of the connections from the cerebellar cortex to the cerebellar nuclei are still in doubt. All parts of the cerebellar cortex project to all the cerebellar nuclei and to the vestibular nuclei. There is, however, some preference for particular parts of the cerebellar cortex to project to selected cerebellar or vestibular nuclei. The pontocerebellum projects preferentially to the lateral cerebellar nuclei and thence both upwards and downwards. The upward projection is to the precentral cortex, with a synaptic relay in the thalamus. The downward projection is via the red nucleus contralaterally and hence, mainly via the reticular formation in man, via the rubro-reticulo-spinal tract to the spinal cord. This involves a double decussation.

Connections of the spinocerebellum

Input

This is from the spinal cord and from the basal ganglia. The basal ganglia, being more primitive than the cerebral cortex, project to the medial parts of the cerebellar hemispheres (Figure 12.7), mainly via the olive (Walberg, 1956).

Input via the spinocerebellar tracts from the spinal cord relays proprioceptive information. The afferent nerve fibres form synapses in the posterior horn on entering the cord. The second-order neurones form the spinocerebellar tracts which project directly to the cerebellum

Proprioceptive input to the cerebellum is also received via the cuneate nucleus, relaying information from the dorsal columns. Another group of fibres ascends from the cord as the spino-olivary tract, relaying to the cerebellum in the olive (Cajal, 1911).

Output

The spinocerebellum projects to the intermediate and roof nuclei of the cerebellum. Thence, they project via the lateral vestibular nucleus and vestibulospinal tract to influence the activity of motoneurones in the spinal cord.

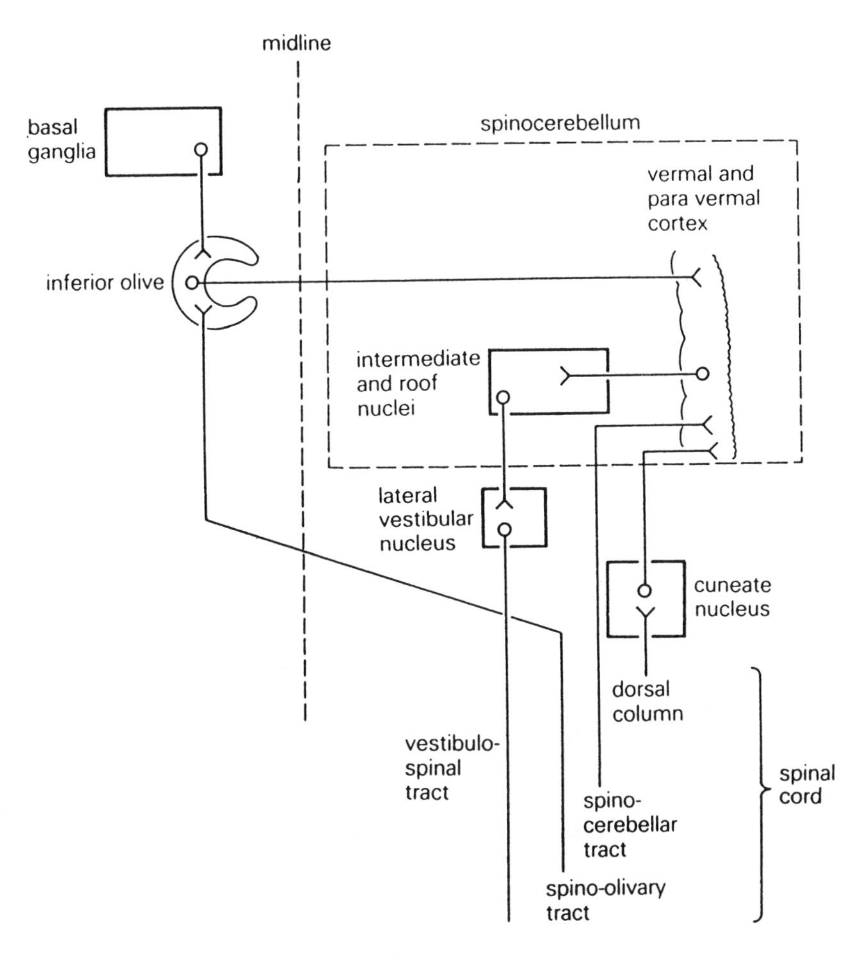

Figure 12.7 The connections of the spinocerebellum.

Connections of the vestibulocerebellum

The flocculo-nodular lobe, having to-and-fro connections with the vestibular system, is equivalent to a vestibular nucleus. Input to the flocculo-nodular lobe is both direct from the vestibule and also indirect via the lateral vestibular nuclei (Figure 12.8). The flocculo-nodular lobe projects directly to the vestibular nuclei without traversing the cerebellar nuclei.

The vestibular input to the vermal part of the vestibulocerebellum is mainly indirect via the lateral vestibular nucleus. The output is principally via the lateral vestibular nucleus, with an additional minor contribution mediated by the direct pathway.

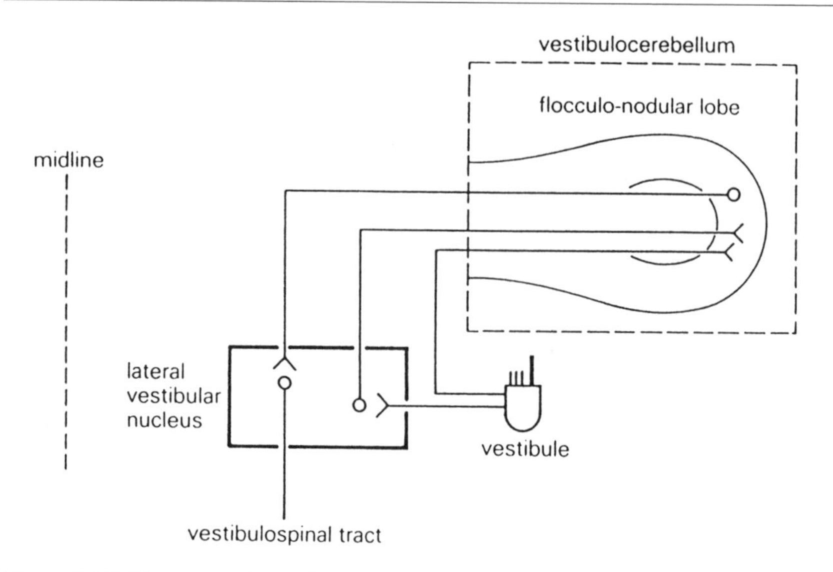

Figure 12.8 The connections of the vestibulocerebellum.

The paleocerebellum and the neocerebellum work together, as evidenced by the continuance up into the hemispheres of the fissures of the vermis. Despite the overlapping of projection, it is possible to detect some anatomical topical organization in the projection from the cerebellar hemispheres to the lateral vestibular nucleus. The latter in turn has a topical projection to the spinal cord. Its rostral part projects to the cervical segments and its caudal part to the lumbar segments of the cord. This is an anatomical substrate for a somatotopical projection from the cerebellar hemispheres to the cord (Voogd, 1964).

Structure of the cerebellar cortex (Figure 12.9)

The surface of each hemisphere consists of grey matter. The deep surface of this grey matter connects with the white matter, which consists of nerve fibres to and from the cortex. The white matter contains a series of cerebellar nuclei. The flow of information in the cerebellum is from other parts of the central nervous system to the cerebellar cortex, thence to the cerebellar nuclei, and finally from these nuclei to extracerebellar regions.

The cerebellar cortex contains cells of several types, the largest of which are the **Purkinje cells**. These cells have flask-shaped bodies and dendrites which ramify profusely. This ramification is confined to a plane transverse to the folium. The Purkinje cells lie in a sheet in the middle of the cortex and can be seen in a transverse section of the cerebellum as a row of cells. They are the only cells of the cerebellar cortex which project away from the

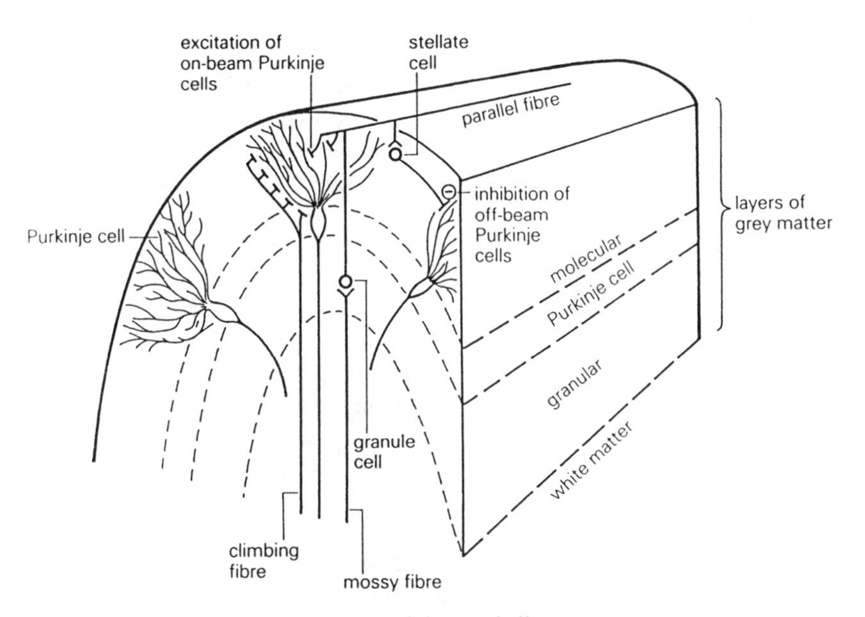

Figure 12.9 Diagram of the structure of the cerebellar cortex.

cortex. They are inhibitory and end synaptically on cells of the cerebellar nuclei (Figure 12.6b). The inhibitory transmitter chemical is γ-aminobutyric acid (GABA). There are cortical layers superficial and deep to the Purkinje cell layer, giving the cerebellar cortex a three-layered appearance. The outermost layer is called the **molecular layer** because of the large number of fine nerve fibres therein. The innermost layer is called the **granular layer**, because it contains small neurones with the appearance of granules. With the exception of the Purkinje cells, all the neurones of the cerebellar cortex are interneurones, i.e. neurones whose axons terminate within the cerebellar cortex itself.

The afferent nerve fibres to the cerebellar cortex are of two types, climbing fibres and mossy fibres.

Climbing fibres

These arise mainly from the cerebellar nuclei. Climbing fibres also originate in the olive. Each climbing fibre winds around its target Purkinje cell, making between 60 000 and 100 000 synapses; the neurotransmitter is glutamate and the post-synaptic receptors are of the n-methyl-D-aspartate variety (Chapter 17). Because of the intense synaptic depolarization, a single action potential in the climbing fibre evokes a so-called 'compound spike' which is a train of action potentials.

Mossy fibres

These arise from the pontine nuclei which are in turn influenced by many extracerebellar centres. Mossy fibres are so called because they terminate in moss-like synapses on the dendrites of granule cells. The axon of a granule cell ascends to the outermost layer of the cerebellar cortex and divides into two terminal branches which run in opposite directions along the long axis of the folium. All of the axons of the granule cells of a particular folium are parallel with each other: they are therefore called parallel fibres. These fibres form excitatory synaptic connections with the apical dendrites of Purkinje cells. For each Purkinje cell, the apical dendrites form a dense ramification across the folium, rather like a fan. Hence the parallel fibres and the apical dendritic trees of Purkinje cells are at right angles to each other. One parallel fibre forms a single synapse on the apical dendrite of the Purkinje cell. An action potential in one parallel fibre gives a single Purkinje cell action potential.

Granule cells are the only excitatory neurones of the cerebellar cortex. Activity in a granule cell results in excitation of a row of Purkinje cells on the 'beam' of parallel fibres from that granule cell. The parallel fibres also excite basket cells and stellate cells, which lie in the outer layer of the cerebellar cortex. These latter two types of cell exert inhibitory synaptic effects on Purkinje cells off-beam; activity in a granule cell thus monosynaptically excites the Purkinje cells on-beam and polysynaptically inhibits the Pukinje cells off-beam. The mutually perpendicular arrangement of parallel fibres and Purkinje cell dendrites would be propitious for learning, and models based on this arrangement are being investigated. Despite our detailed knowledge of the anatomical layout of the cerebellum, little is known of its functional significance.

In the absence of information about how the cerebellum performs its functions, speculations abound. Mossy fibres carry information from many sources in the central nervous system, and their role is likely to be to modulate the activity of Purkinje cells in a fine discriminative fashion according to afferent input from proprioceptors and motor commands from the cerebral cortex. An action potential in a climbing fibre (from the cerebellar nuclei) elicits a burst of action potentials from the Purkinje cell which it innervates. The number of action potentials in the burst is thus regulated via the mossy fibre input by the balance of activity throughout the central nervous system. It is as if the cerebellar nuclei are saying 'How are you Purkinje cells up there getting on?' If there is a background of excitation in the Purkinje cells, they generate a long train of impulses. If the Purkinje cells are relatively unresponsive, the output is a single impulse. In this way activity initiated by the climbing fibres is modulated in the cerebellar cortex by a wide variety of integrative effects from the periphery and from the cerebral cortex.

Function of the cerebellum

The Purkinje cells of the cerebellum only start to discharge towards the end of a sustained movement. It has been suggested that the cerebellum has the function of calculating the difference between the position of the limb and the target, and provides a signal related to this 'error', or the amount of movement still needed. The cerebellum sends this information up to the movement generators in the motor region of the cerebral cortex for limb movements, to the vestibular nuclei for eye movements, etc., and changes their gain appropriately. The cerebellum is thus acting as a calibrating centre.

The clinical corollary is that it is towards the end of a voluntary movement, when the limb needs final finely controlled guidance, that the movement disorder in the form of an intention tremor (see below) becomes obvious.

The effects of deficiency of the neocerebellum

Deficiency of neocerebellar function is characterized by disorders of coordination, particularly for fine, skilled movements (Table 12.4). Since many of these fine skilled movements, such as writing, are performed by the fingers, it is the distal limb musculature which is most affected in cerebellar lesions.

Table 12.4 Cerebellar deficiency

Effects of deficiency of the neocerebellar function
Disorders of coordination, mainly in the distal musculature
Reflex movements
Hypotonea and hyporeflexia
Pendular tendon jerk
Cerebellum is an anti-hunting device
Voluntary movement
Dysmetria: the range of movement is inappropriately large e.g. past pointing
Intention tremor
Dysdiadochokinesis: difficulty with alternating movements.
Nystagmus
Effects of deficiency of vestibulocerebellar function
Cause: medulloblastoma, a childhood malignant tumour
Effect: trunk ataxia

Muscular contractions are weak and easily fatigued. There is muscular hypotonia, and diminished tendon reflexes; this is called **hyporeflexia**. The tendon jerk reflexes may show evidence of incoordination in that a diminished jerk is followed by a series of oscillations; such a response is called **pendular** tendon jerk since it has similarities to the swinging of a pendulum. In a normal individual such oscillations are damped by the cerebellum.

Spinal cord circuitry involves many feedback loops. One is the loop from spindle stretch receptors to the α motoneurones and back to the extrafusal fibres. Any system including feedback is liable to oscillate. A domestic example is the refrigerator, the temperature of which oscillates around the value to which the thermostat is set. This oscillation is often called 'hunting': The system 'hunts' for the value at which the thermostat is set. In the central nervous system, it is the cerebellum which damps out such oscillations; the cerebellum is regarded as an anti-hunting device.

Effects on voluntary movement

Voluntary movement is affected in many ways by neocerebellar deficiency. There is **dysmetria**, the range of movement being inappropriately large in relation to its objective. A particular example is **past-pointing**; when the eyes are closed and the patient is asked to put his finger on his nose, the finger goes right past. Even with the eyes open, there is **intention tremor**, a tremor accompanying voluntary movement and making movement clumsy. Intention tremor becomes most prominent as the termination of a goal-directed motor activity is being approached. There is no tremor at rest.

Dyssynergia is a decomposition of movements. For example, when raising a cup of tea to the lips, the subject is liable to raise the elbow and flex the arm in the wrong order. The force and rate judgement are lost. During walking, the foot strikes the ground hard. A related disorder is **dysdiadochokinesis**, in which the subject finds it very difficult to perform rapidly alternating movements such as alternate pronation and supination of the forearm.

If the patient flexes his elbow against the observer's hand and this resistance is removed, the normal halting of the flexion does not occur; the patient may hit himself in the face; this is called **rebound**. Movements associated with normal strong voluntary effort may be exaggerated, for instance there may be grimacing during speech.

Ocular disturbances also occur. Movements of the eyeballs are amongst the finest movements in the body. In common with other finely controlled movements, co-ordination of the extrinsic ocular muscles depends on cerebellar function. When the cerebellum is defective, there is nystagmus. This may occur with the subject looking ahead, or it may be initiated by asking the subject to look to one side. Unilateral lesions of the cerebellum cause nystagmus which is more severe if the subject looks towards the side on which cerebellar activity is deficient than if he looks to the opposite side. Speech is also a highly complex motor act requiring fine control. With impairment of cerebellar function, speech is jerky and explosive; syllables are separated. Gait is ungainly; the subject tends to stagger towards the side of the lesion.

Immediate and long-term effects of neocerebellar insufficiency

Immediately after damage to the neocerebellum, the signs just described are pronounced. However, they do not persist. Even following irreversible damage to the cerebellum, the signs of cerebellar deficiency slowly subside. After a few weeks, there are minimal signs. The mechanism of this recovery of function is unknown. It has been suggested that the cerebellum teaches another part of the brain, probably the cerebral cortex, how to do skilled acts and that this information is stored in a latent form in the cerebral cortex. When the cerebellum is no longer acting to instruct other centres in the production of a skilled motor act, this information stored elsewhere gradually becomes available. Whatever the mechanism, this recovery of function suggests that other parts of the nervous system can compensate for loss of neocerebellar cortical function.

The effects of deficiency of the flocculo-nodular lobe

A lesion of the flocculo-nodular lobe may be caused in infants and children by a malignant growth called a medulloblastoma. It grows rapidly and is almost uniformly fatal.

The flocculo-nodular lobe is functionally a vestibular nucleus and impaired function of this part of the cerebellum results in disturbances in the motor control of the muscles of the vertebral column and trunk, muscles that are primarily responsible for the reflexes of balance. The effect is known as **trunk ataxia**. There is swaying and staggering as the child tries to walk (Ruch and Patton, 1979). Characteristically, the child learns to wedge his trunk into a corner and, from this support, can operate his limbs in a relatively normal fashion.

The effects of the cerebellum on tonus in skeletal musculature

This will be explained in the context of the interactions of nuclei which were considered in explaining the genesis of decerebrate rigidity. This layout has been copied onto Figure 12.10 and the cerebellar input has been added. Electrical stimulation of the cerebellar cortex causes a reduction in tonus in skeletal musculature in the limbs. This is particularly pronounced when the level of muscle tonus is abnormally high, as in decerebrate rigidity. In such subjects, electrical stimulation of the cerebellar cortex reduces or even abolishes the rigidity.

The mechanism for this reduction in tonus is that activity of the cerebellar cortex leads to reduction of firing of the cells of the cerebellar nuclei. There is an excitatory projection from certain of these cerebellar nuclei to the lateral vestibular nucleus. Stimulation of the cerebellar cortex causes a reduction of activity in the cerebellar nuclei and hence a reduction in activity

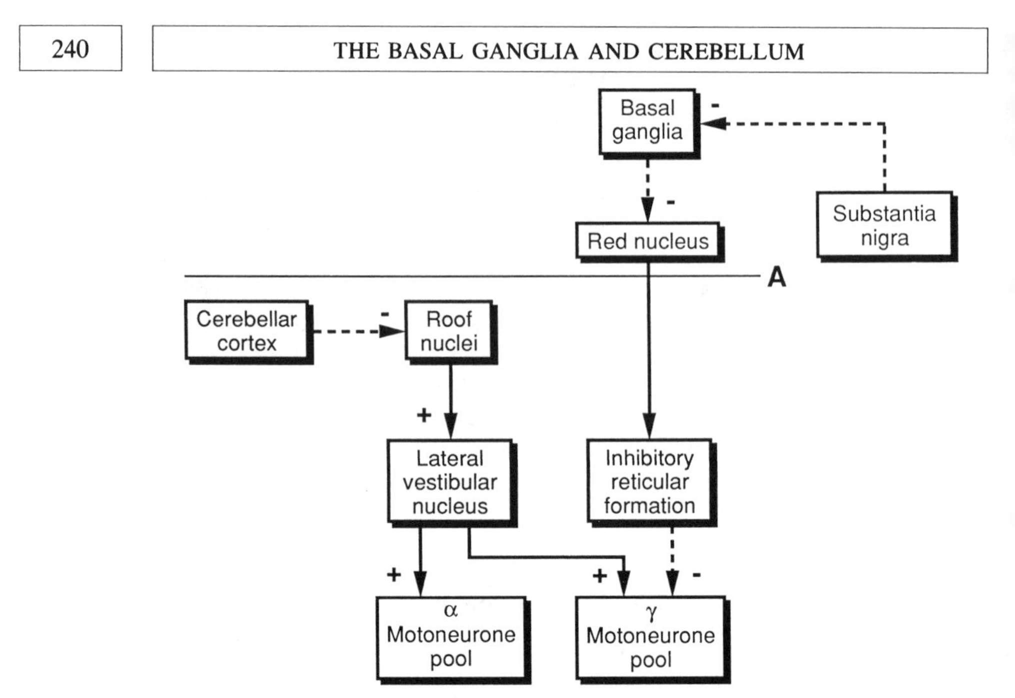

Figure 12.10 Diagram derived from Figure 11.2 to illustrate the effects of the cerebellum on activity in motoneurones. The line A indicates the level of a mid-collicular decerebration; this produces a predominantly γ rigidity due to removal of excitation projected from the red nucleus to the hindbrain inhibitory reticular formation. When in addition the cerebellar cortex is rendered non-functional, the removal of tonic inhibition from the cortex to the cerebellar nuclei, particularly the roof nuclei, results in unleashing of spontaneous activity in the latter nuclei, over-excitation of the lateral vestibular nucleus and the occurrence of α rigidity.

in the lateral vestibular nucleus. This depression of activity in the lateral vestibular nucleus is the origin of the reduced level of activity in the spinal cord.

Cerebellar lesions and muscle tonus

When the cerebellar cortex and nuclei are both damaged, the most common situation in naturally occurring disease, hypotonia and hyporeflexia result, due to removal of the normal excitatory bombardment of the lateral vestibular nucleus from the cerebellar nuclei (Figure 12.10). In rare cases, however, only the cerebellar cortex is damaged, leaving the cerebellar nuclei intact and active. Inhibition of spontaneous activity of the cerebellar nuclei by the cerebellar cortex is then lost and the cerebellar nuclei discharge is uninhibited. This causes excitation of the lateral vestibular nucleus, as shown in Figure 12.10. The result is an excessive excitation of α and γ motoneurone pools with a consequent hyper-reflexia and stiff-leggedness as part of an increase in the supporting reaction. This is the exact converse of the effects of loss of all cerebellar function. This rigidity persists even if conduction in the dorsal roots innervating a limb is blocked. This identifies the rigidity as being α rigidity and emphasizes the difference between the rigidity due to removal of inhibition from the inhibitory reticular formation (projecting principally to γ motoneurones) and the rigidity due to overactivity of the lateral vestibular nucleus, projecting directly to α motoneurones.

This is one of the very few situations in which a rationale can be advanced to explain opposite effects occurring in two patients with disease processes involving approximately the same part of the brain.

12.4 APPENDIX A: NIGRAL INSUFFICIENCY AND RIGIDITY

The neurones projecting from the globus pallidus to the red nucleus are inhibitory. Figure 12.10 shows the pathway from the substantia nigra to the motoneurones. In traversing this route, there are three inhibitory links in the chain. This means that activity in the substantia nigra tends to inhibit the motoneurones, this accounting for the clinical observations of spasticity accompanying lesions of the substantia nigra.

12.5 APPENDIX B: THE CEREBELLUM AND DECEREBRATE RIGIDITY

In Chapter 11, it was explained that decerebrate rigidity is primarily a γ rigidity due to removal of the tonic descending inhibition from inhibitory reticular formation to the spinal cord. The midbrain lesion leading to decerebrate rigidity leaves intact many of the connections of the cerebellum and the spinal cord and we have just seen that the cerebellum makes its own contribution to muscle tonus. If in addition to the destruction of the brain stem in the mid-pontine region (causing decerebrate rigidity) there is also destruction of the cerebellar cortex, the rigidity becomes an α rigidity (Matthews, 1972). This is shown by the fact that the rigidity is no longer abolished by interruption of the dorsal roots. The mechanism of this conversion from γ to α rigidity is the removal of the tonic inhibitory influence of the cerebellar cortex on the cerebellar nuclei (the roof nuclei are particularly important in this context). As a result, these nuclei discharge excessively, excite the lateral vestibular nuclei and they in turn excite α motoneurones.

There are other ways of blocking the activity of the cerebellar cortex, thereby allowing the expression of α rigidity. One way is to cool the cerebellar cortex, thus putting it temporarily out of action but leaving the cerebellar nuclei active. Another situation in which this combination of effects is produced is called **anaemic decerebration**. If the carotid artery on either side and the basilar artery are occluded, most of the brain is destroyed through deprivation of its blood supply. The structures cephalad of the midbrain are destroyed, as is the red nucleus and the cerebellar cortex, but the cerebellar nuclei survive. In anaemic decerebration, there is an α rigidity, in contrast with the γ rigidity seen in the classical decerebrate rigidity resulting from a lesion confined to the midbrain.

The supraspinal control of skeletal muscle

In Chapter 7 it was noted that all voluntary movements which have been investigated have been shown to involve coactivation of both α and γ motoneurones. There are situations in which supraspinal structures produce opposite, instead of similar, effects on α and γ motoneurones. The best known example is the situation just described in which the cerebellar cortex is rendered inoperative. Cooling or destroying the cerebellar cortex increases the frequency of discharge of α motoneurones in the cord, but decreases the frequency of discharge of γ motoneurones there. This example shows that the neuronal connections exist for a central structure to excite one of the types of motoneurone while inhibiting the other. This leads to the speculation that centres such as the cerebellum may beam their commands to one or other of the types of motoneurone rather than commanding coactivation of both types of motoneurone.

The pathway of this reciprocal innervation of α and γ motoneurones is not known. The projection from the lateral vestibular nucleus is, as we have seen, excitatory to both types of motoneurone. There are other pathways from the cerebellum to the cord, via the nuclei of the reticular formation in the lower pons and it is possible that some of these nuclei project in an excitatory fashion to the α motoneurones and in an inhibitory fashion to the γ motoneurones.

FURTHER READING

Evered, D. and O'Connor, M. (1984) Ciba symposium 107 *Functions of the Basal Ganglia*. London, Pitman.

The cerebral hemispheres: general features, localization of function, the limbic system

13

13.1 THE STRUCTURE OF THE CEREBRAL HEMISPHERES

Figure 13.1 shows the lobes and fissures of the cerebral hemispheres. The main subdivision are due to the **central fissure** (or fissure of Rolando) and the **lateral fissure** (or fissure of Sylvius). These divide the hemisphere into four major lobes: frontal, parietal, temporal, and occipital. In the cerebral cortex, the cortical functional areas are organized into the different gyri; these are shown in outline in Figure 13.1b and c. In order to expose the auditory areas, the temporal lobe has been displaced laterally in Figure 13.1b.

The two hemispheres are connected by a thick bundle of nerve fibres called the **corpus callosum** (Figure 13.1c). Note that in Figure 13.1b and c all the sensory areas (apart from the olfactory area) lie behind the central fissure. The somatosensory area lies just behind the central fissure, the visual area lies in the occipital lobe, the auditory areas on the in-folded aspect of the superior temporal gyrus, and olfaction is situated in the most medial parts of the cortex, which are phylogenetically very old.

Different areas of the cortex perform different functions. The **primary sensory areas** (S1, V1, and A1) are those receiving direct input along the specific afferent pathways. Next to each primary area is a **secondary** area (S2, V2, and A2) to which the primary area projects. These areas are involved in analysing the information received by the primary sensory areas. The motor areas are similarly subdivided into the primary motor area (M1), which projects to the motoneurones and the secondary motor areas (M2), which are connected to M1 and are involved in integrating motor activity.

The remainder of the cortex, and this is the majority of cortex in the human brain, is called **association cortex**. This includes the speech areas, which are considered in the next chapter.

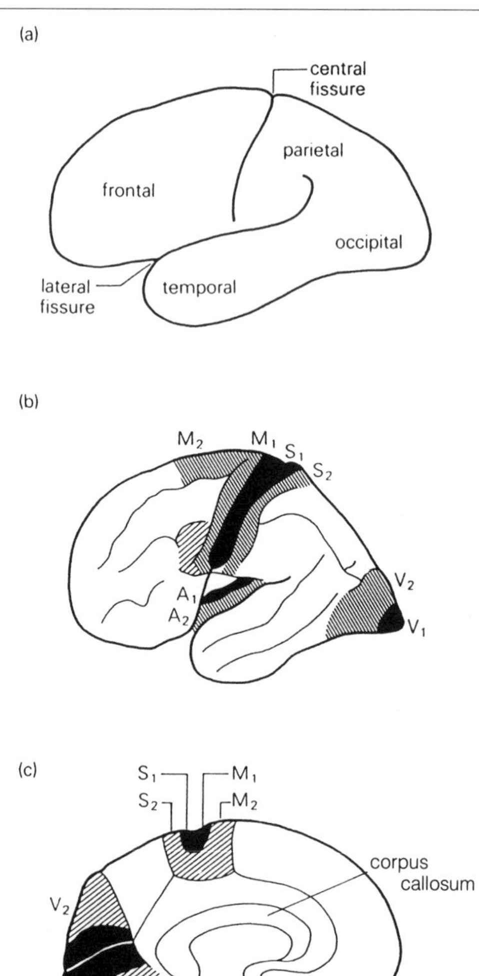

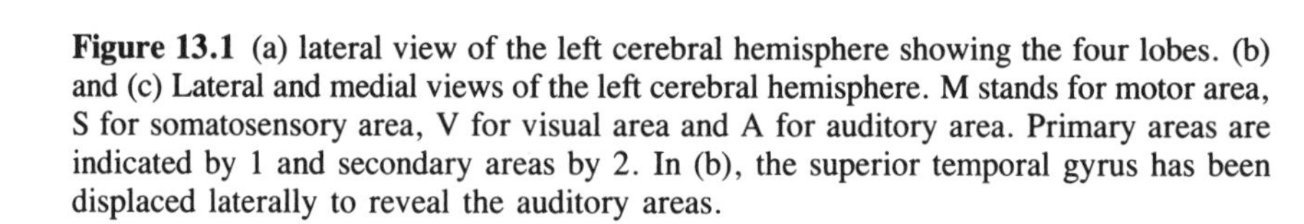
Figure 13.1 (a) lateral view of the left cerebral hemisphere showing the four lobes. (b) and (c) Lateral and medial views of the left cerebral hemisphere. M stands for motor area, S for somatosensory area, V for visual area and A for auditory area. Primary areas are indicated by 1 and secondary areas by 2. In (b), the superior temporal gyrus has been displaced laterally to reveal the auditory areas.

Cerebral cortex: evolutionary considerations (Table 13.1)

Lower animals have no cerebral cortex. When the cerebral cortex first appears in evolution, it is flat (lissencephalic). As more and more cortex develops, its thickness alters little but its surface area proliferates resulting in in-folding to yield **gyri** which are separated by **sulci**.

Table 13.1 The cerebral cortex

Paleocortex (old cortex)
- Rhinencephalon (nose brain)
- Part of limbic system

Neocortex (new cortex)
- Primary sensory and motor areas
- Secondary sensory and motor areas
- Association areas:
  - Anterior frontal
  - Parietal
  - Temporal

Myelination follows the phylogenetic sequence.
Myelination of the association areas is not complete until the age of 20 or more.

The cerebral cortex has developed as outgrowths of various older parts of the central nervous system. Initially, this development was primarily concerned with the sensation of smell. In animals at this stage, smell was by far the most important of the special senses. These oldest parts of the cortex are alternatively known as **paleocortex**, meaning old cortex, or **rhinencephalon**, (literally 'nose-brain'). These parts of the cortex are close to and functionally intimately related to other old parts of the brain, such as the hippocampus and the amygdaloid nucleus of the basal ganglia. Altogether, this system of nuclei and of tracts joining the nuclei is called the limbic system, to which a later section of this chapter is devoted.

All the parts of the cortex that developed subsequently to the paleocortex are called **neocortex**. At this stage, other senses were replacing smell as the most important sense. Vision in particular became dominant: vision gives much more precise information about localization of stimuli than does smell; it also gives information about distant features of the environment. The first neocortical regions to develop were connected with the primary sensory input (other than smell) and motor output. It is as if subcortical structures needed extra computing power and appended to themselves some more circuitry. So, initially, the **primary somatosensory** and **primary motor** areas appeared. Next to appear were the secondary sensory and motor areas. In mammals up to and including the cat, most of the cortex is sensory or motor cortex.

Gyri and sulci

As more and more cortex became needed by the evolving brain, its surface area was increased by it growing around the rest of the brain. A stage was reached when the surface area of the cortex matched that of the whole of the available inner aspect of the skull. Then in-folding began. We do not know why the cortex cannot increase its computing capacity by becoming thicker rather than by becoming more extensive. The increase in surface area could only be achieved by in-folding and at this stage, the cortex was no longer flat (as in the rat and rabbit); it became convoluted (as in the cat). The higher one ascends the evolutionary tree, the more convoluted the cortex becomes.

Association cortex

Very recently in phylogenetic development, the great increase has been in the association areas. This name derives from the assumption that the function of these areas is to form associations amongst features of sensory input and to convey the information that they have extracted to the motor areas. This nomenclature is not approved of by all, because it implies functions of which we are not certain. The areas comprise the anterior frontal, parietal, and temporal regions. We know less of the mechanisms of association cortex than we do for sensory and motor areas. The size of the association areas and the complexity of cortical neuronal circuitry are the most obvious distinguishing features between humans and non-human species.

Primary receiving and motor areas, which bulk so large on the convexity of the monkey brain, are crowded down into the fissures in humans. Of the total surface of the human hemisphere, two-thirds of the cortex is infolded into the sulci, with only one-third occupying the convexity of the cortex. This compares with a mere 7% being infolded in the lowest monkey. The proliferation of the association areas in evolution is exemplified by the fact that the prefrontal area of the association cortex constitutes 3.5% of the cat brain, 17% of the chimpanzee brain (probably the most intelligent primate next to man) and 29% of the human brain.

Embryological development of the human cerebral cortex

Ontogeny recapitulates phylogeny: during development, an embryo goes through stages which are very similar to stages in the embryonic development of its evolutionary forebears. Up to the fourth month of human embryological development, the cortex is flat. It then proliferates and convolutions develop. Myelination also follows the phylogenetic sequence. The corticospinal tracts become myelinated relatively late; they are not fully myelinated until between 6 and 12 months of extrauterine life. The Babinski response assumes the normal adult form at the same stage, reflecting the fact that the pyramidal tract is late to develop in evolution.

The sequence of myelination in the neocortex also follows that of evolution, the primary sensory and motor areas acquiring myelin sheaths first, followed by secondary sensory and motor areas, and finally by the association areas. Myelination of the association areas is not complete until the age of at least 20 in man (Weinberger, 1987). Mental prowess is influenced by two competing factors: it is enhanced as myelination proceeds and it is depressed by neuronal death. In every one of us, brain cells die at the rate of about 30/min or 2000/h from the earliest childhood years onwards; the total deficit is 10% by 60 the age of years in normal subjects. In patients with cerebral arteriosclerosis, the rate of death of cortical cells is much higher. In normal people until the age of about 20, the effect of myelination outweighs that of cell death but, when myelination ceases, mental prowess starts to decline. From the age of 20 onwards, it is downhill all the way, with our mental faculties progressively deteriorating.

Microscopic structure

The neurones of the cortex are of two types, pyramidal and stellate.

Pyramidal cells

Pyramidal cells have cell bodies which are roughly conical in shape (Figure 13.2). Arising from the point of the cone is a long dendrite, called the *apical dendrite*, which projects up almost to the pial surface of the cortex. The apical dendrite of cell bodies lying deep in the cortex spans most of the thickness of the cortex. This dendrite receives synaptic input in all layers of cortex through which it runs. Around the base of the conical cell body arise shorter *basal dendrites* which ramify around the soma. From the centre of the base of the cell body arises the axon, which projects down towards the subjacent white matter. The axon gives rise to many recurrent collateral branches which project to adjacent regions of cortex. Many pyramidal cells send axons to the white matter, whence the axons travel to other regions of cortex, to subcortical nuclei, or to the spinal cord. The pyramidal cells are the output cells of the cortex.

The apical dendrite and axon of a pyramidal cell lie along a line perpendicular to the cortical surface. The axes of adjacent pyramidal cells are parallel so that pyramidal cells in one region of cortex are all lined up.

Stellate cells

There are many varieties of stellate cells, but they all have star-shaped cell bodies, which give them their name. They are of many sizes; a large one and a small one are shown in Figure 13.2. Their dendrites ramify less than those of pyramidal cells. There is no long dendrite like the apical dendrite

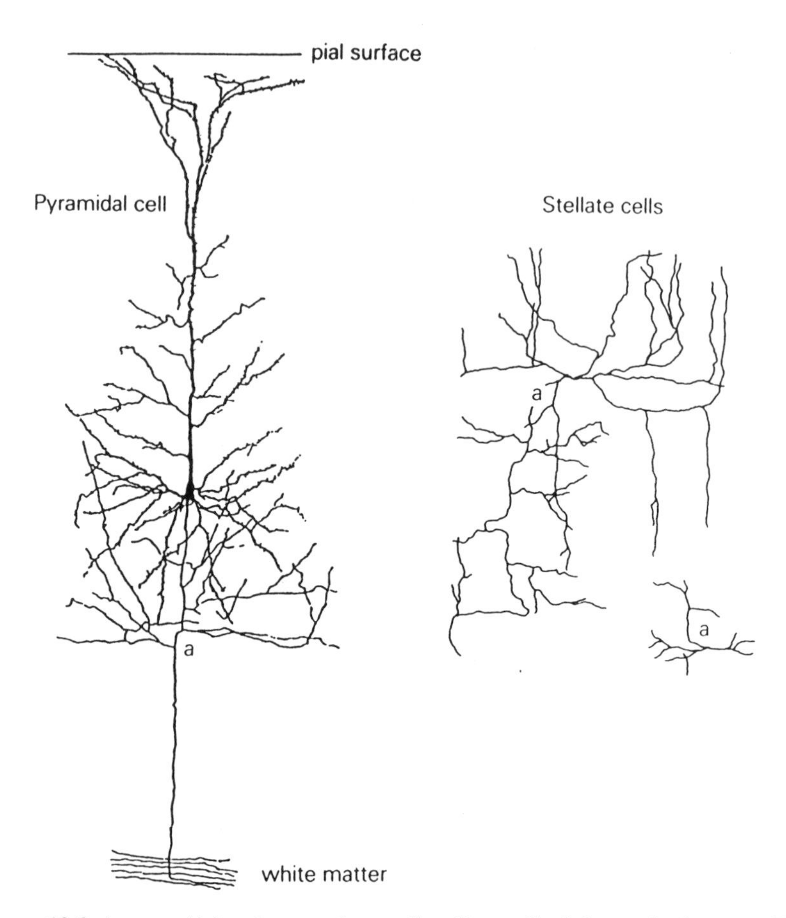

Figure 13.2 A pyramidal, a large and a small stellate cell of the cerebral cortex. For each cell, a indicates the axon.

of the pyramidal cell and the processes of these cells are not lined up, as are those of pyramidal cells. Their axons travel relatively short distances.

As we have had occasion to observe before, the feature that most obviously separates man from the non-human species is the size and development of the neocortex. The human neocortex has a far greater density of stellate cells than does other species. The histological variety and complexity of individual stellate cells is also much greater in humans; the pyramidal cells of the human brain can, by contrast, scarcely be distinguished from those of the higher apes. It is to our stellate cells that we owe our supremacy!

Cortical lamination

The neocortex is laminated: six layers are identifiable, and these are labelled

I to VI, starting at the pial surface. The layering reflects differences in distribution of the different cell types. Pyramidal cells occur most abundantly in layers II to III and V to VI. The apical dendrite runs up to layer I – the pyramidal cells of layer V have a bifurcation of the apical dendrite in layer IV. Layer IV is packed with stellate cells but is free of pyramidal cells. Pyramidal cells of layers II and III send axons mainly to other areas of cortex, those of layers V and VI project to subcortical structures. Different deep layers project to different subcortical sites.

Connections

In the human brain, 99% of the cells in the cortex send their axons only to cortex, either close to their cell bodies or far away. Only 1% send axons to subcortical structures such as the spinal cord. The cerebral cortex is largely a self-contained computer, with a small proportion of cells concerned directly with output.

Gross histological differences amongst various cortical regions

The primary receiving areas, particularly the primary visual cortex, have a thickened layer IV which is packed with stellate cells. As a result, these areas of cortex are called *granular*. The primary input in the receiving areas is onto stellate cells of layer IV and thence to pyramidal cells. There is also direct connection of afferents onto the pyramidal cells. By comparison the motor cortex has no layer IV; this is called *agranular* cortex.

Association cortex is intermediate in histological structure between receiving and motor cortex, and has a moderately developed layer IV. Most of the cortex in humans is association cortex and its structure is regarded as 'typical' cortex. This is the origin of the name *homotypical* cortex as an alternative to 'association' cortex. Some neuroscientists prefer the term 'homotypical' because it does not prejudge the function of the cortex.

Modular organization

The vertical stratification into laminae was known to anatomists and histologists of old. Mountcastle (1957) was the first to find evidence of a mosaic of organization across the surface of the cortex. He found that cortical cells responding to a particualr type of stimulation were grouped together in what he termed *columns* within the primary somatosensory areas. For instance, all the cells in one column would be excited by touch applied to a localized area of the skin. These cells would not respond to hot or cold. In a neighbouring column, all the cells would respond to changes in temperature.

When microelectrode penetrations are made normal to the cortical surface, all neurones encountered during one penetration respond to stimuli of the same modality because the electrode stays within the boundaries of one column. An oblique penetration detects cells from different columns. Physiologically, this is shown by the fact that for a few tenths of a millimetre during the electrode penetration all the neurones respond to stimuli of one modality but that suddenly, as the tip of the electrode advances across the boundary between columns, there is an abrupt change in the modality of stimulation to which the neurones respond. By measuring the obliquity of penetration and the travel of the microelectrode when recording from cells in a given column, Mountcastle was able to estimate the diameter of a column as being between 100 and 500 μm. Later, Hubel and Weisel (1977) showed that the primary visual area of the cortex has a columnar arrangement in that cells with similar response characteristics occur together. This is described in Chapter 14.

13.2 THE DISCOVERY OF THE MOTOR FUNCTIONS OF THE PRECENTRAL GYRUS

Hughlings Jackson, a London neurologist, recognized a century ago that there was a correlation between the site of a cortical lesion and the site of resultant malfunction of musculature. This information was obtained from the character of epileptic fits and the site of the epileptic focus (the site where the abnormal discharge starts). About 50% of patients who have cortical epilepsy experience an aura or forewarning before the full-blown fit occurs. The aura gives a clue as to where the epilepsy is originating. When the precentral gyrus is the seat of the focus, the aura consists of involuntary and uncontrollable contractions, initially confined to the group of muscles appropriate to the site of the focus. A lateral lesion causes twitching of the mouth and tongue; a lesion nearer the vertex causes twitching of the thumb; whereas a lesion on the medial aspect of the hemisphere is associated with twitching in the foot. The convulsions then spread to involve neighbouring musculature until the whole of the side of the body is showing convulsions. Jackson attributed the spread to successive involvement of cells in fields of neurones representing different muscles or movements. Some patients can abort the attack by vigorous rubbing of the region involved in the jerking. The sensory projection from the periphery to the motor strip can therefore inhibit elements of the pathway whose activity is responsible for the pathological synchonization.

At about the same time, the German neurologists Fritsch and Hitzig (1870) first described that direct electrical stimulation of the precentral cortex of the dog caused movements of musculature on the opposite side of the body. Similar stimulation of other parts of the cortex was without discernible effect. A few years later, Ferrier showed that removal of the precentral area resulted in a paralysis of contralateral musculature in monkeys.

This animal work and the observations of Jackson were the basis on which the Glasgow neurosurgeon Sir William MacEwen performed the world's first elective operation on the brain. Until this time, neurosurgery had been confined to cases where the skull had already been cracked open and the brain exposed as a result of a traumatic accident. No surgeon had deliberately opened the cranial cavity to operate on the brain, partly because he did not know where to expect pathology.

Armed with the information, new at the time, about localization of function, MacEwen took the courageous step of deliberately opening the skull. A patient presented to MacEwen who had suffered a penetrating wound of the skull, followed by the development of epilepsy of the type that Jackson had described as originating in motor cortex. On the basis of this evidence, MacEwen argued that there must be an abscess which had tracked to the precentral region. He opened the cranial cavity, confirmed the site and nature of the pathology, removed the abscess, and saved the patient's life.

Wilder Penfield in Canada then took the step of studying the excitable cortex in conscious man. He used local electrical stimulation of different cortical sites, to investigate motor responses and subjective sensory experiences of patients under local anaethesia. Application of the stimulating electrode to the precentral gyrus produced involuntary movements of the opposite parts of the body. These movements showed an unvaried order of sequence from one end of the gyrus to the other, the mouth and face moving when the electrode was placed laterally, the arms and hands with more medial placements, the legs and feet with placements at the most medial part of the gyrus and its extension onto the medial aspect of the hemisphere (Penfield, 1958, p.16). This 'mapping' between the siting of muscles in the body and the position on the motor cortex is called **somatotopic** representation. More informally, we may call it the **body map**.

The result of precentral stimulation was movement involving coordinated contraction and relaxation of muscles operating at a joint; stimulation here never caused the contraction of a single muscle in isolation. Another effect observed during the stimulation of the precentral gyrus was paralysis: the subject was unable to make any voluntary movement in the part of the body corresponding to the stimulated region of cortex. In some cases, precentral stimulation did not produce any involuntary movement, but it always produced this transient paralysis.

Stimulation of the postcentral gyrus produced a feeling of numbness or tingling of the opposite side of the body. The correlation between cortical site and the region where the patient experienced the sensation was the same as that between site on the cortex and region where movement occurred for the adjacent precentral gyrus. During stimulation, the subject was unable to feel a natural stimulus applied to the peripheral site where sensations were being produced by the cortical stimulus. Penfield mapped wide areas of the cerebral convexity in this way and drew up his classical diagrams of the

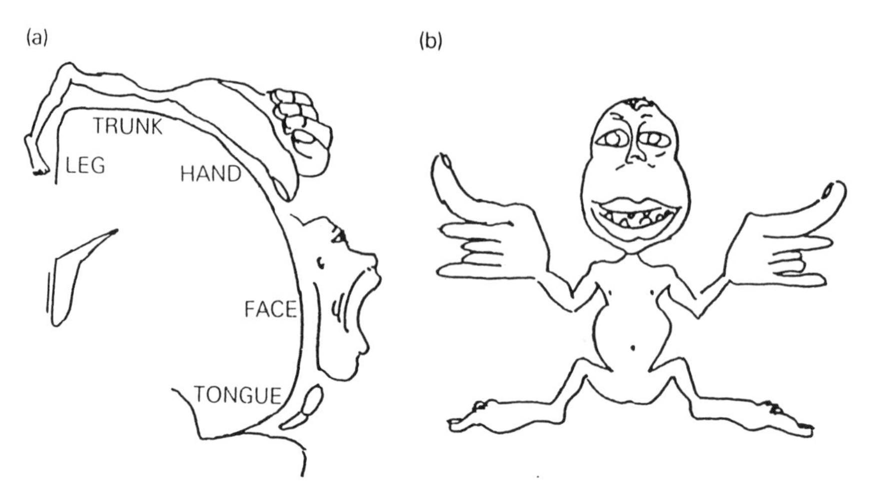

Figure 13.3 (a) The distribution of representation of the musculature on the motor strip. (b) A drawing of a human with different regions scaled to correspond to the area of cortex devoted to it. This figure is known as the homunculus.

distribution of sensory and motor functions on the cortex. Figure 13.3a indicates the amount of motor cortex devoted to different parts of the body's muscles. The sensory map is very similar. The unequal representation of the different parts of the periphery is striking. The lips, fingers, and toes command relatively large areas of cortex. The large area devoted to the thumb and forefinger reflect the control of **apposition** of these two digits. Apposition is the movement of bringing the tips of the thumb and forefinger together to form a letter O. Primatologists regard this movement as important in the advance of these species to the top of the evolutionary tree.

Although one might quite reasonably question how a neurosurgeon can operate on a human brain under local rather than general anaesthetic, it is in fact in the patient's interest. The primary motor and sensory areas of the cortex are reasonably constant from one individual to the next. This is not so for the speech areas described in the next chapter. The sensorimotor and speech areas are indispensable to the patient; the neurosurgeon must avoid them at all cost. The neurosurgeon may wish to remove cortex if, for instance, it is harbouring epileptic foci resistant to pharmacological treatment. During the operation, the neurosurgeon can map the speech areas by local electrical stimulation. If a point within a speech area is electrically stimulated, speech is interrupted. To test this, the surgeon obviously requires the patient to be conscious in order to speak.

13.3 DECUSSATION IN THE CENTRAL NERVOUS SYSTEM

Most of the higher centres are connected to the contralateral periphery. Incoming sensory information decussates completely before it reaches the sensory areas of the cerebral cortex. On the motor side, the motor strip of the cerebral cortex controls the somatic musculature on the opposite side of the body. The extrapyramidal system controls contralateral somatic musculature. Only the cerebellum is related to musculature on the same side of the body. We now consider why there should be this extensive decussation.

It seems to be general principle that nearby points on the body surface project to nearby points in the central nervous system. This has obvious advantages for the integration of influences. For instance, spatial summation in the cord depends on this geometric relationship. Similarly, in the processing of visual information described in the next chapter, it is necessary for neighbouring parts of the visual field to map to neighbouring points in the central nervous system. Clues to the reason for the existence of decussation are given by the study of comparative anatomy. In evolution, decussation first appears in the visual system. The optic nerves of fish and birds decussate completely. This is despite the fact that there is almost no overlap of the visual fields of the two eyes, indicating that decussation does not depend, as one might suppose, on such an overlap. The decussation of the optic nerves first appears, phylogenetically, with the evolution of the simple eye. Such an eye provides an image on the retina which is much better defined than that of lower animals. Due to the optics of the eye, the lens throws an inverted image on the retina. Let us consider a fish or a bird, with a simple eye on each side of its head and with little or no overlap of the visual fields of the two eyes.

In Figure 13.4a, the visual field is shown by the points numbered 1 to 6 on the circumference of the large circle. The task is to retain this order as the points project back to the visual centres of the brain. Figure 13.4a shows the situation that would arise if there were no decussation. Contiguous points 1 and 2 in the peripheral visual field project to contiguous points on the projection pathway, as they should. However, points 3 and 4, contiguous in the visual field, project to widely separated non-contiguous points, and points 1 and 6, far apart in the visual field project to neighbouring points. This may have been the stimulus to decussation, which results in the relationships shown in Figure 13.4b. There is now complete inversion of the visual image on the central nuclei and contiguous points in the peripheral visual field map to contiguous points on the projection pathway.

Other inputs need to be integrated with visual input, such as the sense of touch. These other inputs follow the visual decussation. This is probably the origin of the near total decussation of higher centres. With the forward rotation of the eyes during evolution entailing overlap of the visual fields, the visual decussation of the optic projection has regressed to being

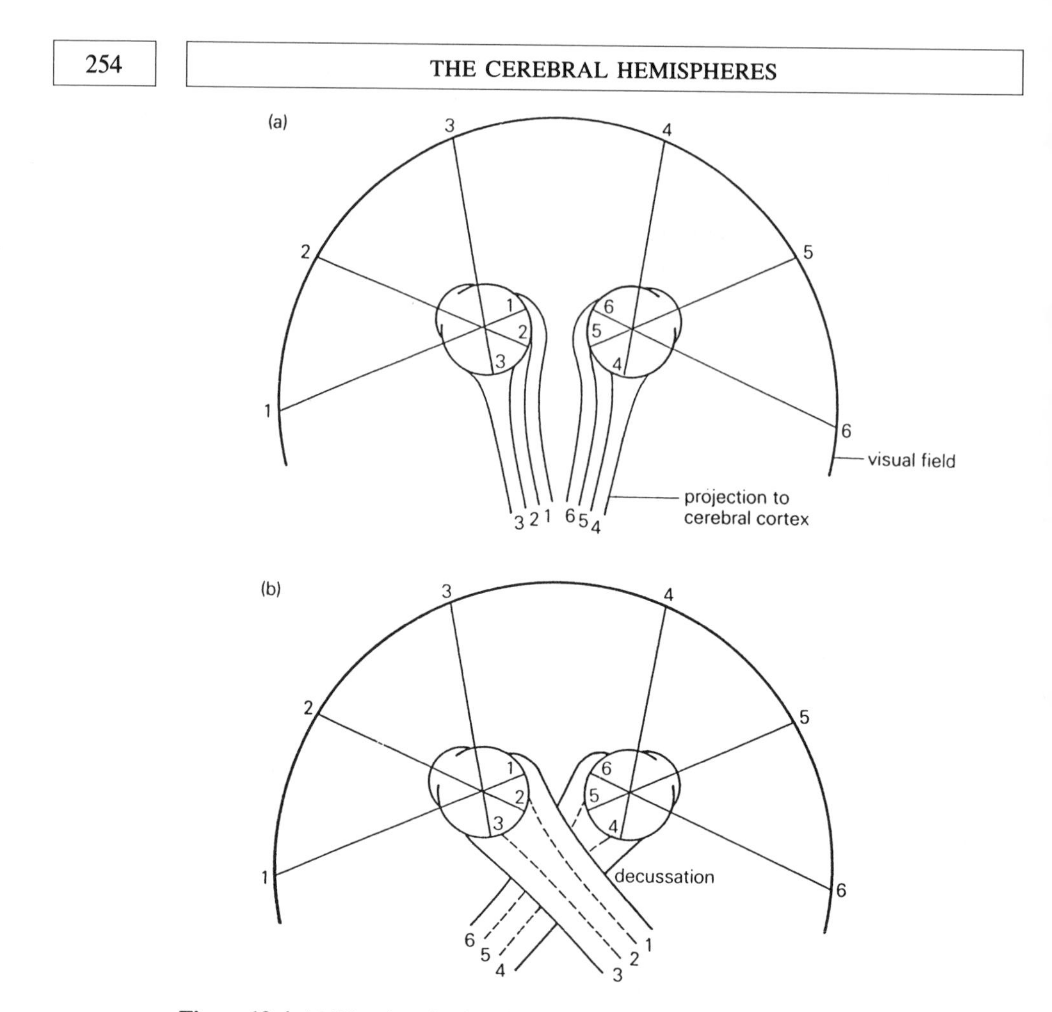

Figure 13.4 (a) Visual projection from simple eyes if there is no decussation of the visual pathways. (b) The situation with decussation.

partial. However, the overall pattern of decussation in non-visual parts of the system has been established and persists.

13.4 LIMBIC SYSTEM

The word **limbic** means border (Isaacson, 1982, p.1) and the term 'limbic system', is applied to a series of phylogenetically ancient nuclei and tracts that border the neocortex. It is a 'limbo' system that we do not understand but which is clearly essential in behaviour. Stimulation of the structures in the limbic system produces a change in behaviour such as attack or flight, or it produces an emotional response. There is always an autonomic component to the response such as changes in respirations and heart rate. As a result, the limbic system has come to be known as the **visceral brain**. The criterion for deciding whether or not a particular structure should be included in the limbic system is that stimulation should elicit such a 'behavioural' response.

The tracts and nuclei of the limbic system were well developed in primitive animals in which the cerebral cortex was rudimentary. Each hemisphere arises around the lateral ventrical, which is a lateral extension of the third ventricle. The lateral ventricle connects with the third ventricle via a foramen called the **foramen of Monro**. The limbic system lies around the foramen of Monro. The expansion of the neocortex has resulted in the limbic system becoming distorted so that its anatomy is very complex.

The components of the limbic system

There are components of the limbic system in the cortex and in subcortical nuclei (Table 13.2). The limbic lobe of the cerebral cortex consists of paleocortex. Anatomically, this limbic cortex occupies the gyri on the medial aspect of the hemisphere surrounding the third ventricle and the corpus callosum, as shown in Figure 13.5a. The cingulate and hippocampal gyri are two of these. One important subcortical component of the limbic system is the **hippocampus**. It is layered, but its structure is simpler than that of cerebral neocortex in being only three-layered. The hippocampus lies on the medial wall of the temporal lobe of the cerebral hemisphere and just behind the amygdaloid nucleus. The cerebral cortex of the limbic lobe runs into the hippocampal cortex, as shown in Figure 13.5b.

Other important subcortical structures in the limbic system are the **hypothalamus**, the **non-specific thalamic nuclei** and the **amygdaloid nucleus**. The various subcortical nuclei are joined by bundles of nerve fibres. Important amongst these is the **fornix**, a band of fibres that forms the efferent pathway

Table 13.2 The limbic system

Components (structures round the foramen of Monro):
- Limbic lobe of the cerebral cortex (paleocortex), including the cingulate and hippocampal gyri
- Hippocampus
- Hypothalamus
- Amygdaloid nucleus

Functions
- Higher autonomic control
- Emotion: rage and pleasure

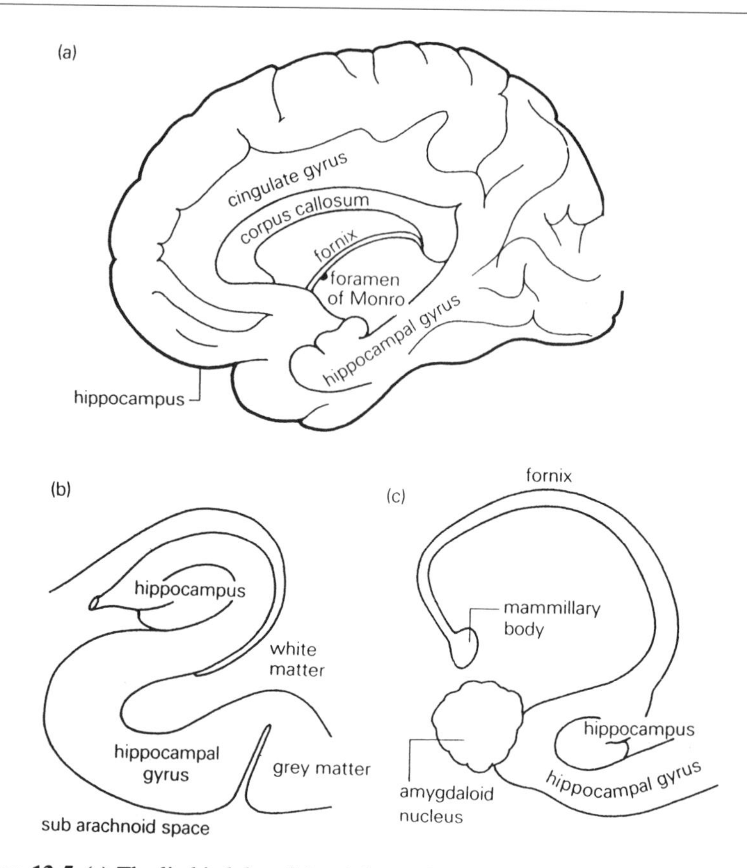

Figure 13.5 (a) The limbic lobe of the right cerebral hemisphere. (b) A coronal section through the hippocampus. (c) Connections of the limbic system. The structures have been dissected out and laid on a flat surface.

from the hippocampus to the mammillary body (Figure 13.5c). Connections between the neocortex and the limbic system are scanty.

The functions of the limbic system

The limbic system arises from structures that were originally concerned with olfaction. In animals which rely on olfaction for much of their information about the environment, this sensation is the origin of many reflexes and emotional responses. As olfaction has become replaced in importance by vision and hearing, the olfactory function of the limbic system has declined, but the role in emotion has become more important. This is illustrated by observing the organism when the limbic system is allowed to express itself free from influences from the cerebral cortex, as in a cat or dog that has lost both cerebral hemispheres (Bard, 1928). Such an animal is liable to show episodes of behaviour that in its motor expression, both somatic and autonomic, appear to be the manifestations of rage. This reaction is called **sham rage** because it has been regarded as the motor expression of an experience that the animal without a cortex cannot possibly experience.

An attack of sham rage may be set off by a stimulus which ordinarily would cause no concern, such as gentle handling. Other attacks may occur without demonstrable external stimulus. Similar attacks of sham rage can be elicited in the intact animal by stimulation of certain parts of the limbic system. In different species, different anatomical sites must be stimulated to give a particular effect.

The limbic system is not only the seat of rage. It is central in all emotions such as that of the feeling of fulfilment. Evidence for this comes from observations of the effects of stimulation at sites in the limbic system other than those which elicit rage. If, in an experiment, an animal is given the option of pressing a bar to deliver stimuli to itself, with the electrode in certain sites, the animal will deliberately stimulate itself even if the experiment is arranged so that the stimulation also carries a penalty, such as an electric shock to the skin. So pleasurable are the effects of the stimulation that the animal may stimulate itself to exhaustion.

The limbic system houses the circuitry important in the physiological basis for emotional behaviour. It determines how we relate to each other.

The limbic system and psychosis

Disorders of function of the limbic system have been implicated as a basis for psychotic disease. The introduction of antipsychotic drugs revolutionized the treatment of psychosis. In a ward of psychotic patients who were previously uncontrollable, many of them not having slept for as long as 3 months, the introduction of antipsychotic drugs made the patients calm and manageable. Many psychotic patients are now so effectively treated by these drugs that they lead relatively normal lives in society instead of having to be confined in an institution. These drugs probably exert their effects by stimulating or suppressing activity in different sites in the limbic system.

13.5 THE HIPPOCAMPUS AND MEMORY

The hippocampus plays a central role in memory by organizing the formation of the memory image (the **engram**). This engram does not itself lie in the hippocampus; it probably lies distributed through the cerebral cortex, particularly in the temporal lobe. The hippocampus acts as the central computer which gathers incoming information, analyses and organizes it, and then sends the information out for storage in the cortex.

Observations have been made on humans in whom the hippocampus on each side has been destroyed. Neurosurgical removal of the temporal lobe is a treatment for temporal lobe epilepsy. In this operation, the hippocampus is removed with the rest of the temporal lobe. As long as the operation is performed on just one side and the remaining hippocampus is functional, the subject experiences little memory defect. However, in an early operation of this nature on a patient with bilateral temporal lobe epilepsy, a neurosurgeon in good faith removed the temporal lobes and hippocampi on both sides. A similar effect is produced when the healthy temporal lobe is removed from a patient in which the laterality of temporal lobe epilepsy has been misdiagnosed. Here the remaining hippocampus is diseased and non-functioning. Bilateral destruction of the hippocampus also occasionally follows viral encephalitis.

If both hippocampi are destroyed, the effect is devastating. The subject can no longer lay down any record of his experiences. He lives at the present instant of time and can recall nothing of the past, even the recent past. Intellectually he is unimpaired. He can speak, read and, if he is a musician, can play his instrument and conduct a choir with the same sensitivity as before the damage to the hippocampus. To do this, he must use images established before the damage. He also recognizes his friends and uses their names; despite this, he denies ever having seen them before. He believes that each instant is the first instant of his own consciousness and existence. If his attention is drawn to evidence to the contrary, such as to some writing of his own from the previous day, he becomes very disturbed and aggressive. Since he can lay down no new memory images, he is a prisoner of the present, with no awareness of the past and no ability to plan for the future. Such is the importance to us of this phylogenetically ancient remnant of neuronal circuitry.

Amnesia is caused by bilateral destruction of the mammillary bodies or by bilateral interruption of the fornix, which is a band, one on each side, between the mammillary body and the hippocampus (Gaffan and Gaffan, 1991). The neuropharmacology of the hippocampus and the subject of memory are discussed in Chapter 17.

Hippocampal neurones signal spatial information

The hippocampus receives information from the eyes, cochleae, vestibules, and somaesthetic sense organs. This information comes via the relevant regions of cerebral cortex, where it has already been processed. Consequently, the hippocampal cells carry highly organized information about the animal relating to the animal's position in the environment.

O'Keefe and Recce (1992) have made simultaneous recordings from several hippocampal cells in the free-running rat. They found that these cells signal the animal's position. When the animal is in one region of its pen, a particular neurone fires (at about 1 Hz). When the animal leaves this specific region, the cell stops firing and another takes over. The first neurone starts to fire again as the animal returns to the original region. It does not matter in which direction the animal traverses its specific region. The boundaries of the region are well defined and stable over long periods of time. These cells are called 'place cells'.

A cell previously identified as a place cell behaves differently in a maze situation when the animal has a choice of only two directions of transit through a particular region. Then the cell may show strong direction sensitivity, firing when the rat enters the specific region from one side but not when entering from the other. This is another indicator that the activity in the cell is influenced by many regions of the brain and that the nature of its activity may be switched according to the type of activity in which the animal is engaged.

13.6 THE PHYSIOLOGICAL ASSESSMENT OF NEURAL FUNCTION

In recent years various physiological methods have been developed to investigate the function of nerve cells and nerve paths within the central nervous system in humans. This section considers some of these.

Primary cortical evoked potentials

A widely used method of investigating sensory function is the recording of **evoked potentials. Somatosensory cortical evoked potentials** are obtained by applying an electrical stimulus percutaneously to a peripheral nerve and recording from electrodes on the scalp overlying the somatosensory area contralateral to the stimulus. **Visual evoked potentials** are produced by stimulating the eye with a flash of light or a patterned light stimulus. The evoked potential is recorded from scalp electrodes over the visual cortex. **Auditory evoked potentials** are recorded as the response to a click stimulus. Since the primary auditory cortex lies in infolded cortex, the potential recorded at the scalp is very weak. A recording electrode on the vertex picks up activity from various sites along the auditory pathway. As explained in Chapter 11, a component attributable to the action potentials in the cells of the midbrain auditory nuclei is depressed in brain stem lesions.

A measurement of importance in neural function is the velocity of conduction along the sensory pathway. This is assessed from the delay between the stimulus and the evoked response. The usefulness of recording somatosensory evoked potentials is limited because of unknown delays introduced by synaptic relays en route, and because of the long distances travelled by the nerve impulse along peripheral nerves. The conduction velocity in these peripheral nerves is sensitive to the temperature of the nerve, and large variations in temperature occur in the limbs. This introduces large variations in the delay between the stimulus to the periphery and the response in the brain. Such variations swamp any changes in conduction velocity as the impulse travels in the brain itself. (The temperature of the brain is held constant by powerful homeostatic mechanisms.) To overcome this problem, investigators have developed methods of recording the central conduction time.

Central conduction time

An electrical stimulus is applied percutaneously to a peripheral nerve, such as the ulnar nerve. Recordings are made from a skin electrode at the back of the neck in the midline in the upper cervical region. A transient voltage is recorded as the volley of nerve impulses travels along the central afferent pathways in the cervical cord. The voltage is small and mixed with unwanted background electrical activity or noise; the recording has to be stored in an electronic averager, which averages the voltages elicited by many stimuli. This allows the low-voltage signal which is time-locked to the stimulus to be separated from the noise, which is not time-locked to the stimulus and averages to zero. Records are derived simultaneously from an electrode on the scalp overlying the primary somatosensory cortex of the opposite cerebral hemisphere, and the cortical evoked potential so recorded is also averaged. The time interval between the evoked potential at the cervical cord and at the cerebral cortex is the **central conduction time**. This measurement is clinically useful. It can be used, for instance, during a neurosurgical operation as a guide to the adequacy of oxygenation of brain tissues. Cerebral hypoxia prolongs the central conduction time. This effect is first observed when the cerebral blood flow falls below 30 ml/100 g tissue/min. Irreversible brain damage does not occur until the cerebral blood flow falls to half this value.

In some neurosurgical operations, the blood flow in a cerebral vessel, the basilar artery for instance, is temporarily interrupted so that the neurosurgeon can operate on brain tissue without the occurrence of too much bleeding. A prolongation of the central conduction time is an indication to the neurosurgeon that he must re-establish the blood supply. If he does so, the brain sustains no permanent damage.

Stimulation of motor tracts

This is like the evoked potential method of testing sensory projections but tests motor projections. A very brief high-voltage stimulus applied to the skin overlying the cervical cord excites the central motor pathways and elicits a twitch from the limb musculature; this can be recorded from electromyographic electrodes (Figure 13.6a). This response cannot be modified volitionally by the subject. A similar shock applied to scalp electrodes overlying the motor cortex elicits a muscle twitch, but this requires volitional reinforcement by the subject; this is shown in Figure 13.6b. If the subject does not provide a background of excitation by producing a sustained voluntary contraction, no muscular activity can be evoked.

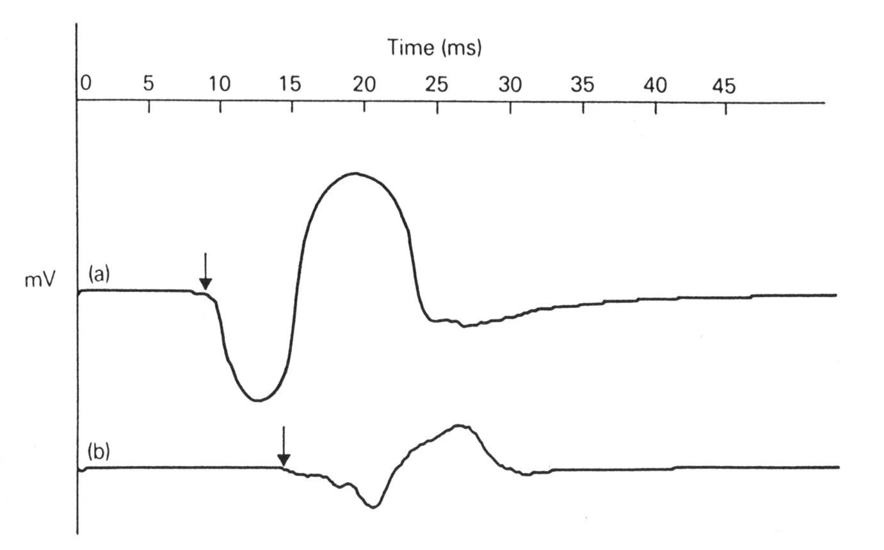

Figure 13.6 Electromyograms from forearm flexor muscles elicited by central stimulation. (a) Stimulation of the cervical cord. (b) Stimulation of the precentral region of the cerebral cortex. Stimuli were delivered at zero time on the time axis. The times of onset of stimulated electromyographic activity are shown by arrows.

The central motor conduction time, from motor cortex to cervical cord, is given by the time interval between the two arrows in Figure 13.6. This represents the time between the onset of electromyograms yielded by stimulation at the two sites.

The electrical stimulus excites structures other than the motor fibres, and the shocks are unpleasant. One side-effect of stimulation is a powerful contraction of the jaw muscles with forceful closing of the teeth. In an uncooperative subject who protrudes his tongue, the tongue is bitten when the brain is stimulated. A recent technique avoids all unpleasantness for the subject. It consists of inducing a rapidly changing magnetic field in the brain by passing a massive current through a coil of wire, about 9 cm in diameter, held over the scalp. The changing magnetic field induces sufficient current in the motor cortex to initiate a volley of nerve impulses travelling along the corticospinal projection. The outcome is a muscular twitch in the limb. Apart from the muscle twitch, the subject is unaware of the stimulus, which is therefore entirely benign.

A disadvantage of both these methods of stimulation is that a relatively large amount of brain is stimulated and the site at which action potentials are initiated cannot be determined exactly.

FURTHER READING

Penfield, W. (1958) *The Excitable Cortex in Conscious Man.* The Sherrington lectures. Liverpool, The University Press.

Seeing and hearing

14

14.1 VISION AND EYE MOVEMENTS

Various aspects of function of the central nervous system are concerned with vision. In addition to the analysis of the sensory input, the motor control of the extrinsic ocular muscles play an important role in the analysis of the visual image. The extrinsic ocular muscles, innervated by cranial nerves III, IV and VI, have their motoneuronal cell bodies in the oculomotor nuclei in the brain stem. As previously described (Section 10.2), these are influenced by the vestibular nuclei. Influence from the cortex is directly from the frontal eye fields of the cerebral cortex and from the superior colliculi (see below). The regions receiving visual input are the lateral geniculate bodies and the occipital lobe of the cortex; processed information is passed thence for further analysis to two regions of association cortex, the posterior parietal area and the inferotemporal region.

14.2 THE CONTROL OF EYE MOVEMENTS

The neural control of the extrinsic ocular muscles has features which are different from those of the control of other skeletal muscles in the body. The precentral gyrus, the primary motor area for all other skeletal muscles in the body, sends no projection to the motoneurones controlling the extrinsic ocular muscles. Instead there is a special region of the cortex, the **frontal eye field** (Figure 14.1) that controls eye movement. This lies in front of the motor area and is sometimes called the **prefrontal area**. This region projects down directly to the oculomotor nuclei and also via the superior colliculus, to be described later. The motor control system for eye movements does not allow us to produce a smooth movement of the eyes voluntarily. If we try to move our gaze smoothly to scan the visual field, our eyes move in a series of jumps. Each jump is called a **saccade**, and is analogous to a ballistic movement. Whereas voluntary movements of the eyes are saccadic, the tracking

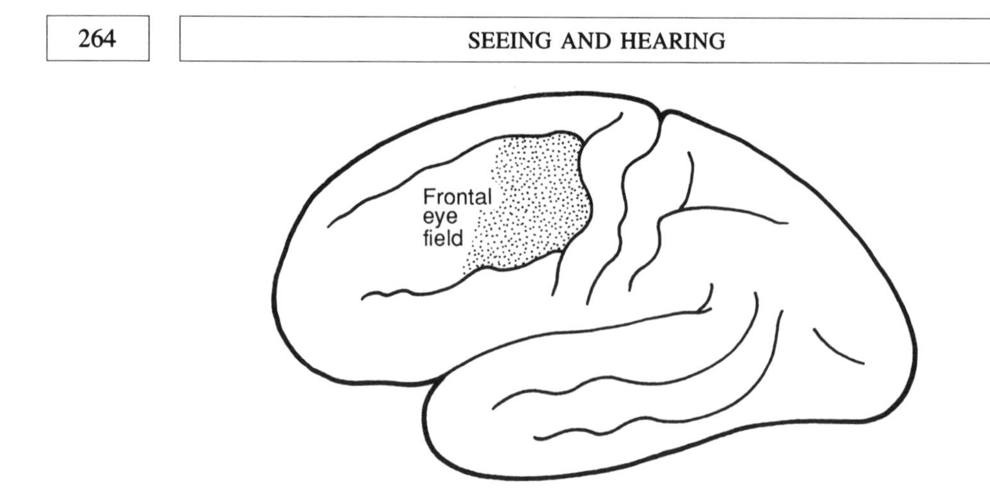

Figure 14.1 The siting of the frontal eye fields.

movements that we involuntary make if the visual world moves past when we look out of the window of a vehicle moving at constant velocity are smooth and controlled by brain stem mechanisms (Chapter 11).

14.3 THE VISUAL PATHWAY

The capture of light by the photosensitive cells (rods and cones) in the retina was described in Chapter 9. The pathway of information from the eyes to the visual cortex is shown schematically in Figure 14.2a. From the retina, the axons of the retinal ganglion cells project as the optic nerve back to the lateral geniculate body, which is part of the thalamus. After a synaptic relay there, the projection continues as the optic tract to the primary visual cortex.

Visual cortex

The primary visual cortex lies at the posterior pole of the occipital lobe; most of it lies on the medial aspect with only a small part overlapping onto the lateral aspect. The secondary visual cortex lies immediately in front of the primary cortex. The primary visual cortex receives the visual input from the lateral geniculate bodies; this input is so dense that layer IV appears as a white line when a section of area 17 is viewed by the naked eye. The whiteness is due to the geniculo-cortical afferent nerve axons.

The retino-cortical map

In a normal person, a clear image of the outside world is formed on the retina. The details of this image are captured by the retinal receptors (rods and

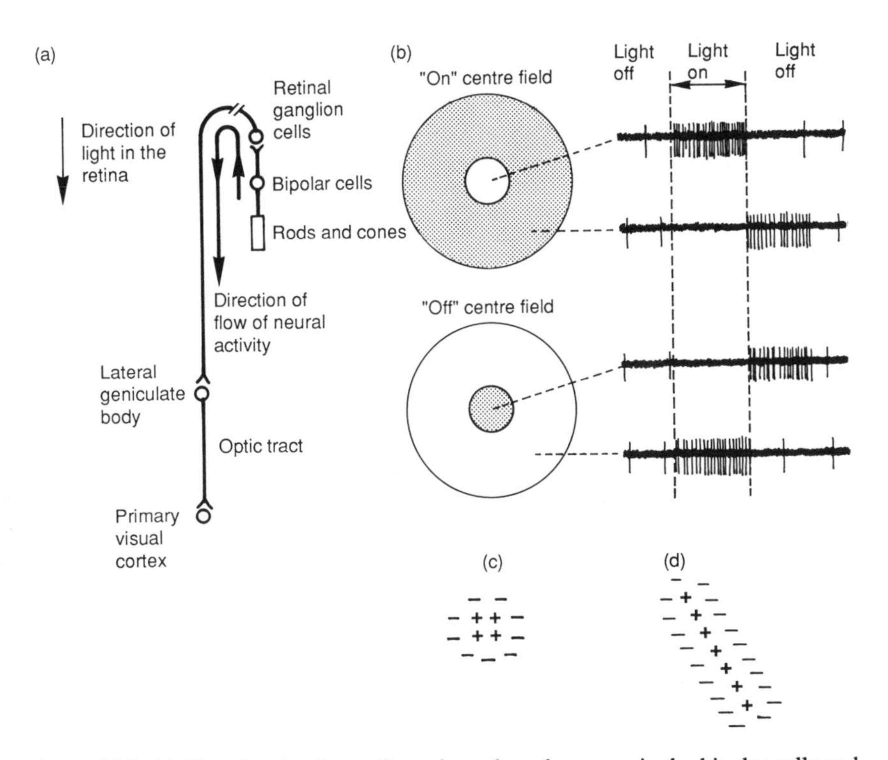

Figure 14.2 (a) The visual pathway from the rods and cones, via the bipolar cells and ganglion cells in the retina, back to the lateral geniculate body and thence to the primary visual cortex. (b) The responses of retinal ganglion cells to spots of light in the visual field. For the cell labelled 'on' centre field, a spot of light shone within the central circle evokes a high-frequency discharge, shown in the uppermost electrical record of (b). A spot of light in the surrounding shaded region gives the effect shown in the second electrical record, with cessation of ongoing activity during the time when the light is on and a burst of impulses immediately after the light is switched off. The cell labelled 'off' centre field shows the converse behaviour, with inhibition of activity when the spot of light is in the centre of the field but excitation when the spot of light is in the surround region. (c) Symbolic representation of an 'on' centre cell. (d) The receptive field of a 'simple' cell in the visual cortex.

cones); each site on the retina projects back to a small region of the primary visual cortex, and a map of the visual world is produced. A given point in the visual field is focused at one point in each eye. As described in Section 9.1, the visual pathways from the two eyes project back in such a way that the information about one point in the visual field projects from both eyes along pathways that come together at a point in the contralateral primary visual area in the occipital lobe.

Just as with the somato-cortical map of Penfield shown in Figure 13.3, the retinocortical map is distorted. The fovea has a much greater share of the cerebral cortex than do peripheral regions of the retina (Figure 14.3). This correlates with the greater visual acuity (the ability to discriminate two points of light which are close together) of the foveal region compared with peripheral retina. The visual acuity of the various regions of the visual field is directly proportional to the amount of cortex devoted to those regions of retina.

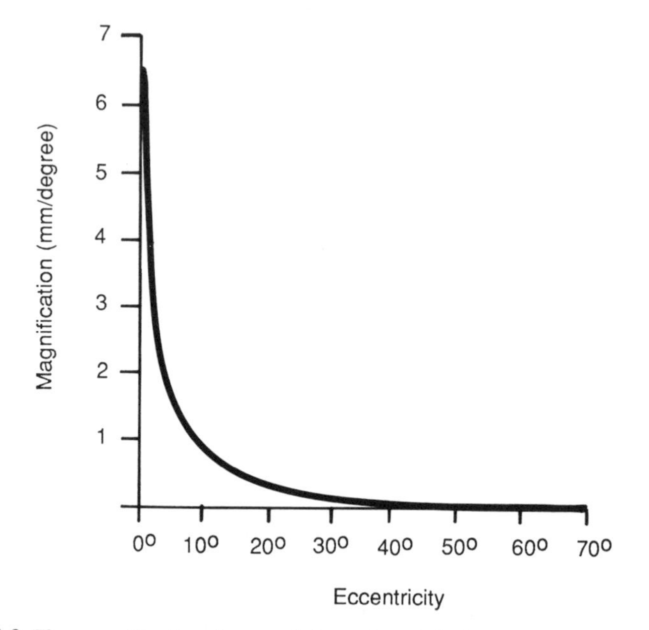

Figure 14.3 The magnification factor of the retinocortical map. This is plotted on the y axis as mm cortex per degree of visual field. The x axis is the eccentricity i.e. the distance of the visual stimulus from the visual axis.

Visual field

The region of the periphery which can be seen by a subject is called the 'visual field'. It is measured by seating the subject with the chin resting on a support. One eye is covered and the subject is asked to fixate on a point in front. A mobile object is moved in from the periphery along a radian towards the fixation point until the subject signals that the object is visible. This point is noted and the procedure repeated along radians at intervals around the whole of the 360°. This determines the outer boundary of the visual field. Gordon Holmes investigated the visual fields of casualities in World War I. The number of casualties from flying metal was horrendous; doctors had the opportunity of examining people with many and varied lesions. Destruction of localized regions of the primary visual cortex produced characteristic defects of vision. Within a region the visual field, contralateral to the site of the cortical damage, the subject was blind. The contour of the edge of this blind area, or **scotoma** was well defined and often irregular in shape, corresponding to an irregular region of destruction of the cortex. The scotomata for the two eyes mapped separately corresponded exactly, confirming that points on the two retinae responding to light in the same part of the visual field project back to a single region of cortex. The subject was entirely unaware of visual objects presented within the bounds of the scotoma.

Research since those days has demonstrated that, although the subject has no conscious access to visual information presented within the scotoma, subconscious regions of the brain do have access to some features of the visual stimulus. If a point of light is flashed within the blind area of the visual field, the eyes move involuntarily towards the point, thus tending to bring the object towards the visual axis. The movements are saccades. This is the phenomenon of **blind sight**. The central nuclei initiating this effect are probably the superior colliculi (Tusa *et al.*, 1986). The afferent visual pathway to the superior colliculi diverges from the geniculo-striate projection before this reaches the cortex (Figure 9.6).

The superior colliculi

The roof of the mesencephalon consists of four small elevations, two on each side. These are called the **superior** and **inferior colliculus**; together the four are called the **corpora quadrigemina**. These are phylogenetically ancient regions of the brain connected with vision (superior colliculi) and hearing (inferior colliculi). In lower vertebrates such as the frog, which has the most rudimentary of cerebral cortices, the main centre for analysis of visual information is the tectum. ('Tectum' means roof; the tectum is the roof region of the mesencephalon.) As one ascends the evolutionary tree, many of the functions of this region are taken over by the cerebral cortex. The superior colliculi in humans are concerned with orientating responses. There is a retino-collicular mapping, and activity in one region of the superior colliculus results in movement of the head and eyes towards an object presented in the visual field. The superior colliculi also project to the cerebral cortex to alert it to the appearance of a novel visual stimulus. Lesions of the colliculi in humans do not significantly impair visual function or eye movements, implying that they are essentially vestiges from phylogenetic ancestors.

Processing of visual information

Visual information leaving the eye via the optic nerve is already processed. The most peripheral element accessible to the recording of action potentials is the retinal ganglion cell; this is the cell body from which an optic nerve fibre arises (Figure 14.2a) and even by this stage, processing has taken place. The retinal ganglion cells are spontaneously active in the absence of light falling on the retina. According to their response to light, these cells are classified into two types, 'on' centre field or 'off' centre field (Figure 14.2b). For an 'on' centre field ganglion cell, a small spot of light falling on the photoreceptive cells that project directly back to the ganglion cell causes the cell to discharge a brief burst of impulses. If the spot of light falls on photoreceptive cells just surrounding the central region, switching on the light inhibits the spontaneous on-going activity in the ganglion cell (an example of surround inhibition). When the light is switched off, the cell gives a vigorous burst of impulses. Behaviour of this type is shown in Figure 14.2c. For 'off' centre cells, a spot of light falling on receptors that project directly back to the ganglion cell blocks its spontaneous firing, but when the light is switched off the ganglion cell gives a burst of impulses. For such a cell, illumination of the surround region is excitatory. The observation that light falling on different regions of the retina causes opposing effects shows that several receptors influence the activity of a single ganglion cell.

Table 14.1 Visual pathways

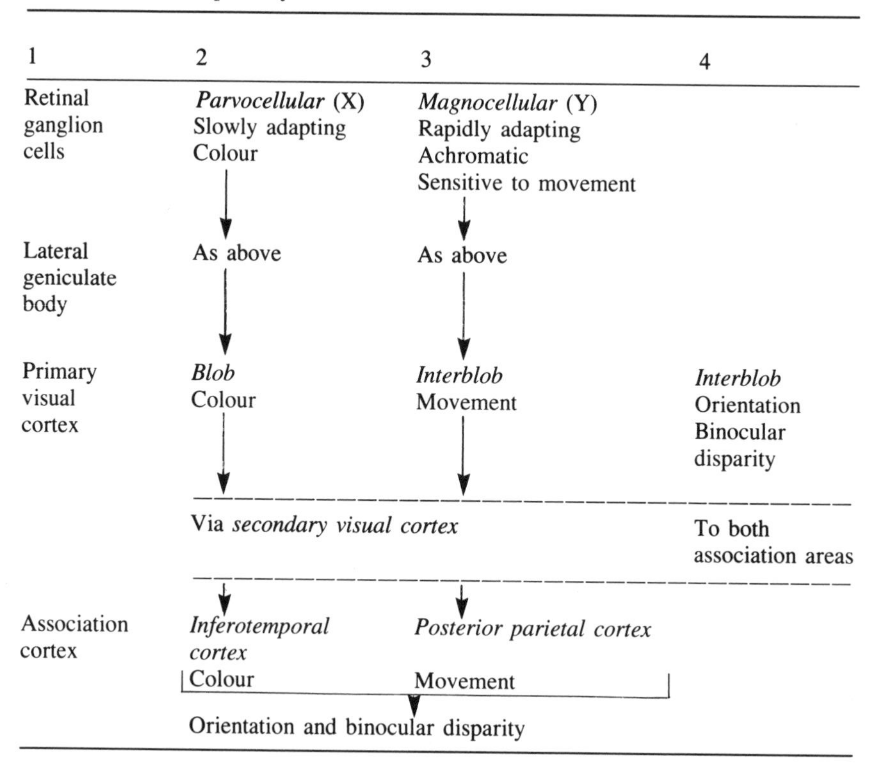

1	2	3	4
Retinal ganglion cells	*Parvocellular* (X) Slowly adapting Colour	*Magnocellular* (Y) Rapidly adapting Achromatic Sensitive to movement	
Lateral geniculate body	As above	As above	
Primary visual cortex	*Blob* Colour	*Interblob* Movement	*Interblob* Orientation Binocular disparity
	Via *secondary visual cortex*		To both association areas
Association cortex	*Inferotemporal cortex* Colour	*Posterior parietal cortex* Movement	
	Orientation and binocular disparity		

In addition to being classified as 'on' or 'off' centre, retinal ganglion cells can be classified according to their size: small X cells (**parvocellular** means 'small cells') and large Y cells (**magnocellular**). Each of these groups contains both 'on' and 'off' centre cells. In other respects, each of these two groups of cells responds in a characteristic fashion (Table 14.1). Parvocellular cells respond with relatively long latent periods to light stimuli, show sustained responses to prolonged stimuli, and show simple summation of effects when different parts of the receptive field are simultaneously illuminated. In primates they are colour sensitive. They are more numerous in the region of the fovea than peripherally. Magnocellular cells are less numerous, they are found in the peripheral regions of the retina, are connected with nerve fibres which conduct at high velocity and respond briskly to movement or changes in light intensity, but cease to respond if constant illumination is maintained. They do not show simple summation when different parts of the visual field are simultaneously illuminated. They are achromatic; they respond to movement of the visual stimulus.

Lateral geniculate body cells show behaviour characteristics similar to those of retinal ganglion cells, and each cell is influenced by only one eye. The cells of the visual cortex show a different pattern of behaviour indicating further convergence of influences from several input nerve fibres and from both eyes onto a cortical neurone.

The primary visual cortex

This is the first stage at which binocular interaction takes place. Most cells are affected by both eyes. They respond to subtle features of the stimulus. For instance, cortical cells called **simple cells** by Hubel and Weisel respond to a bar of light of appropriate orientation (Figure 14.2d) and not to bars of other orientations. Simple cells are orientation detectors. They behave as if each were influenced by a series of lateral geniculate body cells. There are also cells called **complex cells**; these require a bar of light with a particular orientation to be moving in a particular direction. **Hypercomplex cells** demand in addition a bar of particular width and length.

At each stage, several cells of one type converge onto one cell of the next stage in the sequence. This system is hierarchical, since each type of cell is influenced by output from cells earlier in the chain. Each cell passes on only selected features of the information that it receives; information which it filters out is not available to cells beyond it in the sequence.

The visual cortex is arranged as a mosaic of columns lying perpendicular to the cortical surface and extending through all the layers of the cortex. All the neurones in a given column are sensitive to the same orientation of a bar in the visual field. In a neighbouring column, none of the cells respond to that orientation; they all respond to a different orientation specific for that column.

Modern histochemical methods, notably a stain for the enzyme **cytochrome oxidase**, have shown a modular organization of the primary visual cortex on a coarser scale than the columnar arrangement based on the orientation sensitivity of the cells, as had previously been shown by Hubel and Wiesel.

In the monkey, the primary visual area shows a lattice of blobs of a high density of staining for cytochrome oxidase, each blob surrounded by a region of lower density of staining (interblob zone). The blob of high density staining marks the centre of a module.

Three features of each segment of the visual field activate different groups of cells in one module: the colour, the orientation sensitivity and the sensitivity to the direction of movement of the object in the visual field. One module is about 4.5 mm across and serves about 3° of the visual field. The region of high cytochrome oxidase activity contains colour sensitive cells. Orientation- and direction-sensitive cells are in the surrounding areas of low cytochrome oxidase activity. Despite the clear correlation between the different density of cytochrome oxidase staining and physiological characteristics of cortical cells, the significance of cytochrome oxidase levels is not known.

The secondary visual cortex

Because of the complexity of the secondary visual areas, they are divided into regions labelled V2, V3, V4, and V5. Most of the projection from primary visual cortex is to the region immediately in front of it; this is called 'prestriate cortex' or V2.

Different parts of the primary visual cortex project to different regions of V2. These regions in turn project to V3, 4 and 5 in an organized fashion: each of these regions responds to a different feature of the visual stimulus. One responds mainly to colour, another to orientation, and the third only to the direction of movement of the visual target. The details of these projections have still to be determined, and there is currently disagreement amongst those researching the subject. Whatever the details, the overall effect is that different features of an object seen in the visual field are analysed by different parts of the brain.

The projections of visual information away from primary and secondary visual cortex are to regions of the association cortex. There are two parallel lines of analysis. One projection is to the inferotemporal region of cortex for interpretation of the 'what?' of the visual stimulus, and the other is to the posterior parietal area, which is concerned with determining the 'where?' of the stimulus.

Characteristics of the neurones in the inferotemporal region of the cortex

This path for analysis of visual information is the 'what?' path. This originates in the parvocellular part of the retina and it is chromatic. These inferotemporal cells are responsive to the features of the stimulus, but are insensitive to the whereabouts of the visual stimulus in the visual field. Complicated features of the visual input are extracted. For instance, in the monkey there are neurones in inferotemporal cortex that fire impulses only when a monkey's face seen from a particular angle is presented in the visual field. As an exploring recording electrode is moved from site to site in the cortex in this region, the profile of the monkey's face to which the neurones are particularly responsive gradually changes, so that several views of the monkeys head are represented in inferior temporal cortex (Perrett *et al.*, 1992).

In this region of cortex the decision is made as to whether a visual stimulus is of interest or not to the subject. If it is interesting, inferior temporal cortex excites the limbic system which in turn switches on the analytic activity of the posterior parietal neurones. These then provide the brain with the information of the whereabouts and movement of the interesting object.

Characteristics of the neurones in the posterior parietal area of the cortex

Input is derived from the magnocellular component of the retina. This is the 'where?' of the central analysing system of visual information. It is achromatic. Cells in the posterior parietal area have inputs from the visual cortex and from other regions of brain. These cells respond to movement in a particular direction of any shape whatsoever in the visual field. They receive input from the prefrontal region of the cortex, the area which controls saccadic eye movements. They use this information to take account of the direction of gaze; when the direction of gaze changes, this is allowed for in such a way that a particular cell responds to the position of the visual stimulus in the environment and not on the position of the stimulus on the retina.

The perception of the visual image

We end this section with a consideration of what the whole organism perceives when an interesting type of visual input is presented. The detail of how the visual system works varies between species. In a frog, for instance, objects in the visual field are only seen if they move vis-a-vis the background. So long as a fly stands still, a frog cannot see it. Although it may be hard for us to imagine what it is like to be a frog and not to be able to see stationary objects, the human visual system is not so very different. When we think that we are keeping our gaze still, there is a tendency for the eyes to drift. This is compensated for by saccades occurring about 3 times a second. During each saccade, there is a momentary rise in visual threshold which suppresses our awareness of the jump in projection of the peripheral visual field onto the retina.

Our vision depends on the saccades. This can be shown by mounting a tiny mirror on a contact lens which moves with the eyeball. If the picture from a projector is thrown via the mirror onto the retina in such a way that the image on the retina is stabilized and the saccades no longer cause the siting of light incident on the retina to jump from one instant to the next, the perception of the visual image fades completely in a second or so. So just as a frog can see only moving objects, so in a way can we. The difference is that the human causes the visual image to make tiny jumps by the expedient of tiny movements of the eyeball whereas the frog must wait for the fly to jump.

14.4 AUDITORY CORTEX

The primary auditory cortex lies deep on the upper aspect of the infolded superior temporal gyrus. It receives input from the cochlea, relayed by the medial geniculate body. There is a 'tonotopic' organization: different frequencies of sound project to different parts of the primary auditory cortex. As the frequency of a note is gradually increased (i.e. the pitch progressively rises) the projection site moves progressively back across the primary auditory cortex.

Secondary auditory cortex is situated immediately lateral to the primary auditory cortex. In this region there are alternating bands of cortex which respond to sounds presented to one ear or the other. This allows the cortex to integrate and compare input from the two ears and to compare phase relationships and intensity relationships to compute the direction from which a sound is coming (stereophonic sensitivity).

The inferior colliculi

These perform comparable functions in audition with those of the superior colliculi in vision. A novel auditory stimulus excites the inferior colliculi and cause the head to be moved so that the eyes look towards the source of the sound.

The cerebral hemispheres: motor functions

15

15.1 THE MOTOR FUNCTIONS OF THE CEREBRAL CORTEX

Movement is controlled by many higher centres, including the cerebral cortex, the basal ganglia and the cerebellum. We start this chapter with the motor functions of the cerebral cortex and the initiation of voluntary movement.

The corticospinal projection passes via the pyramidal tract. (The pyramidal tract consists of those axons which originate in the cerebral cortex and pass to the spinal cord through the medullary pyramids.) In layers V and VI of the primary motor strip lie the biggest pyramidal cells of the cortex; these are called **Betz cells** and their axons project down via the corticospinal projection. These axons are the thickest in this projection; they project directly to form synaptic connections with the motoneurones of the spinal cord. Such axons constitute a small minority of the nerve fibres in the corticospinal projection; the majority are the axons of small pyramidal cells sited in various regions of the cerebral cortex. At least half the pyramidal tract neurones are unmyelinated and therefore conduct very slowly: it takes more than half a second for nerve impulses to traverse the corticospinal pathway.

The corticospinal tract is only one of many tracts which project down to the cord and influence the activity of motoneurones. Other paths from higher centres down to the cord are often grouped together as the **extrapyramidal pathways**. These consist of rapidly conducting nerve fibres. The motor area of the cerebral cortex projects down to the cord via the extrapyramidal as well as via the pyramidal pathway. One route for this is its projection to widespread regions of the reticular formation and thence to the cord. Since this projection conducts quickly, information passing along it reaches the cord before the slow information passing along the pyramidal tract. Most of the axons descending in the pyramidal and extrapyramidal projections make synaptic contact with interneurones in the cord and act by modifying reflex activity.

A voluntary movement is performed on a background of posture; both the postural adjustments and the voluntary movement itself are important. A drunk trying to insert a key into a keyhole may have little difficulty with fine hand movements controlling the key but sways about so much that he can never put the key in the hole.

Corticospinal tract

The monosynaptic corticospinal tract connection with cord motoneurones in primates is of paramount importance in the control of fine hand and finger movement which is so important to these animals. Although only about 5% of corticospinal tract fibres end monosynaptically on cord motoneurones, each fibre of this type sends branches to exert a strong excitatory synaptic influence on many motoneurones. The influence of this pathway is therefore strong, despite the small proportion of pyramdial neurones in this group. The motoneurones involved lie laterally in the cord (see below) and innervate the fine muscles of the hand and fingers.

If the pyramidal tracts are cut, coarse movements of the arms and hands can still be made. However, the fingers can no longer be used independently of each other. The hand has to be used with all the fingers flexing or extending together at any one time.

In a monkey up to the age of 4 months, there are no monosynaptic connections from corticospinal tract fibres to motoneurones. Monkeys at this age cannot use fingers independently of each other. Independent finger movements become possible later, as the connections are made. An adult monkey can winkle out nuts and other food fragments from nooks and crannies, whereas a young monkey can only use movements of the whole hand; it can only pick up food which is readily available on a smooth surface. The independent use of fingers by adults is also in evidence when monkeys groom each other, catch fleas, and explore the environment by feel. Mammals such as cats cannot use their digits independently of each other. The claws are all extended or flexed together and the paw acts as a whole. These animals do not possess direct corticospinal tract projection to their motoneurones. This direct control is of paramount importance in man for the performance of such highly controlled activities as writing. As a spin-off, it also allows humans to accomplish achievements such as playing the violin. The direct corticospinal projection is a necessary condition for the independent finger control.

A single corticospinal tract in the primate sends branches to many motoneurones and often to motoneurones in several (up to six) motoneurone pools and over several segments of the cord. The precentral output is thus in terms of multiple muscle contacts. This connectivity is such that action potentials in the corticospinal tract fibre cooperate to produce movement by contracting agonist muscles. In many cases the corticospinal tract axon also inhibits motoneurones of antagonist muscles, via the interneurone on the

Group Ia reciprocal innervation inhibitory pathway. This implies a task-related pattern of organization in the motor cortex, with one Betz cell commanding a movement, not contraction of an individual muscle.

The 'pyramidal tract neurones' (PTNs) of the precentral gyrus are of two types, fast and slow, according to the conduction velocity of the axons. Both fast and slow PTNs project down monosynaptically to the motoneurone, but in other respects they have different physiological properties. When a primate performs a long-lasting grip with the hand, the fast PTNs fire a burst of impulses at the start of a movement but then return to a normal irregular firing pattern as the movement proceeds. The slow PTNs, whose on-going firing is regular, increase and maintain their firing throughout the movement. The slow PTNs are the only neurones in supraspinal nuclei whose discharge closely corresponds to the discharge of the motoneurones which comprise the final common path.

The projection of pyramidal cells is primarily to motoneurone pools lateral in the ventral horn of the cord innervating distal musculature; this is important, as already described, in the performance of fine skilled movements, particularly of the digits. From the brain stem motor nuclei (reticular formation, vestibular nuclei), the projection is primarily to motoneurone pools lying medially in the cord grey matter and innervating axial musculature. This is important in stabilizing the trunk during movement and in maintaining posture.

The corticospinal tract projects both to postero-lateral regions of the cord grey matter and to the anterior region. In a 4-day-old rhesus monkey, the corticospinal tract projection to the postero-lateral regions is well developed but the animal at this age is unable to perform independent finger movements. Later, the connections directly to the anterior horn cells develop and, pari passu, the animal is able to perform independent finger movements. This emphasizes the requirement of the direct corticospinal tract projection for independent finger movement.

Reorganization of corticospinal projection

The propensity for plasticity is greater in the central nervous system when it is developing than when it is mature. An example is afforded by patients who suffer major damage to one cerebral hemisphere early in life. An infant may be born with hemiplegic cerebral palsy: this is usually due to a congenital abnormality. Alternatively, one hemisphere may suffer damage in the early years of life. When such a child moves the hand or forearm on the side innervated by the healthy hemisphere, these are invariably accompanied by mirror movements occuring on the palsied side. This is due to the existence in these individuals of abnormally branched corticospinal nerve fibres which innervate motoneurone pools on both sides of the cord. This arrangement can be demonstrated by recording simultaneous electromyograms from the forearm and hand muscles on either side of the body. In a normal person, such records show no correlation with each other, whereas in an individual with mirror movements, there is a high probability of the muscle units in muscles on the two sides of the body being activated simultaneously. This linkage is confined to homologous motoneurone pools on the two sides; a particular dorsal interosseous muscle on one side is linked with its fellow on the opposite side but not with a forearm muscle either on the opposite side or on the same side. This indicates a high degree of specificity of innervation of branching neurones, a given corticospinal neurone being targeted specifically at motoneurones innervating a single muscle. When motoneurones on one side of the cord lack normal corticospinal innervation, a replacement corticospinal innervation develops from the opposite side but only from axons destined for that particular muscle.

15.2 REFLEX TIME AND REACTION TIME

For the human biceps muscle, the reflex time is about 20 ms (Chapter 6). This means that 20 ms elapses between the tap to the biceps tendon and the onset of electromyographic (EMG) activity in the muscle. We contrast the reflex time with the **reaction time**. The subject is asked to make a movement as quickly as possible after he feels a stimulus. The reaction time is the time between the stimulus being applied and the start of electromyographic activity signalling the volitional act. The shortest reaction time is around 100 ms.

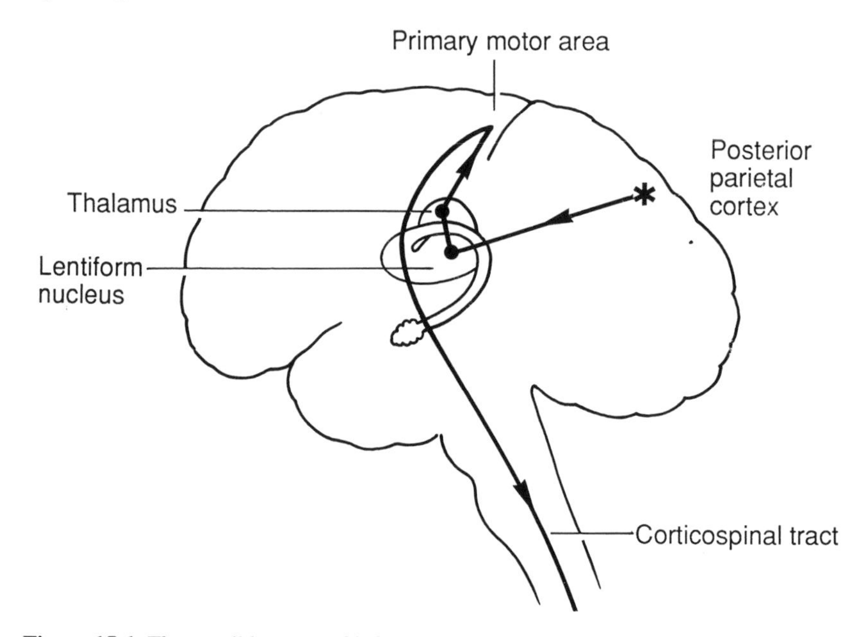

Figure 15.1 The possible route of information flow for a volitional movement. Volition starts in the parietal cortex, indicated by a * Thereafter, there is consecutive involvement of the striatal portion of the lentiform nucleus, the thalamus, the primary motor area of the cerebral cortex and the corticospinal tract.

The neuronal circuits in volitional movement

Records from individual cortical cells show that some cells in the posterior parietal cortex fire specifically in association with voluntary but not with involuntary movements. Their firing precedes the motor response by about 90 ms in the case of movement of the biceps muscle. The sequence of activation thereafter is: basal ganglia, thalamus and, finally, the cells in the primary motor area. The probable circuit is shown in Figure 15.1 (see appendix for more detail).

15.3 'LONG LOOP' OR TRANSCORTICAL REFLEXES AND VOLUNTARY MOVEMENT

Although the response time for a voluntary movement is at least 100 ms, it is possible to influence motor activity much earlier than it is possible to initiate voluntary acts. The tendon jerk can itself be influenced. If you say to a subject 'give a big tendon jerk reflex' and then tap his tendon, there is a much larger bigger response than if you instruct the subject to give as feeble a response as possible. These are the phenomena of facilitation and inhibition of spinal reflexes by higher centres. The effect is achieved by tonic excitatory or inhibitory discharge descending from the higher centres so that the afferent volley finds the reflex pathway facilitated or inhibited.

This leaves unanswered the question of how quickly we can voluntarily change the degree of contraction of a skeletal muscle. (Such an effect is called **phasic** to distinguish it from tonic effects described in the previous paragraph.) The answer depends on the particular type of muscular contraction that one is trying voluntarily to modify. The earliest phasic effect of volition is shown in Figure 15.2. A subject is asked to hold a handle which is originally at rest but can be rotated in either direction and at random times by a motor. Initially, the subject is asked merely to hold the handle and not to resist movements. In this situation, when the handle is rotated, muscles that are stretched exhibit a reflex contraction due to spinal mechanisms. This is shown in Figure 15.2, the onset of EMG activity occurring after a delay of about 20 ms. The experiment is repeated except that the subject is asked to resist the movement of the handle. The subject does not know in which direction the handle will rotate and the rotation must therefore start before the subject can determine in which direction resistance is to be applied. The start of the motor response coincides with the full line because the spinal reflex response is the same whether or not the subject attempts to keep the handle still. At a delay of about 45 ms from the onset of rotation, the traces diverge. This is the first component of a response that is under volitional control; it is

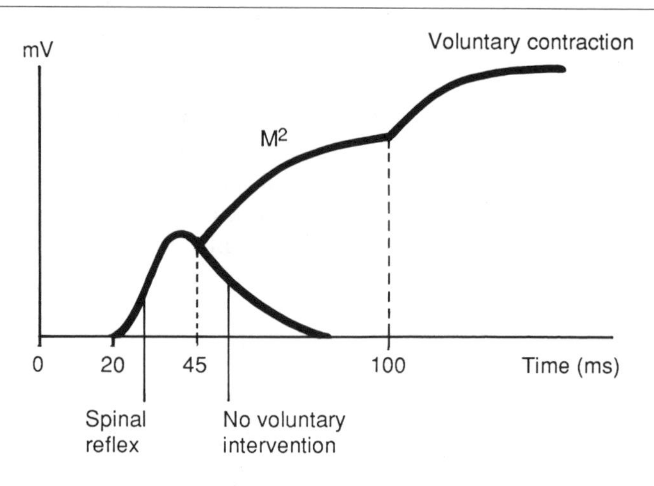

Figure 15.2 Electromyogram records from human biceps muscle. The subject is asked to hold a handle which is originally at rest but can be rotated in either direction and at random times by a motor. Initially, the subject is asked merely to hold the handle and not to resist movements (lower electromyogram). Then the experiment is repeated except that the subject is asked to resist the movement of the handle as soon as the perturbation is felt.

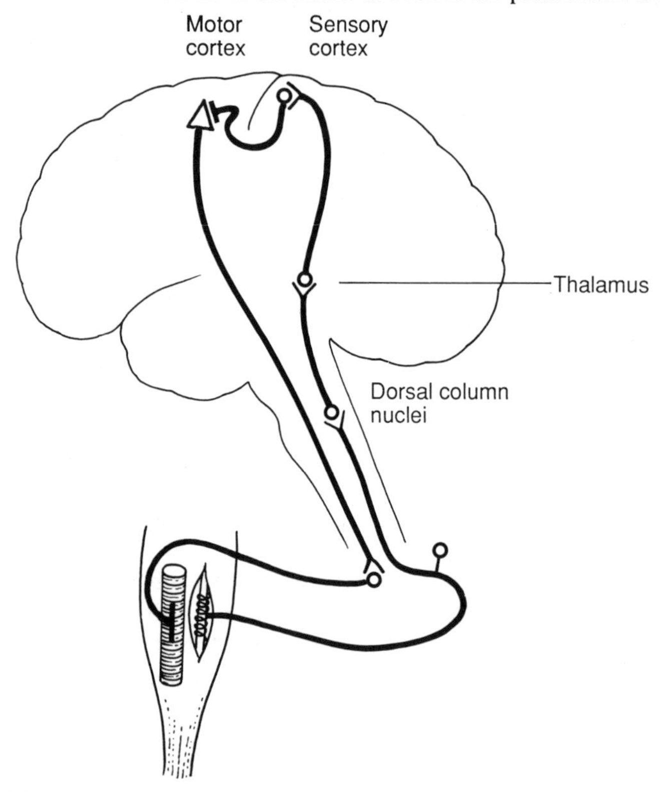

Figure 15.3 The probable circuit of the transcortical 'long loop' reflex.

called the M2 component to contrast with M1, which is the spinal reflex that precedes M2. True voluntary contraction starts with a latent period of around 100 ms.

The latent period of the M2 response is consistent with a pathway up to the cerebral cortex and down again (Matthews, 1991). This pathway is shown in Figure 15.3. The rotation of the handle stretches the muscle and hence the muscle spindle stretch receptors. Activity initiated here initiates nerve impulses which project along the dorsal columns to the postcentral gyrus, then to precentral gyrus and down again. This is called a **long loop reflex**. The pathway passes through the cerebral cortex, where volitional control is exerted. The pathway through the cortex operates if the subject is attempting the keep the weight still but does not operate if the subject is holding the handle passively, that is to say, the circuit can be switched on and off by volition. The most likely site for the volitional control is at the synapse in the postcentral gyrus. The mechanism of volitional control may be by facilitating or depressing the efficacy of synaptic transmission at this site.

In subjects displaying 'mirror movements' because of abnormally branched corticospinal nerve fibres, described earlier in this chapter, ipsilateral M2 responses are accompanied by a long-latency response of the equivalent contralateral muscle. In this situation, the long loop reflex involves action potentials in corticospinal neurones and the abnormal branching provides the anatomical basis for the bilateral long loop reflex contractions.

In some respects, the long-latency reflexes differ between arms and legs. In normal subjects stretching a muscle in one hand produces no reflex contractions of muscles in the opposite hand; this is appropriate since we usually want to move our hands and fingers independently, for instance in undoing a screw. In a person standing on the two feet, however, stretching of muscles in one leg evokes a reflex contraction which involves not only the muscles stretched but also homologous muscles in the contralateral leg. This provides symmetrical postural responses in both legs, an entirely appropriate linkage. When a person stands on one leg, this crossed stretch reflex does not occur.

As we have seen, in a normal person, the M2 component occurs only if the subject is attempting to maintain a constant position of a limb, as in the example of holding a handle which is suddenly displaced. In Parkinson's disease, the M2 component occurs even if the subject is not attempting to maintain a constant position of a limb.

Table 15.1 summarizes the various motor control loops which we have considered. The latent periods refer to the biceps brachii muscle and would be different for muscles closer to or further away from the spinal cord and brain.

Table 15.1 Summary of motor control loops, human biceps muscle

Latent period (ms)	*Name of component*	*Path*
20	Segmental reflex M1, involuntary	Cord
45	Long loop reflex, M2, under voluntary control	Postcentral gyrus Precentral gyrus
100	Voluntary movement	Posterior parietal lobe Basal ganglia Thalamus Precentral gyrus

15.4 STATUS OF THE PRECENTRAL MOTOR STRIP

In the movements considered so far in this chapter, the precentral gyrus is the last centre to be excited before the projection down to the spinal cord; the precentral gyrus is the **penultimate final common path** projecting down to the motoneurones which constitute the final common path. In the evolution of the higher primates and man, the cerebellum, basal ganglia and other subcortical nuclei concerned with finely controlled movement have progressively lost their direct projections to the lower motoneurones and instead project to the precentral gyrus, which acts as a funnel for flow of information. Other motor nuclei discharge earlier than the precentral gyrus. These considerations lead to alternative views of status of the motor strip. One is that the precentral gyrus is demoted to the status of the 'penultimate final common path', operating at the behest of other motor centres which, on this view, take precedence over the precentral gyrus. The highest volitional centre is the parietal cortex; next in authority are the basal ganglia, leaving a subordinate role to the precentral gyrus. The other perspective is that the precentral gyrus is the highest level of motor control, with all other centres subservient and sending requests accompanied by information, but leaving the final decision as to whether or not to act to the precentral gyrus. Either way, all motor information is funnelled from throughout the brain to the precentral gyrus and thence the motor command is launched. This pattern of funnelling has become more and more pronounced as one ascends the evolutionary scale.

Lesions of the precentral gyrus

If the motor cortex is destroyed, for example, by a vascular accident, there is paralysis of voluntary movement; the subject can perform no voluntary movements of the contralateral musculature. However, the mechanism for movement is not lost and involuntary movements occur. When the patient wakes in the morning and stretches himself, both arms move symmetrically, despite the fact that the subject cannot move the affected arm voluntarily. Stretching is an involuntary movement. The side of the face is not accessible to volitional movement but when the subject smiles, this again is an involuntary movement and the mouth moves almost symmetrically.

15.5 THE NEURAL CONTROL OF SKILLED HUMAN MOVEMENT

The principal higher centres of the brain intimately concerned with the control of movement are the motor area of the cerebral cortex, the basal ganglia, and the cerebellum. The basal ganglia and the cerebellum produce their effects primarily by influencing the motor cortex.

Whereas much is known about the types of neurones, the input and output connections, and the details of the intrinsic wiring in the higher neural centres controlling movement, very little is known about what these centres actually do. Movement is produced by coordinated muscular contraction, which in turn is commanded by activity in the nerve cells in the spinal cord that send nerve fibres to innervate the muscles. The higher centres produce their effects via these spinal cord motor neurones. The only region amongst these higher centres which contains neurones discharging synchronously with muscular contraction is the motor area of the cerebral cortex. Even here, it is only the slow pyramidal neurones whose discharge pattern indicates that they directly command muscle contraction. For the discharge of neurones in the cerebellum and basal ganglia, there is a very poor and inconstant relationship between discharge pattern and muscular contraction.

The **basal ganglia** receive profuse input from the posterior parietal region of the cortex, where volitional movements are probably initiated. They project profusely to the supplementary motor cortex, where motor commands are formulated. A minority of neurones in the basal ganglia discharges impulses before a voluntary contraction starts. This, together with the connectivity of the basal ganglia, is taken as indicating that the basal ganglia lie on the pathway for the production of voluntary movement.

The neurones of the cerebellum show no change in their discharge pattern early in a movement; it is only towards the end of a movement that these neurones start to discharge. This indicates that perhaps these neurones are are providing an error signal. Consider a movement to bring the forefinger to a particular position. Early in the movement, the motor cortex launches a signal to produce contraction of the relevant muscles. As the movement proceeds, progressively more feedback is required based on information about the difference between the actual and required positions of the finger. The cerebellum receives profuse proprioceptive input and, as the movement proceeds, the cerebellum is well equipped to provide the error information. The situation is analogous to a liner which has sailed across the ocean by commands from the captain. As the liner approaches a difficult entrance to a harbour, it takes aboard a pilot to provide detailed error signals. The motor cortex is analogous to the captain and the cerebellum analogous to the pilot, who provides guidance as the target of a skilled movement is approached. These ideas begin to provide a crude model of the overall organization of neural control of skilled movement.

15.6 APPENDIX

The cerebral cortex and movement

A decorticate cat or dog is capable of many complex motor activities, such as righting itself if its balance is disturbed, walking, and running. There is loss of placing and hopping reactions (Bard and Rioch, 1937). A primate is much more seriously incapacitated by decortication (Travis and Woolsey, 1965).

Volitional movement

In monkeys, there are neurones in the second somatosensory area and the association area of the parietal lobe of the cerebral cortex whose discharge seems to signal the volitional command of movement. These cells discharge at high rates when a monkey uses his arms and hands for particular voluntary movements; there are cells, for instance, which fire when the animal moves arms and hands to obtain food and drink. These particular cells do not fire during other movements in which the same muscles are used (Mountcastle *et al.*, 1975). Neurones with this pattern of behaviour comprise 10% of all neurones encountered in electrode penetrations in the second somatosensory area (Brodman's area 5) and 30% of all neurones encountered in the neighbouring area of parietal association cortex just behind, a region called the posterior parietal cortex, or Brodman's area 6.

If area 6 is cooled so as to put it out of action reversibly, the monkey is much slower in pressing buttons to obtain his reward (Stein, 1978). Area 6 sends profuse projections to the cerebellum, a modest projection to the prefrontal area, but very little projection direct to the motor cortex (Kuypers *et al.*, 1965; Jones and Powell, 1970).

All parts of the cerebral cortex project to the corpus striatum (Kemp and Powell, 1970); as previously described (Section 11.2), the projection from the posterior parietal cortex is particularly notable. The corpus striatum in turn projects to the globus pallidus, the ventrolateral nucleus of the thalamus, and back to motor cortex. This is the basis for the speculative circuit involved in voluntary movement shown in Figure 15.1.

FURTHER READING

Matthews, P.B.C. (1991) The human stretch reflex and the motor cortex. *Trends in Neuroscience*, **14**, 87–91.

Rothwell, J.D. (1993) *Control of Human Voluntary Movement*. London, Chapman and Hall.

The cerebral hemispheres: association areas, hemispheric asymmetries, speech, split, brain, memory

16.1 ASSOCIATION AREAS

The association areas (i.e. the areas of neocortex not directly concerned with sensation and motor control) comprise the majority of the human neocortex. They lie in the prefrontal region, the temporal lobe (excluding the auditory areas), and the posterior parietal region. Each of these areas includes a speech centre, consideration of which will be deferred until the next section of this chapter.

The association areas are involved in vision as described in Chapter 14; the prefrontal cortex houses the frontal eye fields, involved in the higher control of eye movements, the temporal and posterior parietal lobes are the regions where visual images are analysed.

The prefrontal cortex

This is the entire region of the frontal lobe lying anterior to the motor cortex. This area was originally called 'silent' because no-one knew what it did. It subsequently became clear that the area is far from silent. It has to-and-fro connections with the dorsomedial nucleus of the thalamus which in turn receives fibres from the limbic system.

The absolute and relative size of the prefrontal cortex constitute the human's greatest difference from the apes. This led to the speculation that the prefrontal cortex might be the seat of the intellect. Such ideas were disproved by an experiment of nature involving a certain Phineas Gage. The case was reported by Harlow (1848). Gage had put some explosive into a hole preparatory to blasting. He had tamped the powder slightly with an iron tamping bar. He turned to look to some men working with him when the tamping iron jarred the rock. The powder exploded, driving the iron through Gage's face and the frontal lobes of his brain. The iron then left the skull through the forehead and, covered with blood and brain tissue, landed many metres away. Many years later at post-mortem the injury was shown to have destroyed most of the frontal cortex.

The patient was thrown on his back and gave a few convulsive motions of the extremities, but spoke in a few minutes. His wounds subsequently healed. There was no detectable deterioration of his intellectual facilities and he did not become like an ape. There were, however, dramatic changes in his personality. He became rather impatient, obstinate, fitful, and irreverent, and his language became vulgar. However he was happier, more carefree and less inhibited. After this, experiments were performed on monkeys confirming that removal of the prefrontal areas reduced anxiety, for instance when the animals made mistakes in conditioning tests.

This reduction in anxiety following damage to prefrontal areas suggested the possibility that patients who were unduly anxious might benefit from deliberate frontal lobotomy (isolation of the frontal lobes) or leucotomy (cutting the fibre tracts connecting the frontal lobes with other regions). This 'psychosurgery' began to be practised around 1935 and remained popular until the introduction of pharmacological agents which had similar but reversible effects in the early 1960s. The operation cured the anxiousness. There was little interference with the subjects' intelligence but features of personality such as foresight, imagination, and social behaviour patterns were impaired. Sometimes the changes in personality went too far; side-effects included euphoria, tactlessness, and lack of social inhibitions, the latter exemplified by the tendency to urinate or defaecate on the carpet. Frontal lobotomy or leucotomy also used to be practised for pathological pains which could not be controlled by other means. The pain threshold was not altered but the emotional distress accompanying the pain was alleviated.

The prefrontal areas are also important in the temporal analysis of sensory data. An animal whose prefrontal areas have been destroyed has a greatly impaired ability to remember in which box a reward is hidden. The animal **perseverates**; it always chooses the same box. There is an inability to coordinate previous experience.

The temporal association cortex

This receives input from the visual areas, which lie immediately behind it. This region of association cortex is particularly involved in the interpretation of visual images. Lesions here cause the condition of **psychic blindness**. The subject can see thoroughly well; when walking around the room he will avoid bumping into objects such as a table, but he cannot name the table which he has avoided. He does not even seem to be consciously aware of the existence of the object. The role of the temporal association cortex seems to be in the spatial selection of visual clues. When monkeys' eye movements are monitored, normal monkeys are found to be highly selective for significant objects in the visual field. After temporal lobectomy, eye movements are random.

The anterolateral zone of the temporal lobe was shown by Penfield (1958) to be involved in remembered sequences of events. A patient with an epileptic focus in this region was liable to have an aura which consisted of a vivid memory, or indeed a reliving, of some sequence of events which had occurred in the remote past. Penfield was able to reproduce this experience for the patient by local stimulation of this region of cortex.

The posterior parietal cortex

Lesions of the **parietal** association cortex result in disorders of high level sensory and motor performance. The functions of this part of the brain include assessment of size, shape, weight, texture, and pattern: the 'synthesis' of sensory signals. Lesions of the posterior parietal cortex result in **astereognosis**, which is a condition characterized by an inability to recognize an object such as a pen felt by the hand, although the threshold at which a stimulus is perceived is unimpaired.

On the motor side, defects in the functioning of the parietal association cortex include **apraxia**, which is a defect of high level motor performance with no disturbance of primary motor mechanisms. A subject with a parietal lobe lesion may not be able to perform a particular movement, although he may perform other unrelated movements with the same muscles without any difficulty. A moment later, he may spontaneously make the movement which he had not been able to do voluntarily.

These disabilities are worse when the parietal lobe of the non-dominant hemisphere (see next section) is damaged than when the dominant hemisphere is damaged.

16.2 HEMISPHERIC ASYMMETRY

There are important respects in which the two cerebral hemispheres do quite different complementary things; this phenomenon is referred to as **hemispheric asymmetry**. It is in the functioning of the association areas of the cortex that hemispheric asymmetry is pronounced. Most of us are right-handed; motor coordination emanating from the left hemisphere is superior to that from the right hemisphere. In most people, speech is also a function of the left hemisphere. Even in left-handed people, speech is usually a function of the left hemisphere, although in a small proportion of left-handed people speech is a function of the right hemisphere. The fact that the speech areas are located in the left cerebral hemisphere in most adults can be shown by injecting a short-acting barbiturate into the internal carotid artery on one or other side. Consciousness is scarcely disturbed because the reticular formation is largely supplied by branches of the vertebral arteries. Anaesthesia of one hemisphere has no obvious effect on consciousness. The reticular formation lying in the brain stem and supplied by the basilar artery, together with one functioning hemisphere, is quite sufficient to support consciousness. If you ask the patient to count out loud while you make the injection, then, in most cases, if the injection is into the left internal carotid artery, the patient's speech will temporarily stop, whereas anaesthesia of the opposite hemisphere is without effect on speech. The hemisphere important for speech is sometimes called the **dominant** hemisphere. It contains several **speech centres** as we shall see. The speech areas are regarded as lying in association cortex.

As a result of the siting of the speech areas on the left, the human cerebral hemisphere on the left is significantly bigger than the right. The Sylvian fissure has different angles on the two sides being horizontal on the left but curving up and back on the right. Such an anatomical arrangement is not seen in any subhuman mammals. However it is in evidence on the skull markings of prehuman forms. Some species of birds show cerebral asymmetry associated with singing.

16.3 SPEECH

Normal speech depends on the speech centres in association cortex and on a whole array of other central and peripheral structures. Speech depends, for instance, on the parts of the primary motor area of the precentral gyrus devoted to the lips, tongue, soft palate, and larynx, on the relevant motor nuclei, on the peripheral nerves, and on the muscles themselves. The control of airflow by the respiratory muscles is also important for speech. Interference with different components of this complex results in different speech disturbances.

In all disorders except those involving the speech areas, the structure of language is unimpaired. The subject may have great difficulty in articulation but, when words are spoken, the grammatical structure and the use of language is normal. This difficulty with the mechanism of producing words is called **dysarthria**, which means difficulty with articulation. By contrast, when there is interference with the speech areas, the motor act is unimpaired; any words that are spoken are articulated perfectly. It is the structure of the language which is disordered. Disorders of function in each speech area tend to result in a characteristic speech disorder, although there is much overlap. These disorders of speech are called **aphasia** or **dysphasia**. Aphasia is a misnomer; it means literally 'no speech'. Usually, the subjects can speak but the construction of the spoken language is disordered.

The three principal speech areas (Geshwind, 1970) are shown in Figure 16.1a, in which the temporal lobe has been displaced laterally to expose the in-folded aspect of the superior temporal gyrus which houses the auditory areas. Broca's area is a motor speech area, Wernicke's area is an auditory speech area, and the angular gyrus is a visual speech area.

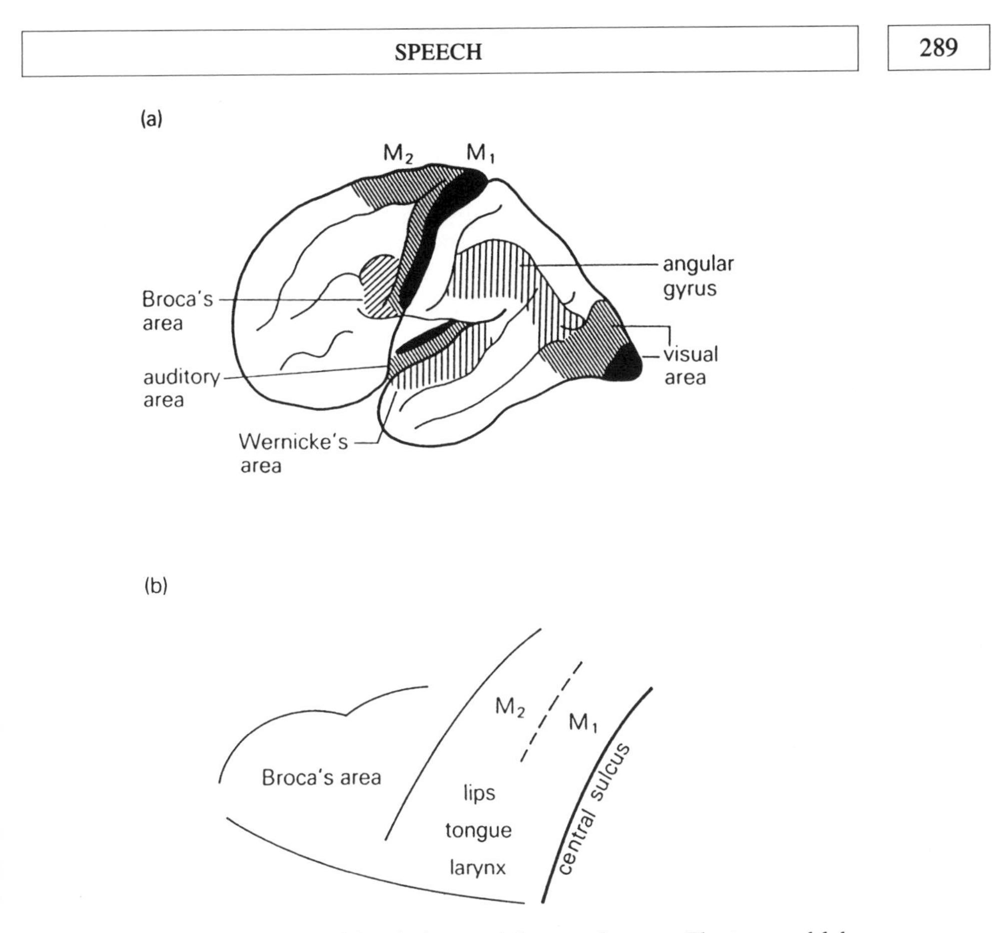

Figure 16.1 (a) The left cerebral hemisphere and the speech areas. The temporal lobe has been displaced laterally to expose the in-folded aspect of the superior temporal gyrus which contains the auditory areas. (b) Broca's area and the contiguous part of the motor strip on a larger scale.

Disorders of speech are almost always accompanied by disorders of reading and writing. However, the disorder of one or other of these activities may predominate.

Broca's area (the motor speech area)

Broca was the first to recognize a discrete speech disorder associated with a focal cortical lesion. The speech area involved is named after him. It lies in the inferior frontal gyrus just in front of the region of the primary motor area connected with lips, tongue, and pharynx (Figure 16.1b). Broca's area projects back to the primary motor area via axons that loop down into the subcortical white matter and back up to the motor cortex. The overall result of activity in Broca's area is discharge of Betz cells in the motor cortex the effect of which is to produce the movements of speech. This pathway from Broca's area to the motor cortex is sometimes interrupted as a result of a small localized lesion in the white matter. In this situation, Broca's area and the motor cortex are disconnected and the motor control of speech is lost.

Broca's area is concerned with co-ordination of motor control and with grammatical construction of speech and writing; it is the centre for memory of the movements that produce speech. It produces integrated commands needed by the motor cortex lying just behind. A lesion in Broca's area results in a defect called **Broca's dysphasia**. Severely affected patients understand everything, but are mute or employ only a few simple words, used mainly for the designation of objects. In milder cases, speech is slow. There is a specific difficulty with saying short words. For instance, it is harder to say 'that is good' than 'parliament'. Short words are omitted, as are participle endings. The result has been described as **telegram speech**.

Broca's area is essential for all linguistic communication, not just verbal communication. A deaf and dumb person can learn to communicate by sign language using muscles of the hand and arm. If such a person suffers a lesion of Broca's area, he can no longer use his hand and arm muscles in this way and can no longer communicate. He finds no difficulty in using the same muscles and movements in, for instance, drinking a cup of tea; there is no paralysis even of voluntary movement in the muscles previously used for communication and it is specifically the communication aspect of hand movement that is lost.

Wernicke's area (the auditory speech area)

Wernicke described another speech area which now carries his name. It is directly adjacent to the primary auditory area, which itself is completely infolded in the Sylvian Fissure. Wernicke's area overlaps out a little way onto the cortical convexity and lies behind and lateral to the auditory area. It is much larger on the left than on the right.

As suggested by its proximity to the auditory cortex, Wernicke's area is important in the interpretation of the speech heard by the subject. The other association regions of the temporal lobe are important in the formation of memory images. Hence, lesions in these regions lead to disturbances of auditory memory images. The patient lacks comprehension of the spoken language. The apparatus for formation of words is intact and is released from the normal control of Wernicke's area. Consequently, the patient speaks continuously.

Mildly affected patients exhibit a continuous flow of rapid speech with the wrong choice of words, neologisms (malapropisms) and circumlocutions. Broca's area is released from inhibition and works overtime. In severely affected patients the speech is so fast and incomprehensible that the condition is often wrongly diagnosed as a confusional state.

Angular gyrus (the visual speech area)

This lies on the course of the tract from Wernicke's area to Broca's area, and is located in the parietal lobe. The angular gyrus lies just in front of the visual cortex of the occipital pole and is important in the visual aspects of speech. Lesions here may specifically impair the ability to read and write. In severe cases, there may be **word blindness** – inability to recognize the meaning of written words. Speech may not itself be disturbed.

Lesions in and around the angular gyrus also result in difficulty in finding names for things. This is called **nominal aphasia**. In its mildest form, it is a common disturbance of function of the normal brain from which many of us suffer; it is exacerbated by fatigue or inebriation. When it is the result of a cerebral lesion, the subject's speech is perfect except for the loss of the occasional word. If he is shown a series of familiar objects and asked to name them, he will perform perfectly for most objects but occasionally will be faced with an object that he recognizes thoroughly well but cannot name. For instance, when shown a pair of spectacles and asked to name it, the subject may say 'It's what you put on – to see better – I wear them myself –'. When he hears one of the words that has been punched out of his repertoire, he cannot recognize it and is unable to attach any meaning whatsoever to it.

Table 16.1 Characteristics of the speech areas

Name	*Siting*	*Functions*	*Deficiency*
Broca's area (the motor speech area)	Anterior to primary motor area; sends commands back to motor area	Grammatical construction of speech, Memory of movements that produce speech	Difficulty with construction of speech, omission of small words, omission of participle endings, little speech, telegram speech
Wernicke's area (the auditory speech area)	Lateral to auditory cortex; projects to and controls Broca's area	Interpretation of speech heard by the subject. Forms and retains memory images of speech	Lack of comprehension of the spoken word. Broca's area is released from control. Speech is fast, incomprehensible, continuous
Angular gyrus (the visual speech area)	Anterior to primary visual cortex	Interpretation of the written word	Lack of comprehension of the written word, difficulty with writing. Nominal aphasia – loss of occasional words

Table 16.1 summarizes the characteristics of interference with the different speech areas.

Speech areas and age

Until the age of about 4 years, both hemispheres participate in speech even though the left superior temporal gyrus is larger than the right in the fetus. After the age of 4, the left takes over. The value of the bilateral potential is that the right may take over if the left hemisphere is atrophic, or if it is damaged by a birth injury or infection. If this occurs before the age of 4, the right hemisphere can take over. Although the subject is liable to be a person of few words, there is no overt disability of speech.

Other hemispheric asymmetries: the right hemisphere

Although speech is the function that is most readily attributable to one hemisphere, many other higher brain functions are also lateralized. Logical and mathematical tasks are performed primarily by the left hemisphere. From observations such as these, the left hemisphere is regarded as the dominant hemisphere. The right hemisphere, however, has its own essential part to play in normal brain functioning. In most humans, singing is a function of the right hemisphere. A singer with a lesion in a speech area can sing perfectly so long as the song is without words. A previously healthy singer who suffers a lesion of the right parietal region can no longer sing, although speech may be perfect. The appreciation of music and reading a music score are both functions of the right hemisphere and are associated with an increase in blood flow in the superior temporal region in the right hemisphere. This and other evidence suggest that the right hemisphere is the seat of the artistic soul whereas the left hemisphere is concerned with handling facts, with rational matters, and with speech.

The right hemisphere is also essential for spatial awareness. Lesions of the parietal association area of the right hemisphere are very disabling and characteristically result in bizarre disturbances of body image. The subject may be quite unaware of the existence of the left side of his body and deny that it exists. When an observer attempts to direct the subject's attention to the subject's own left hand, he disclaims that it is his own, saying that it belongs to someone else. The subject is also unaware of the left side of his environment. If asked to draw a clock face with numbers, he only draws numbers on the right side, squeezing 1 to 12 into the right semicircle. When asked to tell the time from the ward clock, the subject covers half his eyes in an attempt to make sense of the visual input but is quite unable to tell the time. The subject cannot say why he cannot tell the time.

In summary, there are many ways in which the two hemispheres perform different functions that, in the intact subject, are complementary to each other. The name 'dominant' which is often applied to the hemisphere housing the speech centres is probably a misnomer. The hemisphere which does not house speech has its own vital contribution to normal brain function. Some people prefer to use the name 'categorical' for the side responsible for speech and 'representational' for the side responsible for interpretation of spatial form and essential for the artistic aspects of our existence.

16.4 THE SPLIT BRAIN IN MAN

The cerebral hemispheres are richly interconnected by the corpus callosum, which is a massive bundle of around 200 million nerve fibres. In some individuals, it is congenitally absent. It is also deliberately surgically divided to control epilepsy refractory to drug treatment. These individuals have their brains effectively split into two. This produces minimal effects on the subject's temperament, personality, or intelligence.

A subject with a split brain copes well with everyday life. There are times when the isolation of the two hemispheres gives rise to problems. It can happen that the subject tries to tie up shoe-laces with commands from one hemisphere, which wants to go out, and simultaneously tries to untie shoe-laces with commands from the other hemisphere, which wants to stay at home. Another example is of a split-brain subject who angrily and forcibly reached for his wife with his left hand while his right hand grabbed the left in an attempt to stop the aggression (Springer and Deutsch, 1985, p. 33). These are, however, rare incidents, viewed as strange and isolated even by the subject, in an otherwise apparently normal existence.

Special tests, designed to require the hemispheres to cooperate, reveal interesting features. Most of these tests involve providing the two hemispheres with separate visual input which the subject must put together. In every-day situations, an object appearing in the peripheral field causes the subject to turn the head towards the object and to position his gaze so that the object falls on the axis of vision. The object will now project to the occipital poles of both cerebral hemispheres and is thus available for analysis separately by each hemisphere. In order to 'catch out' the split brain, it is necessary to ask the subject to keep the direction of gaze fixed on a point so that the two sides of the visual field project separately to the two hemispheres.

Since the speech areas are on the left, a light presented to the left visual field projects back entirely to the non-speaking right cerebral hemisphere. The split-brain subject 'sees' the light with the right hemisphere, but the speech centres, on the left, know nothing about it. The subject verbally denies its existence. The subject can, however, obey a command to point to the light. Both hemispheres 'heard' the command and the right hemisphere can obey.

A similar phenomenon is found with cutaneous sensation. An object placed in the left hand cannot be named by the subject when asked to identify it without looking at it. It can, however, easily be identified by matching by selection from a series of test objects presented to the left hand.

In another test, the word 'cup' was presented in the left visual field, and so to the right hemisphere. The subject could not say what the word spelt. When a series of objects, including a cup, was presented in the left visual field, the cup was correctly identified by pointing. This shows that the right hemisphere could read and understand the written word, although it could not initiate speech.

The word 'heart' was presented in the visual field so that 'he' was in the left field and 'art' in the right. The subject was asked to pick from a set of cards the word that had been presented. With the left hand, the subject chose 'he' and with the right he picked 'art'. When asked what he had seen, he said 'art'.

When a patterned stimulus or the picture of a face is presented to a split brain subject, each hemisphere operates on the information that it receives and attempts to complete the perception by filling in the unseen part of the

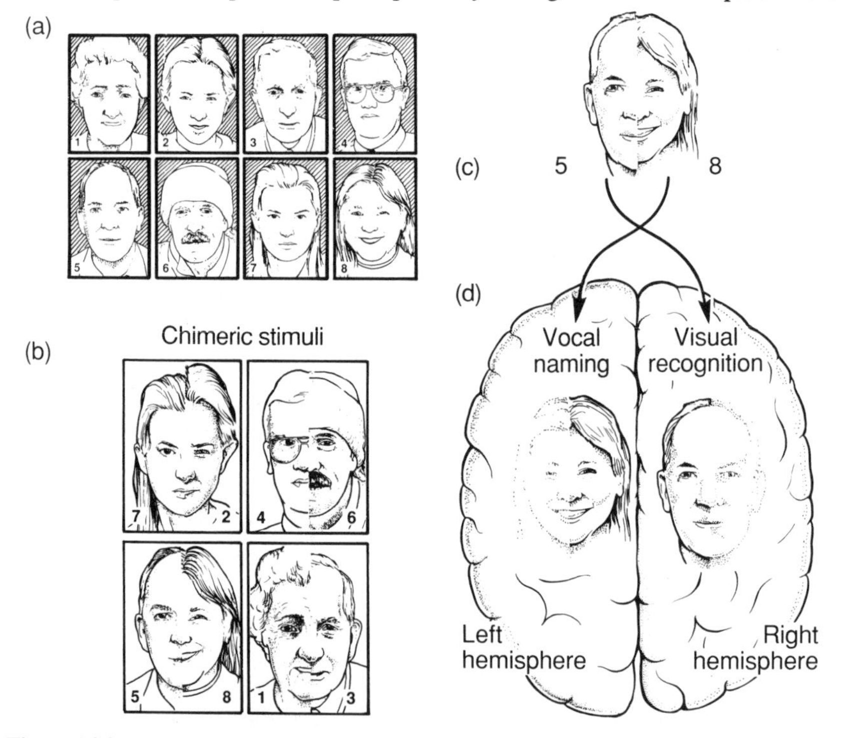

Figure 16.2 The perception of a visual image by a split brain subject. (a) Eight photographs of faces were selected. (b) Photographs were cut in half and different pairs of half faces were mounted together to produce 'chimeric' images. (c) With the subject fixating on the centre of the screen (corresponding to the nose of the chimeric image), the chimeric image was presented for 150 ms. (d) The two cerebral hemispheres received different half faces; each hemisphere interpreted the image as a whole face, filling in the missing half, as indicated by the shadowy half of each face shown in (d).

visual field for itself. This was demonstrated in an investigation of split-brain subjects by Levy *et al.* (1972). These investigators arranged for subjects to be presented with a visual image consisting of the left half of one person's face and the right half of another person's face, as shown in Figure 16.2b. They called this a 'chimeric' stimulus after a mythical monster made of parts of different animals and named Chimera. The two halves of the chimeric image were selected from the series of proper faces shown in Figure 16.2a. The subject was asked to fixate on a spot of red light at the centre of the visual field and the chimeric image was flashed in the visual field for 150 ms, which is too short a time for reflex eye movements to shift fixation to a new point. The subject was then asked to identify the face. The first point of interest is that the subjects were entirely unaware of the fact that the two halves of the visual image were of different faces. The subjects reported no strangeness, discrepancy, or incompleteness, even when there was a gross discrepancy between the two halves of the chimeric image. Each hemisphere completed its own image and since the two hemispheres received different images, they had different ideas of what they had seen. In trials when the subject was asked to name the face, the one corresponding to the image in the right visual field was chosen. In trials when the subject was shown the chimera, then shown the faces in Figure 16.2a and asked to point at the matching face, the one corresponding to the half face in the left visual field was chosen.

In making the choices, the left hemisphere found the task of identifying a face extremely difficult; it hesitated and latent periods were very long and the percentage of errors was high. The right hemisphere made its decisions more rapidly and more accurately, whether this response was pointing with the right or the left hand. This indicates that the right hemisphere has a greater capacity than the left for the apprehension of shape, supporting the suggestion that the kind of cerebral organization suited for language is distinctly inferior when required to deal with apprehension of shape. This reinforces the impression gained from observations described earlier in this chapter, that hemispheric asymmetry provides the possibility of specializations required for language and related symbolic functions in the left hemisphere separated from the different specializations for spatial perceptual abilities in the right hemisphere. In the normal subject, these two faculties complement each other and operate harmoniously together; in the split-brain subject, the two are isolated from each other and cannot co-operate.

To return to the tests on split-brain subjects, in some situations, information may be obtained by a **cross-cue** from one hemisphere to the other. The subject is told that a red or blue light will be presented in the left visual field and that he must choose between it being red or blue. The experimenter then presents the colour and asks 'What colour was that?' The subject's first attempt is a random guess, with equal chances of being right or wrong. When this first guess is wrong (e.g. a red light was presented and the subject reports it as blue) the subject hears with both ears his own verbal response. The right hemisphere, the one which received the sensory input, recognizes the fault and the subject immediately corrects himself and says 'Oh no, I meant red'. This indicates that the disconnection between the hemispheres is not absolute; some sort of communication between them is still possible. This communication is not of precise information but is of an emotional kind. When there is a wrong choice of colours, the right hemisphere signals disapproval to the left which, on receiving this non-specific cue, is able to correct itself. The remaining connections between the two sides of the brain are via the **anterior commissure** and the **hippocampal commissure**. These are two commissures, much smaller than the corpus callosum, which connect the rhinencephalon on the two sides.

Another indication of transfer of emotional information in the split-brain human occurs when the left visual field is presented with a pornographic picture. The subject blushes and, on being questioned, will say that there is something embarrassing happening. But he cannot say what.

There are certain tasks in which a split-brain monkey can perform better than a normal monkey. An example is when an array of panels is presented, 8 to the left visual field and 8 to the right. A random selection of the panels briefly lights up (for 0.5 s) and the animal is rewarded according to how many of the previously illuminated panels he touches. The normal monkey cannot remember the whole array; he presses correctly one panel after another until, about ¾ of the way through the task, he forgets which panels were illuminated. The split brain monkey can act more quickly. By operating the left panels with the left hand and the right panels with the right hand he can complete the whole task. Each cerebral hemisphere has stored the information from the appropriate half of the environment; each hemisphere handles information from 8 of the 16 panels.

Whether a split-brain subject has two brains and two independent sets of consciousness is a matter for speculation. The subject certainly cannot tell you himself, because his speech centres are on the left and can only communicate the awareness of the left hemisphere.

16.5 CONDITIONED REFLEXES, MEMORY AND LEARNING

This is a group of related phenomena, the scientific study of which was pioneered by Pavlov. The story is told of Pavlov that he instructed his technician to feed the dogs in his animal house every day at midday. The technician was to do this when the bells of St. Petersburg struck twelve. One day, Pavlov was in the animal house and his technician bringing the food was delayed. The bells struck twelve and Pavlov noticed that, despite there being no food, the dogs started to salivate. Thus was born the subject of conditioned reflexes.

A conditioned reflex is the association of a sensory signal which originally produces no reflex effect with an unconditioned reflex. In the case of Pavlov's dogs, the unconditioned reflex is salivation reflexly occurring when the dog sees food. The conditioning stimulus is the ringing of bells. Originally the ringing produces no response. If however on many occasions the ringing of bells is followed within a short period of time by the unconditioned reflex, then the ringing of bells becomes associated with the presentation of food. Then the bells alone produce the reflex salivation.

Another type of conditioning is **operant conditioning**. Here the animal must itself take the initiative to be rewarded; for instance it must press one particular lever to be rewarded with food.

Humans exhibit the ability to learn and to remember. The performance of a conditioned reflex in an animal obviously requires processes analogous to learning and memory. However the relationship between these processes in the human, who is consciously trying to acquire knowledge, and the animal in the conditioning situation, who is behaving entirely without introspection, is difficult to ascertain.

There is a phenomenon at the level of the synapse which is regarded by many as a likely mechanism for learning. It is called **long-term potentiation** or LTP. It is more prominent in some central structures than in others: it is particularly pronounced in the hippocampus, which is important in memory. If a piece of hippocampus is maintained *in vitro*, a sequence of test stimuli applied at 1 Hz to a presynaptic nerve bundle results in consistent responses as recorded in the postsynaptic cell. If a period of stimulation at 100 Hz lasting for a few seconds is interposed, thereafter the infrequent stimuli elicit a greatly potentiated postsynpatic response. The potentiation lasts indefinitely, for as long as the slice remains viable. Hence the name 'long-term potentiation'. Such a modification in synaptic efficacy has attracted much attention as a possible physiological basis for conditioning and memory. Its mechanism is considered in Chapter 17.

Phases in memory

When something is remembered, the memory mechanism involves at least two stages. There is **short-term memory**, a mechanism of retaining information for a few minutes. During this time the information may be consolidated into **long-term** memory, a mechanism to retain information for long periods, perhaps for the remainder of the subject's life. Short-term memory does not lead automatically to long-term memory. An example is remembering a telephone number between the time when one looks it up in the telephone directory and dials the number. In this example, most of us are not able to proceed to transfer this to the long-term store.

There is evidence to suggest that short-term memory is based on activity in reverberating neuronal circuits. By contrast, consolidation into long-term memory involves an increase in synaptic efficacy, perhaps by synapses growing larger or by some intracellular structural change in pre- and post-synaptic regions in the neurones storing the information. Evidence is provided for the first by the phenomenon of retrograde amnesia. Consider a motor cyclist who at night crashes head first into a lorry approaching in the opposite direction. The blow to the head results in loss of consciousness. When consciousness is recovered, the subject may be unable to remember occurrences in a period of time up to and including the accident, although memory of occurrences long before the accident and after recovery of consciousness are unimpaired. This is called 'retrograde amnesia'. The more serious the injury the longer the period of retrograde amnesia. As time progresses, the subject may start to remember occurrences of which he was previously unaware. The sequence of recall is such that the duration of retrograde amnesia shrinks but the subject never comes to remember the accident itself. The subject may report 'I am told that I was riding without lights and crashed recklessly into the lorry, but I can remember nothing of it'.

The interpretation of this observation is that, in the reverberating circuits in which the nerve impulses comprising short-term memory is held, the shock of the blow stops the nerve impulses so the information is lost. Occurrences earlier than the period of retrograde amnesia are remembered because consolidation had had time to occur and the information was stored in the form of enhanced synaptic connectivity, a form which is not destroyed by the damage produced by the blow to the head. There is also evidence to suggest that the temporary memory storage system is located in the hippocampus, whereas the long-term store is probably distributed through the cerebral cortex.

Transfer of information from the temporary store to long-term store is powerfully promoted by emotional factors. Most people remember the details of the situation in which they first met the person with whom they subseqently fell in love and married. Another striking example (Carpenter, 1990, p. 328) is that almost everyone who was adult at the time of the assassination of President Kennedy can remember exactly where they were and what they were doing at the time they heard the news, even though those things were of themselves of no significance and would certainly not have been remembered except for the shock of the news.

Hebbian synapses

Hebb (1949) was the first to consider the modification of synaptic transmission which he hypothesized as underlying the acquisition of conditioned reflexes. He formulated rules which a synapse must obey for adaptive behaviour to be exhibited. They are

a. If a presynaptic input is active at the same time as the postsynaptic cell is active, then the synaptic efficacy is increased.

b. If a presynaptic input is quiescent when the postsynaptic cell is active, then the synaptic efficacy is diminished.
c. If the postsynaptic cell is quiescent there is no effect on efficacy.

In recent years, synaptic mechanisms underlying these postulates have been discovered, as will be described in Chapter 17.

FURTHER READING

Springer, S.P. and Deutsch, G. (1985) *Left Brain, Right Brain* (2nd Edn). New York, W.H. Freeman and Co.

Neurotransmitter chemicals in the brain

17

17.1 NEUROTRANSMITTERS IN GENERAL

Various chemicals act as neurotransmitters, and different regions of the brain employ different neurotransmitters. The centres that possess the greatest number of different neurotransmitters are the basal ganglia. New neurotransmitters are discovered frequently in this region, although the reason for the existence of so many neurotransmitters is obscure. In the higher centres, there appear to be fewer neurotransmitters and some order can perceived in the subject.

Neurotransmitters in the cerebral cortex

In the higher neural centres, including the cerebral cortex, there are pathways of many speeds from very fast to very slow. They can be divided into three groups, according to their speed and mechanism. The pathways which are best understood and with which this book is primarily concerned are the fast pathways. Here is the synaptic delay is around 1 ms and the neurotransmitters are amino acids (Table 17.1). The mechanism is the attachment of neurotransmitter to recognition sites (receptors) on the surface of the post-synaptic membrane and the opening thereby of ionic channels which

Table 17.1 Cortical pathways

	Delay	*Mechanism*
Fast	1 ms	Direct linkage of receptor and channel
Medium	100 ms	G-coupled receptors
Slow	5 s	Second messengers (e.g. cyclic AMP)

are immediately adjacent to and directly linked with the receptors. For these reasons, the whole process is extremely fast.

The pathways next in terms of speed have delays of the order of 100 ms. These involve indirect coupling of the receptor on the surface membrane with ion channels via the intermediary of G proteins. These proteins are attached to the inner aspect of the cell membrane and connect receptor sites with ion channels that are some distance away. They act as cytoplasmic shuttles moving between receptor and effector molecules on the internal surface of the membrane (Chabre, 1987).

The slowest pathways introduce delays of the order of 5 s. The mechanism here is the attachment of neurotransmitter to a receptor which in turn activates an enzyme that generates an intracellular messenger, or second messenger, such as cAMP or calcium ions. The second messenger concept was originally introduced to describe the role of intracellular cAMP acting to couple hormone action on the cell membrane with intracellular effects. In the present context, the second messengers diffuse via an intracellular pathway to the cell membrane and open or close ionic channels. It is this intracellular pathway which introduces the delay of seconds into the time between attachment of neurotransmitter to the cell surface and the change in membrane channel conductance.

The synaptic pathways classified according to speed use different neurotransmitter chemicals. The fast paths use amino acids, soon to be considered in more detail. The slower pathways involve large numbers of neurotransmitters such as acetylcholine, noradrenaline, histamine, 5-hydroxytryptamine and neuropeptides, including endogenous opioids involved in the supraspinal control of pain.

Identification of neurotransmitter chemicals

Dale and coworkers established that acetylcholine is the transmitter at the skeletal neuromuscular junction. Skeletal muscle contains a single type of neuroeffector junction. For synapses in the central nervous system, the task of identifying transmitters is rendered difficult because any one neurone receives synapses of many different types and, in any one region of the brain, neurones of many different types are intermingled. Experimenters wishing to identify the neurotransmitter at a particular synapse may inject a putative neurotransmitter, but it is difficult to be sure that an observed effect is due to activation of the synapse being studied.

What types of evidence would one wish for establishing the identity of the neurotransmitter at a particular synapse? This subject has already been discussed briefly in Chapter 4, but now we proceed to a more extended consideration. Appeal is made to the classical work which established acetylcholine as the neurotransmitter at the end-plate but it will subsequently become apparent that one cannot expect all these criteria to be met for central neurotransmitters.

1. The neurotransmitter is present specifically in neurones suspected of using it.
2. The enzymes required for the synthesis of the neurotransmitter are present in the neurone. For acetylcholine, choline acetylase is present in the terminals of the motor nerve fibres.
3. The neurotransmitter is found in the blood or other fluid draining the region when the synapse in question is operating. For acetylcholine acting as neurotransmitter as the vagal terminals in the heart, this was demonstrated by Loewi in his famous crossed-perfusion experiment.
4. Specific recognition sites should exist on the postsynaptic membrane; neurotransmitter released by the presynaptic terminal should attach specifically to these so-called 'receptors'. These receptors are protein macromolecules and methods are available to raise antibodies that specifically bind to any given configuration of antibody. One powerful technique is to label the antibody with a radioactive atom and identify the localization of antibody by pressing thin sections of prepared neural tissue against a photographic plate. This is subsequently developed photographically and the receptor siting is identified as regions where the plate has been exposed; this is the technique of **autoradiography**.
5. Enzymes to break down the neurotransmitter are present. The presence of cholinesterase at the neuromuscular junction was one of the important pieces of evidence in Dale's work.
6. Application of the chemical from a microelectrode onto the synapse should yield a postsynaptic potential.
7. **Pharmacological experiments**: the application of the putative neurotransmitter to the synapse should obviously have the same effects as activation of the synapse by impulses in the presynaptic nerve fibre. Pharmacological agents that influence transmission at the synapse should produce similar effects on both modes of activation of the synapse; agonists should enhance, and antagonists should block both similarly. In particular, the existence of a specific antagonist for a putative neurotransmitter provides a powerful tool for investigating the role of that neurotransmitter in the functioning of the brain.

Neurotransmitter chemicals for fast pathways

These are probably amino acids, that is, they possess an amino (NH_2) group and a carboxyl (COOH) group. Amino acids differ in the nature of the R group. Figure 17.1 shows the chemical structure of the most important amino acid neurotransmitters. α-Amino acids, with the carboxyl group on the α-carbon atom, are involved in general amino acid metabolism, being the building blocks of the body peptides and proteins. They are ubiquitous and this makes for difficulties in identifying those that also play a role as neurotransmitters.

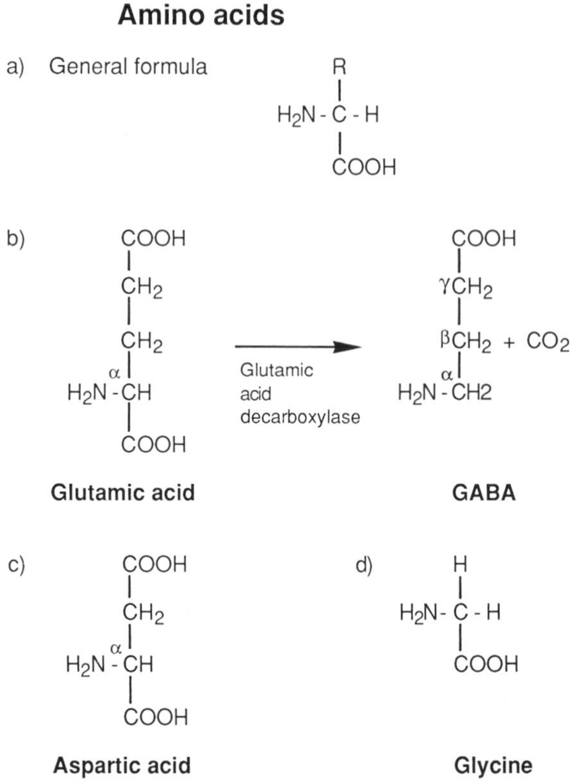

Amino acids

a) General formula

R
|
H_2N - C - H
|
COOH

b)

COOH
|
CH_2
|
CH_2
α |
H_2N - CH
|
COOH

Glutamic acid

→ Glutamic acid decarboxylase

COOH
|
γCH_2
|
βCH_2 + CO_2
α |
H_2N - CH2

GABA

c)

COOH
|
CH_2
α |
H_2N - CH
|
COOH

Aspartic acid

d)

H
|
H_2N - C - H
|
COOH

Glycine

Figure 17.1 (a) The general formula of an amino acid. Different amino acids differ in the nature of the R grouping. (b) The formation of GABA from glutamic acid, a reaction catalysed by glutamic acid decarboxylase (GAD). (c) Aspartic acid. (d) Glycine.

Inhibitory neurotransmitters

GABA

For inhibitory synapses in the cerebral cortex and other higher centres, the principal inhibitory transmitter is **γ-aminobutyric acid (GABA)**. Gamma amino acids are not involved in general amino acid metabolism, nor are they constituent amino acids of peptides and proteins. For this reason, the evidence for GABA being a neurotransmitter is relatively secure.

GABA is present in high concentrations in the central nervous system but not elswhere in the body, and, within the nervous system, it is found only in specific central neurones. GABA is synthesized from glutamic acid, as shown in Figure 17.1b, in a reaction catalysed by a specific enzyme **glutamic acid decarboxylase (GAD)**. This enzyme can be detected readily by histochemical means. Also, it is possible to raise antibodies against GAD for autoradiographic identification of binding. With such techniques, it has been established that any one neurone contains either a high concentration of GAD or almost none. The presence of GAD in a neurone is good evidence that the neurone is GABAergic. GABA agonists and antagonists are available and have been used to establish a strong case for GABA being the inhibitory neurotransmitter. Altogether, these several lines of evidence have established beyond reasonable doubt the role of GABA as the primary inhibitory neurotransmitter in higher centres including the cerebral cortex.

Glycine

GABA is not the only inhibitory neurotransmitter in the cortex, as evidenced by the fact that blockade of GABAergic transmission does not abolish inhibitory action altogether. A candidate for the neurotransmitter subserving non-GABAergic inhibition is glycine. This is the simplest of all amino acids, with the R group being hydrogen (Figure 17.1d). Unlike GABA, glycine is involved in the general metabolic processes of the body. Consequently, the presence in a neurone of glycine is not a specific indicator that the neurone cell uses glycine as its transmitter. The spinal cord and brain stem contains cells which stain heavily with stains specific for glycine; these are candidates for glycinergic neurones. In neocortex, there is no dramatic staining of cell bodies and no staining of dendrites and axons: cell body staining may merely be due to the presence of glycine in normal amino acid metabolism. Unlike GABA, there is no specific enzyme in the production of glycine, so it is not possible to raise an antibody for cells suspected of being glycinergic. These problems in establishing that glycine, rather than another amino acid, is responsible for non-GABAergic inhibitory synaptic transmission have led to the introduction of the name 'non-GABAergic' inhibition to describe the phenomenon.

The most promising approach for establishing a role for a chemical such as glycine as a putative neurotransmitter is to demonstrate specific postsynaptic receptors. Glycine acts as an inhibitory neurotransmitter and is blocked by strychnine. Histological sections of neural tissue may be incubated with tritiated strychnine to localize the ligand by autoradiography. This method reveals receptors in abundance in the cord and sparsely distributed in the cerebral cortex. The existence of glycinergic inhibition in the cerebral cortex is thus still in doubt. The balance of evidence is against the existence of glycinergic neurones intrinsic to the cortex. There may, however, be glycinergic input to the cortex from subcortical structures (extrinsic glycinergic inhibitory mechanisms).

As will be described shortly, there are also glycine receptors as part of the receptor complex for the NMDA receptor. This binding site behaves differently in that it is not blocked by strychnine and it does not bind antibodies raised to the strychnine-sensitive sites.

Comparison of inhibitory neurotransmitters in the cord and in higher centres

GABA is the principal inhibitory neurotransmitter in the higher centres. Postsynaptic inhibition in the spinal cord is mediated by glycine, whereas presynaptic inhibition is mediated by GABA. Presynaptic inhibition does not occur in the cerebral cortex.

Excitatory neurotransmitters

Evidence is accumulating to suggest that the dicarboxylic α-amino acid glutamate is the most widely distributed fast excitatory neurotransmitter, with aspartate probably attaching to receptors on the presynaptic terminal. Both of these are common amino acids involved in metabolism and are present in all tissues. There is no difference in average concentration of these amino acids between inhibitory and excitatory neurones. However there is recent autoradiographic evidence that certain presynaptic nerve terminals are rich in glutamate, adding weight to the claim that this chemical is acting as a neurotransmitter. There may be barriers separating the synaptic cleft from the surrounds, and isolating neurotransmitter glutamate within from metabolic glutamate around.

There are techniques for collecting fluid superfusing neural tissue and detecting glutamate. A related technique which samples fluid from around neurones is to insert two cannulae, to allow injection of physiological saline through one and its extraction through the other, a 'push – pull' technique. When the relevant synapses are stimulated, glutamate overflows into the fluid that is collected. A rise in the yield of glutamate is evidence of its role as a neurotransmitter.

Uptake and release of glutamate by synaptic terminals

At least half the synapses in the mammalian central nervous system are glutamatergic. At these synapses, the glutamate required for neurotransmission is taken up from extracellular fluid (concentration 1 μM) into the cytosol of the presynaptic terminal (concentration is 1 mM). Transport into the terminal is by a specific coupled carrier, glutamate entry being coupled to sodium ion entry. The energy for transferring glutamate against its concentration gradient is derived from sodium ions flowing down their electrochemical gradient. There is a separate carrier mechanism for transferring glutamate from the cytosol to the synaptic vesicle; inside the vesicle the glutamate concentration is 100 mM.

Lowering of the membrane potential of the terminal reduces the energy drive for sodium entry and the glutamate pump runs in reversed direction. This is the origin of glutamate release from the terminals in cerebral hypoxia (Section 17.4 below).

17.2 THE POSTSYNAPTIC RECEPTORS FOR AMINO ACID NEUROTRANSMITTERS

The postsynaptic receptor for GABA

The attachment of GABA to its postsynaptic receptor results in the opening of channels that allow the passage of chloride ions (Section 4.4). The GABA receptor complex has various components, shown diagrammatically in Figure 17.2, taken from Krogsgaard-Larsen *et al.* (1986). The site at which GABA itself attaches is adjacent to the chloride channel which is opened by GABA. There is another attachment site, spatially distinct from the GABA attachment site, to which the benzodiazepine drugs attach. There is yet a third site to which barbiturates attach.

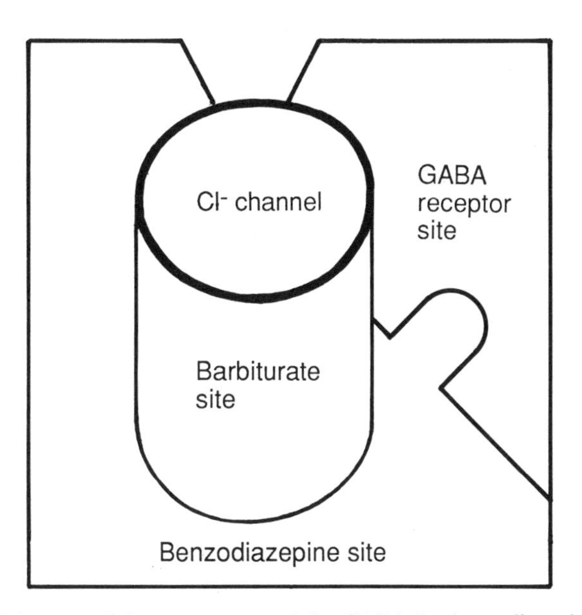

Figure 17.2 Diagram of the components of the GABA (or benzodiazepine) receptor.

The most common benzodiazepine is diazepam, a drug widely used as an anxiolytic agent, particularly in America. When a benzodiazepine attaches to the GABA complex, it does not itself influence the conductivity of the chloride channel. However when benzodiazepine and GABA are both present, the potency of the GABA is increased. This is due to alteration of the shape of the GABA receptor site to enhance its affinity for GABA; the name given to this type of effect is an **allosteric** relationship. Just as the affinity of GABA binding sites is increased by benzodiazepines, GABA also enhances the affinity of the benzodiazepine binding site for its ligand.

By virtue of binding to the barbiturate site on the receptor complex, the sedative barbiturates increase the binding of both GABA and benzodiazepines. By this means, these barbiturates increase the sensitivity to GABA without themselves opening the chloride channel. These allosteric effects are blocked by certain convulsants such as picrotoxin. These observations provide a rational basis for the therapeutic actions of benzodiazepines as anti-anxiety drugs at low concentrations and as anaesthetics at higher concentrations. They probably also explain the anaesthetic effects of barbiturates in terms of enhancement of inhibitory transmission at central synapses.

The GABA receptor complex is often referred to as the 'benzodiazepine' receptor complex. In order to study the receptor protein macromolecule, the molecular biologist attaches a ligand to it, purifies the ligand-bound receptor by separating from other cell membrane constituents and then dissociates the ligand–receptor complex to yield the pure receptor complex. This complex can then be studied by the methods of molecular biology to reveal its molecular structure. Central to this process is the requirement of a ligand which binds specifically and tightly to the complex to act as a handle for extraction. When the GABA receptor was first isolated, benzodiazepine was used as the ligand, hence this receptor became called the benzodiazepine receptor.

The benzodiazepines and the barbiturates are exogenous pharmacological agents and there is speculation about whether their binding sites have any function in the normal physiological activity of the brain.

Receptors for excitatory neurotransmitters

The excitatory neurotransmitter glutamate attaches to postsynaptic receptors of several different pharmacological types. This is reminiscent of the situation of nicotonic and muscarinic receptors for acetylcholine. In the case of the receptors for acetylcholine, the pharmacological agents muscarine and nicotine specifically attach to one or other type of cholinergic receptor and this forms the basis for distinguishing between them. The different types of receptor are found at different junctions and have different physiological properties.

In the case of the excitatory synapses in the brain, the receptors are named after the three agonist agents which bind preferentially to them: n-methyl D-aspartate (NMDA), α-amino-3-hydroxy-5-methyl-4-isoxazoleproprionate (AMPA), and kainate. These three chemicals are exogenous pharmacological agents which specifically attach to one or other group of receptors and mimic the effects of endogenous neurotransmitter. As with the cholinergic receptors of different types, these different post-synaptic receptors seem to be involved in different physiological functions. The AMPA and kainate receptors are considered together as 'non-

NMDA receptors' and they seem to be responsible for the lower neurophysiological functions of transferring information from place to place; they mediate spinal reflexes and cortical evoked potentials, for instance. The NMDA receptors are responsible for the more advanced functions of the brain, such as learning and long-term potentiation.

The effect of membrane potential on NMDA-operated channels

Before describing the behaviour of the NMDA-operated channel, we will review that of the GABA-operated channel, described in Chapter 4. As with most channels operated by a neurotransmitter, for GABA-operated channels the conductance is voltage insensitive. The conductance measures the amount of opening of channels. For GABA, this membrane conductance (Figure 17.3a) is the same irrespective of the membrane potential at the time of arrival of the inhibitory input. The synaptic current depends on two factors, the magnitude of the conductance of the channels and the driving force. For the GABA-operated channel, the driving force is the difference between the membrane potential and the chloride equilibrium potential (−70 mV). With a membrane potential of −60 mV, the driving force is for chloride to move into the cytoplasm and this makes the potential more negative, towards the chloride equilibrium potential. With a membrane potential of −70 mV, there is no driving force, no net current passes and there is no synaptic voltage excursion recorded. With a membrane potential of −80 mV, the driving force is reversed, chloride leaves the cytoplasm and the synaptic potential is depolarizing. In all three cases, however, the amount of opening of the GABA-operated channels is the same; only the driving force is varied.

The non-NMDA operated channels opened by glutamate exhibit a similar simple behaviour. Attachment of glutamate to the receptor opens channels irrespective of the ambient membrane potential. For this type of synapse, the equilibrium potential is close to zero, so that the driving force is approximately equal to the membrane potential.

The NMDA-operated channels are different, in that their conductance is voltage dependent. The membrane potential influences their conductance rather similarly to its influence on the voltage-sensitive sodium channels responsible for the upstroke of the action potential in nerve. At the resting potential of neurones, the NMDA-operated channels allow very little ionic current to flow (Figure 17.3a). This is because they are plugged with magnesium ions, as shown in Figure 17.3b. These magnesium ions fit into the NMDA-operated channels but can scarcely squeeze through. They are pulled into place by the membrane potential, the intracellular negativity attracting the postively charged magnesium ions. If the membrane is depolarized towards zero membrane potential, the attracting force on the magnesium ions weakens, the magnesium ions tend to diffuse away and leave the channels open. Now the channels are truly open and, since the channels are large, sodium, potassium, chloride,

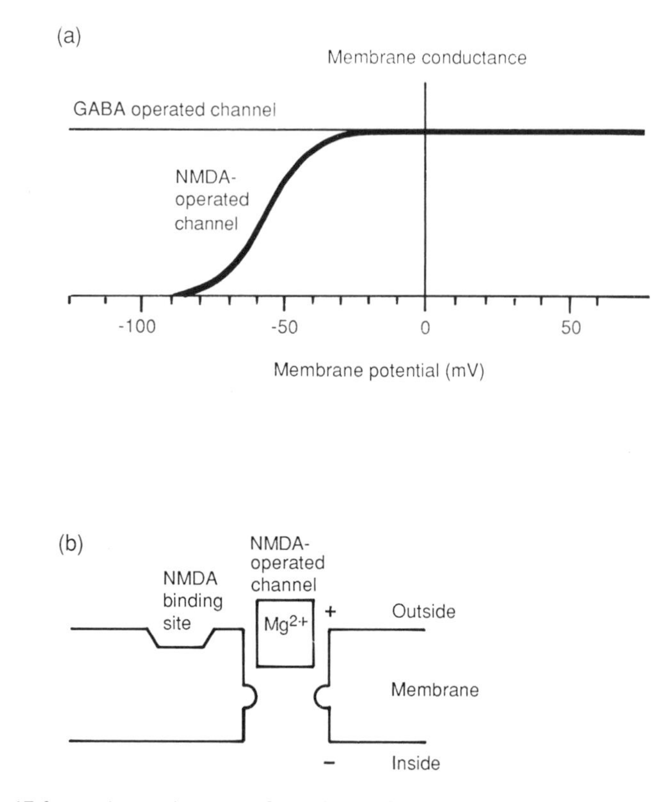

Figure 17.3 (a) The conductance of membrane channels operated by GABA and by NMDA as a function of membrane potential. (b) The NMDA-operated channel, with a magnesium, ion plugging the channel. The resting potential, outside positive and inside negative, provides an electrostatic force sucking the magnesium ion into the channel thereby blocking it.

and calcium ions can move through. Since the channels allow these ions through indiscriminately, they act as a short circuit and ion movement brings the membrane potential towards zero.

For NMDA-operated channels to produce membrane depolarization, the channels must be opened by neurotransmitter and the neuronal membrane must be already depolarized due to some other effect. If these synaptic channels are activated by neurotransmitter at the resting potential, very little synaptic current flows. In its normal functioning, the NMDA operated channel usually cooperates with non-NMDA channels. Synaptic activation of both types of channel occurs simultaneously. The non-NMDA channels cause a depolarization of the membrane, allowing the NMDA-operated channels conduct. This causes further depolarization and further opening of

NMDA-operated channels. The system slides into a positive feedback destabilizing loop.

Calcium and NMDA-operated channels

The size of the channels operated by NMDA receptors is greater than that of channels operated by the non-NMDA receptors. Whereas non-NMDA operated excitatory channels allow only the small inorganic ions sodium, potassium and chloride to pass, NMDA-operated channels additionally allow calcium ions to pass. Because of the huge inward driving force for calcium ions, the opening of these channels results in an insurgence of calcium ions. Such ions play an important role as intracellular messengers and influence the biochemical processes within the postsynaptic cell. This provides a mechanism for heterosynaptic interactions, as in Hebbian synapses, described below.

NMDA receptors, the hippocampus and memory

The hippocampus has the highest concentration of NMDA receptors in the whole brain. A pharmacological agent, amino-phosphono-valeric acid (APV), specifically blocks the NMDA receptors but not the other glutamate receptors. If this is administered to rats, their normal activity such as running, feeding, and grooming, is unaltered. However they can no longer learn tasks involving spatial memory.

In one group of experiments performed by Morris, rats were put in a pool of water so deep that, if they stood on the bottom, their noses were below the surface of the water. Rats do not like swimming and they try to find a way out of the water. They were provided with a submerged platform that was obscured from view by adding some milk to make the water opaque. The rats had the task of learning the whereabouts of the platform from visual clues in the environment. Normal rats quickly learn to find the submerged platform. Rats treated with the NMDA-blocking agent could not learn to find the platform; they swam about at random until eventually they reached the platform by chance.

This is just one of many observations pointing to the importance of this particular type of receptor in learning.

How does the NMDA-operated channel provide a mechanism for heterosynaptic interaction?

A neurone with two synaptic inputs can be regarded as a model of conditioning. Input 1 represents the unconditioned stimulus, such as the presentation of food. Input 2 represents the conditioning stimulus, such as the bell which chimes just before food is presented. Input 1 provides intense depolarization of the postsynaptic cell, which fires impulses. These produce the unconditioned response, such as salivation, in response to the presentation of food. Initially, the bell by itself activates Input 2 but this is insufficiently intense to cause the postsynaptic neurone to fire. We suppose that the postsynaptic receptors at synapse 2 are of the NMDA variety, so that their activation at normal resting potential produces very little synaptic current because of the magnesium blockade.

Simultaneous or near simultaneous activation of Inputs 1 and 2 must occur repeatedly for conditioning to be established. The strong depolarization of the postsynaptic cell by Input 1 relieves the magnesium blockade of the NMDA operated channels of Input 2, so now synaptic current flows at synapse 2. Calcium enters into the postsynaptic cell through synapse 2 and this activates intracellular processes in the region of synapse 2 to potentiate synaptic transmission there. After repeated pairing of stimuli, Input 2 has become sufficiently potentiated for the bell alone to trigger impulses in the postsynaptic cell and hence establishment of the unconditioned reflex.

The NMDA receptor–ion channel complex

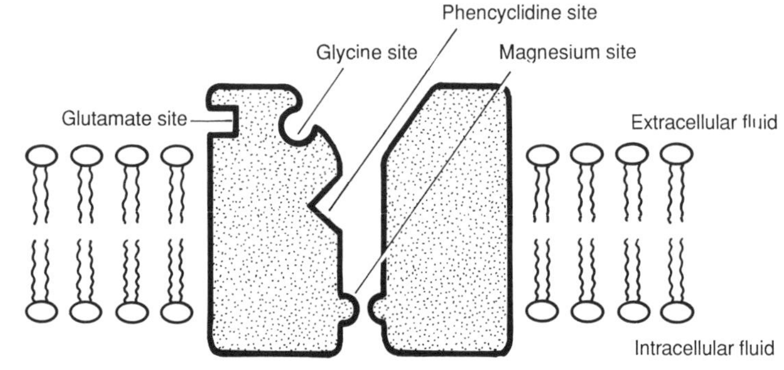

b) Activated (channel open)

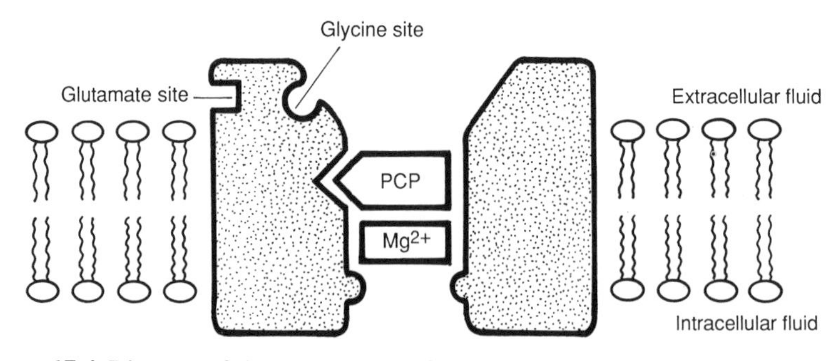

Figure 17.4 Diagram of the components of the NMDA receptor.

As shown in Figure 17.4, this complex has several distinct sites (Bullock and Fujisawa, 1992). **Magnesium** fits into the ion channel, as already described. The complex also has a **glycine-binding site** which is insensitive to strychnine; this pharmacologically distinguishes this receptor from the postjunctional receptors for strychnine at inhibitory synapses in the spinal cord. The affinity of the glycine binding site on the NMDA complex is so great that the concentration of glycine normally present in the interstitial fluid of the brain saturates the site; glycine does not therefore act as a neuromodulator since its concentration is always supramaximal.

Channel binding site for phencyclidines

Phencyclidine, or angel dust, was a drug of addiction, used primarily in America. It dissociates the individual from reality. Phencyclidine receptor ligands block NMDA responses in a non-competitive, activity dependent manner, probably by binding within the ion channel. The drug is therefore an open channel blocker.

Metabotropic receptors

These are receptors for glutamate separate from the NMDA, AMPA, or kainate receptors. Attachment of glutamate to NMDA, AMPA, or kainate receptors leads to the opening of membrane channels that allow the passage of inorganic ions. Attachment of glutamate to metabotropic receptors does not cause membrane channels to open. Instead, it operates via G-proteins and intracellular messenger systems, resulting in an increase in intracellular calcium ion concentration released from intracellular stores.

Thus glutamate increases intracellular calcium ion concentration both by allowing calcium entry through NMDA-operated membrane channels and release from intracellular stores by operation of metabotropic receptors.

Neuroactive peptides

Neuroactive peptides are widely distributed in the nervous system. Many of them also occur elsewhere in the body. They range in size from dipeptides such as carnosine to polypeptides as large as hormones. There are probably at least 50 neuropeptides. They are synthesized in the cell body and transferred by axoplasmic transport to the nerve terminal. This is a point of contrast with neurotransmitter chemicals such as amino acids and acetylcholine, which are synthesized by cytosolic enzymes at synaptic sites.

The postsynaptic receptors for neuropeptides have to date escaped precise characterization, and the raising of antibodies for binding studies has not yielded useful results at presumed peptidergic synapses.

Peptides usually produce their effects not by acting as ligands for ligand-gated channels but by interacting with neurotransmitters. Many neurones release neuropeptides in addition to the primary neurotransmitter chemical. Such peptides may modulate the postsynaptic response to neurotransmitter by binding to a component of the receptor for the neurotransmitter and thereby increasing or decreasing the receptor's affinity for its ligand.

These peptides may also bind to their own receptors on the postsynaptic membrane and these, via intracellular messenger systems, may modulate the postsynaptic response. The effects of a neuropeptide may be sensitive to the situation in which they are released; it is no mean task to determine the role of neuropeptides in normal neuronal functioning.

Second messengers and retrograde messengers

cAMP is the classical second messenger; novel ones are arachidonic acid and nitric oxide (Bockaert, 1992). These two are unusual in being both intra- and extracellular messengers. Nitric oxide diffuses across cell membranes with ease. It is produced within the cell from arginine when NMDA receptors are activated. It diffuses out of the cell in all directions, thus acting as a retrograde messenger. Nitric oxide depresses the NMDA sensitivity of the postsynaptic membrane and the system thus acts as a negative feedback loop to help prevent over-activity of the NMDA system. Nitric oxide is a free radical and at high concentrations is toxic.

Arachidonic acid is released from a component of the membrane on the intracellular aspect of the membrane. It is also released as a consequence of activation of NMDA receptors. The NMDA-operated channels allow the entry of calcium, and this acts in its capacity as a second messenger to release the arachidonic acid into the cytoplasm close to the cell membrane. The arachidonic acid can then leave the cell by passing through special membrane channels. The ways in which it may operate as a retrograde messenger are described in the next section.

17.3 LONG-TERM POTENTIATION

This phenomenon was described in the previous chapter. It was originally observed in the hippocampal slice. Long-term potentiation in the hippocampus has been widely studied since the mechanisms involved in its induction and maintenance are believed to underlie the fundamental properties of learning and memory. At synapses that exhibit long-term potentiation, the neurotransmitter is glutamate and NMDA type postsynaptic receptors are present. Long-term potentiation is blocked by NMDA antagonists.

Long-term potentiation requires a strong depolarization of the postsynaptic cell because of the need to unveil NMDA operated channels. This strong depolarization can be achieved by a variety of means. Most simply, it is achieved by activating multiple inputs to the cell. If the postsynaptic element is prevented from being depolarized by a polarizing current applied during the period of rapidly repeated stimulation, long-term potentiation does not occur. This demonstrates the need for the postsynaptic cell to be depolarized. Again, this is because NMDA-operated channels must be activated for long-term potentiation to be established.

At the other extreme, it is possible to produce long-term potentiation by stimulation of a single presynaptic neurone if, concurrently with the repetitive stimulation, a depolarizing current is passed into the postsynaptic cell. A single presynaptic neurone can induce long-term potentiation in its target if the experimenter is prepared to help by artificially depolarizing the postsynaptic cell. Long-term potentiation does not occur if the depolarization is given before or after the tetanus.

There is currently heated debate about whether long-term potentiation is a presynaptic or postsynaptic phenomenon. It is likely that both mechanisms operate and that one or other may predominate depending on the exact experimental circumstances and the nature of the tissue being studied. An increase in release of neurotransmitter glutamate is associated with long-term potentiation. At first sight this seems to indicate a presynaptic mechanism, but there is evidence that the effect is due to arachidonic acid acting as a retrograde messenger. With the strong activation of NMDA receptors needed for the induction of long-term potentiation, calcium enters the cytoplasm and this leads to the release of intracellular arachidonic acid, as previously described. This arachidonic acid then leaves the cell and diffuses back to act on the presynaptic terminal. One mode of its action is to increase intra-synaptosomal calcium levels and thereby enhance the release of neurotransmitter. Another effect is to block the uptake by the presynaptic terminal of released glutamate and this results in a prolongation of depolarization of the postsynaptic membrane. The increased release of neurotransmitter associated with long-term potentiation thus involves cooperation between pre- and postsynaptic cells.

We are still a long way from understanding the cellular and molecular basis of learning and memory but this type of experimentation promises to be a starting point for progress in this subject.

17.4 EXCITOTOXINS

This name is applied to chemicals, both exogenous and endogenous, that stimulate neurones by acting on excitatory synapses; in some circumstances they generate so much activity that the neurones are damaged or killed. Glutamate acting as a excitatory neurotransmitter is innocuous at normal concentrations but at higher concentrations it is an excitotoxin. This action involves all the different types of postsynaptic receptor. The action of excess glutamate on AMPA and kainate receptors results in sodium entry followed by water entry. Glutamate acting at both NMDA and metabotropic receptors causes an increase in concentration of calcium in neuronal cytoplasm. The effects of excitotoxins are commonly mediated by an excessive increase in intracellular calcium concentration. The effects of excessive activation gives neuronal death with a characteristic histological appearance. In two common degenerative neurological conditions, Huntington's disease and Alzheimer's disease, degeneration with these characteristic features is present. This suggests that these diseases may have as a component excessive activation of NMDA receptors by endogenous chemicals such as glutamate.

Glutamate toxicity

Hypoxia, hypoglycaemia, and epilepsy result in the release of glutamate in most mammals, including man. Certain marine vertebrates that dive for long periods can withstand cerebral hypoxia: resistance to hypoxia is due to an adaptation which means that hypoxia does not cause glutamate release. In the human brain, the release of glutamate that accompanies cerebral hypoxia contributes drastically to the brain cell death which accompanies stroke. Pharmacological agents which block calcium channels in cell membranes are currently the most effective way of protecting cells against death in stroke patients. They have a dual mode of support. First, they relax the smooth muscle of blood vessels and thus assist in perfusing hypoxic neural tissue with blood. Second, they impede the calcium entry into neurones and thus help neurones to resist excitotoxicity. NMDA blockade is also a promising tool for protection of hypoxic brain against excitotoxic damage.

FURTHER READING

Chiari, G.D.C. and Gessa, G.L., eds (1980) Glutamate as a neurotransmitter. *Advances in Biochemical Psychopharmacology*, **27**, New York, Raven Press.

Hartmann, A. and Kuschinsky, W., eds (1989) *Cerebral Ischaemia and Calcium*. New York, Springer-Verlag.

Watkins, J.C. and Collingridge, G.L. eds (1989) *The NMDA Receptor*. Oxford, IRL Press.

Pain

18

18.1 THE SENSATION OF PAIN

Pain is an unpleasant sensory and emotional experience associated with actual or potential tissue damage (Wall and McMahon, 1986). It is produced if the skin is cut, compressed, burnt, or frozen. Pain is a non-specific sensation in that it alerts the subject to the fact that something dangerous is happening to a part of the body without supplying much information about the nature of the danger.

Pain makes the subject aware of danger and forces him to do something about it. Pain is a sensation which we cannot ignore; pain we obey. Pain is a common cause of human suffering and doctors spend much of their professional life concerned with its alleviation.

When we have pain, we feel that man should not have been born to suffer and we are likely to think that pain is an outright curse. This is not true and pain has several useful functions. First, it is a warning of threat to tissue or to the organism as a whole. Pain occurring before serious injury can prevent further injury. For instance, if one steps lightly on a sharp object such as broken glass, pain produces immediate withdrawal or other appropriate reaction to prevent further injury. Second, pain serves as a basis for learning not to touch dangerous things. Third, pain forces the subject to rest an inflamed limb or, if the pain is widespread, it enforces the whole body to rest. This gives the defence mechanisms a chance to contain and repair the damage.

The absence of pain and its effects

Individuals have been described in whom the sensation of pain is absent. No consistent anatomical abnormality has been found, although in some cases an obvious cause is present. Lack of pain is sometimes familial and sometimes a result of acquired disease when central mechanisms subserving pain are destroyed.

In our formative years when we are acquiring a strategy for survival, we learn whether or not a particular activity is dangerous by whether or not it is painful. People with no pain sensation lack this essential indicator. They do not learn that it is dangerous to fall while climbing because it does not hurt them to fall. They fear nothing. They do not learn to protect themselves against injury.

In subjects devoid of the sensation of pain, reflexes elicited in normal people by noxious stimuli are also absent. Cutting and burning of the skin occurs without the subject being aware of anything untoward. This leads to chronic skin ulceration, tissue damage, and infection. The weight-supporting joints also become damaged. The normal human is continuously receiving a low level of noxious stimulation signalling that articular surfaces are being compressed and ligaments stretched. This stimulates minor adjustments which distribute the strains of everyday life over the different parts of the supporting system. In subjects who do not receive this input, the adjustments are not made. The subject does not shift weight when standing, does not turn in sleep. Individual components get too much wear and they break down. As a result, joint surfaces become damaged and ligaments stretched so that the joint becomes unstable and unable to carry weight. This condition is known as a **Charcot joint**.

Infection is not drawn to the attention of the patient since the pain usually generated by infection of tissues is absent. A relatively minor condition such as acute appendicitis may go unnoticed until peritonitis has developed and reached such an advanced state that it is a hazard to life.

Variation in the perception of pain depending on circumstance

Different individuals have a similar threshold to pain, but their reactions may be very different. Soldiers severely wounded in battle often feel little or no pain for hours or days after an injury as severe as amputation. This may be due to the relief of escaping alive. These same soldiers complain just as bitterly as a civilian at an inept venepuncture, when back in the field hospital.

In some cultures, initiation rites and rituals involve procedures that an onlooker would expect to be very painful and yet the participants experience no discomfort, only exaltation. An example is an African ritual called trepanation. This involves cutting scalp and muscles to expose a large area of underlying skull. The subject sits holding a pan under the chin to catch the blood, aware only of the exultation of the ceremony in which he is participating and for which he has had the honour to be selected for special treatment. The inbuilt pain-suppressing mechanism is a downward projection of neurones to the cord from the brain stem; the transmitters involved are opioids, described later in the chapter. Via this projection, the mind can reduce pain.

If a person's attention is focused on a potentially unpleasant experience, the pain perceived is more intense. Apprehensive people find that pain produced at the dentist seems much worse than that produced by a similar amount of trauma in other circumstances.

Stimuli which elicit pain

Pain is usually provoked by injury. The potential for producing pain depends on the nature of the damaging energy. Infection and mechanical damage are usually painful. Ultraviolet or X-rays cause much tissue destruction, but no pain until irremediable damage has occurred. This is also true in some cases of malignant disease, because of the late occurrence of pain the condition may not be recognized until it is too late for effective treatment.

At the other extreme, severe pain may be associated with relatively minor tissue damage. An example is the passing a renal stone. This is described by the patient as painful beyond any expectation that pain can reach such an intensity, although the innervation of the ureter is sparse by comparison with an equal amount of skin. In some conditions, pain occurs in the absence of any detectable tissue damage. The majority of people who suffer from low back pain have no apparent injury.

Pain from different tissues

The stimuli that give rise to the sensation of pain are those which signal impending tissue damage. In almost all living tissues, inflammation causes pain. Apart from inflammation, the type of stimulus which is perceived as painful depends on the nature of the tissue being stimulated. For the skin, cutting elicits pain, this being a common cause of injury to the skin. For skeletal muscles, things are different since, lying as they do inside the body, they are rather rarely cut in nature and therefore do not yield the sensation of pain when they are cut. In muscle, ischaemia is painful, this being the signal that oxygen supply is inadequate. The stimulus effective in eliciting pain in viscera is distention, since this is the potentially damaging condition which commonly occurs in life. We now consider muscle and visceral pain in more detail.

Ischaemic muscle pain

This occurs in active skeletal muscle without adequate blood supply. If the circulation to a healthy limb is completely occluded by applying a sphygmomanometer cuff at a pressure above the systolic arterial pressure, exercise of the limb results in severe pain after about 30 s of exercise; this becomes intolerable after 70 s. It is not due to vascular spasm since blood vessels lose their tone after occlusion. It is not due to muscle tension because the pain is continuous, not accentuated during contraction. It is related to the amount of exercise and is probably due to accumulation of a chemical called substance P. The pain disappears within 4 s of removal of the occlusion.

A clinical example of ischaemic muscle pain is **intermittent claudication**, an intense pain during muscular exertion which occurs in subjects as a result of narrowing of the blood vessels supplying the limb. The circulation is adequate at rest but on exertion, muscle pain develops. The pain makes the patient rest, substance P is washed away, and the pain subsides.

In the heart of a subject whose coronary arteries are narrowed, the blood supply fails to match demand when the subject exercises. This gives a severe pain in the chest, called **angina of effort**. As with intermittent claudication, the pain subsides when exercise is stopped and the pain has the useful function of inhibiting the subject from taxing his heart to the level of ischaemic damage. If a coronary vessel is completely occluded, as in coronary thrombosis, there is severe pain in the chest at rest, due to ischaemia of the affected area of heart muscle. The pain persists for as long as nerve endings remain alive.

Visceral pain

The nature of the stimuli that give rise to the sensation of pain varies widely in the different parts of the viscera. The parietal layers of the pleura and of the intestine are painful if they are cut, as are the mesenteric arteries. Apart from these components, all parts of the alimentary tract are insensitive to cutting, burning, or clamping and it is possible to operate on the exposed colon in the conscious patient without anaesthesizing it. As we have already noted, distenstion of the gut is very painful.

Colic

This is the name applied to an intermittent pain in which periods of a few minutes of intense pain alternate with periods of complete freedom from pain. This pain accompanies obstruction of a hollow organ (viscus) such as the intestine; it is due to stretching of the deformation receptors in the wall of the viscus. The time pattern is due to the alternate contraction and relaxation of the smooth muscle in the wall of the viscus in the movements of peristalsis.

Renal colic occurs when a stone which has formed in the renal pelvis travels along the ureter. Renal colic is the only true colic, with complete cessation of pain in the periods of intermission. Intestinal colic (occurring when the lumen of the bowel is obstructed) and biliary colic (occurring when a stone passes from the gall bladder along the common bile duct to the duodenum) give a pattern of pain similar to that of renal colic but pain does not disappear completely in between the spasms.

Localization of visceral pain; referred pain

Pain from a viscus is often poorly localized by the subject. In other cases, the pain is interpreted by the subject as coming from the surface of the body. Pain being perceived as coming from a site different from the site of the pathology is called **referred pain**. Many dorsal horn neurones respond to stimulation either of a somatic receptive field or of a viscus. They are called viscero-somatic neurones and their axons travel up in the spino-thalamic and spino-reticular tracts. Referred pain arises because of convergence of visceral and cutaneous input on to common neurones on the central afferent pathway.

The brain more often receives impulses from parts of the body of which we may be aware, such as the skin, than from other parts such as the viscera. This is why the brain interprets the arriving impulses as coming from the body surface and not from the viscera. Referred pain is felt in the region of the body surface that receives the same segmental nerve supply as the viscus. This in turn means that the pain is felt at the site close to the embryological origin of the viscus. A classical example is the gall bladder, which originates as part of the fifth cervical segment. Pain originating from the gall bladder is felt as coming from the tip of the right shoulder, which is innervated by the fifth cervical segment.

Another structure which has a similar segmental innervation is the diaphragm, which is innervated by cervical segments 4 and 5. Inflammation of the diaphragm may also cause pain interpreted by the patient as coming from the tip of the shoulder.

The pain of appendicitis is usually initially a midline pain localized above the umbilicus, reflecting the origin of the gut as a midline structure and the segmental level of the appendix as lower thoracic. As the infection spreads to involve the peritoneal covering, the pain becomes lateralized and is felt in the right iliac fossa.

The pain of angina pectoris is distributed in the lower cervical and upper thoracic segments. It is a frightening choking pain, felt as a constricting band round the chest which often radiates to the inner aspect of the left arm. Accompanying referred pain, there is a disturbance in normal sensation from the area of skin to which the pain is referred. The skin is more sensitive than usual to painful stimuli, a phenomenon known as **hyperalgesia** (see Section 18.2). Rubbing hyperalgesic skin can alleviate the pain by mechanisms to be described next.

18.2 NOCICEPTIVE MECHANISMS

When certain points on the skin are pricked with a fine needle, the subject reports the sensation of pain. These points are separated by regions which can be pricked without eliciting pain. The pain spots do not coincide with the distribution of other modalities of sensation. This proves that there are receptors in the skin which are responsible for the sensation of pain and which are distinct from the receptors for pressure etc. Since these receptors are excited specifically by stimuli which are damaging or potentially damaging i.e. noxious stimuli, they are called **nociceptive** receptors or **nociceptors**. Nociceptors are slowly adapting free nerve endings; they are unique amongst skin receptors in that they penetrate into the epidermis. In addition to exciting nociceptors, noxious stimuli also powerfully excite mechano- and thermoreceptor systems; this provides the brain with ancillary information for localizing and assessing the nature of the painful stimulus. These latter systems are driven to maximal activity by intensities of stimulation which do not reach the level of causing pain. Such systems therefore cannot themselves mediate pain.

Inflammation and hyperalgesia

Injury to the skin causes inflammation. A stimulus such as a light touch, which in normal circumstances would cause no discomfort, causes intense pain when applied to inflamed skin. This is called **hyperalgesia**. It is caused by the release of endogenous chemicals which stimulate nociceptors. This may be demonstrated by forming a skin blister, removing the loose skin and applying fluids or chemicals to the base of the blister. The application of fresh serum, causes no pain indicating that it contains no algogens. Serum which first touches glass is painful when subsequently applied to the blister base. This is due to the disintegration of blood platelets with the release of 5-hydroxytryptamine and by the formation in the plasma of bradykinin. A chemical which causes pain when so applied is called an **algogen**. An algogen has a specific sensitizing action on nociceptors; it has no effect on the threshold of other receptors such as mechanoreceptors. These chemicals, together with histamine and prostaglandins, are produced widely in inflamed tissues and this is the explanation of the hyperalgesia which occurs as a result of sunburn or rheumatoid arthritis.

Prostaglandins alone do not cause pain (Ferrieira, 1979) but they potentiate the effects of the other algogens. Anti-inflammatory agents such as aspirin and indomethacin relieve the pain of inflammation. They block the action of the enzyme prostaglandin synthetase, and the consequent reduction in release of prostaglandins in hyperalgesic tissue contributes to the reduction of pain (Iggo, 1982). These agents do not alter the threshold of nociceptors in uninflamed skin. For more information on hyperalgesia, see Appendix A, Section 18.4

First and second pain

In a normal subject, a noxious stimulus gives an immediate sensation of sharp, well-localized pain and, after a second or so, a second more intense less well-localized pain. These are called 'first and second' pains.

The two types of pain are mediated by different groups of peripheral nerve fibres, as the following observations indicate. Cocaine, or other local anaesthetic injected into nerve blocks the smallest fibres first and slow pain is lost before fast. Ischaemia or asphyxia blocks fast fibres before slow and blocks fast pain before slow. The latent period for the second pain increases with distance from the cord suggesting that it is carried by slowly conducting nerve fibres.

The two groups of nerve fibres connect to types of nociceptor ending which have differing physiological properties. The fast-conducting fibres are Group A δ and the slow are C fibres. The Group A δ fibres are 2–6 μm in diameter and conduct at 5–25 m/s. The receptors with which they connect subserve prickling, bright, well-localized pain. In addition to being stimulated by frankly noxious stimuli such as strong acids applied to the skin, these receptors are stimulated by firm pressure or cutting; they are uninfluenced by brushing the skin or light mechanical stimulation. They are similarly unresponsive to temperature. They are thus high-threshold mechanical nociceptors. Because they respond to strong mechanical stimuli and to noxious stimuli, they are called **polymodal receptors**. They are **mechanical nociceptors**.

The non-medullated fibres are 0.5–1 μm in diameter and conduct at around 0.5–2 m/s. The receptors with which they connect subserve the second wave of pain, which is burning or aching and is poorly localized. These receptors are called **mechanothermal receptors** because, in addition to responding to strong mechanical stimuli, they give a vigorous response to severe thermal stimuli. So slowly do the sensory fibres from these receptors conduct that, for noxious stimuli applied to the hand or foot, it takes around 1 s between the damage and the first appreciation of the dull pain. We have all experienced the feeling, immediately after cutting or burning ourselves, of knowing that we have hurt ourselves, from information travelling along the fast pathway, but of having to wait a second or so before assessing whether it is a bad injury or merely a minor one. That second of waiting is occupied by the time taken for information to pass along the slow pathway.

It is slow pain that is distressing and which, if it becomes persistent, forms the basis of pathological pains, to be described in a later section.

Central pathways for nociception (injury messages)

The Group A δ fibres form synapses in the dorsal horn on neurones which decussate to join the lateral spino-thalamic tract. Many ascend or descend several segments before decussating. They project to the specific thalamic relay nuclei and thence somatotopically to the postcentral region of the cerebral cortex. It is this projection that allows a subject to localize pain.

In the peripheral nerves, non-medullated (C) fibres are far more numerous than the Group A δ fibres. They form synapses in the dorsal horn with neurones that project along multisynaptic pathways through the grey matter of the spinal cord up to the brain stem reticular formation and medial nuclei of the thalamus (not the specific relay nuclei). From here, they project to the prefrontal areas of the cerebral cortex. Here the somatotropic organization is poor but the region is involved in the emotional stress associated with pain. Frontal lobectomy removes this distress, without altering the pain threshold.

As it ascends through the reticular formation, the pathway sends collateral branches to the hypothalamus. This structure mediates the autonomic effects of pain including increased heart rate, increased respiration, and protein catabolism with nitrogen excretion.

The control of nociceptive pathways

The transmissivity along the pathways signalling pain is controlled at the spinal cord level. This control can be classified into two groups. The first is due to impulses entering the cord from the periphery along thickly myelinated nerve fibres. The second group of control mechanisms is due to descending influences from higher centres.

Control due to input from the periphery

Many lines of evidence indicate that the interpretation of the intensity of pain depends on activity in several groups of afferent nerve fibres. Pain is perceived when there is activity in fibres from nociceptive endings. For a given level of discharge in these fibres, the intensity of pain is reduced if there is concurrent activity in thick afferent nerve fibres. This is the basis for the common experience that rubbing or warming a painful area alleviates pain, the phenomenon of **counter irritation**. The rubbing initiates impulses in thick afferent nerve fibres and hence reduces the intensity of the pain.

Consistent with this, ischaemia of a limb, which blocks thick nerve fibres before it blocks thin ones, impairs sensations such as pressure and position sense before it impairs pain. At the time when sensation mediated by the thickly myelinated nerve fibres is abolished and nociceptive mechanisms are operating, a given noxious stimulus is experienced by the subject as being more painful than it would normally be. This is because the role of thick fibres in holding pain at bay is interrupted at a time when the nociceptive pathway is still unimpaired.

Using the technique of microneurography described in Chapter 3, researchers have found that a single nerve impulse in an afferent nerve fibre from a nociceptive ending is never perceived. It takes a stimulus adequate to cause the afferent nerve fibre to discharge briskly before pain is perceived. (You will remember from Chapter 3 that, for a sensation such as touch, a single afferent impulse may be perceived in some circumstances.) This indicates that it requires the central summation of input before pain is perceived.

If heat or pressure at high intensity are applied to a point on the skin, they both give the same sensation of pain. With microneurography, it is possible to demonstrate that the intensity of discharge in afferent nerve fibres for nociceptive endings is quite different in the two cases, although the intensity of perceived pain is similar. The firing rate when the stimulus is just intense enough to be painful is typically 0.5 Hz for the heat stimulus and 10 Hz for the pressure stimulus. This is because the pressure stimulus, unlike the heat stimulus, also activates large nerve fibres and these reduce the intensity of pain.

It is possible to exploit this pain-reducing effect of activity in thick nerve fibres to help patients suffering from pain. One way is to use a stimulator adjusted to excite large nerve fibres. This is attached to the skin or to the peripheral nerve supplying the region in which the pain is felt; switching the stimulator on alleviates the pain. Another manoeuvre is to apply stimuli through electrodes implanted in the dorsal columns. These elicit orthodromic impulses travelling up the cord and antidromic impulses travelling down. These latter constitute thick fibre input to the nociceptive pathway in the cord. They depress the intensity of pain.

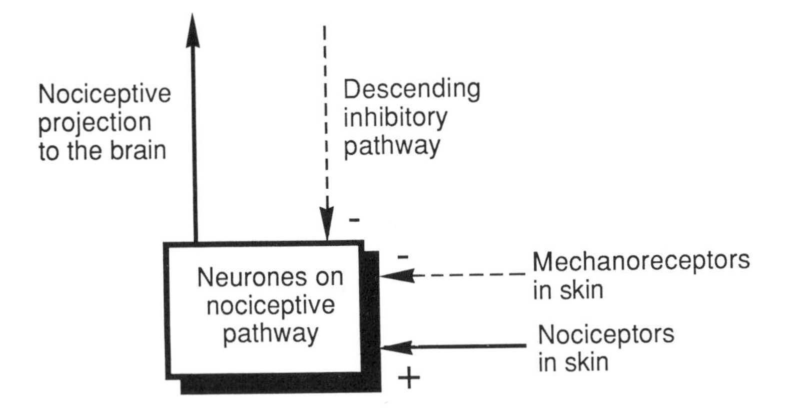

Figure 18.1 The control of transmission along the nociceptive pathway.

The interaction of thick and thin fibre inputs to the cord is shown diagrammatically in Figure 18.1. When a noxious stimulus is applied to a region of the skin, nociceptors there generate impulses which enter the cord and excite neurones on the nociceptive pathway. These same calls receive inhibitory input from the thick afferent nerve fibres innervating cutaneous mechanoreceptors in and around the stimulated region of skin. With this arrangement, the frequency of impulses travelling up the nociceptive projection to the brain is increased if there is a predominance of activity in small afferent nerve fibres and reduced if large activity predominates (Appendix B, Section 18.5). There is physiological evidence for the existence of such interactions in the spinal cord. Responses of a nociceptor-driven dorsal horn neurone are completely blocked by activation of mechanoreceptors in the skin. The effect is specific, since activity in afferent nerve fibres from muscle mechanoreceptors is without effect on the discharge rate of these dorsal horn neurones (Iggo, 1982). Figure 18.1 also shows a descending control, to be described next.

Control due to descending influences from higher centres: opioid peptides

In addition to the control of transmissivity along nociceptive pathways in the cord by thick afferent nerve fibres, there is descending control from the brain stem. This control is mediated by opioids. One of the most ancient treatments of pain is poppy extract, the active agent of which is morphine. Morphine exerts its effects by binding to specific receptor sites on the surface of neurones. This suggested to Hughes *et al.* (1975) that these receptors were more likely to have evolved to recognize endogenous morphine-like chemicals than to recognize exogenous morphine. This led them to search for morphine-like effects of extracts of brain. As a sensor of the presence of opioid activity, they used the smooth muscle of the vas deferens. If a strip of vas is electrically stimulated *in vitro*, the smooth muscle contracts. Morphine reduces or abolishes this stimulated contraction. A specific antagonist called naloxan blocks this effect of morphine. Extracts of brain tissue were found to have similar effects to that of morphine and the effects were blocked by naloxan, indicating that brain contains chemicals with actions similar to those of morphine. The chemicals are called **opioids**, and they are peptides (unlike morphine, which is an alkaloid). Subsequent work showed that there are three groups of endogenous opioids. Each group is derived biologically from a long-chained precursor. The three groups are known as **encephalins, endorphins** (derived from 'endogenous morphines') and **dynorphins** (Goodman Gilman, 1990, p. 486).

The opioid receptors function by exerting inhibitory modulation of synaptic transmission. They are located on presynaptic nerve terminals, where their action results in a decreased release of excitatory neurotransmitter when the nerve terminal is invaded by an action potential. The endogenous opioids are involved in the descending control of spinal circuitry. The nuclei involved are shown in Figure 18.2. Stimulation of the periaqueductal grey matter of the midbrain and the medullary nuclei generates analgesia. Microinjection of minute amounts of morphine into these same regions has similar effects. These regions are rich in endogenous opioid receptors. The action of opioids is to excite the descending inhibitory pathway. This descending pathway causes presynaptic inhibition of the primary afferent nerve fibres coming in to the cord and reporting noxious stimulation. As a result, the message from these primary afferent fibres is attenuated. The inhibitory transmitter is an opioid. The region in the cord where these cell bodies are found is rich in opiate receptors and microinjection of opiates here also generates analgesia. All these observations add up to the diagram shown in Figure 18.2, with descending control of nocioceptive pathways using opiods as inhibitory transmitters. The figure is oversimplified, but it illustrates the essentials.

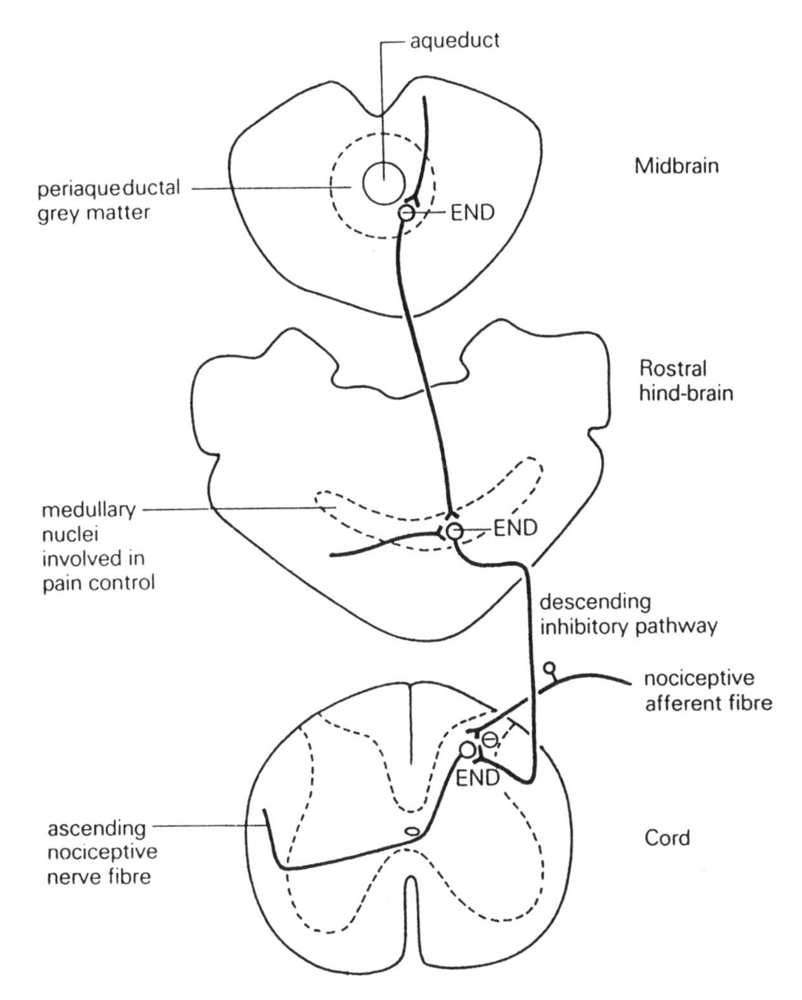

Figure 18.2 Diagram of descending pathways controlling the nociceptive pathways through the cord. END represents 'endogenous opioid'. The synapses are all excitatory except for the ending of the descending inhibitory pathway, which is inhibitory.

18.3 PATHOLOGICAL PAINS

When pain is severe enough to interfere *per se* with a patient's well-being, it is a pathological pain. In many cases, the mechanism of the pain is a matter for speculation. Neural circuitry subserving pain shows a strong tendency of becoming self-perpetuating once it has been set into activity. Such pains often resist treatment. When pain takes over, it ceases to inform. It dominates the existence of the sufferer.

Pathological pains are slow pains. They are caused by injury to nerves or to the central nervous system and not by injury to non-nervous tissue. A few examples are given below.

The neurophysiological basis of pathological pains

When a peripheral nerve is cut across, regeneration will occur if the circumstances are propitious, as we saw in Chapter 6. The sprouting ends of regenerating neurones respond to certain biologically active chemicals applied to them. Regenerating C fibre sprouts, for instance, are stimulated to generate impulses if bradykinin or histamine are applied to them at concentrations similar to that found in plasma in pathological conditions. This response is reminscent of the excitatory effect of these chemicals on nociceptors in inflamed tissues and the mechanism may be similar in the two cases.

The excitation of the sprouting terminals is enhanced by adrenaline or noradrenaline. This may form part of the pathophysiological mechanism by which pain due to nerve lesions is enhanced in states of sympathetic activation described in the next paragraph (Nathan, 1983). Regenerating postganglionic sympathetic fibres are contained in a neuroma and the catecholamines released there during sympathetic activity would sensitize the C-fibre sprouts to circulating bradykinin and histamine.

Causalgia

Subjects who have suffered an injury in which a nerve has been violently and rapidly deformed by a missile experience a severe pain with a burning character thereafter. This pain is called **causalgia**, which means 'burning pain'. It is usually accompanied by abnormal sympathetic activity in that limb. The skin is cold, dripping sweat, the fingernails are brittle and shiny, and the skin discoloured due to vascular changes. If treatment is initiated early enough, injection of local anaesthetic into the sympathetic ganglia may abolish these signs and the pain for long periods, sometimes indefinitely. One factor is the removal of sympathetic release of catecholamines, which are known to irritate sprouting neurones. Another is that, as a result of the disordered sympathetic activity, the blood supply to the somatic nerves is impaired and this contributes to causalgia. These pains are the worst known to man. Pain of this nature which has persisted for a long time may become resistant to all treatment. Even rhizotomy (section of the dorsal nerve roots) is ineffective. This is an example of the creation within the central nervous system of a self-perpetuating neural circuit which discharges indefinitely.

A particularly bizarre example of pain following damage to nerve fibres is when a subject who has lost a limb in an accident subsequently develops severe pain which he feels to be coming from the absent limb. This **phantom limb** pain is common in cases where loss of the limb was associated with avulsion of the nerves.

Amputation

If general anaesthesia alone is administered for amputation, phantom limb pain is a common sequel. This is because the injury discharge of action potentials reaching the cord cause the cord to remember the pain. If in addition to the general anaesthesia, local anaesthetic is injected around the cord or the nerves in the limb before cutting, the injury discharge to the cord at the time of nerve section is interrupted and this dramatically reduces the incidence of phantom limb pain.

Another factor contributing to the mechanism of phantom limb pain may be denervation hypersensitivity of neurones in the central nervous system to which the lost limb previously projected. Partial denervation of a neurone is known to result in the development of hypersensitivity of the neurone to excitatory neurotransmitter. When such a transmitter chemical is released at synapses near to the denervated neurone, a little escapes and reaches the denervated neurone. The result is excitation of this neurone in the absence of normal physiological input. The neurone then spuriously fires impulses which are interpreted by the higher centres as painful.

Another example of pain persisting long after the stimulus for pain has ceased is **post-herpetic neuralgia**. After herpes zoster, a virus infection of the cells in the dorsal root ganglia, has been apparently completely cured, pain felt in the affected region may persist for months or years. This post-herpetic neuralgia is particularly likely to occur if the trigeminal nerve is the site of the herpes.

Thalamic syndrome is a condition in which some pathways through the thalamus are destroyed and others left intact. The condition is most frequently caused by occlusion of the thalamo-striate artery. The threshold for sensation is raised, as expected with destruction of nerve fibres on the specific afferent pathway. When a stimulus reaches the threshold for perception, it evokes a peculiarly unpleasant pain which may become very hard for the subject to tolerate. This may occur with minor everyday stimuli to which the average person pays no attention. This suggests that the upper centres interpret any input as painful if the balance of activity in different groups of nerve fibres subserving sensation is upset.

Other examples of pain occurring when the normal balance of afferent bombardment of the cord is disturbed are the intractable pain of carcinoma and syringomyelia. The latter condition is a cavitation in the central canal of the cord which leads to partial interruption of the spino-thalamic tract.

There is a phylogenetic interpretation of certain types of pathological pain, such as the thalamic syndrome and the pain which is experienced by a subject who has suffered transection of a peripheral nerve and who experiences pain as the nerve regenerates (Chapter 5). For afferent activity associated with modalities of sensation which, on a phylogenetic scale, have been acquired relatively recently, the correct interpretation by the central nervous system of this input depends on a spatio-temporally balanced pattern. When this balance is upset, impulses arriving along afferent pathways can no longer be interpreted. The central nervous system, faced with input that it cannot interpret, takes a 'fail-safe' strategy and interprets the input as pain. So when, in the thalamic syndrome, the central afferent pathway has been partially interrupted, the pattern of impulses reaching the cerebral cortex is jumbled; the brain interprets this as pain. When the regenerating peripheral nerve fibres are initially not properly and fully linked with the receptors, the afferent activity is jumbled; again this is interpreted as pain.

Activity in spinal neurones during nociception

There are three regions in the grey matter of the spinal cord to which nociceptive afferents project. As described in Chapter 8, the primary projection of these afferents is to the outermost border of the dorsal horn of the grey matter. These cells have small receptive fields, such as one digit (Basbaum, 1992). The nociceptive input also projects to cells in the intermediate region of the dorsal horn in a region called the 'neck' region of cells; the cells here are rather more dispersed and have larger receptive fields, indicating that each of these cells receives input from several cells in the posterior grey matter. In the ventral part of the cord is the third group, each neurone having a huge receptor field, such as both hind limbs and the back half of the trunk, indicating input to each cell from many cells in the intermediate zone.

Some neurones exhibit a rise in concentration of a messenger RNA called *c-fos* as a consequence of repetitive firing of action potentials. Staining for *c-fos* thus provides a marker for cells which have recently been very active. This technique has been applied (Basbaum, 1992) to the study of the activity of spinal cord neurones resulting from noxious stimuli applied to the integument. A discrete nociceptive cutaneous stimulus gives activation of cells in all three regions of the cord and over many (up to 10) segments.

As already noted, morphine injection i.v. operates by stimulating the periaqueductal grey matter, giving descending inhibition to switch off the nociceptive neurones in the cord. This gives complete analgesia behaviourally. Cells in the neck region and in the ventral grey now no longer stain for *c-fos*, although there remain cells in lamina I which do still stain for *c-fos*. This indicates that the descending control of nociceptive input is operating in the cord at interneuronal levels of integration of the nociceptive projection rather than at the first synapse.

Neuropathic pain is pain caused by injury to nerves or the CNS and not due to injury to non-nervous tissues. As already observed, such pain is often unresponsive to analgesics. When pain is produced by damage to peripheral non-neural tissues, the *c-fos* levels rise to a maximum in 0.5 h and decline, approaching zero a further 1.5 h later. Neuropathic pain stimuli cause *c-fos* levels to persist for weeks.

18.4 APPENDIX A: PRIMARY AND SECONDARY HYPERALGESIA

If a small area of skin is exposed to an intense source of ultraviolet light, the skin is damaged. A gentle pin prick to this area evokes a sensation of pain which is far more severe than if the same stimulus is applied before the damage. Moreover, the threshold for pain is lowered at the damaged site. This is called **primary hyperalgesia**.

Surrounding such a localized region of skin damage there develops an area of **secondary hyperalgesia**. In this secondary region, the threshold for pain is unaltered, but when a pin prick is sufficient to cause pain, the intensity of pain which the subject experiences is far greater than before the damage. This secondary hyperalgesia is abolished if the passage of nerve impulses from the primary area is blocked, indicating that the secondary hyperalgesia is due to facilitation of pathways within the central nervous system by impulses in nerve fibres from nociceptors in the damaged region.

18.5 APPENDIX B: THE GATE THEORY OF PAIN

Of the many theoretically possible neuronal circuits which would behave like the black box diagram of Figure 18.1 Melzack and Wall (1965) advanced one which became famous as the 'gate theory of pain'. The theory supposes a symmetrical arrangement of input to the dorsal horn from mechanoreceptor and nociceptor afferent nerve fibres. There is physiological evidence that the theory may not be accurate in its entirety (Iggo, 1972; Nathan, 1976). Detailed consideration of this subject is beyond the scope of this book.

FURTHER READING

Basbaum, A.I. and Fields, H.L. (1984) Endogneous pain control systems. *Annual Review of Neuroscience*, **7**, 309–338.

Bond, M. (1979) *Pain, Its Nature, Analysis and Treatment.* London, Churchill Livingstone.

Holden, A.V. and Winlow, W. eds. (1983) *The Neurobiology of Pain.* Symposium of the Northern Neurobiology Group. Manchester, Manchester University Press.

Nathan, P.N. (1988) *The Nervous System* (3rd edn) Oxford, Oxford University Press.

Autonomic nervous system 19

19.1 PERIPHERAL AUTONOMIC NERVOUS SYSTEM

The autonomic nervous system is a special part of the nervous system concerned with the efferent innervation of cardiac muscle, smooth muscle and glands – all of the body's effectors excluding skeletal muscle. Its function is to coordinate the body's vegetative functions, such as supply of energy and oxygen to tissues, so that different regions of the body receive nutrients appropriate to their requirements. These requirements are continuously varying; for instance as the subject changes from rest to muscular exercise, the blood flow to active muscles increases whilst that to the viscera is reduced. The autonomic nervous system contributes to the adjustments that are required.

On the whole, the structures controlled by the autonomic nervous system are not under voluntary control; they are controlled 'automatically' by the nervous system, hence the name 'autonomic nervous system'. There are exceptions to this generalization. For instance, breathing is involuntary and is brought about by skeletal muscle. Micturition and defaecation are under voluntary control but are due to activity of smooth muscle, innervated by the autonomic nervous system. Originally, the term autonomic nervous system was applied only to the efferent nerves, but nowadays it is applied also to the afferent nerve fibres primarily concerned with autonomic function (e.g. baroreceptors and chemoreceptors) and to the central nuclei that control autonomic function.

The autonomic nerves arise from three regions of the neuraxis: certain of the cranial nerves, the thoraco-lumbar region, and the sacral region. As described in more detail later, the effects of stimulation of the cranial and sacral outflow are generally similar and these effects are opposed by stimulation of the thoraco-lumbar outflow. The thoraco-lumbar outflow was called the **sympathetic** nervous system, the name being derived from the fact that this part of the nervous system acts so that the tissues operate in harmony with each other. The cranio-sacral outflow was called the **parasympathetic** nervous system.

Comparison of the anatomical layout of the somatic and automatic nervous systems

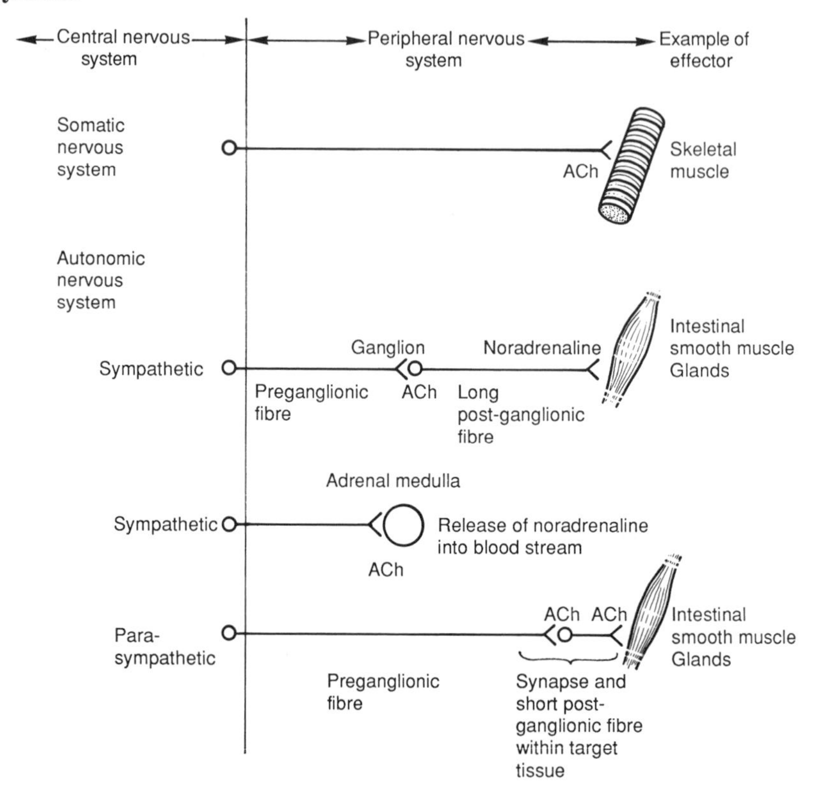

Figure 19.1 The layout of the somatic and autonomic nervous systems.

The motor nerve fibres in the somatic nervous system arise from cell bodies in the central nervous system. The axon projects all the way to the target tissue, such as skeletal muscle, without any synaptic relay en route. The transmitter chemical at the neuro-effector junction is acetylcholine (ACh). This is shown at the top of Figure 19.1. The autonomic nervous system is different in that for each nerve fibre there is a synaptic relay in a ganglion outside the central nervous system. The nerve fibre arising in the cord and projecting to the ganglion is called the 'preganglionic' fibre and that projecting from the ganglion to the target cell is called the 'postganglionic' fibre. The synaptic relay in the ganglion is cholinergic.

Outline of anatomy of autonomic nervous system

The preganglionic fibres arise from a series of homologous nuclei in the brain and spinal cord. These nuclei lie between the sensory and motor nuclei. In the spinal cord they cause a bulge in the grey matter which is called the **lateral horn**. The preganglionic nerve fibres leave the neuraxis with the motor roots of the somatic nervous system. These fibres are B fibres, i.e. thinly myelinated fibres. A nerve bundle containing preganglionic fibres appears white, because of the myelin surrounding each nerve fibre. The postganglionic nerve fibres are C fibres. A bundle of such fibres, being unmyelinated, appears grey.

The parasympathetic nervous system

The parasympathetic nerves arise from the cranial and sacral ends of the central nervous system but are distributed peripherally very widely. For the cranial outflow, the preganglionic nerve fibres arise from cells in the central nervous system and pass peripherally. The autonomic nerve fibres run with certain of the cranial nerves; embryological factors determine which particular cranial nerves act as carriers. After travelling with the cranial nerves, the parasympathetic nerve fibres form synapses in ganglia close to the tissue that is innervated. The neuroeffector transmitter is acetylcholine. The cranial parasympathetic nerves innervate the constrictor pupillae muscle and the ciliary muscle in the eye, and the salivary and the lachrymal glands. Via the vagus nerve, they innervate the heart, the lungs, and the alimentary tract as far as the splenic flexure. The parasympathetic relays lie in Auerbach's plexus, a plexus of nerves in the wall of the gut. The anatomy of the cranial parasympathetic outflow is given in the appendix to this chapter.

The sacral parasympathetic nerve fibres arise from segments S2 and S3 and form the pelvic splanchnic nerves. They innervate the descending colon, rectum, bladder, uterus, and the erectile tissue of the penis or clitoris. The relay is in the wall of the target structure.

The sympathetic nervous system

This arises from segmental levels T1 to L3. The preganglionic fibres form synapses in sympathetic ganglia, which lie distant from the organs which the postganglionic fibres innervate. The sympathetic postganglionic nerve fibres run with the mixed peripheral nerves and also along blood vessels. There is a chain of sympathetic **collateral ganglia**, one pair corresponding to each segment of origin of the preganglionic fibres. These ganglia lie close to the junction of the posterior and anterior roots of each segment. There are also more peripheral sympathetic ganglia. In the neck, the sympathetic system projects upwards as a nerve called the **cervical sympathetic trunk** which lies in the carotid sheath. On its course lie three cervical ganglia, inferior, middle, and superior, on each side of the body.

The sympathetic supply to the head arises from segmental levels T2 and T3. It innervates the eye (dilator pupillae muscle and the smooth muscle fibres of the retractor muscle of the upper eyelid), the salivary glands, the vasculature, and the skin. The sympathetic nervous system also innervates the heart (segmental level T2 and T3) with branches from all three cervical ganglia and from the upper thoracic segmental sympathetic ganglia. The larynx, trachea, bronchi and lungs, the abdominal and pelvic viscera, and the external sex organs all receive sympathetic innervation.

The sympathetic system has aggregations of nerve fibres and ganglia into three plexuses, cardiac, coeliac, and hypogastric, sited in the thorax, abdominal, and pelvic cavities respectively. From these plexuses, branches run to the viscera. The regions of the neuraxis which innervate the limbs do not give rise to autonomic outflow. In these two regions the cord is fatter than elsewhere because of the large numbers of neurones needed for innervation of the limbs and this has squeezed out the autonomic component.

Most sympathetic postganglionic nerve terminals release a mixture of chemicals, noradrenaline being one of importance. Some, notably the sympathetic postganglionics supplying sweat glands, are cholinergic.

A special part of the sympathetic nervous system is the adrenal medulla (third row in Figure 19.1). This is a modified sympathetic ganglion, receiving input from preganglionic nerve fibres. The cells of the adrenal medulla are modified nerve cells which do not give rise to axons. They release their transmitter chemicals, noradrenaline and adrenaline, direct into the blood stream. Thence these chemicals are distributed throughout the body to influence many effectors such as the heart, the blood vessels and the gut.

Muscarinic and nicotinic receptors

Acetylcholine is the transmitter at all autonomic ganglia and at the parasympathetic neuro-effector junction. The recognition sites for acetylcholine (called **receptors**) are different at the two types of junction, as can be shown pharmacologically. The ganglionic receptors are called **nicotinic**, because nicotine attaches to the receptors and, if the nicotine is present in small doses, it stimulates the receptors just as acetylcholine would. Nicotine is said to be an **agonist** of acetylcholine at this junction. The receptors at the parasympathetic neuro-effector junction are called **muscarinic**, since muscarine is a specific agonist at this site. Some pharmacological agents specifically block either nicotonic or muscarinic receptors. A widely used muscarinic blocker is **atropine**.

Innervation of blood vessels

Many blood vessels, and particularly arterioles and veins, receive a sympathetic innervation. The target for this innervation is the smooth muscle in the vessel wall. Parasympathetic innervation of blood vessels is much rarer; only the erectile tissue of the external genitalia and the salivary glands have been shown to receive such innervation. There is evidence, however, that blood vessels in the brain receive autonomic inhibitory innervation which is probably purinergic.

The discharge of peripheral sympathetic nerve fibres may be phasic. In a limb nerve, sympathetic fibres supplying arterioles show modulation of firing frequency in time with the arterial pulse. This indicates rapid responsiveness of autonomic reflexes driven by presso-receptors in the carotid sinus and aortic arch.

Sensory nerve fibres

Both the somatic and autonomic fibres of sensory nerves pass from the nerve terminal at the receptor into the central nervous system without relaying. Both have cell bodies in the dorsal root ganglion or, in the case of cranial nerves, in ganglia which are homologous with the dorsal root ganglion.

The physiological effects of activity in the autonomic nervous system

In circumstances when the existence of an individual is threatened, as when it is necessary to fight or flee to avoid destruction, the complex of physiological reactions is called the **fight or flight** reaction. There is usually a massive discharge throughout the sympathetic nervous system, coupled with a dramatic reduction in activity in the parasympathetic system (exceptions to this are described later). Respiration is stimulated in preparation for greater oxygen usage by the body. The heart rate and contractility of cardiac muscle are increased, so that the circulation of the blood can proceed more quickly than usual, to provide vital tissues such as skeletal muscle with extra oxygen and wash away the metabolic products of muscular activity (CO_2, lactic acid, potassium, heat). The blood flow to areas such as the gut and kidney, which survive with a reduced supply, is cut down. Glucose is released into the blood and tissue fluids to provide the energy for fight or flight. The reduction in parasympathetic activity means that intestinal activity is depressed, and micturition and defaecation are inhibited.

By contrast with this involvement of the sympathetic system in the fight or flight reaction, the parasympathetic system stimulates the body's mechanisms that carry on when the individual is resting (e.g. digestion, defaecation, formation of urine, micturition) and has been labelled the 'housekeeping' system. These labels can be misleading, as we shall see.

A corollary to these overall differences in function is the fact that, in many organs, the two divisions of the autonomic nervous system have opposing effects. For instance, in the heart, activity in the sympathetic system causes an increase in the heart rate whereas activity in the parasympathetic system causes a decrease. Conversely in the gut, activity in the parasympathetic nerve supply increases activity of the smooth muscle of the gut wall (excluding the sphincter musculature), whereas activity in the sympathetic system decreases intestinal motility. However, there are many exceptions to this rule. The sympathetic nervous system innervates all the blood vessels (except for capillaries) whereas the parasympathetic system influences only restricted regions. The limbs and the skin receive only sympathetic and no parasympathetic innervation. Regions that are innervated by the sympathetic and not by the parasympathetic system can obviously only be affected by the former.

In certain regions which are innervated by both divisions of the autonomic nervous system, the two systems act together instead of in opposition. For instance, in the male genital system erection is primarily a function of the parasympathetic nervous system. Parasympathetic discharge to the vasculature of the penis causes dilatation of arterioles. The engorged vessels compress the venous drainage from the organ and hence bring about the accumulation of blood at high pressure. Ejaculation requires participation of the sympathetic system. In ejaculation, there is movement of the semen from the seminal vesicles into the urethra; this is achieved by contraction of the smooth muscle of the vas deferens and the seminal vesicles and is controlled by activity in the sympathetic nerves from the lumbar cord. Finally there is propulsion of the semen out of the urethra by contraction of the bulbocavernosus muscle, a skeletal muscle. So in the sexual act in the male, participation of both divisions of the autonomic nervous system and the somatic nervous system is essential.

The designation of the sympathetic system as the mediator of the fight or flight reaction must not be taken to mean that the sympathetic nervous system is inactive in more normal circumstances. At all times in a healthy human, both the sympathetic and the parasympathetic nervous components of the autonomic nervous system are discharging as an essential part of the complex homeostatic mechanisms for maintaining the vegetative functions of the body such as maintenance of a fairly constant core temperature.

We have seen that, in the fight and flight reaction, there is generalized activity throughout the sympathetic nervous system. This involves both the release of noradrenaline as the neurotransmitter of the sympathetic postganglionic nerve endings and the release into the blood stream from the adrenal medulla of adrenaline, with consequent distribution of this hormone and activation of adrenergic receptors throughout the body. Nevertheless, the sympathetic nervous system is capable of patterned activity much subtler than this massive generalized activity, as an example will illustrate. A human in a hot environment needs to lose heat from the skin. The physiological mechanisms include vasodilatation of the vasculature in the skin so that blood is transported in large amounts to the most superficial layers of the dermis carrying heat as close as possible to the body surface. This maximizes the loss of heat by conduction and convection. The deepening of the colour of the engorged skin maximizes the loss of heat by radiation. The channelling of blood to the surface layers of the dermis is achieved by a reduction in the activity in sympathetic vasoconstrictor nerves supplying the arterioles in the skin. Sweating, which allows the loss of heat by evaporation of water from the skin surface, depends on an increase in activity in sudomotor nerves, which are also part of the sympathetic nervous system. So in this circumstance the hypothalamic centres which control the fine patterning of activity in the autonomic nervous system send down inhibitory influences to sympathetic neurones innervating skin arterioles but excitatory influences to those sympathetic neurones in the same segments of the cord which supply the sweat glands.

When there is generalized discharge of the sympathetic nervous system, this subtle patterning is abandoned, sometimes with adverse consequences. There is a massive reflex discharge of the sympathetic nervous system in a human who has suffered a severe haemorrhage. In this case the receptors responsible for the reflex are the pressure receptors, in the aortic arch and the carotid bodies, which sense the fall in arterial blood pressure resulting from the blood loss and consequent circulatory collapse. Part of the massive sympathetic response is vasoconstriction of skin vessels (an appropriate response since this tends to increase the peripheral resistance of the vascular bed and hence buffer the fall in arterial blood pressure) and stimulation of the sweat glands (an inappropriate effect in these circumstances since it results in the loss of even more fluid). These two sympathetic effects on the skin account for the cold clammy feel of the skin of a patient in haemorrhagic shock, cold because of peripheral vasoconstriction and clammy because of sweating.

All in all, the sympathetic nervous system is a mixture of a fine grain of control over the balance of activity in some neurones with inactivity in others (as in loss of heat from the skin) with examples of dissemination of sympathetic influence over an inappropriately wide range of tissues (as in the sweating accompanying hameorrhage).

History of the discovery of the autonomic nervous system

It was known to early anatomists that some nerves have lumps on them. Gaskel (1886–1916) showed that the lumps were ganglia with synapses and that the system was essentially efferent. He called it the vegetative nervous system.

Langley, between 1898 and 1905, found that, when he stimulated autonomic nerves, the cranial and sacral outflows had similar effects that were opposed by those of the thoraco-lumbar part. He designated these the parasympathetic and sympathetic respectively. Coupled with this was the concept of double innervation of viscera with opposite effects.

Loewi and Dale showed that the effects of sympathetic and parasympathetic stimulation were mediated by neurotransmitters which differed and which affected different receptors. In most cases the sympathetic nerves released adrenaline at the nerve terminals whereas the parasympathetic nerves released acetylcholine.

By the 1950s 'adrenergic' and 'cholinergic' came to be used as synonyms for 'sympathetic' and 'parasympathetic'. Intracellular recording from smooth muscle with sympathetic and parasympathetic innervation revealed that there was usually depolarization for excitation and hyperpolarization for inhibition. Structures innervated by autonomic nerves are spontaneously active and this is consistent with both excitatory innervation, to increase the spontaneous activity, and inhibitory innervation, to reduce the spontaneous activity. This is a feature in which smooth muscle (with its dual innervation) and skeletal muscle (with its simpler somatic innervation) differ. Skeletal muscle shows no spontaneous activtiy. Its innervation is purely excitatory.

Autonomic nervous system – modern modifications of classical views

The straightforward idea that adrenergic and cholinergic transmission accounted for all the efferent effects of the autonomic nervous system was already being doubted at the time of Dale's pioneering work. Since then it has been shown that other chemicals play an important role as neurotransmitters. There are peptides that were originally discovered in parts of the alimentary tract and that coexist with acetylcholine in parasympathetic nerve fibres. Synaptic transmission involves the release of acetycholine and peptides, both contributing to the response, a phenomenon called **co-transmission**. Stimulation of the parasympathetic nerve supply to the parotid gland causes profuse secretion of a salivary fluid rich in enzymes and a concomitant vasodilatation. These effects are mediated respectively by acetylcholine and one of the neuropeptides, called vasoactive intestinal peptide (VIP), as shown in Figure 19.2. This can be shown by adding atropine, which is a specific blocker of cholinergic transmission at parasympathetic neuro-effector junctions. In the parotid gland experiment, atropine blocks the secretion but not the vasodilatation elicited by nerve stimulation. Antisera to VIP added to the perfusion fluid abolish vasodilatation but not the secretion. This is just one example of a widespread phenomenon in exocrine glands in which a single nerve fibre releases two transmitters, acetylcholine and a peptide. Similarly, noradrenaline and the purine adenosine triphosphate (ATP) often act as a synergic pair.

In other situations, chemicals modify synaptic transmission without themselves being transmitters. These are called **modulators**. Their effect is to facilitate or inhibit (i.e. modulate) the response of the postsynaptic membrane to the specific transmitter chemicals. A modulator may come from the same neurone or from a second neurone. The latter situation occurs in the gut which receives both parasympathetic and sympathetic innervation. Except at the sphincters, the parasympathetic innervation is excitatory to the gut's smooth muscle, and the sympathetic supply antagonizes this excitation. The parasympathetic nerve fibres penetrate right to the smooth muscle fibres to form neuro-effector junctions. The sympathetic nerves do not penetrate as far; they terminate in Auerbach's plexus of nerves in the gut wall.

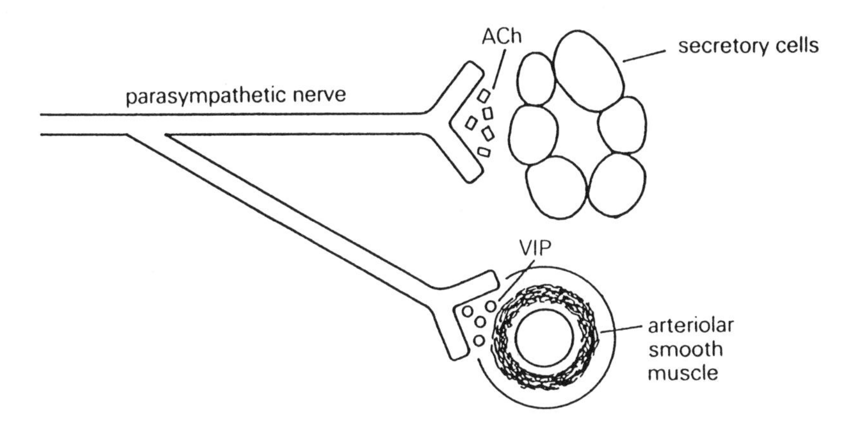

Figure 19.2 Cotransmission. The parasympathetic nerve fibre releases both acetylcholine and vasoactive intestinal polypeptide (VIP). At the gland cell, there are receptors for acetylcholine and at the smooth muscle of the arteriole, there are receptors for the VIP.

These adrenergic nerves release noradrenaline which reduces the responsiveness of the gut to the parasympathetic cholinergic pathway. This is the mechanism of antagonism of the sympathetic and parasympathetic systems in the gut.

19.2 CENTRAL AUTONOMIC CONTROL

Two regions of the brain are primarily concerned with central autonomic control, the integration of the vegetative functions. They are a group of centres located in the medulla extending up into the pons and another group of centres in the hypothalamus. The activity of the two groups of centres is co-ordinated. Both are exceptional regions of the brain in that they contain neurones which are sensors of the cerebral environment. The existence of these receptors has been established by experiments in which localized regions of the brain have been exposed to changes such as heating, cooling, or change in pH. The receptor cells have not been identified histologically.

Both centres directly influence the level of activity in the lateral horn cells of the sympathetic system. The medullary and pontine centres are phylogenetically the more primitive. They exert their influences through neurally controlled effector mechanisms, primarily the autonomic nervous system. The hypothalamus is more advanced and performs highly complicated integrative tasks, producing its effects by a harmonious interaction of neural and endocrine effector mechanisms.

The medullary and pontine centres

The medulla and pons contain regions where stimulation causes cardiac effects (an increase in heart rate and in myocardial contractility) and vasoconstriction throughout the blood vessels of the body. Stimulation of neighbouring regions of neural tissue results in inhibition of sympathetic activity. Observations of this nature have led to the concept of medullary cardiovascular centres. 'Centre' is to be understood to mean a system of neurones with similar functions and not an anatomically defined nucleus. Neurones of excitatory and inhibitory centres are intermingled and the effect of stimulation at any one site depends on which type of neurone predominates at that site.

In physiological circumstances these medullary centres are a link in the chain of baroreceptor reflexes. Discharge of arterial baroreceptors (in the carotid sinuses and aortic arch) reflexly depresses sympathetic tonus. Discharge of baroreceptors in the right atrium has the opposite effect. The cardiovascular centres are the integrating centres for these reflexes. The medulla and pons also house the respiratory centres which are the central generators of the rhythmic respiratory movements. As with the cardiovascular centres there are inspiratory and expiratory centres, neurones of the different centres being intermingled. The most important sensors in the control of respiration are neurones situated superficially on the ventral surface of the medulla near the origin of the IXth and Xth cranial nerves. These receptor neurones sense the hydrogen ion concentration in the cerebrospinal fluid in which they are bathed. Their axons project to the respiratory centres which lie in the substance of the brain stem.

The hypothalamus

In magnitude, the hypothalamus is relatively insignificant, occupying a mere 1 cm^3. It extends from the third ventricle medially to the internal capsule laterally. In some parts, cells are aggregated into groups that form readily identifiable nuclei; other regions are not so convincingly organized.

Despite its small size, the hypothalamus is essential for survival. The homeostatic mechanisms that it controls are temperature regulation, the water content and osmolarity of the body, food and energy balance, and sexual function and reproduction. For many physiological variables (e.g. temperature), the hypothalamus is the most important central integrator of autonomic action. To some extent the different functions of the hypothalamus are located in different anatomical sites. An example is the suprachiasmatic group of nuclei described below. In other respects, early hypotheses about hypothalamic 'centres' have been abandoned. For instance, the idea that there were specific centres for hunger and thirst have not withstood the test of experimental investigation since no neural structures have been identified which, when stimulated locally, reproducibly cause the behavioural response, and ablation of which abolishes the response.

The hypothalamus plays a central integrating role in autonomic control. It samples the status of the body by means of input from baroreceptors and chemoreceptors. There are also specialized receptor cells in the hypothalamus itself, including osmoreceptors and temperature receptors. It receives input from the rest of the brain notably from the limbic system.

The 'function' of the hypothalamus can be observed if all the higher centres are destroyed leaving the hypothalamus and its connections with the pituitary cord and lower neural centres intact. Most of the homeostatic mechanisms continue; the animal eats and drinks, it exhibits an alternating pattern of waking and sleeping, it maintains a normal body temperature, and it can reproduce. If the hypothalamus is destroyed, none of these mechanisms survive. Biological rhythms, such as waking and sleeping, depend on the integrity of part of the hypothalamus, notably a group of nuclei called the suprachiasmatic nuclei. Their name is derived from their siting above the optic chiasma. This group of nuclei is also essential for other biological rhythms such as the oestrus cycle. An individual with an intact hypothalamus but no higher centres lacks anticipatory, exploratory, and social behaviour.

19.3 FEEDBACK LOOPS

Stability versus the need for change

There are many mechanisms in the body which incorporate negative feedback loops. An example is the feedback from stretch receptors in the muscle spindle to anterior horn cells. Its effect is a tendency to stabilize muscle length by matching muscular contraction to the effort needed to maintain muscle at a given length (the length being commanded by the central nervous system).

More generally, homeostatic mechanisms (which often depend on neural and endocrine effects) by definition involve negative feedback to stabilize the internal environment. These mechanisms are so important that the student sometimes gains the impression that all physiological mechanisms involve negative feedback. This is wrong because if every factor in the body were stabilized all the time, the individual would be immobile and capable of no change. There would be no adaptation to changing circumstance. In the realm of motor activity any movement would be rigidly opposed by a reflex loop involving negative feedback. The ability to move and to adapt to changing circumstance, which is cardinal in the dominance of man among living things, depends on the ability to interrupt the stabilizing mechanisms when the occasion arises. In many instances when change is demanded, positive feedback is involved. An example in the nervous system is the opening of sodium gates in the nerve membrane in response to depolarization of the membrane. The action potential involves a mechanism in which stabilization of the resting potential must be abandoned if the system is to operate in its capacity of transmitting information.

An example in the endocrine system is the release of progesterone from the corpus luteum and luteinizing hormone from the anterior pituitary. This positive feedback loop is responsible for the abrupt hormonal changes responsible for menstruation.

We must therefore realize that negative feedback loops stabilize a system and that initiation of change involves different types of mechanism, some of them involving positive feedback. Homeostasis is one side of the coin; the other side, which is just as important, is the propensity for change.

19.4 APPENDIX: THE ANATOMY OF THE CRANIAL PARASYMPATHETIC OUTFLOW

The **oculomotor nerve (III)** carries parasympathetic fibres arising from the cells of the Edinger-Westphal nucleus in the midbrain. The parasympathetic preganglionic nerve fibres leave the parent nerve as the short ciliary nerve. The nerve fibres form synapses in the ciliary ganglion and the postganglionic nerve fibres course back to the oculomotor nerve to innervate the constrictor pupillae muscle of the iris, and the ciliary muscle, contraction of which causes relaxation of the lens and hence accommodation for near vision.

The **facial nerve (VII)** carries parasympathetic nerve fibres travelling to the submandibular and lingual salivary glands and also to the lacrimal glands. The nerve fibres to the submandibular and lingual salivary glands arise from the superior salivary nucleus. They leave the facial nerve as the chorda tympani, which traverses the tympanic cavity to reach the lingual nerve, by which route they travel to the submandibular ganglion where they relay before travelling to the two salivary glands. The nerve fibres to the lacrimal gland also leave the facial nerve as the greater superficial petrosal nerve. This leads to the spheno-palatine ganglion, where the synaptic relay is situated.

The **glossopharyngeal nerve (IX)** carries the parasympathetic innervation of the parotid salivary gland. The preganglionic fibres arise in the inferior salivary nucleus. They leave the glossopharyngeal nerve as the lesser superficial petrosal nerve and relay in the otic ganglion.

The vagus nerve (X) contains a large proportion of parasympathetic nerve fibres. The preganglionic fibres arise from the dorsal nucleus of the vagus. They travel to the lungs, heart, oesophagus, stomach, small intestine, and large intestine as far as the splenic flexure. They relay in ganglia in the walls of the structures that they innervate. There are nerve fibres which are distributed peripherally with the vagus but which leave the hindbrain in the rootlets of the accessory nerve, subsequently to leave the accessory nerve and join the vagus.

The cerebral environment 20

20.1 THE CONSTANCY OF THE CEREBRAL ENVIRONMENT

It is the complexity of brain function that most dramatically separates the human from the most advanced subhuman species. The advanced human brain requires a very stable environment in which to operate, more stable than that required by any other system. When the chemistry of the body is disturbed by disease, the central nervous system ceases normal function before other systems. For instance, if the hydrogen ion concentration falls, the subject becomes unconscious long before other vital body functions, such as heart, liver, and kidney activity are impaired. Similarly, with variations in body temperature, the hypothermic survivor of a shipwreck becomes unconscious, whereas the patient with a high fever is delirious.

The human brain contains 10^{10} neurones. As a consequence, the human brain is extremely bulky. This carries its own dangers, such as the liability to inertial damage, discussed in Chapter 11. To avoid an even larger brain, during evolution there has been a relative reduction in the volume of the extracellular fluid of the brain. On average throughout the body extracellular fluid is some 30% of the whole; in the brain it is only 15%.

Powerful homeostatic mechanisms have evolved to stabilize the composition of the cerebral interstitial fluid against variations in composition. Several groups of these mechanisms act in concert. First, there are the homeostatic mechanisms of the body as a whole which serve to keep the composition of the blood fairly constant, so that the brain is perfused by a fluid whose composition varies little. The remainder of this section is devoted to other mechanisms that are specific to brain.

The blood–brain barrier and cerebrospinal fluid (CSF)

The capillaries supplying the brain are specialized to form a structure called the **blood–brain barrier**. In most other tissues, there is free communication between the blood plasma and the interstitial fluid for the interchange of all small molecules and ions. In the brain, the blood–brain barrier impedes the movement of all water-soluble substances, such as glucose. (Fat-soluble substances interchange readily between blood and cerebral tissues.) Protein in cerebral interstitial fluid is 0.4% of its level in plasma; in other tissues it is about 2% of its plasma level. So even if the composition of water-soluble substances in the plasma of the blood perfusing the brain wavers, the composition of interstitial fluid of the brain can be maintained almost constant. The interstitial fluid of the brain is in some ways different from that of other tissues. It is drained into the cerebrospinal fluid or CSF, which is the fluid filling the cavities within the brain.

The cerebrospinal fluid is formed at two types of capillary in the brain, those in the choroid plexus (described in Chapter 21) and those within the brain tissue. These capillaries have different structures. In the choroid plexus the capillaries have fenestrations, or windows, where the two layers of the membrane of the endothelial cells become closely apposed. The fenestrations house a regular array of pores large enough for water and

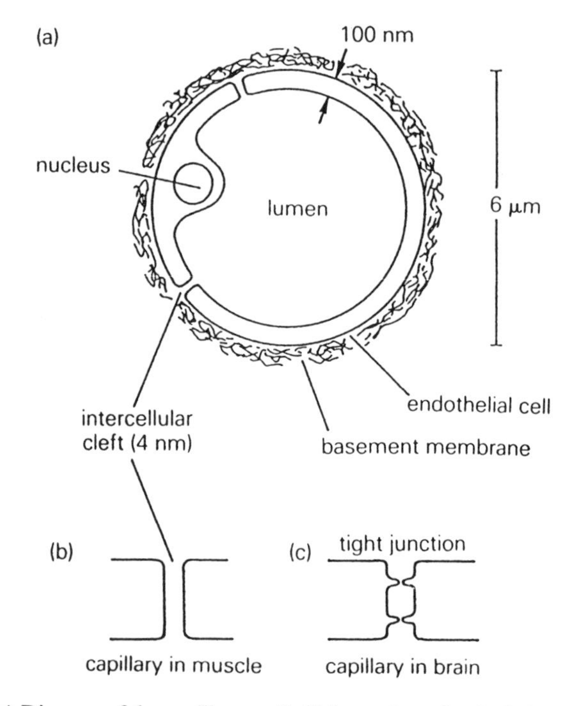

Figure 20.1 (a) Diagram of the capillary wall. This consists of a single layer of endothelial cells surrounded by a basement membrane. (b) Intercellular clefts between endothelial cells in capillaries of skeletal muscle; these clefts allow the passage of molecules with diameters of 4 nm or less. (c) Intercellular clefts between endothelial cells in capillaries of capillaries in the parenchyma of the brain; the tight junctions allow the passage only of molecules with diameters of 1.2 nm or less.

small chemicals to pass. Here the capillary wall is leaky, but the capillaries are separated from the cerebrospinal fluid by a continuous sheet of epithelial cells which are connected to each other with tight junctions. This epithelial lining selectively retains many solutes. The epithelial cells constitute the blood–brain barrier in these capillaries.

The capillaries supplying the substance of the brain have no surrounding layer of epithelial cells; the endothelial cells are surrounded by a basement membrane and a sheath of astrocytes. Endothelial cells and glia both contribute to the blood–brain barrier. The endothelial cells show several separate specializations connected with their contribution to the blood–brain barrier.

Tight junctions

The intercellular clefts found between adjacent endothelial cells in most tissues are sufficiently wide to allow all water-soluble substances to pass readily (Figure 20.1b). The endothelial cells of cerebral vessels are special in that these intercellular junctions are modified to form **tight** junctions, one close to the capillary lumen and one at the basement membrane aspect of the endothelial cells, as shown in Figure 20.1c. These narrowings in the intercellular clefts limit the free diffusion of water-soluble substances.

Absence of pinocytotic vesicles

In the brain, pinocytotic vesicles (the vesicles in the cytoplasm of endothelial cells in the capillaries of many tissues such as skeletal muscle, vesicles which are thought to act as ferry boats for large molecules between the plasma and the interstitial fluid) are rare or absent.

Lack of contractility of endothelial cells

Endothelial cells of cerebral blood vessels do not contract. They have no actomyosin, as do capillaries elsewhere in the body. Associated with this is a lack of effect of histamine on cerebral capillaries. Elsewhere histamine acts by cellular contraction and separation. The lack of contractility in cerebral capillaries ensures that tight junctions are not pulled open.

Lack of amoeboid motion

In capillaries of most tissues other than brain, the endothelial cells are amoeboid. Cerebral capillary endothelial cells are much more sedentary.

The blood–brain barrier imposes an important constraint on the physician who wishes to treat a patient with a drug that must reach the brain substance. Many drugs taken by mouth or injected intravascularly do not cross the blood–brain barrier and so do not reach the parenchyma of the brain. One way round the problem is to use precursors of the active agent; if the precursors cross the blood–brain barrier, then an effective way of administering the drug is available. We encountered an example of this in the use of L-dopa in the treatment of Parkinson's disease (Chapter 12). Another way round the problem is to inject the agent itself directly into the cerebrospinal fluid. The latter is in direct communication with the interstitial fluid of the brain; the cerebrospinal fluid is on the brain side of the blood–brain barrier.

Glia

Having seen that the tissues of the brain are insulated from free interchange of substances with the plasma, we now consider the mechanisms which the brain possesses within the fortress provided by the blood–brain barrier, to stabilize the cerebral environment. An important contribution is made by the supporting cells of the brain, which are called the **glial cells** or **glia**. These supporting cells are derived, as in other tissues, from mesoderm. The stabilizing effect of glia on the composition of the internal environment contrasts with the lack of any such function of supporting cells in other organs in the body.

The glial cells make up a large proportion of the brain, their total volume being approximately equal to that of the neurones themselves. This large volume is associated with a special function, not found in supporting cells elsewhere, of regulating the composition of the cerebrospinal fluid. If the composition of the cerebrospinal fluid alters, the glial cells transfer chemicals into or out of their cytoplasm to cancel the changes in composition of the extracellular space. For instance, when neurones conduct impulses, potassium ions leave the neurones and accumulate in the restricted extracellular space. If there is a rise in the extracellular concentration of potassium, this results in hyperexcitability of the neurones and spontaneous synchronized discharges, giving the clinical condition of epilepsy. In normal brain, this rise in extracellular potassium concentration is buffered by the glia, which absorb the excess potassium into their cytoplasm. Subsequently, the neurones themselves take potassium up again and at this time the glia release potassium. In this way the glial cells hold the extracellular concentration of potassium constant, so that neurones can operate normally. For such a system to work, the glia need a considerable capacity for temporary storage or depletion of potassium. This is probably why during evolution the glial cells have come to comprise such a large proportion of the brain's volume. A small extracellular space in the brain is efficiently held at a constant composition by active participation of a glial system which has of necesssity to be relatively bulky.

20.2 CEREBRAL BLOOD FLOW

By comparison with the body average, the brain uses up more oxygen. It therefore has a relatively high blood flow. The global cerebral blood flow is 17% of the cardiac output whereas the brain comprises only 2% of the body mass.

The difference in blood pressure in the cerebral arteries and cerebral veins provides the driving force for the perfusion of the brain. This pressure is therefore called the **cerebral perfusion pressure** . The pressure in the cerebral veins is close to the intracranial pressure and, as we shall see in a subsequent section, an important effect of a rise in intracranial pressure is to compromise the cerebral blood flow.

Autoregulation of blood supply

Autoregulation means that the system regulates itself according to the needs of the situation. Autoregulation is a feature of various vascular beds but it operates with greatest efficiency in the central nervous system. There are two mechanisms for autoregulation, mechanical and chemical. Their roles are rather different. The mechanism for mechanical regulation is that smooth muscle, when it is stretched, responds by contracting. If the mean arterial blood pressure rises, this rise stretches the walls of the cerebral arterioles. The stretching of the smooth muscle lying in the arteriolar wall brings about an arteriolar vasoconstriction. This counteracts the rise in cerebral blood flow which the rise in arterial pressure would by itself entail. So a rise in arterial blood pressure results in little change in cerebral blood flow.

A fall in arterial blood pressure results in the converse sequence of responses to that described in the last paragraph. There are of course limits of variation in arterial blood pressure which can be compensated in this way; these limits are typically cerebral perfusion pressures of 60–160 mm Hg. Within these limits, changes in arterial blood pressure result in only minor changes in cerebral blood flow. This maintains an adequate blood supply to each region of brain independent of other regions. It does not compensate for variations in tissue requirement for blood supply; this is the function of chemical autoregulation. Where the cerebral perfusion pressure falls below 60 mm Hg, the cerebral blood flow falls.

The chemical autoregulation of cerebral blood flow operates to allow a regional match between tissue needs and blood supply. A feature of the central nervous system is that at different times, different portions exhibit high activity. A highly active region of brain requires a high blood flow to supply oxygen and glucose and to wash away metabolites such as CO_2.

The brain requires a constant supply of oxygen, for reasons to be considered shortly. Any fall in oxygen content of arterial blood puts the brain at risk. If the compensating mechanism were to depend on sensing oxygen tension, it would require a fall in this tension as a prerequisite to being called into operation and this fall would jeopardize brain function. This problem has been solved by the evolution of mechanisms which, instead, depend on other chemical factors to initiate arteriolar dilatation; these factors are a rise in tension of carbon dioxide and a rise in hydrogen ion concentration. Both these factors increase as a result of increased tissue metabolism. This mechanism is rendered extremely sensitive by the fact that the chemical buffering of the cerebrospinal fluid is much inferior to that of blood. The release of CO_2 and other metabolites into the cerebral interstitial fluid results in an increase in an unbuffered hydrogen ion concentration. This results in vasodilatation of the vessels, an effect confined to the active region of brain; the blood supply is increased locally to the regions of brain that require it. The poor chemical buffering power of the brain's interstitial fluid is an advantage in the maintenance of an adequate oxygenation in regions of active brain. So effective is this mechanism that sometimes the P_{O_2} in an active region of brain is above normal.

There are circumstances in which the mechanism of CO_2 and $[H^+]$ vasodilatation does not match oxygen supply to requirement. An example is when the gas being breathed by a subject contains an amount of oxygen so much less than normal that the blood cannot carry enough oxygen for the demands of the tissues. In such a situation, cerebral arterial hypoxia does occur and this elicits a powerful dilatation of the cerebral arterioles.

Relevance to clinical practice

Various brain injuries including stroke cause autoregulatory failure. In these circumstances even a small fall in arterial blood pressure will reduce cerebral blood flow to damaged regions of brain in addition to those due to the primary pathology itself. For a patient on artificial ventilation, it is important not to over-ventilate because the fall in arterial P_{CO_2} which this produces will cause cerebral vasoconstriction and again exacerbate the hypoxia in already damaged regions of brain.

In hypertension of long standing, the limits of arterial blood pressure over which autoregulation operates are adjusted to a higher level, a compensation to allow the subject to operate at the new blood pressure. Typically the range for autoregulation in a hypertensive subject is 90 to 210 mm Hg. The middle of the range corresponds approximately to the subject's mean arterial blood pressure. If the well-meaning doctor prescribes hypotensive therapy, the lowering of the blood pressure which this produces to a value which would be normal for a normotensive subject may reduce the cerebral perfusion pressure to below the lower limit for autoregulation and the brain is damaged by hypoxia as a result.

Cerebral metabolism

The brain depends on a continuous supply of oxygen and glucose. Global anoxia of the brain causes loss of consciousness in 8 s. Irreversible damage to brain cells starts after 3 min and death of the whole brain in 6 min.

In normally functioning brain, the change of metabolism accompanying neural activity is far less than when a tissue such as skeletal muscle contracts. If the metabolism of brain undergoing maximal physiological activity is taken as 100%, then the basal metabolic processes require 60% of this metabolism, 20% for biosynthetic processes and 40% for pumping ions across neural membranes. The final 40% is the excess metabolism available for increased neural activity. In pathological conditions of overactivity of the brain, notably epileptic cerebral dysrhythmias, the oxygen consumption of neural tissues may increase several-fold. This is beyond the capabilities of supply of oxygen by the blood and hypoxia of the hyperactive regions of the brain ensues, with the danger of brain cell death.

The brain can only metabolize aerobically. This contrasts with tissues such as skeletal muscle which, if they do not receive enough oxygen for their ambient requirements, can metabolize anaerobically. Muscle can in this way incur an **oxygen debt** but brain cannot. This is one of the reasons for the brain requiring an adequate blood supply all the time.

There is another factor which operates in a similar way. Brain cells, unlike skeletal muscle cells, contain only a tiny reserve of glycogen. If the blood supplying the brain is deficient in glucose, the neuronal reserve of glygogen and glucose is exhausted in about 2 min. Hypoglycaemia is almost as damaging to brain as hypoxia. The results of both are brain cell death. The most sensitive region within the brain in this respect is the cerebral cortex, as observed in Chapter 11.

20.3 INTRACRANIAL PRESSURE

The brain is protected from mechanical damage by the skull. In this respect it differs from most of the other organs in our bodies. As always, this advantage carries its own penalties. There is no spare space inside the skull and if anything which takes up space is added to the contents of the skull, something else must move out. The only component which can move out easily is the blood. So the blood supply of the brain, which is so vital, is compromised

The normal intracranial pressure is between 0 and 10 mmHg. As the volume of the intracranial contents increases, the intracranial pressure rises. At first this rise is slow because the skull has some compliance. Once this stretching has reached its limit, the pressure increase becomes very steep, as shown in the brain compliance curve in Figure 20.2. As an intracranial space-occupying lesion (such as a haematoma) expands, the unit rise in pressure is greater when the haematoma is larger. When the intracranial pressure rises, the arterial blood pressure is no longer capable of forcing the blood into the cranial vascular bed. This is another factor reducing the blood supply to the brain when the volume of the intracranial contents increases. A homeostatic mechanism has evolved to combat this. A rise in intracranial pressure is accompanied by

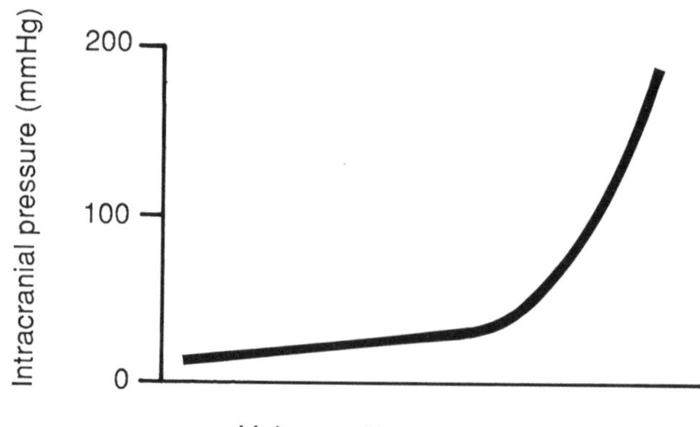

Figure 20.2 The intracranial pressure as a function of the volume in the intracranial contents.

a reflex rise in arterial blood pressure to a level above that in the cranial cavity over a wide range. This restores the ability of the arterial blood to perfuse the brain.

The intracranial pressure is unequivocally raised at 20 mmHg. With raised intracranial pressure, there is the potential for the development of a vicious circle. Consider a man with a head injury and a haematoma. As a result of the head injury he is unconscious. He has an airway obstruction. The hypoventilation involves hypercapnia and hypoxia which cause cerebral vasodilation and the haematoma expands. The intracranial pressure rises, leading to further reduction of cerebral perfusion with the development of cerebral oedema. Intracranial oedema fluid is space-occupying; there is a further rise in intracranial pressure and a progressive fall in cerebral perfusion.

20.4 THE MEASUREMENT OF CEREBRAL BLOOD FLOW AND METABOLISM: BRAIN 'IMAGING'

Mean cerebral blood flow

Cerebral blood flow was first measured by the nitrous oxide method (Ketty and Schmidt, 1945). The method is based on the Fick principle. The Fick principle is the law of conservation of matter; for an inert chemical, the amount which accumulates or is lost from a tissue or organ is equal to the amount entering or leaving from the blood.

Over the course of 10 min, the subject breathes a gas mixture containing 15% of the inert gas, nitrous oxide. The concentrations of nitrous oxide in blood entering and leaving the brain are measured. For measuring the arterial concentration, blood may be obtained from any systemic artery, since arterial blood is well mixed and the concentration of nitrous oxide is uniform. For the venous blood samples, blood is taken from the internal jugular vein, the large vein which drains the brain. Sufficient samples are taken to allow the investigator to plot the time-course of the concentrations, as shown in Figure 20.3.

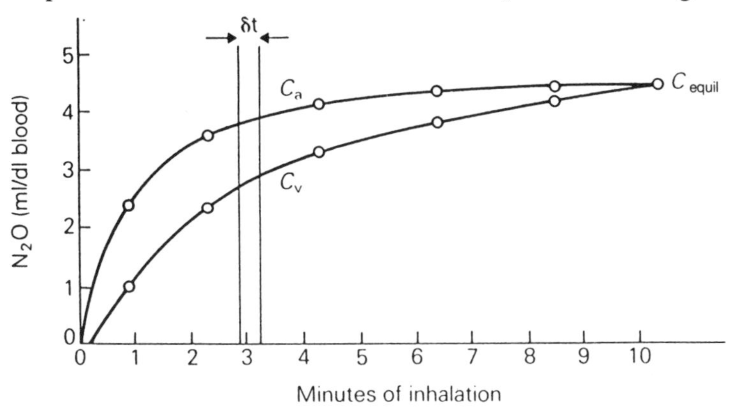

Figure 20.3 The nitrous oxide method of measuring cerebral blood flow. The subject inhales nitrous oxide at a constant concentration in inspired gas, starting at time zero. The graph shows C_a and C_v, the arterial and cerebral venous blood concentrations of nitrous oxideduring the inhalation period. $C_{(equil)}$ is the equilibrium concentration when there is no further net movement of nitrous oxide from blood to brain. Other features are described in the text.

For any interval of time, the amount of nitrous oxide leaving the blood to enter 100 g of brain is

$$F(C_a - C_v) \times \delta t$$

where F is the blood flow (ml blood/min/100 g brain), C_a and C_v are the concentrations of nitrous oxide in the arterial and venous blood, and δt is the interval of time.

To measure the amount of nitrous oxide entering 100 g of brain over the whole of the 10 min period, one measures the equilibrium concentration $C_{(equil)}$ of nitrous oxide in arterial and venous blood when nitrous oxide has been inhaled for the 10 min period. At this time, the concentration of nitrous oxide in brain equals that in the blood.

The Fick principle then yields

Amount present = amount which entered

$$C_{equil} = F \int_{t=0}^{10\ \text{min}} (C_a - C_v)\, \delta t$$

The integral is the area between the curves C_a and C_v. The blood flow F is the only unknown and is calculated from the measurements.

Although the possibility of measuring the cerebral blood flow was a great step forwards, the method suffers from two drawbacks. Firstly, it measures the cerebral blood flow averaged over the 10 min inhalation period, this making it impossible to use the method to measure rapid short-lived changes in blood flow. Secondly, the method gives an average of the blood flow to different regions. Any differences between the flow to different regions, e.g. differences in blood flow to grey and white matter in the cerebral cortex, are ironed out by the method. In a quest to overcome this latter shortcoming, investigators sought a new method which allowed the measurement of flow to different regions of the brain.

Regional cerebral blood flow

Measurement of regional blood flow involves the administration by inhalation or intracarotid injection of a radioactive indicator, such as ^{133}Xenon gas, which readily diffuses from blood to brain and which emits high energy rays; the energy must be sufficiently high for them to be able to penetrate the cerebral tissues and to reach extracranial detectors. The concentration of the indicator is monitored with directionally sensitive extracranial detectors. Readings are made of the intensity of radiation with the detector at many different sites and orientations around the head. The information is fed into a computer, which then calculates the three-dimensional distribution of ^{133}Xenon in the brain. It displays this information as a contour map of blood flow projected into a surface diagram of the brain, or as a map of the density of ^{133}Xenon at a series of different planes through the brain.

This technique has shown that a given amount of grey matter has four times the blood flow as the same amount of white matter. At rest, the frontal lobes have a higher blood flow than temporal or occipital lobes. Sensory stimulation on one side of the body produces an increased perfusion of the somatosensory region of the opposite hemisphere. Auditory stimulation etc. similarly cause increased blood flow to the relevant regions of cortex so that these now have the greatest blood flow of all the cortex. Speech is associated with an increase in blood flow in the left hemisphere. When a person hears a piece of music which he appreciates, there is an increase in blood flow in the superior temporal region in the right hemisphere.

Measurement of cerebral metabolism

Just as important as measuring cerebral blood flow is the measurement of the level of metabolic activity of the brain. If cerebral blood flow does not match metabolism, the neurones run out of energy and are liable to damage. This subject is particularly important to the clinician since, in certain pathological conditions, the increase in blood flow that normally accompanies increased metabolism is blocked and the brain is liable to anoxic injury. A major advance in this field was due to Sokoloff *et al.* (1977), who introduced the deoxyglucose method.

Deoxyglucose and the mapping of metabolic activity

The method as originally developed can be used only in experimental animals, since it involves removing the brain for histological sectioning. Radioactively labelled 2-deoxyglucose is infused intravenously at a constant rate over the course of 20 min. 2-Deoxyglucose is an analogue of glucose which, when injected into the blood, is taken up by cells in just the same way as glucose. Once it is inside the cell, it is phosphorylated but it cannot be further metabolized by the enzymes that metabolize glucose. Neither can it leave the cells, because the phosphorylated form is impermeant. It accumulates inside the cells at a rate proportional to the rate of metabolism of the cells. The deoxyglucose concentration in the cell thus reflects the rate of glucose uptake and hence the metabolism of the cell. It measures the metabolic activity averaged over the 20 min period during which deoxyglucose is in the circulating blood. At the end of the 20 min infusion period, the animal is killed, the brain removed and sectioned. The histological sections are laid on photographic emulsion. The radioactive emissions from the labelled deoxyglucose cause exposure of this emulsion. Photographic development then results in a photographic plate on which darkness is proportional to the concentration of deoxyglucose. The investigator has a slide which looks like a histological section, but the density at each site is a measure of the rate of metabolism of the neural tissue at that site.

For many years it was not possible to apply the method to humans but this problem has now been solved. The method is known as **positron emission tomography** or **PET** scanning. This measures the distribution in the brain of an injected or inhaled isotope that emits positrons. Positrons are like electrons but positively charged; they come from the nucleus of the atom. Within a millimetre, each positron collides with and annihilates an electron and the masses of these particles are converted into high energy γ-rays. Because of their high energy, these γ-rays can be detected by direction-sensitive detectors outside the skull. The readings are fed to a computer which builds up the three-dimensional distribution of the positron emitting density.

Deoxyglucose can be bonded to the positron-emitting isotope ^{18}F and this allows imaging of the metabolic activity of the living brain. With this method, it is possible to show, for instance, that in the normal subject a sensory stimulus entails an increase in metabolism of the receiving area of the cerebral cortex. Pathologically, the method has demonstrated that epileptic brain shows increased metabolism which outstrips the increase in cerebral blood flow found to occur during epileptic seizures.

The PET scan method can obviously be adapted for use with chemicals other than analogues of glucose. It has been exploited, for instance, by bonding positron-emitting atoms with neurotransmitters, their agonists and antagonists, to map the density of the receptors for neurotransmitter chemicals in the brain, as described in Chapter 17.

20.5 PHARMACOLOGICAL PROTECTION AGAINST CEREBRAL ISCHAEMIA

If the blood supply to the brain is interrupted for 4 min or more, brain cells die. The longer the duration of anoxia, the greater the consequent neurological deficit. An important factor in causing brain cell death is the release of excitotoxins from hypoxic neurones. The protection afforded in stroke patients by calcium channel blockers and NMDA blocking agents has been described in Chapter 17.

Neuroanatomy

21

21.1 THE BRAIN AND ITS COVERINGS

The central nervous system consists of the brain and spinal cord. The spinal cord is a long cylinder lying in the vertebral canal. At the top it is continuous with the brain. The entire central nervous system is protected by bony casing, the brain by the skull and the cord by the vertebral column. Peripheral nerves connect with the brain through foramina in the skull and with the spinal cord via intervertebral foramina. Figure 21.1a shows an outline diagram of the brain. The brain consists of five parts, the forebrain, the midbrain, the pons, the cerebellum and the medulla oblongata.

The forebrain (**prosencephalon**) is subdivided into the cerebral hemispheres (Figure 21.b; see also Chapter 13), the basal ganglia (Chapter 12) and the **diencephalon**, the latter comprising the thalamus and the hypothalamus (Chapter 19). The midbrain (**mesencephalon**), the pons, and the medulla connect the forebrain with the cord and together are called the brain stem (Chapter 11).

The brain and spinal cord are enveloped by the **meninges** which are three membranes called the dura mater, the arachnoid mater and the pia mater. The **dura mater** (meaning tough membrane) is tough connective tissue. It lines the bony cavities in which the central nervous system lies. In the cranial cavity, the dura extends into the cavity in sheets to divide the cavity with a series of interconnecting spaces for housing the different anatomical parts of the brain. One of these sheets is called the **tentorium cerebelli**, the tent of the cerebellum. In the upright head it lies approximately in the horizontal plane and lies on the cerebellum. It provides a support for the posterior (or occipital) lobes of the cerebral hemispheres. The tentorium is attached behind to the bones of the skull. Its front border is free and the midbrain lies in front.

The **subdural space** is a very narrow space between the dura mater and the arachnoid mater. It is not in continuity with the subarachnoid space to be described shortly. It is not infrequently the site of an accumulation of blood

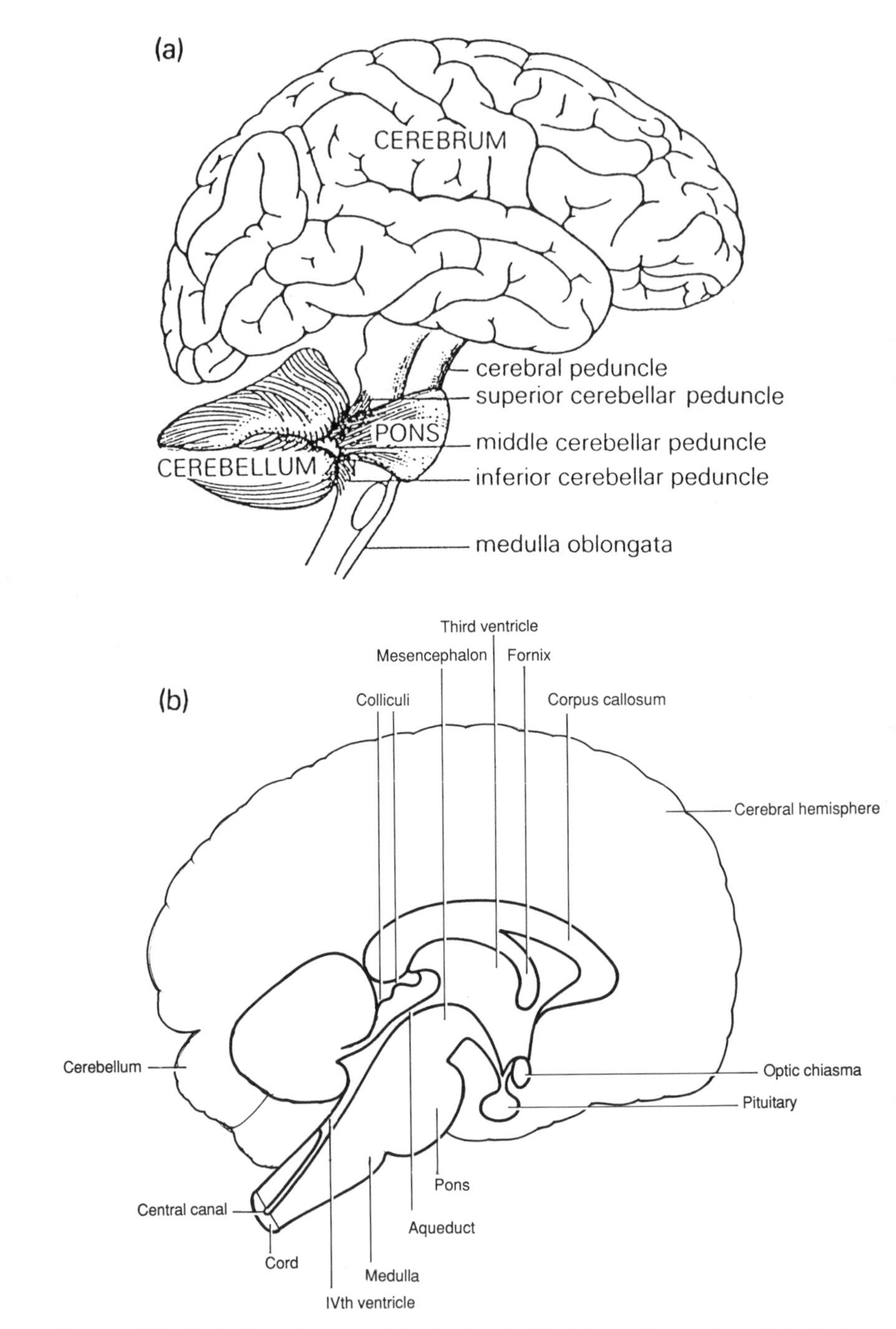

Figure 21.1 (a) Outline diagram of the brain. (b) Medial saggital section of the brain.

when a blood vessel is torn as a result of head injury. This condition is known as a subdural haemorrhage.

The **arachnoid mater** is a continuous membrane within and much more delicate than the dura mater. Between the arachnoid mater and the pia mater is the subarachnoid space which contains a fluid of special composition called the cerebrospinal fluid (CSF). The capacity of the subarachnoid space is much greater than that of the subdural space. The subarachnoid space is traversed by blood vessels.

The brain develops embryologically as a tube. The **telencephalon** develops as a pair of outward expansions from the front end of the tube. The central cavity of the tube develops into a complex of cavities. These cavities are filled with cerebrospinal fluid. On each side there is a lateral ventricle within the cerebral hemisphere. Each lateral ventricle consists of a series of interconnecting cavities which project into the frontal, temporal and occipital lobes. Medially, each lateral ventricle communicates with the third ventricle which lies in the midline. This connects through a narrow tube called the aqueduct (in the midbrain) with the fourth ventricle which lies between the medulla anteriorly and the cerebellum posteriorly. The roof of the fourth ventricle is deficient at a median aperture and two lateral apertures. Through these apertures, the ventricles and the subarachnoid space are in continuity.

The **choroid plexuses** are two plexuses (or networks) of blood vessels that invaginate the roof of the fourth ventricle, one on either side. They are covered with a secretory epithelial lining. Cerebrospinal fluid is formed at the choroid plexuses and also in the capillaries of vessels supplying the brain tissues. The cerebrospinal fluid circulates through the ventricular system to the subarachnoid space. It finds it way back to the circulating blood through the **arachnoid villi**. These are minute projections from the subarachnoid space into the large venous sinuses which drain blood from the brain. Many such villi are aggregated together to form an arachnoid granulation, a macroscopic granule protruding from the subarachnoid space into the venous sinuses. Some arachnoid granulations are so large that they cause absorption of bone producing pits in the inner aspect of the skull-cap.

21.2 THE SPINAL CORD

The spinal cord lies in a canal formed by the vertebrae; these latter are the bones which form the spinal column. The vertebrae are named according to where they lie; there are seven cervical vertebrae, 12 thoracic vertebrae; five lumbar vertebrae and three sacral vertebrae. The sacral vertebrae are fused together but the others articulate with each other. The nervous connections between the cord and the extravertebral tissues is made through nerve roots, which are bundles of axons. There is one set of nerve roots for each segment of the spinal cord (i.e. for each vertebra) and the roots emerge from the vertebral canal by passing through foraminae between adjacent vertebrae.

The uppermost roots are called the first cervical roots or C1. They pass between the skull and the first cervical vertebra. The second cervical roots C2 pass between the first and second vertebrae and so forth. The roots passing between the seventh cervical and first thoracic vertebrae are called C8. From there down, the roots are numbered according to the vertebrae below which they pass.

At each segment, all the afferent (sensory) nerve fibres from one side of the body collect together to form the posterior root whereas all the motor nerve fibres leave the cord together as the anterior root. Each segment therefore has a posterior and an anterior root on each side. Outside the cord, the afferent and efferent roots join to form spinal nerves; these in turn join and divide to connect with the peripheral nerves which are usually mixed motor and sensory nerves. The cell bodies of the afferent fibres all lie together to form a swelling called the dorsal root ganglion in the dorsal root. This ganglion lies in the foramina between adjacent vertebrae. There are no synapses in the dorsal root ganglion.

In the first months of life, the vertebral column grows faster than the cord and consequently the length of the roots in the spinal cord increases from the cervical level down to the sacral level. For the first and second cervical segments the roots are short and transverse. For the remainder of the cervical region, the roots become progressively more oblique. The nerve root from the eighth cervical segment emerges from the vertebral canal one segment lower than its level of origin from the cord. In the thoracic region, the nerve roots emerge from the vertebral canal two segments lower than their level of origin from the cord. In the lumbar region, the nerve roots emerge from the vertebral canal two or three segments lower than their level of origin from the cord.

In the adult, the spinal cord does not extend below the level of the second lumbar vertebra. In the lower lumbar region, the nerve roots of the lower lumbar and sacral segments lie within a tube of dura. Each nerve root penetrates the dura at its level of exit from the vertebral column. This anatomical relationship is exploited when a clinician wishes to obtain a sample of cerebrospinal fluid. The procedure is known as **lumbar puncture**. A needle of appropriate length and diameter is introduced between the lower lumbar vertebrae and pierces the dura. The manometer to which the needle is connected indicates the pressure of the cerebrospinal fluid bathing the spinal cord. A sample of the fluid can be removed for analysis. There is no fear of mechanical damage to the cord. The nerve roots slip aside as the needle passes them. A corollary of the anatomical arrangement is that injuries to the lower lumbar and sacral regions of the vertebral column in adults do not directly damage the cord; their effects are limited to damage of nerve roots. In the sacral region, the long nerve roots have the appearance of a horse's tail and are appropriately called the 'cauda equina'.

Cutaneous distribution of spinal segments (dermatomes)

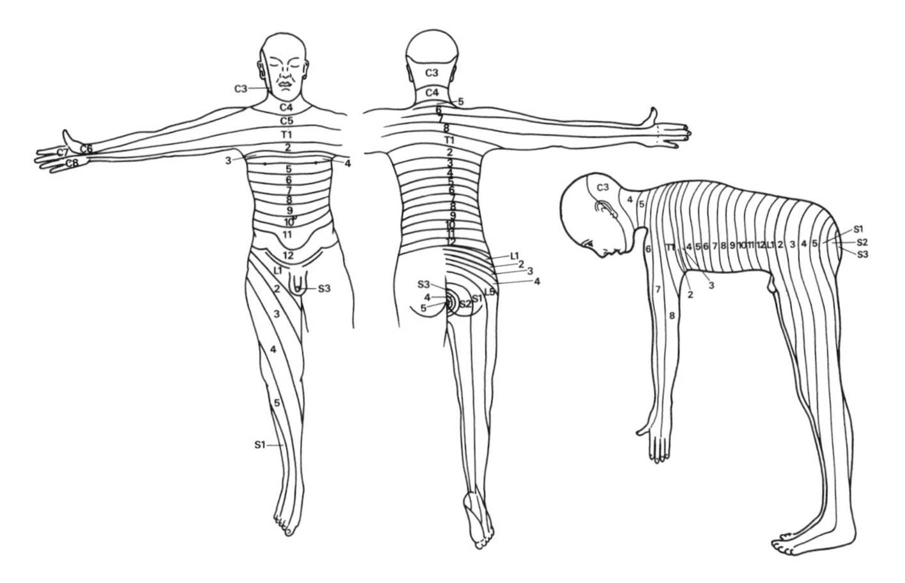

Figure 21.2. The sensory cutaneous distribution of the spinal segments. (a) Front view. (b) Back view. (c) The dermatomes form a sequence of almost straight stripes when the subject assumes the ancestral quadrupedal position.

Figure 21.2 shows the sensory innervation of the skin by the different segments of the cord. For the trunk, each segment innervates a simple band, whereas in the limbs, the pattern is more complicated. This is because of the complex embryological development of the limbs. Figure 21.2c shows that, when man adopts a quadrupedal position, the segmental innervation appears as a series of regular fairly straight bands, indicating our ancestral lineage from a segmented worm. It is the fact that man has pulled himself up to the bipedal position that results in most of the apparent complexities of segmental innervation. There is considerable overlap of innervation by contiguous spinal roots so that cutting one or two roots does not result in a band of complete anaesthesia, only in a band of diminished sensitivity. Three contiguous roots have to be sectioned before a band of anaesthesia is produced.

The medulla oblongata

At its rostral end, the spinal cord increases in size and is continuous with the medulla oblongata, without any sharp line of demarcation (Figure 21.1b). Its lower end is at the level of the foramen magnum and, at its upper end, it is separated from the pons by a groove on the anterior aspect of the brain. It contains the bulk of the respiratory centres and the cardiovascular centre.

The anterior median fissure continues up from the spinal cord but then ends in the medulla at the decussation of the corticospinal tracts, alternatively known as the decussation of the pyramids.

The olivary nucleus is a hollow mass of grey matter with irregular walls and a hilum. It lies in the medulla just dorsal to the pyramids and creates an elevation on the surface of the medulla. The olivo-cerebellar tract consists of fibres that emerge from the hilum, cross the midline, and enter the inferior cerebellar peduncle. The olivary nucleus probably acts as a relay to the cerebellum from the cord, the vestibular nuclei and the cerebral cortex.

On the dorsal aspect, the medulla houses the fourth ventricle, which is in continuity below with the central canal of the spinal cord. The dorsal column nuclei (Section 8.1) lie on the dorsal aspect of the medulla and are displaced laterally by the edges of the fourth ventricle.

From the medulla arise the hypoglossal (XII), glossopharyngeal (XI) and vagus (X) nerves. Other cranial nerves arise from the border between the medulla and the pons; they are the abducens (VI), facial (VII) and auditory (VIII) nerves, named in order from medially to laterally.

The pons

The pons (Figure 21.1b) joins the medulla oblongata below with the midbrain above. It is divided into two parts. The dorsal part of the pons resembles the medulla oblongata, of which it is a direct continuation. The ventral part of the pons is composed mainly of structures which are peculiar to this level; the exception to this is that the corticospinal tracts traverse the ventral pons to become the medullary pyramids below. The pons forms a conduction path uniting the cerebellar hemispheres with the cerebral hemispheres. It contains pontine nuclei, which form the target of corticopontive fibres and also receive input from collaterals of the corticospinal fibres as they pass through. From the pontine nuclei, most fibres traverse the median plane to reach the opposite cerebellar hemisphere through the middle cerebellar peduncle. These fibres are responsible for the most striking feature of the ventral pons in transverse section, which is of fibres traversing from side to side.

The trochlear (IV) nerves arise from the posterior aspect of the pons.

The midbrain (mesencephalon)

This is the upward continuation of the dorsal part of the pons. On the ventral aspect on either side of the midline lies the basis pedunculi, consisting of fibres of cortical origin on their way to the ventral part of the pons. The middle part consists of the corticospinal projection, as yet uncrossed.

On the dorsal aspect of the mesencephalon lie the corpora quadrigemina, alternatively called the superior and inferior colliculi. The colliculi on either side are separated by a shallow groove in which lies the pineal body. There are connections from the lateral geniculate body to the superior colliculus and from the medial geniculate body to the inferior colliculus.

The diencephalon

This connects the cerebral hemispheres with the mesencephalon. The axial portion resembles the upper part of the mesencephalon, with which it is in continuity. The projection from the cerebral hemispheres downwards maintains the same position in the basis pedunculi as in the mesencephalon.

More laterally in the diencephalon, large nuclei have developed in the tendency to telencephalization. These are the thalamus, subthalamus and hypothalamus and from the diencephalon develop the retinae and the optic nerves.

The thalamus

This is a large ovoid mass and consists of a large number of nuclei. The thalamus on either side is connected with its opposite number via a narrow band of neural tissue. The lateral nuclei of the thalamus house the relays on the central afferent projection pathways; all these pathways have their second synaptic relay in the thalamus. Other thalamic nuclei act as relay stations on the routes from the basal ganglia and from the cerebellum to the cerebral cortex. The most medial thalamic nuclei are phylogenetically the most ancient. They make up the cephalad component of the ascending reticular formation.

Basal ganglia

As shown in Figure 21.3, these are deeply placed masses of grey matter within the cerebral hemisphere. The caudate nucleus (Figure 12.1) is throughout its extent closely related to the lateral ventricle. The lentiform nucleus is deeply placed in the white matter of the hemisphere and lies between the insula laterally and the caudate and thalamus medially. It rests on the white matter forming the roof of the inferior horn of the lateral ventricle. The amygdaloid body lies at the very front of the roof of the inferior horn of the lateral ventricle. The subthalamic nucleus lies in the diencephalon, with the red nucleus and substantia nigra lying immediately below in the mesencephalon.

Cerebral hemispheres

These are connected by the corpus callosum, a thick band of nerve fibres. Each hemisphere is divided, rather arbitrarily, into lobes. In addition to the four lobes seen on the cortical convexity, there is the insula, which is infolded cortex in the base of the lateral fissure. It lies over the corpus striatum.

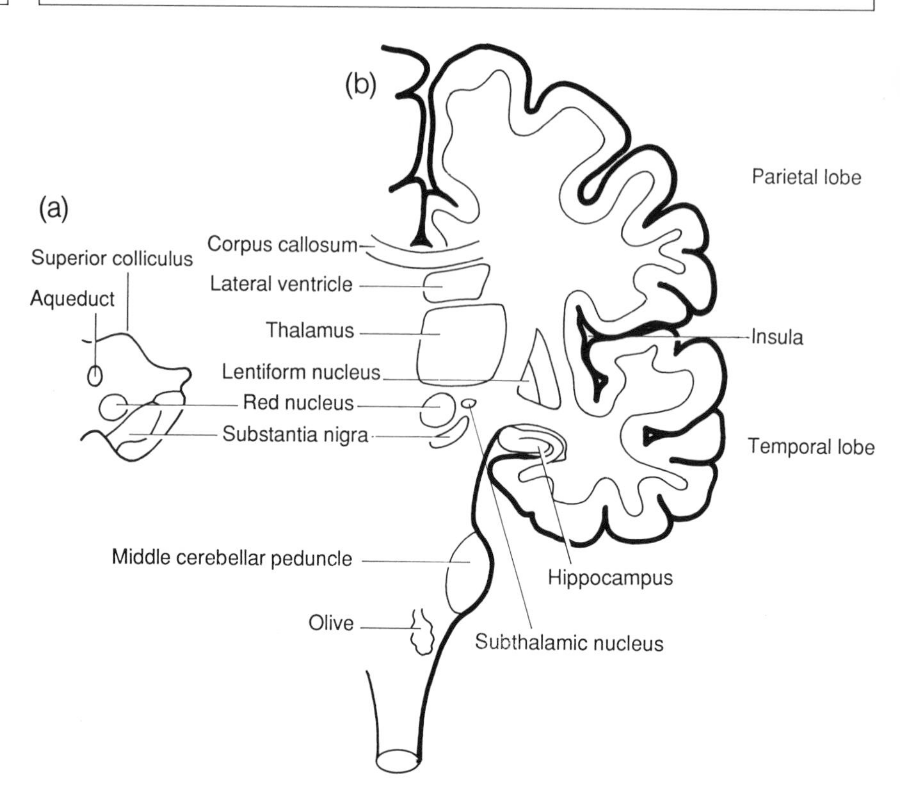

Figure 21.3 (a) transverse section through the mesencephalon. (b) Coronal section through the brain.

Subdivisions of the lobes of the cerebral hemispheres

These are shown in the frontispiece.

The frontal lobe

The frontal lobe lies anterior to the central sulcus. In front of the central sulcus and running parallel with it is the precentral sulcus. The precentral gyrus lies between these sulci. The area of the frontal lobe anterior to the precentral gyrus is subdivided into the superior, middle and inferior frontal gyri.

Parietal lobe

The lateral aspect is subdivided into three areas by the postcentral sulcus and the intraparietal sulcus. The postcentral gyrus lies between the central sulcus in front and the postcentral sulcus behind. The superior parietal lobule lies medial to the intraparietal sulcus. The inferior parietal lobule lies below the intraparietal sulcus.

The inferior parietal lobule is divided into three parts, anterior (the supramarginal gyrus), middle (angular gyrus) and posterior.

The temporal lobe

This is divided into superior, middle and inferior temporal gyri.

The occipital lobe

This is divided into superior and inferior occipital gyri.

Central afferent and efferent pathways

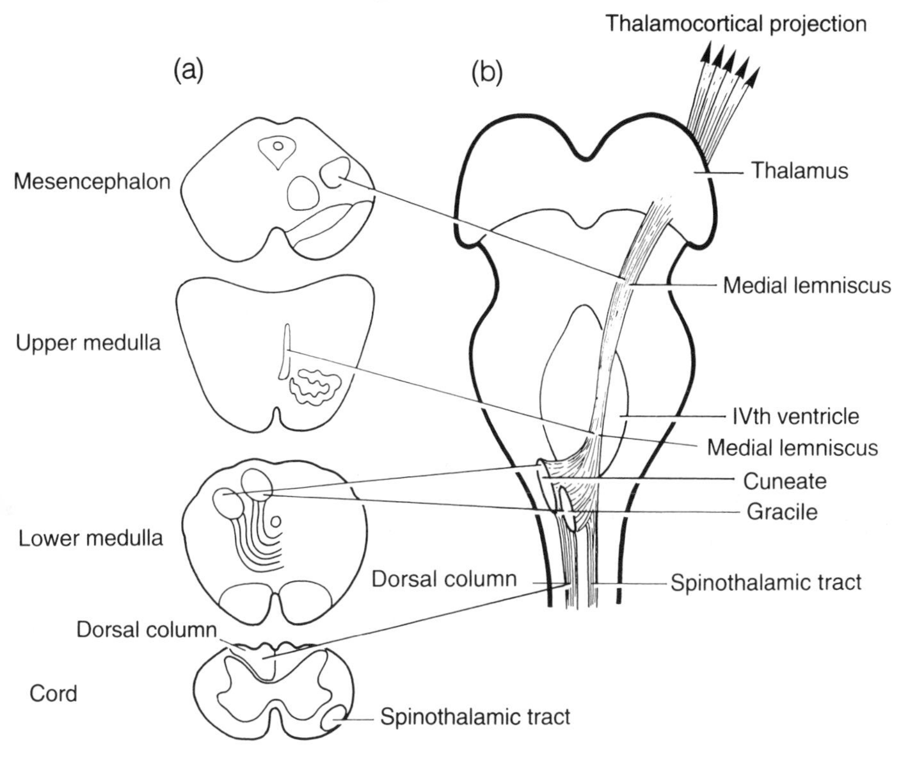

Figure 21.4 Central afferent pathways from the cord to the cerebral cortex. (a) Transverse sections of the brain stem and cord. (b) A dorsal view of the afferent projection as it courses from the cord through the brain stem.

These were described in Sections 6.6 and 6.7. The anatomy of the afferent pathways is shown in Figure 21.4. The first order neurones ascending in the dorsal columns form their first relay in the dorsal column nuclei (the gracile and cuneate nuclei) which lie on the dorsum of the medulla. From these nuclei, the second order neurones immediately decussate, passing ventrally and in front of the central canal to form the medial lemniscus (fillet). Fibres ascending via the spinothalamic tracts join the medial lemniscus which runs up to the lateral relay nuclei in the thalamus. Here the second relay on the pathway is made and the third order neurones project up from the thalamus to the postcentral gyrus of the parietal lobe of the cerebral hemisphere.

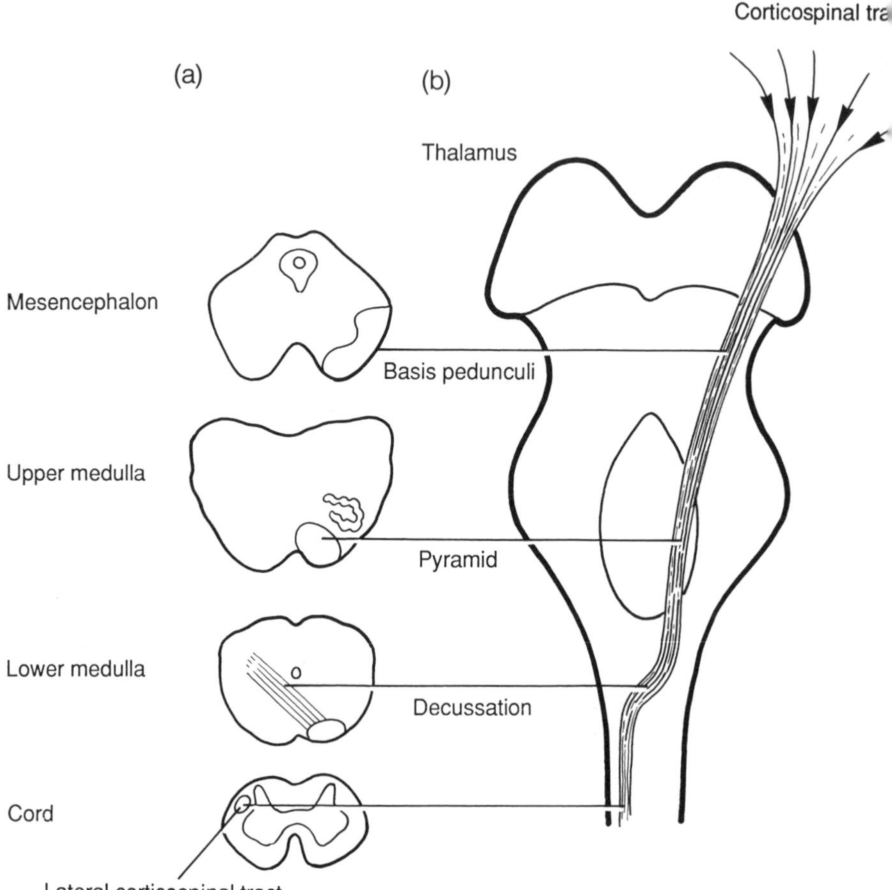

Figure 21.5 The corticospinal projection. (a) Transverse sections of the brain stem and cord. (b) A dorsal view of the projection as it courses from the cortex down through the brain stem to the cord.

The corticospinal projection is shown in Figure 21.5. The tract originates in the pyramidal cells of the precentral gyrus and projects down via the internal capsule to the brain stem, running in the basis pedunculi, through the pons and then forming the pyramids in the medulla. In the lower medulla, the pyramids decussate, and the fibres become the lateral corticospinal tract which projects down through the spinal cord.

FURTHER READING

Barr, M.L. and Kiernan, J.A. (1989) *The Human Nervous System. An Anatomical Viewpoint* (5th edn.) London, Lippincott.
Ranson, S.W.R. and Clark, S.L. (1959) *The Anatomy of the Nervous System* (19th edn). Philadelphia, W.B. Saunders Co.

Self-test questions and answers 22

Many of these questions are multiple choice questions (MCQs). An MCQ consists of five parts, or statements. Each statement is either correct or incorrect. The decision for each statement is entirely separate from the decision for any other statement.

SENSATION AND MOVEMENT

Question 1.1 MCQ

The skin of a finger of a normal human was tested for light touch (with a fine bristle) and for pain sense over a grid of points marked with dots in the figure. At points marked with a cross, the touch stimulus was perceived and at circled points the pain stimulus was perceived.

x = touch

⊙ = pain

5 mm

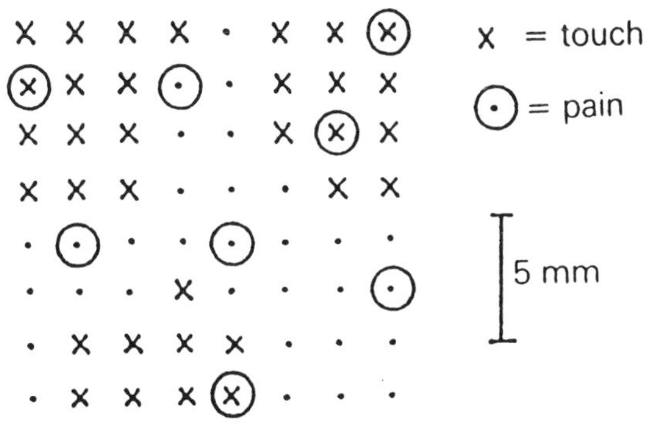

The subject could discriminate three different positions of the light touch stimuli but was unaware of the differences in the positions of the pain-sensitive spots.

From this information it can be deduced that, in the region of skin tested:

(a) There are points where the subject could not perceive the light touch stimulus.
(b) There are at least three touch endings.
(c) All the touch endings may be connected to a single afferent nerve fibre.
(d) All the pain endings may be connected to a single afferent nerve fibre.
(e) The sensation of pain depends on excessive stimulation of touch receptors.

Question 1.2 (MCQ)

Concerning receptor mechanisms in humans:

(a) The frequency of impulses generated in an afferent nerve fibre by a constant stimulus to the receptor may decline progressively.
(b) The receptive field of an afferent nerve fibre may consist of several non-contiguous areas of skin.
(c) There are receptors for every type of energy impinging on the skin.
(d) The perception of burning of the skin depends on the conduction of heat along the afferent nerve fibre.
(e) The speed of conduction of action potentials along afferent nerve fibres equals the speed of light.

Question 1.3 (MCQ)

In a normal adult person, mechanisms involved in producing an increase in muscle effort include increase in:
(a) The amplitude of action potentials in motor nerve fibres.
(b) The frequency of action potentials in motor nerve fibres.
(c) The number of motor nerve fibres conducting action potentials.
(d) Repetitive contraction of muscle fibres in response to a single action potential in the motor nerve.
(e) The safety factor of neuromuscular transmission.

Question 1.4 (MCQ)

Concerning skeletal muscle and its innervation:
(a) Tone is due to action potentials in the motor nerve fibres innervating the muscle.
(b) The peak tension produced by a muscle fibre in response to a single action potential in its motor nerve is only a small fraction of the tension in response to a burst of impulses at high frequency in the motor nerve.
(c) A single motor nerve innervates several muscle fibres.
(d) The cell bodies of the motor nerve fibres lie in the dorsal root ganglia.
(e) A co-ordinated movement involves the relaxation of certain muscles simultaneously with the contraction of others.

NERVE

Exercise 2.1 Structured exercise

The exercise consists of working through the text, filling in the blank spaces and deleting inappropriate expressions enclosed in brackets. In some instances, more than one item within the brackets may be correct.

Nerve fibres

Peripheral nerve fibres are of two types, myelinated and unmyelinated. The myelinated are

(thicker/thinner) than the unmyelinated. In myelinated fibres, the myelin is produced by

(the nerve fibre/supporting cells called Schwann cells). The myelin is interrupted at intervals of about

(0.1/1./10/100) mm at sites called the Nodes of Ranvier. The action potential is generated at the

(internodes/nodes).

Resting potential

In the interstitial fluid, the principal cation and anion are, respectively

(sodium/potassium/calcium); and

(chloride/nitrate/organic anion). Inside the axoplasm, the principal cation and anion are, respectively

(sodium/potassium/calcium) and

(chloride/nitrate/organic anion).

At rest, the inside of the nerve fibre is

(negative/positive) to the outside. The absolute magnitude of the voltage is of the order of

(0.01/0.1/1/10) volts. At rest, the membrane peremeabiltiy to sodium ions is

(lower/higher) than the permeability to potassium ions.

For sodium, the equilibrium potential is typically

(−90/0/+55/+100) mV. For potassium the equilibrium potential is typically

(−90/0/+55/+100) mV. So, for the membrane at rest the membrane potential is close to the equilibrium potential for

(sodium/potassium) ions.

Action potential

The adequate stimulus for initiating an action potential in the nerve axon is

(acetylcholine acting on the membrane/depolarization). This causes transient

(opening/closing) of channels which allow the passage of (sodium/potassium) ions. Consequently these ions

(enter/leave) the axon carrying with them their positive charge and hence making the internal voltage more

(positive/negative). The membrane potential moves towards the equilibrium potential for

(sodium/potassium) the peak voltage is typically .... mV.

After the peak of the action potential the membrane potential returns towards its resting value. Contribution to this is made by

(potassium ions leaving/sodium ions leaving/chloride ions entering) the nerve fibre. It involves ions moving

(down/against) their electrochemical gradient. If the sodium pump is blocked, the downstroke of the action potential is

(little affected in the short term/greatly prolonged).

Characteristics of nerve fibres

The thickest nerve fibres in the human body are typically

(0.2/2/20/200) μm in diameter and conduct at

(1/10/100/1000) m/s. The smallest nerve fibres in the human body are typically

(0.05/0.5/5/50) μm in diameter and conduct at

(0.01/1/10/100) m/s.

Calculation of wavelength

For a mammalian nerve fibre of large diameter, the action potential lasts for about 0.5 ms at body temperature. Its speed of conduction is about 100 m/s. Calculate the length of axon depolarized at a given instant of time.

Question 2.1 (MCQ)

The action potential in nerve:

(a) Is initiated by a depolarization of the membrane.
(b) Is a change in membrane potential towards the equilibrium potential for sodium ions.
(c) Involves a decrease in the membrane permeability to potassium ions.
(d) Is associated with an increase in the electrical resistance of the membrane.
(e) Is propagated along the axon by means of the release of acetylcholine.

Question 2.2 (MCQ)

The conduction velocity at 37°C:

(a) For the thickest nerve fibres in the pyramidal tracts (in the central nervous system) is on average greater than for the thickest peripheral nerve fibres.
(b) In a Group 1a afferent fibre is typically 500 m/s.
(c) In a non-medullated fibre is typically 5 m/s.
(d) In thick myelinated fibres is inversely proportional to the fibre diameter.
(e) In unmyelinated fibres is proportional to the square root of the fibre diameter.

Question 2.3 (MCQ)

In a nerve fibre,

(a) At rest, the intracellular potassium concentration exceeds the intracellular sodium concentration.
(b) At the peak of the action potential, the intracellular sodium concentration exceeds the intracellular potassium concentration.
(c) At rest the intracellular ionized calcium concentration is about 1 mmol/l.
(d) If the sodium pump is poisoned, the action potential mechanism is immediately blocked.
(e) If the sodium pump is poisoned, the axon progressively shrinks.

Question 2.4 (MCQ)

In a nerve fibre

(a) The resting potential depends primarily on the ratio of concentrations of sodium ions inside and outside the ribre.
(b) At rest, the concentration of potassium ions in the axoplasm exceeds that in the interstitial fluid.
(c) The net inward movement of sodium during the rising phase of the action potential is due to transient intense activity of the sodium pump mechanism.
(d) The downstroke of the action potential is due to extrusion of sodium from the axoplasm by the sodium pump mechanism.
(e) The amplitude of the action potential is inversely proportional to the diameter of the fibre.

Receptor mechanisms

Questions 3.1 to 3.5 cover aspects of mechano-electrical transduction and also the conduction of the nerve impulse. The five questions all refer to the figure, which shows (a) the structure of a Pacinian corpuscle, a receptor sensitive to deformation, and (b) the transmembrane potential in the ending of the nerve fibre.

Some of the questions test your physiological knowledge and others test your ability to make deductions from the experimental data presented in the figure.

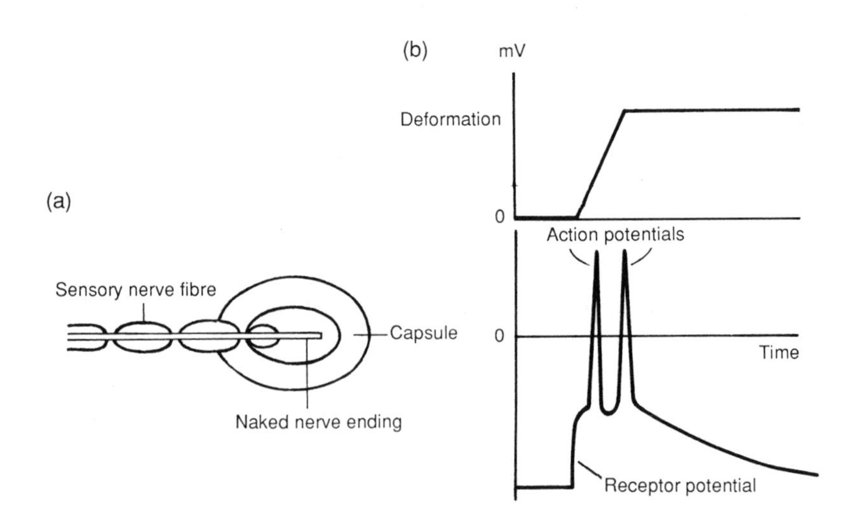

Question 3.1 (MCQ)

(a) The receptor potential is generated in the capsule.
(b) The action potentials would propagate without decrement along the sensory nerve fibre.
(c) The receptor potential would propagate without decrement along the sensory nerve fibre.
(d) If the nodes of Ranvier were blocked by local anaesthetic, the receptor potential would be abolished.
(e) If the nodes of Ranvier were blocked by local anaesthetic, the action potentials would be abolished.

Question 3.2 (MCQ)

(a) The receptor potential and the action potential are both due to opening of common ionic channels (or gates).
(b) The discharge of the sensory nerve fibre shows adaptation.
(c) If the same final deformation had been attained with a slower more prolonged ramp, the sensory nerve fibre would have fired more impulses
(d) As a result of a deformation of this magnitude the receptor fails to transmit information to the central nervous system.
(e) The electrical record may have been derived from the sensory nerve fibre some cm away from the receptor itself.

Question 3.3 (MCQ)

(a) The membrane of the unmyelinated nerve terminal is electrically excitable.
(b) The membrane of the first node of Ranvier is electrically excitable.
(c) Chemical transmission is an essential step in the linkage of the receptor potential and the action potentials.
(d) The receptor potential is a hyperpolarization.
(e) The action potential is a hyperpolarization.

Question 3.4 (MCQ)

(a) At the offset of the stimulus, not shown in the diagram, there would be a receptor potential of the same polarity as at the onset.
(b) At the offset of the stimulus, action potentials would be generated.
(c) Conduction of the nerve impulse along the sensory nerve fibre is saltatory.
(d) The sensory nerve fibre is capable of conducting an action potential towards the receptor.
(e) The cell body of the sensory nerve fibre is in the dorsal horn of the grey matter of the spinal cord.

Question 3.5 (MCQ)

(a) If the receptor were decapsulated, the receptor would adapt more slowly to a maintained stimulus.
(b) A very intense thermal stimulus would stimulate the receptor.
(c) Pacinian corpuscles are found in the skin of the lobe of the ear.
(d) Pacinian corpuscles are found in the cornea.
(e) Pacinian corpuscles are found in the epidermis.

Question 3.6 (MCQ)

With respect to sensation in the human skin:

(a) The sensation of touch can be mediated by free nerve endings.
(b) There is better two-point discrimination in the thumb than in the upper arm.
(c) There are local variations in the sensitivity of skin to touch.
(d) Pacinian corpuscles are temperature receptors.
(e) A single point on the skin can be sensitive to both pain and touch.

Question 3.7 (MCQ)

An increase in strength of stimulus to the skin will evoke, in the sensory nerve fibre, action potentials which:

(a) Are longer in duration.
(b) Are higher in amplitude.
(c) Are more frequent.
(d) Are conducted more rapidly.
(e) Occur in a larger number of nerve fibres.

Exercise 4.1 Structured exercise

Neuromuscular transmission

In a nerve, an action potential can propagate

(only in one direction/in either direction).

At the neuromuscular junction, transmission
(can occur in either direction/is unidirectional).

Chemical transmitter is stored in the cytoplasm

(diffusely/in packets) in the

(presynaptic/postsynaptic) cell. In the absence of action potentials in the motor nerve fibre there is (release/no release) of small amounts of transmitter. An action potential in the motor nerve fibre causes the calcium concentration in the axoplasm of nerve terminal to

(rise/fall). This is

(unrelated to/an essential step in) the process of

transmission. Transmitter chemical

(diffuses/is pumped) across the synaptic cleft and, on arrival at the postsynaptic membrane, the chemical

(attaches to receptor sites of the surface/penetrates into the cytoplasm) of the postsynaptic cell. As a

result of this, there is an opening of ionic channels which allow the passage

(of sodium ions specifically/of all small inorganic ions)

which consequently

(are pumped against/flow down) their electrochemical gradient. The equilibrium potential for the end-plates potential is around

(−90/−50/0/+50) mV. This electrical change results in depolarization of the neighbouring muscle fibre membrane and initiates an action potential there. The linkage between end-plate potential and action potential in the muscle fibres is

(via the intermediary of acetylcholine/by local current flowing out from the end-plate region). The delay

introduced by neuromuscular transmission is around

(0.03/0.3/3/30) ms. For a nerve fibre with a diameter of 10 μm this delay corresponds to conduction along a length of

(1.5/15/150/1500) mm of nerve. An action potential in the muscle fibre

(is confined to the region near the end-plate/spreads the whole length of the muscle fibre). On either side of the end-plate the action potential is propagated in

(the same direction/opposite directions).

Exercise 4.2 Matching exercise

Properties of the action potential, receptor potential and synaptic potential

For each of these three types of potential, decide which characteristics from the list below apply and fill in the table appropriately, e.g. the top left-hand entry should be 'all-or-none' because the action potential is all-or-none.

	A	B	C	D	E
Action potential					
Receptor potential (somaesthetic receptors)					
Synaptic potential					

List of alternatives

(a) All-or none/graded.
(b) Propagates without decrement/localized.
(c) Waveform resembles intensity of excitation (i.e. analogue coding)/digital coding.
(d) One-way conduction only/orthodromic and antidromic conduction are possible
(e) Mechanism is electrical/mechanism is chemical

NEURONAL INTERACTIONS

Exercise 5.1 Matching exercise

Junctional transmission

After each of the following items (1 to 7), indicate which items (a) to (i) in the list are appropriately associated with it.

1. The release of chemical transmitter from a nerve terminal depends on .....
2. Essential in the process of presynaptic inhibition are (in correct order) ....., ... and .....
3. The ionic channels in the postsynaptic membrane that are opened to give an excitatory postsynaptic potential are .....
4. The ionic channels in the postsynaptic membrane that are opened to give an inhibitory postsynaptic potential are .....
5. The ionic mechanism of an end-plate potential is .....
6. Curare blocks neuromuscular transmission by .....
7. Succinyl choline blocks neuromuscular transmission by .....

List

(a) Reduction of the action potential excursion in the presynaptic excitatory nerve terminal.
(b) Reduction in amount of excitatory transmitter released.
(c) Attachment to the receptors in the post-junctional membrane thereby preventing access to acetylcholine.
(d) Chloride channels
(e) Opening of membrane channels that allow all small inorganic ions to pass.
(f) Opening of channels in the presynaptic nerve terminal.
(g) Increase in concentration of free calcium in the cytoplasm of the presynaptic nerve terminal.
(h) Depolarization block of the end-plate.

Question 5.1 (MCQ)

An excitatory post-synaptic potential (EPSP):

(a) Consists of a change of the membrane potential towards the equilibrium potential for potassium ions.
(b) Is produced by ions flowing down their electrochemical gradients.
(c) May summate with other postsynaptic potentials generated on the postsynaptic nerve cell membrane.
(d) Has the same ionic mechanism as the action potential.
(e) Is an essential step in the activation of motoneurones during a stretch reflex.

Question 5.2 (MCQ)

An inhibitory postsynaptic potential (IPSP) in a mammalian motoneurone.

(a) May prevent an excitatory postsynaptic potential (EPSP) from initiating an action potential.
(b) Is a hyperpolarizing potential irrespective of the membrane potential from which it arises.
(c) Is a change in the membrane potential towards the equilibrium potential for sodium ions.
(d) Is directly produced by an increase in activity of the sodium pump.
(e) Is an essential step in the production of a reflex muscle twitch.

Question 5.3 (MCQ)

An inhibitory postsynaptic potential (IPSP) in a spinal motoneurone:

(a) Reduces the excitability of the motoneurone when tested by a direct electrical stimulus to the motoneurone.
(b) Is accompanied by an increase in conductance of the postsynaptic membrane.
(c) Is directly produced by ions flowing down their electrochemical gradients.
(d) Is mediated by a chemical transmitter.
(e) Can block one excitatory input on the motoneurone leaving all others unmodified.

Question 5.4 (MCQ)

Presynaptic inhibition

(a) Occurs in the spinal cord.
(b) Can block one excitatory input to the postsynaptic cell leaving all others unmodified.
(c) Involves an increase in the overshoot of the action potential in the presynaptic terminal.
(d) Has the same ionic mechanism as postsynaptic inhibition.
(e) Is an increase in membrane potential in the postsynaptic cell.

Question 5.5 (MCQ)

Presynaptic inhibition in the spinal cord

(a) Is an increase in membrane chloride permeability of the postsynaptic membrane.
(b) Is mediated by GABA (γ-aminobutyric acid).
(c) Is an essential step in the mechanism of the recurrent inhibitory pathway.
(d) Involves decreased sensitivity of the postsynaptic membrane to neurotransmitter.
(e) Involves decreased release of transmitter by the presynaptic terminal.

Question 5.6 (MCQ)

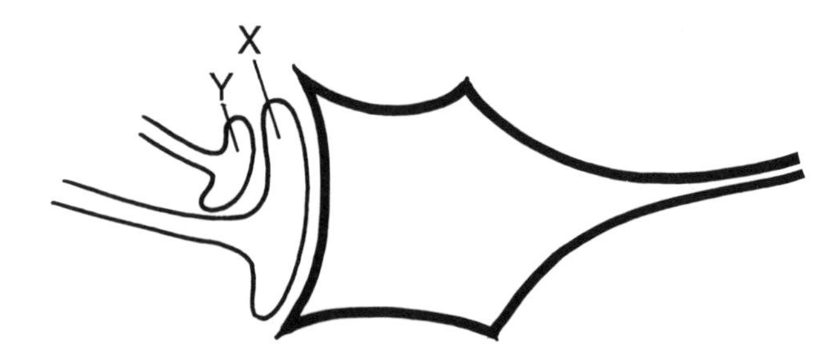

In presynaptic inhibition, the presynaptic terminal X exhibits:

(a) Regions of membrane which are presynaptic.
(b) Regions of membrane which are postsynaptic.
(c) Synaptic vesicles.
(d) Cytoplasmic continuity with the postsynaptic cell.
(e) A nucleus.

Questions 5.7 and 5.8 refer to the figure of a neurone in the spinal cord of a man. Various inputs are shown; + represents postsynaptic excitation and − represents postsynaptic inhibition.

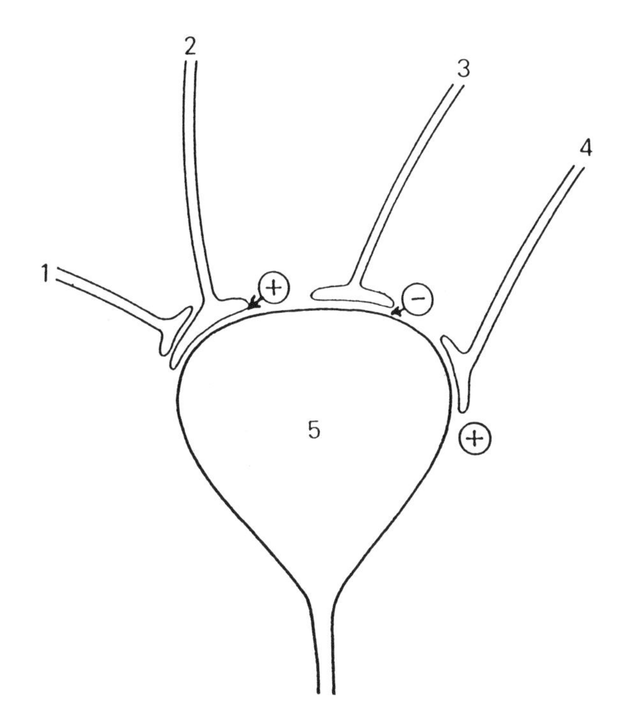

Question 5.7 (MCQ)

(a) Activity in 1 inhibits the excitatory effects of input 2 on neurone 5.
(b) Activity in 1 inhibits the excitatory effects of input 4 on neurone 5.
(c) Activity in 3 inhibits the excitatory effects of input 2 on neurone 5.
(d) Activity in 3 inhibits the excitatory effects of input 4 on neurone 5.
(e) Activity in 1 causes a hyperpolarization of 5.

Question 5.8 (MCQ)

In each case, is it possible that the element shown in the diagram is part of a neurone with a long axon?

(a) 1
(b) 2
(c) 3
(d) 4
(e) 5

SPINAL CORD MECHANISMS

Question 6.1 (MCQ)

Concerning the plantar (Babinski) reflex elicited by a firm stroke to the outer aspect of the sole of the foot

(a) The response in a normal adult is plantar flexion.
(b) An extensor plantar response (dorsiflexion) is a spinal reflex.
(c) An extensor plantar response indicates damage to the cortico-spinal projection.
(d) An extensor plantar response is normal in a newborn baby.
(e) A Babinski reflex may be one of the first reflexes to reappear after the period of spinal shock that follows a spinal transection.

Question 6.2 (MCQ)

Two weeks after section of a cutaneous nerve, the denervated skin is studied with the subject in a warm environment. Will the following be observed?

(a) Insensible perspiration.
(b) Cold skin.
(c) Sweating.
(d) Local oedema formation when the skin is damaged.
(e) Flaring of the skin when it is damaged.

Question 6.3 (MCQ)

An hour after an uncomplicated transection of the spinal cord at the level of C8 in an adult human, there is

(a) Paralysis of the diaphragm.
(b) Paralysis of intercostal muscles.
(c) An atonic bladder.
(d) Sudomotor paralysis at levels below the lesion.
(e) An extensor plantar response.

Question 6.4 (MCQ)

A year after an uncomplicated transection of the spinal cord at the level of C8 in an adult human there is

(a) Rhythmic respiratory contractions of the intercostal muscles.
(b) A reflex rise in blood pressure accompanying filling of the bladder.
(c) Loss of bladder sensation.
(d) Loss of the stretch reflexes in the lower limbs.
(e) Loss of abdominal reflexes.

Question 6.5 (MCQ)

Long-term consequences of complete transection of the spinal cord at the level of C8 include:

(a) Loss of reflex emptying of the bladder.
(b) Flaccid paralysis of the legs.
(c) Plantar flexion when the sole of the foot is firmly stroked.
(d) Defect in blood pressure regulation in response to changes of posture.
(e) Loss of all sensation in the legs.

Question 6.6 (MCQ)

Concerning spinal mechanisms:

(a) An hour after a complete transection of the spinal cord in the lower cervical region of the cord, an electrical stimulus applied percutaneously to a motor nerve in a limb will cause a contraction of the muscles which it innervates.
(b) Spinal shock is due to release of toxins from the cord at the site of injury.
(c) In spinal shock the muscle spindle stretch receptors generate action potentials in the afferent nerve fibres when the muscle is stretched.
(d) The lack of muscle tone in spinal shock is due to interruption of proprioceptive input to the cord.
(e) Recovery from spinal shock (after a transection of the cord) is due to regeneration of peripheral nerve fibres.

PROPRIOCEPTION

Exercise 7.1 Structured exercise

M and H responses

An electrical stimulus is applied to the skin overlying a mixed motor nerve. When the stimulus strength is very small, just sufficient to excite a few nerve fibres the (smallest/largest) nerve fibres are stimulated. The

largest nerve fibres in the body are

(afferent/efferent) nerve fibres. They conduct at about

(1/10/100/1000) m/s and are about

(0.2/2/20/200) μm in diameter. They are

(thinly/thickly) myelinated. Their mode of conduction is called

(continuous/saltatory). These fibres elicit a

(direct/reflex) muscular contraction. The type of spinal reflex with the smallest reflex time involves

(0/1/2/3) synapses in the cord and subserves the

(tendon-jerk/stretch/flexor withdrawal) reflex. This reflex pathway is activated as part of the reflex response to electrical stimulation of the mixed motor nerve. So after about

(0.3/3/30/300) ms there is a reflex contraction of the muscle. This is called the H response, H for Hoffman, the person who first described it.

Increasing the stimulus strength recruits

(larger/smaller) fibres which are

(afferent/efferent/both). Excitation of efferent fibres leads to another deflection in the electromyogram, of

(smaller/greater) latent period than that elicited by a threshold stimulus. This is called the M response, M for

. . . .

As the stimulus strength is further increased, the M response becomes bigger and the H response becomes smaller. Explain this.

Question 7.1 (MCQ)

Concerning skeletal muscle in the adult human:

(a) The gastrocnemius muscle usually has a single muscle spindle.
(b) Muscles involved in the performance of fine movement have a greater density of muscle spindles than those involved in coarse movements.
(c) In the motor nerve supplying a typical skeletal muscle, there are about ten times as many motor nerve fibres innervating the extrafusal muscle fibres as there are innervating intrafusal fibres.
(d) Most of the nerve fibres innervating intrafusal muscle fibres send branches to extrafusal fibres.
(e) The γ efferent nerve fibres typically end in direct association with the sensory terminals.

Question 7.2 (MCQ)

Concerning the muscle spindle receptors and afferent nerves in an human adult:

(a) The response of a primary ending shows a larger dynamic component during stretch than does that of a secondary ending.
(b) When a primary ending is activated, the action potentials in the afferent nerve fibre are initiated in the sensory terminals themselves.
(c) The afferent nerve arising from a primary ending conducts faster than one arising from a secondary ending.
(d) The afferent arising from the primary ending may connect directly (i.e. monosynaptically) on α motoneurones.
(e) The primary ending acts as the principal receptor in the tendon jerk reflex.

Question 7.3 (MCQ)

Concerning innervation of muscle and reflexes:

(a) The knee-jerk reflex involves discharge of α motoneurones.
(b) For the knee-jerk reflex, it takes longer for the nerve volley to travel from the cord to the quadriceps muscle than from the receptors to the cord.
(c) Cooling of a limb will reduce the reflex times in that limb.
(d) Group II afferents arise mainly from tendon organs.
(e) When an intrafusal muscle fibre is activated, the equatorial region is stretched.

Question 7.4 (MCQ)

Concerning skeletal muscle in a healthy human adult:

(a) When the end-plate of an extrafusal muscle fibre is depolarized to threshold, the contraction that is initiated occurs along the whole length of the muscle fibre.
(b) The end-plates of intrafusal muscle fibres are usually in the equatorial region.
(c) The initiation of impulses in afferent nerve fibres by spindle receptors is via the intermediary of a chemical synapse.
(d) When extrafusal muscle fibres shorten, the equatorial region of the intrafusal muscle fibres are stretched.
(e) Contraction of all the intrafusal fibres in a muscle results in the development of a smaller tension than the contraction of all the extrafusal fibres.

Question 7.5 (MCQ)

In the skeletal muscles of an adult mammal *in vivo*, the contraction of extrafusal muscle fibres due to the discharge of:

(a) γ motoneurones is abolished if the dorsal roots are sectioned.
(b) γ motoneurones is abolished if the ventral roots are sectioned.
(c) γ motoneurones is more readily blocked by local anaesthetic injected into the mixed motor nerve than is contraction due to discharge of α motoneurones.
(d) α motoneurones is abolished if the dorsal roots are sectioned.
(e) α motoneurones is abolished by selective blockade of the end-plates of intrafusal muscle fibres.

Question 7.6 (MCQ)

In adult human skeletal muscles, an intrafusal muscle fibre typically exhibits:

(a) More than one motor end-plate.
(b) A sensory receptor applied to its surface.
(c) Nuclei concentrated in the equatorial region.
(d) Striations visible under the interference light microscope.
(e) Innervation by α motoneurones.

Question 7.7

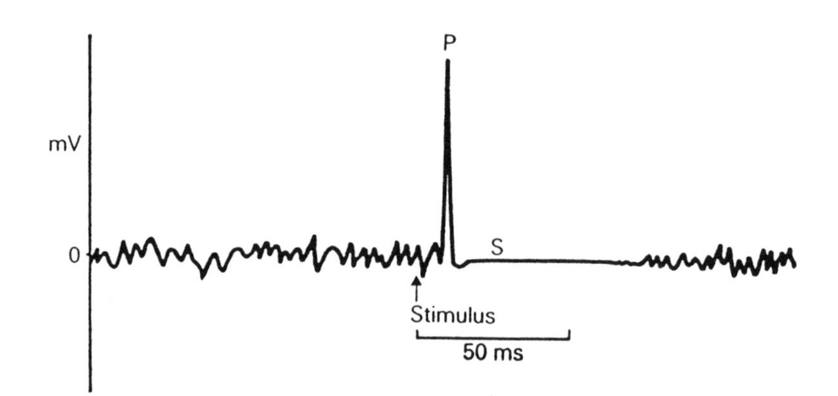

The trace shows the electromyogram recorded from electrodes on the skin overlying a muscle. The subject is voluntarily contracting the muscle throughout the trace. At the arrow, an electrical stimulus to the mixed motor nerve supplying the muscle initiates a volley of impulses in the α efferent nerve fibres supplying the muscle.

(a) The oscillations before the arrow are generated directly by action potentials in the intrafusal muscle fibres.
(b) The large deflection (P) is the compound nerve action potential from the motor nerve.
(c) The 'silent period' (S) may be due to unloading of the spindle stretch receptors.
(d) The 'silent period' (S) may be due to recurrent inhibition in the spinal cord.
(e) In addition to stimulating motor fibres, the stimulus would stimulate sensory nerve fibres.

Questions 7.8 and 7.9 refer to the figure

Electrical stimuli were applied percutaneously below the level of the knee to the mixed motor nerve supplying the gastrocnemius muscle. The records show the electromyograms recorded from skin electrodes overlying the gastrocnemius muscle.

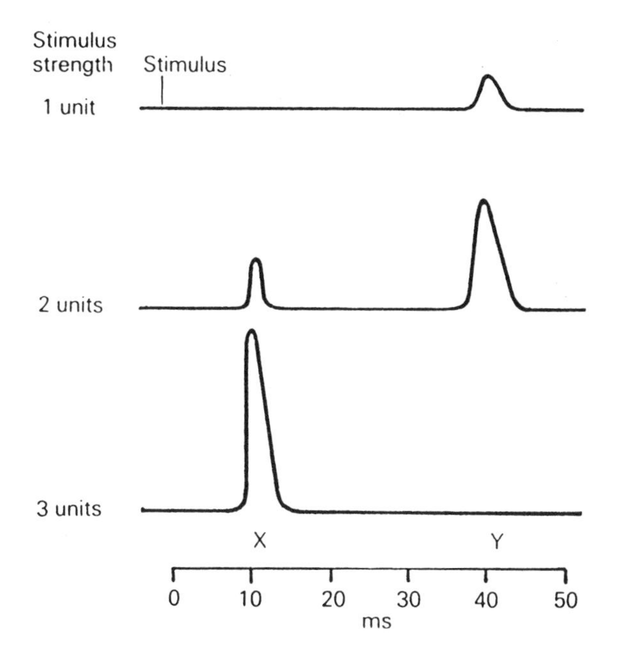

Question 7.8 (MCQ)

(a) In the upper trace, the wave is the response of the muscle following excitation of motor nerve fibres under the stimulating electrode.
(b) In the middle trace, the wave at delay X is part of a monosynaptic reflex.
(c) In the middle trace, the wave at delay Y is initiated by impulses in Group Ia afferent nerve fibres.
(d) In the lower trace, the total number of nerve fibres excited by the stimulus is greater than in the middle trace.
(e) In the lower trace, the lack of response at delay Y is due to collision with mutual extinction of antidromic with orthodromic nerve impulses.

Question 7.9 (MCQ)

(a) If the nerve were blocked by local anaesthetic above the level of the knee-joint, the wave at delay X would disappear.
(b) If the nerve were blocked above the knee joint, the wave at delay Y would disappear.
(c) If the nerve were blocked between the stimulating electrode and the gastrocnemius muscle, both waves X and Y would disappear.
(d) If the stimulus were given at the level of the ischium, the wave X would occur with shorter delay.
(e) If the stimulus were applied at the level of the ischium, the wave Y would occur with shorter delay.

Questions 7.10 and 7.11 refer to the figure of a spinal reflex pathway in a person.

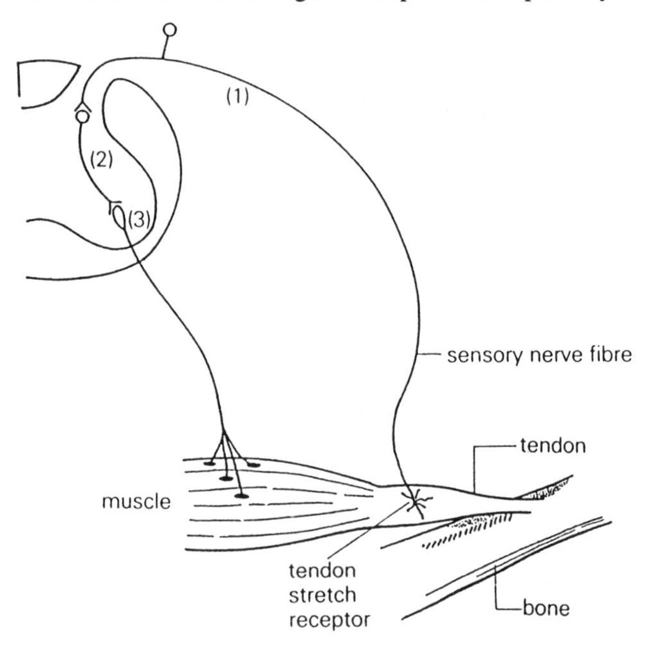

Question 7.10 (MCQ)

(a) It is a diagram of the tendon-jerk reflex.
(b) It shows a disynaptic reflex.
(c) The nerve fibre (1) is non-myelinated.
(d) The nerve fibre (2) is an interneurone.
(e) The transmitter between neurone (2) and neurone (3) is likely to be glycine.

Question 7.11 (MCQ)

(a) The synapse between neurone (1) and neurone (2) is an inhibitory synapse.
(b) The reflex arc shown has the fewest number of synapses of all reflexes.
(c) Other reflex pathways with receptors in the muscle converge on neurone (3).
(d) Influences from proprioceptors on the opposite side of the body converge on neurone (3).
(e) The neurone (3) itself gives off one or more collaterals (not shown in the diagram) which run back to form synapses on other neurones in the cord.

Question 7.12 (MCQ)

Concerning the tendon-jerk reflex in the human

(a) It has the smallest central reflex time of all types of spinal reflexes.
(b) The stimulation of stretch receptors is due to a contraction of the intrafusal fibres
(c) Antagonist muscles relax before the reflex contraction occurs.
(d) If tendon afferents (Group Ib afferents) are activated, this increases the amplitude of the tendon-jerk reflex.
(e) The reflex can be influenced by higher centres.

SPINAL MECHANISMS

Exercise 8.1 Structured exercise

Reflex pathways

1. Draw a fully labelled diagram of the components of the tendon-jerk reflex. Make sure that you include the following, which are deliberately not written down in their correct sequence.

   Motoneurone. Afferent nerve fibre. Cell body of afferent nerve fibre. Efferent nerve fibre. Synapse (or synapses?) Receptor. Stimulus. Motor unit. End-plate(s). (How many for each muscle fibre?).
2. In what ways would an inhibitory reflex pathway (as in reciprocal innervation differ from that for the tendon-jerk reflex?
3. In what ways would the pathway for the flexor withdrawal reflex differ from that for the tendon-jerk reflex?
4. On your diagram of the tendon-jerk reflex pathway, indicate approximate values for the time delays introduced by each component.

Question 8.1 (MCQ)

(a) A reflex may involve the discharge of action potentials from all the motoneurones in a motoneurone pool.
(b) If two weak stimuli are applied to neighbouring points on the skin surface, the magnitude of the reflex response to the two stimuli applied together is likely to be greater than the sum of the responses to the two stimuli given separately.
(c) If two strong stimuli are applied to neighbouring points on the skin surface, the magnitude of the reflex response to the two stimuli applied together is likely to be less than the sum of the responses to the two stimuli given separately.
(d) The phenomenon of occlusion demonstrates that there must be divergence of neuronal pathways.
(e) The magnitude of a reflex response to a constant stimulus is more variable than the magnitude of a muscle twitch in response to electrical stimulation of the motor nerve.

Question 8.2 (MCQ)

(a) A reflex pathway with unmyelinated nerve fibres as the afferent limb is likely to have a longer reflex response time than a pathway with myelinated afferent nerve fibres.
(b) A typical spinal reflex pathway involves a synaptic relay in the dorsal root ganglion.
(c) A subject is usually aware or a stimulus which elicits a spinal reflex.
(d) The motor nerve axons responsible for a somatic spinal reflex have their cell bodies in the dorsal horn of the grey matter of the spinal cord.
(e) The reflex response to stimulation of a cutaneous nerve may be inhibited by activity in other reflex pathways.

Question 8.3 (MCQ)

Concerning the reflex contraction of the quadriceps femoris muscle elicited by tapping the patellar tendon in a normal human:

(a) The afferent nerve fibres involved in the reflex arc are myelinated fibres.
(b) The reflex response time is about 1 s.
(c) Most of the time delay between the tap and the onset of EMG (electromyogram) activity is attributable to conduction along the afferent and efferent nerve fibres.
(d) The muscle goes into a long tetanic contraction.
(e) The mechanical contraction of the muscle follows the same time course as the electromyogram (EMG).

Question 8.4 (MCQ)

With respect to the general properties of somatic spinal reflexes in vertebrates:

(a) The response may involve the contraction of more than one muscle.
(b) If a strong painful stimulus is applied to one leg, the reflex effects are confined to motoneurones in the spinal cord on the same side of the body as the stimulus.
(c) If the dorsal roots are cut, spinal reflexes can be elicited if the stimulus is very strong.
(d) Every reflex involves at least one synaptic relay in the spinal cord.
(e) A maintained stimulus may evoke an oscillatory reflex response.

Question 8.5 (MCQ)

Concerning human spinal reflex pathways:

(a) These pathways may include motoneurones in several segments of the cord.
(b) Some reflex pathways include excitatory interneurones in the cord.
(c) An inhibitory reflex pathway includes at least one interneurone.
(d) An inhibitory reflex pathway has at least two synaptic relays in the spinal cord.
(e) There are inhibitory motoneurones.

SPECIAL SENSES

Question 9.1 (MCQ)

Concerning the human eye

(a) The fovea contains more cones than rods.
(b) There is a greater density of rods around the optic disc than at its centre.
(c) Close to the optic disc, the density of cones exceeds the density of rods.
(d) The photosensitive receptors are expansions of the terminals of optic nerve fibres.
(e) Rhodopsin, the visual pigment in rods, is broken down by smaller amounts of blue light than is photopsin, the pigment in cones.

Question 9.2 (MCQ)

If the left optic nerve is cut,

(a) Light shone in the right eye will cause pupillary constriction of the left eye.
(b) Light shone in the left eye will cause pupillary constriction of the right eye.
(c) Convergence will be accompanied by constriction of both pupils.
(d) The subject will be able to see objects to the right of the midline.
(e) As the right eye moves in scanning the visual field, the left eye will move too.

Question 9.3 (MCQ)

A person with normal vision has both eyes open.

(a) If the person is in a bright environment with pupils constricted, all the light entering his eye falls on the fovea.
(b) If the person wears red spectacles in daylight, his vision is due to stimulation of rods.
(c) In a brightly illuminated environment, the rhodopsin in the rods is completely bleached.
(d) When the person moves from a bright environment to an unilluminated dark room, the process of dark adaptation is complete within about 5 min.
(e) There is a region within each side of his visual field where he is blind.

Question 9.4 (MCQ)

Concerning vision in a normal person:
(a) Convergence of neuronal pathways enhances discrimination of two points of light.
(b) Convergence of neuronal pathways jeopardizes visual sensitivity.
(c) Lateral inhibition enhances two point discrimination.
(d) Visual acuity is greater for cone vision than for rod vision.
(e) At night, a faint point of light in the visual field is more easily seen if the light is on one side of the visual axis than if it is on the visual axis.

Question 9.5 (MCQ)

Concerning audition:

(a) Contraction of the stapedius and tensor tympani muscles increase auditory sensitivity.
(b) The sensitivity of auditory receptors can be decreased by activity in the efferent nerve fibres innervating the receptors.
(c) Sounds of high frequency excite receptors closer to the apex of the cochlea than do sounds of low frequency.
(d) For low frequency notes, action potentials in the auditory nerve fibres fire in phase with the oscillations of the sound stimulus.
(e) The connection between the receptor element and the first-order auditory nerve fibre is via a chemical synapse.

Question 9.6 (MCQ)

Concerning receptors in the vestibular apparatus:

(a) The connection between the receptor and the afferent nerve fibre is mediated by a chemical transitter.
(b) The hair cells in an ampulla of a semicircular canal are randomly orientated.

(c) The response of the receptors in a semicircular canal is greatest when the acceleration is around an axis in the plane of the canal.
(d) The hair cells in the utricle are uniformly orientated.
(e) The deflection of utricular hair cells is greatest when the direction of linear acceleration is perpendicular to the plane of the utricular macula.

BALANCE

Question 10.1 (MCQ)

When a blindfolded normal subject is rotated in a rotating chair:

(a) With the onset of rotation, the direction of movement of the eyeballs in the orbits is opposite from that of rotation of the chair.
(b) For the first few seconds of rotation, the subject exhibits nystagmus.
(c) If the rotation continues at a constant angular velocity for several minutes, the subject continues to be aware of the direction of rotation.
(d) If after a minute the rotation is suddenly stopped, the subject interprets the change as the onset of rotation in the opposite direction.
(e) The subject senses rotations as a result of input primarily from the receptors in the utricle.

Question 10.2 (MCQ)

Concerning the input from the inner ear:

(a) It is possible for a linear acceleration initiate input which is confused with that initiated by a tilt of the body and head.
(b) It is possible for an angular acceleration to initiate input which is confused with that initiated by a tilt of the body and head.
(c) If a person sits in an aeroplane and holds the top of a plumb line still, when the plane performs a properly banked turn, the plumb line points to the centre of the earth.
(d) When a person stands in a vehicle moving at a constant velocity, his centre of gravity is vertically above his support.
(e) A passenger standing in a bus which suddenly stops takes a step in the same direction as the previous motion.

BRAIN STEM

Exercise 11.1 Structured exercise

Vestibulo-occular reflexes and caloric testing

Movement of the endolymph inside the semicircular canals relative to the wall of the canals is sensed by the mechanoreceptors in the ampullae. They initiate reflex movements of the eyes to stabilize the visual image on the retina. Such reflexes are named **vestibulo-ocular reflexes**. There are three semicircular canals, arranged with their planes mutually perpendicular. For the present, we will consider only the lateral semicircular canal; it lies approximately in the horizontal plane when the head is upright. It is also the semicircular canal which lies nearest to the external auditory meatus; the significance of this will emerge shortly.

Physiological vestibulo-ocular reflex

Suppose you are sitting with your head upright and you rotate your head to the right. Due to its inertia, the endolymph in the left lateral semicircular canal moves, relative to the wall of the canal,

(from anterior to posterior/from posterior to anterior).

The reflex contraction of the extrinsic ocular muscles stabilizes the visual image; with respect to the head, the eyes deviate to the

(left/right) i.e. in the

(same/opposite) direction compared with the rotation of the head. This reflex thus

(increases/decreases) the apparent inertia of the eyes. The reflex

(will/will not) occur in a blind subject. The reflex

(will/will not) occur in a subject with defective labyrinthine function.

Testing vestibulo-ocular reflexes in an unconscious subject

The cephalo-ocular reflex

The cephalo-ocular reflex is one type of vestibulo-ocular reflex which can be elicited from a patient who is unconscious, but whose brain stem is functioning. The reflex is elicited by passively rotating the head. With the patient lying supine, the doctor stands at the head of the couch and holds the patient's head with his hands. He then rotates the head quite quickly to one side. The cephalo-ocular reflex results in a deviation of the visual axis in the

(same/opposite) direction compared with the rotation of the head. This response is sometimes referred to as 'doll's head phenomenon' because some dolls are provided with high-inertia eyes which behave in the same way. This response is one of the tests for the integrity of the

(brain stem/cerebral cortex) in an unconscious patient.

Caloric testing

This test is also for the presence of vestibulo-ocular reflexes and can also be performed on the unconscious subject. Imagine that a subject is lying supine and that cold water is infused into his left external auditory meatus. As a result, there is cooling of the endolymph of the nearest semicircular canal. This is the lateral semicircular canal.

Cool endolymph is

(heavier/lighter) than warm endolymph. Consequently,

convection currents are set up in the endolymph, the cold endolymph tending to

(rise/fall). In the semicircular canal, this causes a movement of endolymph in the direction

(occiput to nose/nose to occiput). This is the same relative movement as if the head were rotating to the

(right/left). The resultant vestibulo-ocular reflex causes deviation of

(the eye on the side of the infusion/both eyes) to the

(right/left). The direction of deviation of the visual axis is

(towards/away from) the side of the infused cold water.

If the same procedure is adopted except that warm water is used instead of cold, the directions of movement of endolymph would be

(the same/reversed).

If the subject were sitting with head upright instead of lying, the effects on the movements of endolymph along the length of the lateral semicircular canal would be

(less/exaggerated) and so the reflex effects on eye deviation would be

(less/more).

Question 11.1 (MCQ)

Concerning the eye:

(a) The cephalo-ocular reflex decreases the apparent inertia of the eyeball.
(b) The vestibulo-ocular reflexes serve to stabilize the visual image on the retina.
(c) Reflex deviatons of the visual axis depend on the integrity of the brain stem
(d) The fast phase of nystagmus depends on the integrity of the cerebral cortex
(e) Nystagmus only occurs in a normal person if there is movement of endolymph relative to the membraneous labyrinth

Question 11.2 (MCQ)

In a subject with a normally functioning brain stem but no cortical function:

(a) Cold water infused into the external auditory meatus causes nystagmus.
(b) Hot water infused into the external auditory meatus causes deviation of the visual axis.
(c) If the head is passively rotated by an observer, relative to the orbit, the eyes rotate in the opposite direction from that of the head.
(d) If a bright light is shone in one eye, the pupil constricts.
(e) There is spontaneous breathing.

Question 11.3 (MCQ)

Concerning caloric testing in a normal subject in different postures, the following procedures will cause the visual axis to deviate to the left:

(a) Subject lies supine and cold water is infused into left ear.
(b) Subject lies supine and warm water is infused into left ear.
(c) Subject lies prone and cold water is infused into left ear.
(d) Cold water infused into one external auditory meatus causes deviation of the direction of gaze only of the eye on the same side.
(e) Cold water infused into the external auditory meatus is more effective in producing deviation of the visual axis if the subject is sitting head upright than if the subject is lying supine.

Question 11.4 (MCQ)

Concerning the semicircular canals and their receptors:

(a) The effective stimulus is movement of the endolymph relative to the membraneous labyrinth.
(b) The stimulation of receptors in caloric testing is a direct effect of temperature on the sensitivity of the receptors.
(c) Sensory input from one semicircular canal influences only the extrinsic ocular muscles of the eye on the same side.
(d) A motor nerve innervating extrinsic ocular muscles (cranial nerve III, IV or VI) supplies muscles in the orbits both on the left and right.
(e) It is possible to rotate the head in such a way that the receptors in only one semicircular canal are influenced.

Question 11.5 (MCQ)

In a patient with no brain stem function, it may be possible to elicit:

(a) Reflex withdrawal of the foot when a toe is pinched.
(b) Spontaneous respiratory movements when the arterial P_{CO_2} is raised to 50 mmHg.
(c) Grimacing on pressing over the supraorbital notch.
(d) Ocular deviation in response to infusing cold water into the external auditory meatus.
(e) Intelligible speech.

Question 11.6 (MCQ)

In a subject with no brain stem function:

(a) The subject may be fully conscious.
(b) Reflex swallowing can be elicited.
(c) There are likely to be abnormalities in auditory evoked potentials
(d) Normal evoked potentials may be recorded over the somatosensory region of the cerebral cortex in response to stimulation of the periphery.
(e) Normal evoked potentials may be recorded over the visual cortex in response to flash stimulation.

Question 11.7 (MCQ)

Does increased activity in the structure on the left lead to an increase in activity in the structure on the right?

(a)	Red nucleus	Cells of the descending reticular formation
(b)	Motor strip of cerebral cortex	Cells of the descending reticular formation
(c)	Descending reticular formation	γ motoneurones in the cord
(d)	Lateral vestibular nucleus	γ motoneurones in the cord
(e)	Lateral vestibular nucleus	α motoneurones in the cord

Question 11.8 (MCQ)

Decerebrate rigidity:

(a) Is a rigidity of the anti-gravity muscles.
(b) Occurs after transection above the red nucleus.
(c) Is abolished in the left half of the body after destruction of the right lateral vestibular nuclei.
(d) Is abolished in one limb if the dorsal roots supplying that limb are cut.
(e) Is exaggerated if an electrical stimulus is applied to the cerebellar cortex.

Question 11.9 (MCQ)

(a) If a subject on artificial ventilation with air exhibits no rhythmic activity of his respiratory muscles, then his brain stem must be dead.
(b) The vestibulo-ocular reflexes depend on the integrity of the auditory area of the cerebral cortex.
(c) A small lesion in the cerebral cortex is less likely to cause derangement of consciousness than a similar sized lesion in the brain stem.
(d) The inhibitory reticular formation projects extensively to the cerebral cortex.
(e) Swallowing is abolished with lesions of the motor strip of the cerebral cortex.

BASAL GANGLIA AND CEREBELLUM

Question 12.1 (MCQ)

Features of a patient with a lesion confined to the cerebellar cortex include:

(a) Hypotonia.
(b) Involuntary movement at rest.
(c) Intention tremor.
(d) Paralysis of voluntary movement.
(e) A positive Babinski sign.

Question 12.2 (MCQ)

The axons of the Purkinje cells of the cerebellar cortex.

(a) Constitute the main efferent pathway from the cortex.
(b) May terminate in the spinal cord.
(c) May terminate in the cerebellar nuclei.
(d) May terminate in excitatory synapses.
(e) Produce influences mainly on the contralateral musculature.

Question 12.3 (MCQ)

(a) If the cerebellar cortex is destroyed, the resulting disability recovers to a large extent.
(b) If the substantia nigra is damaged, the resulting disability is permanent.
(c) More nerve fibres carrying proprioceptive information go to the cerebellum than to the basal ganglia.
(d) The direct influence on motoneurones in the cord is greater from the cerebellum than from the basal ganglion.
(e) The disorders of movement produced by lesions of the basal ganglia and of the cerebellum are indistinguishable clinically.

Question 12.4 (MCQ)

Concerning the basal ganglia:

(a) The corpus striatum projects to the motor area of the cerebral cortex.
(b) The putamen is the principal output nucleus of the basal ganglia.
(c) The body is somatopically represented in the corpus striatum.
(d) Overactivity of the substantia nigra results in spasticity.
(e) If the cholinergic dopaminergic balance in the basal ganglia is disturbed such that the dopaminergic influence predominates, this results in Parkinson's disease.

Question 12.5 (MCQ)

Concerning the basal ganglia:

(a) The basal ganglia have a denser innervation from the cerebral cortex than from the central afferent pathways of the spinal cord.
(b) Input to the basal ganglia of information about the internal environment is largely through the hypothalamus.
(c) Cells of the basal ganglia increase their discharge rate during ramp movements.
(d) Cells of the basal ganglia increase their discharge rate during ballistic movements.
(e) There are cells in the basal ganglia whose activity increases before the onset of voluntary movements.

Question 12.6 (MCQ)

Concerning the basal ganglia:

(a) Cells in the substantia nigra react strongly with stain for dopamine.
(b) The nigro-striatal projection is dopaminergic.
(c) The nigro-striatal projection is predominantly inhibitory.
(d) As a neurotransmitter, acetylcholine is predominantly excitatory in the basal ganglia.
(e) The dopamine content of the striatum is increased in Parkinson's disease.

Question 12.7 (MCQ)

Concerning the cerebellum:

(a) The climbing fibres arise mainly from cells in the olivary nucleus.
(b) A nerve impulse in a climbing fibre is insufficient to elicit an action potential from the Purkinje cell which it innervates.
(c) The axons of granule cells form direct synaptic connections with the Purkinje cells.
(d) The granule cells are excitatory.
(e) The parallel fibres run in the same direction as the planes of the dendritic trees of the Purkinje cells.

Question 12.8 (MCQ)

Concerning the cerebellum, does increased activity in the structure on the left lead to an increase in activity in the structure on the right?

(a)	Purkinje cells of the cerebellar cortex	Cells of the cerebellar nuclei
(b)	Cells of the cerebellar nuclei	Spinal α motoneurones supplying extensor muscles
(c)	Climbing fibres	Purkinje cells of the cerebellar cortex
(d)	Parallel fibres	Purkinje cells of the cerebellar cortex
(e)	Axons of basket cells of the cerebellar cortex	Purkinje cells of the cerebellar cortex

CEREBRAL CORTEX, GENERAL FEATURES, LOCALIZATION OF FUNCTION

Question 13.1 (MCQ)

Concerning the cerebral cortex:

(a) A greater proportion is devoted to primary sensory and motor functions in humans than in non-human mammals.
(b) During development, myelination of the association areas antecedes that of the primary sensory areas.
(c) At birth, myelination of the central nervous system is complete in humans.
(d) The primary motor area of cortex lies behind most of the primary sensory areas.
(e) First-order sensory neurones project without relaying to the primary somatosensory area of the cortex.

Question 13.2 (MCQ)

In the human cerebral cortex
(a) Stimulation of the motor strip will usually cause contraction of one muscle in isolation.
(b) Stimulation of the motor strip causes a temporary local paralysis of voluntary movement.
(c) The somatic receiving area is an undistorted scaled-down version of the body surface.
(d) The speech areas are generally on the right.
(e) Stimulation of the cortex can cause pain.

SEEING AND HEARING

Question 14.1 (MCQ)

Concerning the primary visual cortex:

(a) The form of the stimulus that optimally stimulates the neurones is the same as for optimal stimulation of neurones in the lateral geniculate body.
(b) In an electrode penetration perpendicular to the cortical surface, the orientation sensitivities of neurones are intermingled in an apparently random fashion.
(c) The area of cortex serving 1 mm^2 of fovea is less than that serving 1 mm^2 of retina 30° off the visual axis.
(d) A region in the peripheral visual field to the left of the midline projects to the primary visual cortex of both hemispheres.
(e) Layer IV is absent from primary visual cortex.

Question 14.2 (MCQ)

Concerning the eye and vision:

(a) The optic nerve contains axons of retinal ganglion cells.
(b) Rods are associated with colour vision.
(c) Activity in the sympathetic nerve supply to the eye causes pupillary constriction.
(d) The primary visual cortex lies in the parietal lobe of the cerebral hemisphere.
(e) Visual input from an object in the right visual field projects to the right cerebral hemisphere.

Question 14.3 (MCQ)

(a) The cortical projection from one eye is exclusively to the opposite cerebral hemisphere.
(b) The decussation of the visual pathway is via the corpus callosum.
(c) Cortical control of eye movements is from the primary motor area of the cortex (the precentral gyrus).
(d) Slow eye movements can be volitionally made.
(e) During a saccade, there is suppression of the visual image.

MOTOR CONTROL

Question 15.1 (MCQ)

An adult human performs a voluntary limb movement involving shortening of a particular limb muscle.

(a) If afferent activity from a spindle stretch receptor increases during shortening, there must be activation of γ motoneurones.
(b) Movement could be caused by activation of just the γ motoneurones.
(c) Movement could be caused by activation of just α motoneurones.
(d) The delay between a stimulus and a volitional muscular response is typically 30 ms.
(e) The pathway of the projection from the higher centres down to the motoneurones is via the dorsal columns.

CEREBRAL CORTEX: ASSOCIATION AREAS

Question 16.1 (MCQ)

Features likely to be present in a patient with a lesion of the parietal lobe of the cerebral cortex include:

(a) Agnosia.
(b) Anaesthesia.
(c) Apraxia.
(d) A disinclination to voluntary movement.
(e) Dysarthria.

Question 16.2 (MCQ)

Concerning the speech areas in the cerebral cortex:

(a) These areas all lie together in the same lobe of the cerebral hemisphere.
(b) Disordered function of the speech areas results in difficulty with articulation.
(c) The motor speech area (Broca's area) lies just posterior to the primary motor area.
(d) Disordered function of the auditory speech area (Wernicke's area) results in reduction in the amount of speaking.
(e) If in an adult the hemisphere housing the speech areas is destroyed, the remaining hemisphere is able to take over the speech functions.

Question 16.3 (MCQ)

Concerning hemispheric function:

(a) It is mainly the association areas that show asymmetry of function.
(b) Mathematical ability and logic are attributes of the right hemisphere in most subjects.
(c) Creative and artistic ability are attributes of the right hemisphere in most subjects.
(d) A subject with a split brain (section of the corpus callosum) can name an object presented in the right visual field.
(e) It takes longer for a split-brain subject to identify, by pointing, a face presented in the left visual field than if the face is presented in the right visual field.

NEUROTRANSMITTERS

Question 17.1 (MCQ)

Concerning glutamate:

(a) It is the principal inhibitory transmitter in higher centres of the brain.
(b) It is involved in the general amino acid metabolism of the body.
(c) The increase in the intracellular calcium concentration induced by glutamate is due partly to calcium influx from the extracellular fluid.
(d) The increase in the intracellular calcium concentration induced by glutamate is due partly to calcium release from intracellular stores.
(e) In hypoxic brain, there is a decrease in release of glutamate.

Question 17.2 (MCQ)

Concerning GABA (γ-aminobutyric acid):

(a) It is the principal inhibitory transmitter in higher centres of the brain.
(b) It is involved in the general amino acid metabolism of the body.
(c) It is synthesized in the synaptic terminals of GABAergic neurones.
(d) The synaptic vesicles in which GABA is stored are synthesized in the synaptic terminals.
(e) Membrane channels operated by GABA have a conductance that varies with membrane voltage.

Question 17.3 (MCQ)

Concerning synaptic transmission in the higher brain centres in humans:

(a) Repetitive activity of GABA-operated channels tends to result in an increase in synaptic efficacy.
(b) Repetitive activity of glutamate-operated channels tends to result in an increase in synaptic efficacy.
(c) Presynaptic inhibition plays a prominent role in the cortical processing of information.
(d) Peptides mediate fast synaptic transmission.
(e) Neuroactive peptides are synthesized in axon terminals.

PAIN

Question 18.1 (MCQ)

Concerning pain:

(a) A subject experiences more pain if the wall of the intestine is stretched than if it is cut.
(b) The intensity of pain experienced by a subject when a noxious stimulus acts on the skin is increased if thick afferent nerve fibres are simultaneously activated.
(c) Transmission through spinal cord relays may be inhibited by descending influences from brain stem centres.
(d) Tactile stimuli applied to the cornea give rise to the sensation of pain.
(e) A stimulus which is innocuous when applied to normal skin may elicit pain in inflamed skin.T

Question 18.2 (MCQ)

Concerning pain:

(a) Nociceptive endings are free nerve endings.
(b) Nociceptive endings are found in the epidermis.
(c) If the specific nuclei of the thalamus on the central afferent pathway for somaesthetic senses are partly destroyed, a normally innocuous stimulus may be painful.
(d) The afferent nerve fibres mediating the sensation of pain are amongst the largest nerve fibres in the body.
(e) Cutting the somatosensory area of the cortex causes pain.

Question 18.3 (MCQ)

Concerning pain:

(a) The projection of nociception from the periphery to higher centre is partly via polysynaptic pathways in the grey matter of the cord.
(b) A patient with a dorsal column lesion can perceive pain.
(c) In a limb devoid of pain sensation, the joints frequently become damaged.
(d) Ischaemia of a muscle results in pain.
(e) Pain due to inflammation of the appendix may be referred to the midline of the abdominal wall.

Question 18.4 (MCQ)

Concerning pain:

(a) The pleura are devoid of nociceptors.
(b) Pain due to inflammation of the gall bladder may be referred to the tip of the right shoulder.
(c) If the thin myelinated and the unmyelinated nerve fibres in a peripheral nerve are blocked, the sensation of pain is lost whilst the sensation of touch is not lost.
(d) Rubbing a painful area of skin can reduce the intensity of pain.
(e) A noxious stimulus to a limb may cause a reflex extensor thrust in that limb.

Question 18.5 (MCQ)

Concerning pain:

(a) Peripheral unmyelinated afferent nerve fibres are involved in nociception.
(b) Peripheral myelinated afferent nerve fibres are involved in nociception.
(c) Nociception is mediated by nerve fibres which conduct faster than those mediating vibration.
(d) The sensation of pain may persist even when the original cause of the pain has been removed.
(e) There are nerve endings which respond specifically to noxious stimuli.

Question 18.6 (MCQ)

Concerning pain:

(a) Stimulation of the dorsal columns is a method of alleviating pain.
(b) Endogenous opioids increase the firing rate of neurones in the periaqueductal grey matter.
(c) Endogenous opioids increase the firing rate of neurones on the nociceptive pathway in the cord.
(d) Nociceptive pathways project to the postcentral gyrus.
(e) Nociceptive pathways project to the frontal lobes of the cerebral cortex.

AUTONOMIC NERVOUS SYSTEM

Question 19.1 (MCQ)

Concerning the autonomic nerve supply to the skin:

(a) The skin receives parasympathetic nerve fibres.
(b) The sympathetic nerve supply for the whole body arises from the thoraco-lumbar region of the cord.
(c) The cell bodies of the sympathetic nerves concerned lie in the dorsal root ganglia.
(d) The preganglionic sympathetic nerve fibres are thicker than the post-ganglionic ones.
(e) At the synapse between pre- and post-ganglionic sympathetic nerves, the transmitter is acetylcholine.

CEREBRAL ENVIRONMENT

Question 20.1 (MCQ)

Autoregulation of cerebral blood flow to changes in arterial blood pressure:

(a) Is mediated via the arteriolar smooth muscle.
(b) Has a lower limit of about 30 mmHg in a normal adult man.
(c) May be abolished by brain injury, e.g. stroke.
(d) Has limits which are higher than normal in chronic hypertension.
(e) Is achieved by cerebral vasodilatation when the blood pressure rises.

ANSWERS

Answer 1.1

(a) Yes. There are points without crosses.
(b) Yes. The subject could discriminate three different positions of touch.
(c) No. If this were so, the subject would not be able to discriminate different positions of touch.
(d) Yes. The subject was unaware of the difference in the position of the pain sensation spots so the pain endings in the area tested might all be connected to a single afferent nerve fibre.
(e) No, because some of the pain-sensitive spots are where the skin is insensitive to touch.

Answer 1.2

(a) Yes. This is the phenomenon of adaptation.
(b) Yes. Branches of an afferent nerve fibre often innervate receptors in non-contiguous areas of skin, as shown in Figure 1.1.
(c) No. For instance there are no receptors sensitive to radiowaves; we need a radio receiver to transduce radiowaves into a form of energy, such as sound, which we can perceive.
(d) No. The information is conducted in the form of action potentials.
(e) No. The upper limit of speed of conduction in human nerves is around 100 m/s; the speed of light is around 300 000 000 m/s.

Answer 1.3

(a) No. Action potentials are 'all-or-none'.
(b) and (c). Yes. These are the two mechanisms for increasing musuclar effort in skeletal muscle.
(d) No. Every nerve action potential evokes one, and only one, action potential in the muscle fibres which it innervates.
(e) No. The safety factor of neuromuscular transmission in health is high and any change would be without effect on muscular effort.

Answer 1.4

(a) Yes. This is the origin of tone in skeletal muscle; by contrast in smooth muscle tone is also due to excitation by chemicals circulating in the blood.
(b) Yes. The 'tetanic' tension is several times greater than the twitch tension.
(c) Yes. The motor nerve fibre and the muscle fibres that it innervates comprise the 'motor unit'; see Section 1.3.
(d) No. The cell bodies lie in the anterior horn of the spinal cord or, in the case of cranial motor nerves, in a homologous site in the brain stem.
(e) Yes. See Section 1.3.

Structured exercise 2.1

Nerve

thicker
Schwann cells
1 mm
nodes

Resting potential

sodium
chloride
potassium
organic anion
negative
0.1 volts
lower
+55 mV
−90 mV
potassium

Action potential

depolarization
opening
sodium
enter
positive
sodium
+40 mV
potassium leaving and chloride entering
down
little affected in the short term

Characteristics of nerve fibres

20 μm
100 m/s
0.5 μm
1 m/s

Calculation of the wavelength of the action potential

A speed of 100 m/s corresponds to 100 mm/ms. In 0.5 ms, an action potential propagating at 100 m/s travels 100 × 0.5 = 50 mm. The length of nerve depolarized at any one time is thus 5 cm.

Answer 2.1

(a) Yes. See Chapter 2. Section 2.
(b) Yes.
(c) No. There is an increase in membrane permeabilty to potassium that occurs after the increase in sodium conductance.

(d) No. The opening of ionic gates constitute a decrease in the electrical resistance of the membrane.
(e) No. Acetylcholine release is not important in the propagation of the action potential along the nerve but is an essential step in neuro-muscular transmission.

Answer 2.2

(a) No. The conduction velocity in the pyramidal tract fibres is rather less; this is because space inside the central nervous system is limited and the nerve fibres in the central nervous system tend to be thinner than their counterparts in the peripheral nerve.
(b) No. 100 m/s is a typical value.
(c) No. 1 m/s or less.
(d) No. It is directly proportional to the fibre diameter.
(e) Yes.

Answer 2.3

(a) Yes.
(b) No. The amount of sodium that enters the axoplasm during the upstroke of the action potential is insufficient to cause a significant change in the intracellular sodium concentration.
(c) No. It is of the order of 10^{-4} mmol/l.
(d) No. The action potential mechanism involves ions flowing down their electrochemical gradients. If the sodium pump is poisoned, the battery driving the action potential runs down very gradually.
(e) No. In any cell, the only impermeant chemicals are the internal anions. Consequently, poisoning of the sodium pump results in a progressive swelling.

Answer 2.4

(a) No. The resting potential depends primarily on the concentration ratio for potassium.
(b) Yes. The potassium concentration in the axoplasm is about 30 times as great as in the interstitial fluid.
(c) No. Sodium ions flow down their electrochemical gradient.
(d) No. The downstroke of the action potential is due to potassium ions leaving the axoplasm and by chloride ions entering; both move down their electrochemical gradients.
(e) No. The amplitude of the action potential is sensibly independent of fibre diameter.

Answer 3.1

(a) No. The receptor potential is generated across the membrane of the nerve ending.
(b) Yes. A characteristic of action potentials is that they propagate without decrement; they are 'all-or-none'.
(c) No. The receptor potential is a local potential and does not propagate.
(d) No. The receptor potential is generated in the unmyelinated tip; it is not affected by interference with the nodes of Ranvier of the afferent nerve fibre.
(e) Yes. The action potential is generated in the membrane at the nodes of Ranvier.

Answer 3.2

(a) No. The receptor potential is due to opening of channels that allow all ions to traverse; the action potential is due to opening of channels specifically allowing sodium ions to pass.
(b) Yes. A reduction in frequency or cessation of action potentials after the initial discharge is called 'adaptation'.
(c) No. There would have been fewer or no impulses. The receptor is rapidly adapting and therefore responds only to a rapidly changing stimulus.
(d) No. Since action potentials are initiated, this deformation is signalled to the central nervous system.
(e) No. The receptor potential cannot be recorded more than 1 mm or so away from the receptor.

Answer 3.3

(a) No. The membrane of the terminal is a mechano-electrical transducer; it is not excited by electrical stimulation.
(b) Yes. The membranes of all nodes of Ranvier are electrically excitable; this is the basis of their ability to generate action potentials.
(c) No. For somaesthetic receptors in general, the initiation of the nerve impulse is due to electrical spread of the generator potential from the nerve terminal to the first node of Ranvier; there is no chemical junction.
(d) No. The receptor potential is a depolarization.
(e) No. The action potential is a depolarization and reversal of polarization of the nerve membrane potential.

Answer 3.4

(a) Yes. At the offset, the nerve ending is deformed once more and so generates a potential similar to that at the onset of the stimulus.
(b) Yes. The generator potential at the offset elicits action potentials from the nerve fibre.
(c) Yes. The sensory nerve fibre is myelinated.
(d) Yes. Nerve fibres can conduct impulses in either direction.
(e) No. Every first-order somatic afferent nerve fibre has its cell body in the dorsal root ganglion, which lies in the intervertebral canal outside the spinal cord.

Answer 3.5

(a) Yes. Much of the adaptation shown by this receptor is due to the mechanical properties of the capsule.
(b) Yes. Stimulation of any type will, if sufficiently intense, stimulate all receptors.
(c) No. The skin here only has free nerve endings.
(d) No, as in (c).
(e) No, in the areas of skin where they occur, for instance in the fingertips, Pacinian corpuscles are in the dermis.

Answer 3.6

(a) Yes, the skin of the lobe of the ear, for instance, has only free nerve endings and is sensitive to touch, as you can readily confirm on yourself. So the sensation of touch from this area must be mediated by free nerve endings.
(b) Yes, see Section 3.1.
(c) Yes, this was shown in the experiment of question 1.1.
(d) No. Pacinian corpuscles are receptors for vibration.
(e) Yes, the receptive fields of different types of receptor overlap and the field for any one receptor may fall within that for the other. For instance, in the figure of question 1.1 certain points are sensitive to both nociceptive and touch stimuli.

Answer 3.7

(a) No, the wave form of the action potential is stable and not influenced by the strength of the stimulus.
(b) No, see (a)
(c) Yes. This is one mode of transmisson of information about the strength of the stimulus.
(d) No, see (a)
(e) Yes, this is the other mode of transission of information about the strength of the stimulus.

Structured exercise 4.1

Neuromuscular transmission

in either direction
is unidirectional
in packets
presynaptic
release
rise
an essential step
diffuses
attaches to receptor sites on the surface
of all small inorganic ions
flow down
0 mV
by local current flow from the end-plate zone
3.0 ms
150 mm (1 s corresponds to 50,000 mm and 1 ms to 50 mm)
spreads the whole length of the muscle fibre
opposite directions

Matching exercise 4.2

	A	B	C	D	E
Action potential	all-or-none	propagates without decrement	digital	bidirectional	electrical
Receptor potential (somaesthetic receptors)	graded	localized	analogue	unidirectional	electrical
Synaptic potential	graded	lozalized	analogue	unidirectional	chemical

Matching exercise 5.1

1. (g)
2. (f), (a) and (b)
3. (e)
4. (d)
5. (e)
6. (c)
7. (h)

Answer 5.1

(a) No, an EPSP is a depolarizing potential whereas the E_K is more negative than resting potential.
(b) Yes.
(c) Yes, this is an important mechanism for integrative action in the nervous system.
(d) No, the EPSP is due to an opening of channels which allow all small ions to pass whereas the action potential is due to opening of channels specific for sodium ions.
(e) Yes. The EPSP is the basic excitatory synaptic mechanism in spinal reflexes.

Answer 5.2

(a) Yes, this is a basic mechanism in the integrative activity in neurones in the spinal cord.
(b) No, it is a move in membrane potential towards the equilibrium potential for chloride ions; this equilibrium potential is typically -70 mV, so if the resting potential is more negative than this, the inhibitory postsynaptic potential is depolarizing.
(c) No, the equilibrium potential for sodium ions is inside positive; a change in this direction is excitatory, not inhibitory.
(d) No. It is produced directly by the opening of chloride channels.
(e) No, this would be true of an excitatory postsynaptic potential.

Answer 5.3

(a) Yes, chloride channels are open so some of the injected current flows through these channels. Extra injected current is needed to bring the membrane potential to threshold.
(b) Yes, it involves opening of membrane channels that allow ions to cross so the total conductance of the postsynaptic membrane increases.
(c) Yes.
(d) Yes, the IPSP is mediated by a chemical synapse.
(e) No, because the whole of the soma of the postsynaptic cell is hyperpolarized and the efficacy of all excitatory inputs to the soma is reduced.

Answer 5.4

(a) Yes. Presynaptic inhibition was first demonstrated in the spinal cord.
(b) Yes, because presynaptic inhibition directly involves the presynaptic terminal, not the postsynaptic cell.
(c) No, there is a decrease in the overshoot of the action potential in the presynaptic terminal.
(d) Yes. They are both due to opening of chloride channels.
(e) No. The resting membrane potential of the postsynaptic membrane is not altered during presynaptic inhibition.

Answer 5.5

(a) No. There is an increase in chloride permeability of the presynaptic nerve terminal, not the postsynaptic neurone.
(b) Yes.
(c) No. This involves postsynaptic inhibition. See Figure 5.3.
(d) No. The sensitivity of the postsynaptic membrane is unaltered. Presynaptic inhibition is the reduction in the amount of transmitter released by the presynaptic terminal.
(e) Yes. See (d)

Answer 5.6

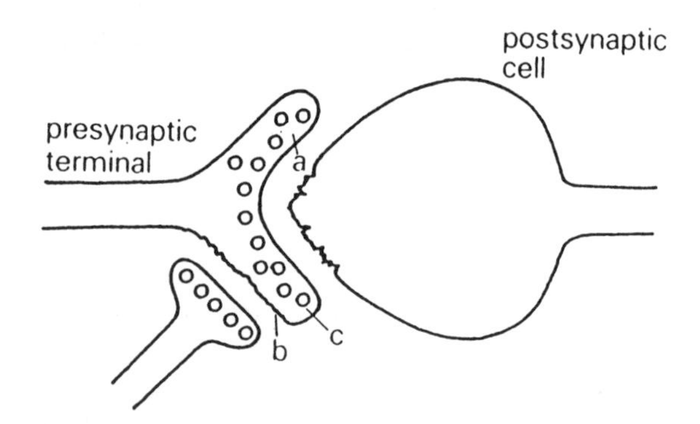

(a) Yes, see Figure – a
(b) Yes, see Figure – b
(c) Yes, see Figure – c
(d) No, the synaptic cleft separates presynaptic and postsynaptic membranes.
(e) No, a neurone contains only one nucleus and this resides in the soma, not in the axonal terminals.

Answer 5.7

(a) Yes, this is presynaptic inhibition.
(b) No, presynaptic inhibiton only influences input along the presynaptic fibre which is involved in the process; input (2) in this case. The excitability of neurone (5) is not altered so the influence of input (4) on the neurone (5) is not affected.
(c) Yes. Activity in input (3) results in postsynaptic inhibition of neurone (5) and the excitability of neurone (5) is reduced; the efficacy of all excitatory synaptic activity to neurone (5) is counter-balanced.
(d) Yes, see (c)
(e) No. The postsynaptic membrane is not directly influenced in presynaptic inhibition.

Answer 5.8

(a) No. Presynaptic inhibition is mediated by terminals of neurones with short axons.
(b) Yes, excitatory synaptic connections are made by neurones with long axons and some neurones with short axons (excitatory inter-neurones).
(c) No, inhibitory neurones in the cord are interneurones and have short axons.
(d) Yes, see (b)
(e) Yes, neurone (5) may be a motoneurone.

Answer 6.1

(a) Yes.
(b) Yes. It is the spinal reflex response that is found after connections with higher centres have been severed, for instance, in a complete transection of the cervical cord.
(c) Yes. Any damage to the corticospinal projection results in an extensor plantar response.
(d) Yes, in the newborn the corticospinal projection is not yet functioning. Consequently the plantar response is extensor.
(e) Yes. In some subjects, it is one of the first reflexes to reappear as spinal shock wears off; it is an extensor plantar response after spinal transection.

Answer 6.2

(a) Yes. Insensible perspiration is the seepage of fluid through the skin and does not depend on innervation.
(b) No. The sympathetic nerve supply to the skin runs in the cutaneous nerves, so after cutting of this nerve, the skin is warm due to loss of sympathetic vasoconstrictor tonus.
(c) No. Sweating depends on activity in sympathetic sudomotor nerve fibres and these fibres are sectioned.
(d) Yes. This depends on histamine release and is not abolished by denervation.
(e) No. This is an axon reflex subserved by sensory nerve fibres. Two weeks after section of a nerve, the nerve fibres in the skin have degenerated and the axon reflex is lost.

Answer 6.3

(a) No. The innervation of the diaphragm is via the phrenic nerve, segmental levels C3, 4 and 5 so the pathway connecting the medullary respiratory centres with the diaphragm is intact.
(b) Yes. The intercostal muscles are innervated by the thoracic segmental nerves. At all levels below C8 there is a state of spinal shock involving paralysis of skeletal musculature.
(c) Yes. There is no reflex activity of the cord at this stage. Lack of bladder reflex results in the bladder being atonic (having no tone).
(d) Yes. This is an autonomic component of spinal shock.
(e) No. The cord exhibits no reflex activity at this stage.

Answer 6.4

(a) No. The contraction of intercostal muscles in normal people in respiratory movements depends on phasic drive from the medullary respiratory centres. This is abolished if the medulla and thoracic cord are disconnected.
(b) Yes. This reflex is more pronounced than in normal people, since effective reflex control of blood pressure depends on connection between the vasomotor centre in the medulla and the sympathetic outflow to the splanchnic areas.
(c) Yes, there is permanent loss of all sensation after a spinal transection.
(d) No, the stretch reflexes in the lower limbs are usually exaggerated in chronic spinal man.
(e) Yes, the abdominal reflexes depend on the integrity of the cortico-spinal projection, which is interrupted in spinal man.

Answer 6.5

(a) No. The bladder of chronic spinal man empties reflexly; it is called a 'reflex bladder'.
(b) No. The paralysis is spastic, not flaccid.
(c) No. Dorsiflexion occurs.
(d) Yes. Because of interruption of central descending pathways from the vasomotor centres.
(e) Yes. The loss of sensation is permanent.

Answer 6.6

(a) Yes. The peripheral apparatus functions normally; spinal shock is a phenomenon of the spinal cord itself.
(b) No. It is due to interruption of descending influences from higher centres.
(c) Yes. As in (a).
(d) No. Proprioceptive input still occurs.
(e) No. Peripheral nerves below the segmental level of a spinal transection do not degenerate and therefore no regeneration occurs.

Key to exercise 7.1

M and H responses

largest
afferent
100
20
thickly
saltatory
reflex
1
tendon jerk (the stretch reflex involves at least two synaptic relays in the central nervous system and the flexor withdrawal reflex more than four).
30
smaller
both afferent and efferent
smaller
motor
Explanation: see text of chapter 7

Answer 7.1

(a) No. There are more than 100 spindles in a muscle of this size.
(b) Yes. A greater density of muscle spindles allows finer control of muscular effort.
(c) No; α and γ efferent axons are usually about equal in number; α may exceed γ, but never by a factor of ten.
(d) No. Most efferent nerve fibres innervate one or the other type of muscle fibres, not both.
(e) No. They innervate the endplates of intrafusal muscle fibres at the poles. The stretch receptors coil around the equatorial region of these muscle fibres.

Answer 7.2

(a) Yes. The primary endings respond to the rate of change of length, the secondaries to the amount of stretch.
(b) No. Action potentials are initiated at the first node of Ranvier of the afferent nerve fibre.
(c) Yes. The primary endings connect with Group Ia fibres, the secondary endings with Group II fibres.
(d) Yes. The Group Ia afferents connect directly with α motoneurones in the cord.
(e) Yes. The Group Ia afferents comprise the afferent limb of the reflex arc of the tendon-jerk reflex.

Answer 7.3

(a) Yes, the alpha motoneurones innervate the motor units.
(b) Yes, because the afferent nerve fibres are thicker than the efferent ones and hence conduct more quickly.
(c) No, it will increase reflex times by slowing conduction in peripheral nerves.
(d) No, Group Ib afferent fibres innervate tendon organs. Group II afferent fibres innervate secondary spindle receptors.
(e) Yes, the polar regions contract. The equatorial region, being non-contractile, is stretched.

Answer 7.4

(a) Yes. For extrafusal fibres, a threshold depolarization at any point initiates an action potential which travels away from that point in both directions to the end of the muscle fibre; the whole of the muscle fibre is involved.
(b) No. End-plates are found at the poles of the intrafusal fibres.

(c) No. The receptor is the terminal of the afferent nerve fibre. The initiation of impulses is by passive (electrotonic) conduction from the receptor to the first node of Ranvier of the nerve fibre.
(d) No. This takes the stretch off the whole of the intrafusal muscle fibre.
(e) Yes. The direct contribution of the intrafusal fibres to the tension developed by a muscle is insignificant.

Answer 7.5

(a) Yes. Gamma motoneurone activity causes a reflex contraction of extrafusal fibres; the path of this reflex traverses the dorsal roots.
(b) Yes. Section of the ventral roots abolishes all contraction of neural origin.
(c) Yes. Small nerve fibres are more readily blocked than large ones by local anaesthetic. So γ efferent fibres are blocked before α.
(d) No. Activity in α motoneurones is conducted to the extrafusal fibres via the ventral roots.
(e) No. Blockade of neuromuscular transmission of intrafusal fibres does not alter contraction of extrafusal fibres activated by α motoneurones.

Answer 7.6

(a) Yes. Intrafusal fibres have at least one end-plate on each of the polar regions.
(b) Yes. The equatorial region of an intrafusal fibre has a receptor applied to its surface.
(c) Yes. Unlike extrafusal muscle fibres, the intrafusal fibres have equatorially placed nuclei.
(d) Yes. Intrafusal fibres are thin striated muscle fibres.
(e) No. Innervation is by γ motoneurones.

Answer 7.7

(a) No. The EMG is generated by action potentials in the extrafusal muscle fibres.
(b) No. The EMG is generated by action potentials in extrafusal muscle fibres.
(c) Yes. Action potentials in the α afferent nerve fibres cause the extrafusal muscle fibres to contract, unload the spindle stretch receptors and this effect, altrui paribus, will cause a pause of the component of voluntary activity initiated by the γ route.
(d) Yes. Antidromic action potentials in the alpha motor nerves cause recurrent inhibition via the Renshaw cell pathway; this causes a pause in discharge of α motoneurones.
(e) Yes. A stimulus strong enough to stimulate motor nerve fibres would certainly stimulate the thicker sensory fibres.

Answer 7.8

(a) No. This response is the reflex activity initiated by stimulation of Group Ia afferents. These latter are the biggest fibres in the mixed motor nerve and so have the lowest threshold to electrical excitation.
(b) No. Wave X is the EMG due to direct electrical stimulation of motor nerve fibres.
(c) Yes. Group Ia afferent nerve fibres are the nerve fibres with the lowest threshold to electrical stimulation. They are stimulated in the upper trace and give rise to the reflex contraction of the muscle. In the second trace, the stronger stimulus stmulates more of these fibres so that the deflection at this delay is greater.
(d) Yes. The stronger stimulus excites more nerve fibres.
(e) Yes. The lack of the deflection at Y is because all the nerve fibres are excited by the stimulus; antidromically conducted impulses in motor nerve fibres collide with the reflexly elicited action potentials and thus extinguish the reflex response at delay Y.

Answer 7.9

(a) No. X is produced by stimulation of motor fibres which connect directly with the muscle end-plates; this pathway is not blocked.
(b) Yes, because Y is a reflex response and the pathway twice traverses the nerve between the stimulus and the cord.
(c) Yes, because both responses involve conduction of impulses along the length of the nerve between the stimulation electrode and the muscle.
(d) No, because the length of the pathway from stimulus to the muscle is increased.
(e) Yes, because the length of the pathway from the stimulus to the cord is decreased.

Answer 7.10

(a) No. For the tendon-jerk reflex, the receptors are in the muscle spindles, not in the tendon as shown in the figure. Also the tendon-jerk reflex involves only one synapse in the cord. The reflex pathway shown in the figure is the clasp-knife reflex.
(b) Yes. There are two synapses in the cord.
(c) No. The afferent nerve fibre from a tendon stretch receptor is a Group Ib fibre, a thickly myelinated axon.
(d) Yes. it lies entirely in the grey matter of the cord.
(e) Yes. The reflex shown is an inhibitory reflex. The interneurone (2) is inhibitory. Postsynaptic inhibition in the cord is probably mediated by glycine.

Answer 7.11

(a) No. The connections of afferent nerve fibres with nerve cells in the cord are always excitatory.
(b) No. The tendon-jerk reflex has only one synapse in the cord.
(c) Yes. Many synaptic influences converge on a motoneurone, which is the final point of integration for contraction of skeletal muscle.
(d) Yes. Influences from proprioceptors on both sides of the body converge on a motoneurone. An example of contralateral influences is the pathway for the crossed extensor reflex.
(e) Yes. In the circuitry of Renshaw cell feedback, for instance.

Answer 7.12

(a) Yes. Central reflex time is largely due to the synaptic delays of the synapses on the reflex pathway. The tendon-jerk reflex is the only monosynaptic reflex in the body and so has the shortest central reflex time of all types of reflex.
(b) No. The stretch of the muscle stimulates the stretch receptors. The intrafusal muscle fibres do not twitch when the tendon joint reflex is elicited.
(c) No. Inhibition of antagonist muscles is polysynaptic and hence its delay is longer than that of the contraction of the agonist muscle, the latter being a monosynaptic reflex.
(d) No. This input inhibits the motoneurone pool of the muscle whose tendon is tapped.
(e) Yes. A subject can voluntarily reinforce a reflex. One way of achieving this is to ask the subject to clench his two hands together and pull on them as hard as possible. This raises the excitability of many spinal reflexes, particularly in the upper part of the cord. This is used by the clinician if he cannot, for instance, elicit a jaw jerk reflex in the resting subject; if the subject clenches his hands, the reflex may become appreciable.

Structured exercise 8.1

Reflex pathways

1. See Figure.
   Points to be noted:
   The stretch is transmitted mechanically from the tendon to the muscle spindles where the receptors lie.
   The tendon jerk reflex pathway has only one synapse in the cord: this is the only monosynaptic reflex in the body.
   There is one, and only one, end-plate supplying each extrafusal muscle fibre.
2. There would be an inhibitory interneurone interposed on the pathway.
3. The receptor would be a nociceptive receptor in skin. The afferent nerve fibres would be A delta and non-myelinated fibres. In the cord there would be four or more interneurones. The effector would be a flexor muscle.
4. See Figure 8.4

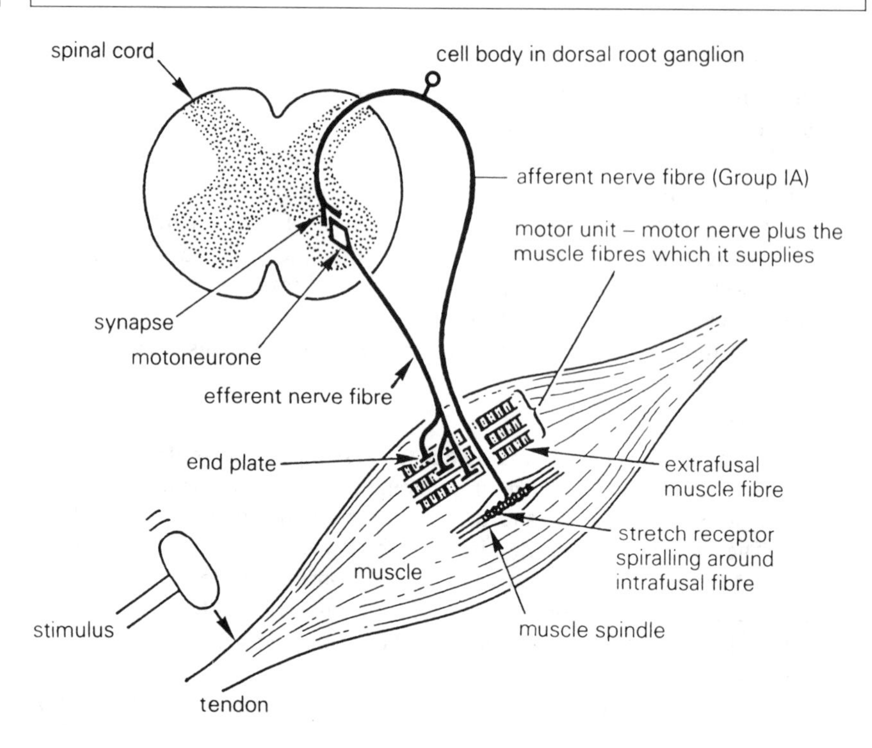

Answer 8.1

(a) No. No reflex commands the whole of a motoneurone pool.
(b) Yes. This is facilitation.
(c) Yes. This is occlusion.
(d) No. Occlusion demonstrates that there is convergence.
(e) Yes. The central excitatory state of a motoneurone pool varies at different times and so the magnitude of reflex responses shows considerable variation.

Answer 8.2

(a) Yes. Because unmyelinated fibres conduct more slowly than myelinated ones and therefore introduce extra delay into the reflex response time.
(b) No. The dorsal root ganglion does not contain synapses.
(c) Yes. For example, the subject is certainly aware of a painful stimulus eliciting a flexor withdrawal reflex.
(d) No. They lie in the ventral horn.
(e) Yes. For instance if an extensor muscle is reflexly contracting in a crossed extensor reflex and a nociceptive stimulus is applied ipsilaterally, this stimulus will evoke inhibition of the extensor muscles by reciprocal innervation.

Answer 8.3

(a) Yes, the Group Ia afferent nerve fibres.
(b) No, 35 ms is a typical value.
(c) Yes, less than 1 ms is due to synaptic delay, the remainder due to conduction along nerve fibres.
(d) No, this reflex contraction is a muscular twitch.
(e) No, the mechanical response occurs after the electromyogram and its time course is much more prolonged than that of the electromyogram.

Answer 8.4

(a) Yes, indeed this is usually the case. A reflex is a movement and usually involves the contraction of several agonist muscles.
(b) No, in addition to the reflex withdrawal of the leg to which the stimulus is applied, there is a postural component with contraction of extensor muscles in the contralateral leg; this is called the crossed extensor reflex.
(c) No, a spinal reflex pathway includes the dorsal roots. If these are cut, the reflexes are abolished.
(d) Yes. This is true for the tendon-jerk reflex. For all other reflexes, there are two or more synaptic relays in the cord.
(e) Yes. For instance, a scratch reflex response to a constant itch of the flank.

Answer 8.5

(a) Yes, the motoneurones supplying any large muscle are usually distributed through several segments of the cord.
(b) Yes, indeed this is true of most reflex pathways.
(c) Yes. Since all the branches of an afferent nerve fibre are excitatory, an inhibitory interneurone is obligatory on every inhibitory reflex pathway.
(d) Yes, this is a consequence of the requirement of an inhibitory interneurone on the pathway.
(e) No, in man all motoneurones are excitatory.

Answer 9.1

(a) Yes. There are only cones, no rods, in the fovea.
(b) Yes. There are no photosensitive receptors at the centre of the optic disc, or blind spot.
(c) No. This is peripheral retina, where the density of rods exceeds that of cones.
(d) No. The photosensitive receptors are of epithelial origin. There is a chain of synapses between the receptor cells and the terminals of the optic nerve fibres.
(e) Yes. Rhodopsin is sensitive to blue light (see Figure 9.4) and rods are more sensitive than cones for light of this wavelength.

Answer 9.2

(a) Yes. The input pathway is the right optic nerve and the output pathway is the parasympathetic nerve supply to the left eye; both these pathways are intact.
(b) No. The left optic nerve is cut so the input pathway for the reflex is interrupted.
(c) Yes. This reflex does not depend on visual input.
(d) Yes. The seeing eye has a visual field to both sides.
(e) Yes. The nerve supply to the extrinsic ocular muscles is intact and both eyes move.

Answer 9.3

(a) No. A small pupil throws a clearer image but the extent of the image thrown on the retina is independent of the pupil size.
(b) No. Rods are not sensitive to red light.
(c) Yes. The rods do not function in bright light.
(d) No. Complete dark adaptation takes about an hour.
(e) No. The blind spot for one eye falls within the visual field of the other eye.

Answer 9.4

(a) No. Convergence results in blunting of two-point discrimination since the excitation generated by stimulation of the two points converges onto a single afferent nerve fibre.
(b) No. Convergence favours sensitivity because weak excitation projecting along two pathways sums as a result of convergence onto the neurone to which the inputs are converging.
(c) Yes. The mechanism is described in Section 5.5.
(d) Yes. There is much less convergence for cones than for rods so that visual acuity, which is the ability to discriminate two points of light, is greater for cones than for rods.
(e) Yes. An object on one side of the visual axis projects an image onto rods which are at highest density just around the fovea. The fovea itself is devoid of rods and sensitivity at low levels of illumination is consequently low for objects directly on the visual axis.

Answer 9.5

(a) No. It decreases auditory sensitivity.
(b) Yes. This is a mechanism for the peripheral control of transmission along a sensory pathway.
(c) No. The parts of the basilar membrane near to the base of the cochlea respond to higher frequencies than the parts further away.
(d) Yes. For low frequencies, this locking of action potentials to the waveform of the sound is one way in which information about the frequency of the incident waveform to the central nervous system.
(e) Yes. This is true of most special sense receptors.

Answer 9.6

(a) Yes.
(b) No. All the receptors in an ampulla are similarly orientated and respond to displacement of the cupula along the axis of its canal.
(c) No. It is greatest when the acceleration is around an axis perpendicular to the plane of the canal and through its centre.
(d) No. Hair cells in the utricle have their directions of maximum sensitivity variously orientated in order to signal displacements in any direction of the utricular macula.
(e) No. The displacement is greatest when the linear acceleration is parallel with the macula.

Answer 10.1

(a) Yes. This stabilizes the retinal image.
(b) Yes. Nystagmus is the alternation of slow deviation and quick flick back.
(c) No. The time constant of the receptor mechanism is about 20 s and, after a minute of rotation, the subject will no longer sense it.
(d) Yes. The receptor mechanism is activated in the opposite direction from that at the start of rotation; this is interpreted as rotation in the opposite direction.
(e) No. The sensors of most importance in this respect are in the semicircular canals.

Answer 10.2

(a) Yes. Both situations are signalled by receptors in the utricle and saccule. A tilt can produce a force on the otoliths acting in the same direction as the force produced by an acceleraton.
(b) No. Angular acceleration is a rotation and tilting can be confused with linear acceleration, not with rotation.
(c) No. The plumb line indicates the vector sum of the gravitational force and the force needed to accelerate the weight towards the centre of turn.
(d) Yes. At constant velocity there is no accelerating force other than gravity.
(e) Yes. This repositions the base under the thrust force.

Structured exercise 11.1

Physiological vestibulo-ocular reflexes

from anterior to posterior
left
opposite
increases
will
will not

The cephalo-ocular reflex

opposite
brain stem

Caloric testing

heavier
fall
nose to occiput
right
both eyes
left
towards
reversed
less
less

Answer 11.1

(a) No. When the head is passively rotated, the eyeball reflexly rotates in the opposite direction, stabilizing the visual image. A similar movement of the eyeball would occur if its inertia were greatly increased, so the cephalo-ocular reflex increases the apparent inertia of the eyeball.
(b) Yes. This is the function of the vestibulo-ocular reflexes.
(c) Yes. The nuclei responsible for these reflexes lie in the brain stem.
(d) Yes. The fast phase of nystagmus depends on the cerebral cortex, the slow phase on the brain stem.
(e) No. When a person moving at constant speed in a train sits still looking out of the window, the relative movement of the landscape causes a physiological nystagmus, so called 'opto-kinetic' nystagmus. In this situation there is no movement of endolymph relative to the membraneous labyrinth.

Answer 11.2

(a) No, the fast phase of nystagmus depends on the integrity of the cerebral cortex. The caloric test will result in conjugate deviation of the visual axes, but no alteration of slow deviation with rapid flicking back of the visual axis, i.e. no nystagmus.
(b) Yes, see (a)
(c) Yes, brain stem reflexes are intact and the cephalo-ocular reflex results in movement of the visual axis in the opposite direction from that of the rotation of the head.
(d) Yes, this reflex is mediated by the Edinger Westphal nucleus in the brain stem.
(e) Yes, the respiratory centres lie in the brain stem.

Answer 11.3

For this question you must realize that the lateral semicircular canal which is approximately in the horizontal plane when the head is upright is nearest to the external auditory meatus so that endolymph in this canal is much more influenced than that in the others by cold or hot water in the meatus.

(a) Yes, the endolymph of the left lateral semicircular canal is cooled and falls. This convection current is in the same direction as the relative movement of the endolymph in the left lateral semicircular canal when the head is rotated to the right. To stabilize the visual image in this situation the visual axis rotates to the left.
(b) No, warm water causes a convection current in the opposite direction from that in (a).
(c) No, the downward movement of cooled endolymph is now in the direction from the back of the head to the front. It simulates movement of the head from right to left. To stabilize the visual image, the visual axis rotates to the right.
(d) No, the eye movements are conjugate.
(e) No, the opposite. If the subject is sitting, the plane of the lateral semicircular canal is approximately horizontal. The cooling of endolymph in the lateral semicircular canal causes only minor convection currents along the axis of the canal.

Answer 11.4

(a) Yes
(b) No. The stimulation is brought about by convection currents of endolymph, mainly in the lateral semicircular canal on the side to which the hot or cold water is applied. These currents move the endolymph relative to the membranous labyrinth and it is this relative movement that stimulates the receptors.
(c) No. The information from the single semicircular canal is integrated by the central nervous system and the reflex response is conjugate movement of both eyes.
(d) No. Each of these cranial nerve innervates muscles on only one side.
(e) No. Any rotaton of the head must influence receptors in at least two semicircular canals; more usually the receptors in all six are influenced.

Answer 11.5

(a) Yes, this is a spinal reflex and does not depend on brain stem function.
(b) No, the respiratory centres lie in the brain stem and, with no brain stem function, spontaneous respiratory movements do not occur.
(c) No, grimacing involves activity in the nucleus of the facial nerve, which lies in the brain stem.
(d) No, vestibulo-ocular reflexes including caloric testing depend on the nuclei of cranial nerves III, IV and VI which lie in the brain stem.
(e) No, the subject cannot speak because he is unconscious due to lack of activity of the ascending reticular formation.

Answer 11.6

(a) No. Consciousness depends on ascending activation of the cerebral cortex by the ascending reticular formation which lies in the brain stem.
(b) No, swallowing depends on hindbrain nuclei and the hindbrain is part of the brain stem.
(c) Yes. Transmission along the auditory pathway involves synaptic relays in the vestibular nuclei which lie in the brain stem. Abnormalities of certain components of auditory evoked potentials are characteristic of brain stem damage.
(d) Yes, these depend on pathways through the brain stem but without synaptic relay in the brain stem. Conduction of the nerve impulse has a high safety factor and so conduction through the damaged brain stem may be normal.
(e) Yes, see (d)

Answer 11.7

(a) Yes.
(b) Yes.
(c) No. The descending reticular formation has predominantly inhibitory effects on spinal γ motoneurones.
(d) Yes. This is an important factor in decerebrate rigidity, when the excitatory effect of the lateral vestibular nuclei is not counterbalanced by inhibition.
(e) Yes.

Answer 11.8

(a) Yes, the lateral vestibular nuclei intensely excite spinal motoneurones innervating antigravity muscles and this makes an important contribution to decerebrate rigidity.
(b) No, the red nucleus excites the inhibitory reticular formation which provides the inhibitory descending influences in spinal motoneurones. Decerebrate rigidity is due to interruption of this effect of the red nucleus on the inhibitory reticular formation. A section above the red nucleus does not result in decerebrate rigidity.
(c) No, the connections of the lateral vestibular nucleus are mainly with the ipsilateral spinal motoneurone.
(d) Yes, because γ activation is important in decerebrate rigidity.
(e) No. Such stimulation increases inhibition by the Purkinje cells of the cerebellar nuclei; these in turn send descending excitation of spinal α motoneurones.

Answer 11.9

(a) No. He may be being overventilated; this would remove the normal stimulus to respiration.
(b) No. They depend on the integrity of the brain stem but can occur when cortical function, including the function of auditory cortex, is widely impaired.
(c) Yes. A small lesion of the cortex does not cause derangement of consciousness whereas a small lesion in the brain stem, by depressing activity of the ascending reticular formation, rapidly depresses consciousness.
(d) No. The inhibitory reticular formation projects primarily down to the cord.
(e) No. Swallowing depends on the integrity of nuclei in the hind brain and persists even if the cerebral cortex, including the motor strip, is not functioning.

Answer 12.1

(a) No. There is hypertonia due to release of the cerebellar nuclei, which are predominantly excitatory to spinal motoneurones, from tonic inhibition exerted by the Purkinje cells of the cerebellar cortex.
(b) No, this would be true of a lesion in the basal ganglia.
(c) Yes, this is a characteristic of cerebellar lesions.
(d) No, voluntary movement is not paralysed although its efficient execution is disrupted by intention tremor.
(e) No, a positive Babinski sign indicates a lesion of the cortico-spinal projection; this projection is not interrupted by a cerebellar cortical lesion.

Answer 12.2

(a) Yes, the Purkinje cells are the output cells of the cerebellar cortex.
(b) No, they terminate in the cerebellar nuclei.
(c) Yes, see (b).
(d) No, they are inhibitory cells.
(e) No, the connections of the cerebellum with the periphery are uncrossed.

Answer 12.3

(a) Yes. This is a characteristic of the cerebellar cortex. The disorder of movement associated with damage to the cerebellar cortex is extreme but largely recovers over the course of weeks even if damage to the cerebellum is irreversible.
(b) Yes.
(c) Yes. The proprioceptive input to the cerebellum is profuse; the input to the basal ganglia is mainly from higher centres, notably the cerebral cortex.
(d) Yes, the cerebellum projects extensively to the cord via the vestibular nuclei. The basal ganglia have less influence via the descending reticular formation.
(e) No. The lesions of the basal ganglia result in involuntary movement and/or rigidity; those of the cerebellum result in intention tremor.

Answer 12.4

(a) Yes.
(b) No. The globus pallidus is the principal output nucleus of the basal ganglia.
(c) Yes. Although the representation is less well resolved spatially than in the somatosensory area of the cerebral cortex.
(d) No. Reduction of activity in the substantia nigra results in spasticity.
(e) No. The opposite is the case; a reduction in dopaminergic influence is associated with Parkinson's disease.

Answer 12.5

(a) Yes.
(b) Yes.
(c) Yes.
(d) No. Characteristically, it is during ramp movements that cells of the basal ganglia increase their discharge rate; during ballistic movements, their discharge is unaltered.
(e) Yes. This implicates the basal ganglia as important in the initiation of voluntary movement.

Answer 12.6

(a) Yes. The cells of the substantia nigra are dopaminergic.
(b) Yes.
(c) Yes.
(d) Yes.
(e) No. The dopamine content of the striatum and substantia nigra is decreased in Parkinson's disease.

Answer 12.7

(a) Yes.
(b) No. The synaptic contact between climbing fibres and Purkinje cells is dense and a single impulse in the climbing fibre evokes a train of impulses in the Purkinje cell.
(c) Yes.
(d) Yes, they are the only excitatory interneurones in the cerebellar cortex.
(e) No. The two axes are at right angles to each other.

Answer 12.8

(a) No. The Purkinje cells are inhibitory neurones.
(b) Yes. Cooling of the cerebellar cortex, with a consequent cessation of firing of Purkinje cells, results in increased excitability of α motoneurones supplying extensor muscles.

(c) Yes.
(d) Yes.
(e) No. Basket cells are inhibitory interneurones.

Answer 13.1

(a) No. It is the proliferation of association areas that characterizes the human brain.
(b) No. The sequence of myelination is the same as the sequence of phylogenetic development; the association areas are phylogenetically more recent than the primary receiving areas.
(c) No. Myelination of association cortex proceeds until the late teens.
(d) No. The primary motor area lies in precentral cortex; most primary sensory areas are postcentral.
(e) No. The specific somatosensory projection pathways, which are the most direct projections, traverse two synapses (see Chapter 6).

Answer 13.2

(a) No, stimulation of the motor strip causes a movement and this usually involves contraction of a group of agonist muscles, seldom of one muscle in isolation.
(b) Yes, the subject cannot voluntarily move the part of the body to which the stimulated area of the motor strip projects.
(c) No, the representation is distorted. Those parts of the musculature that are finely controlled, such as the musculature of the thumb and the forefinger, receive motor projection from a relatively large area of the motor cortex.
(d) No, they are generally on the left.
(e) No, stimulation of brain tissue does not cause pain.

Answer 14.1

(a) No. Neurones in the lateral geniculate body respond best to circles of light or dark; those in the primary visual cortex respond best to bars.
(b) No. The cortex is arranged in a columnar fashion, with all neurones in a column responding optimally to the same orientation.
(c) No. The ratio of area of cortex to area of retina is greatest for the fovea and falls progressively for retinal sites far from the fovea.
(d) No. The decussaton of the optic projection in the optic chiasma results in the lateral peripheral visual field projecting strictly to the contralateral cerebral hemisphere.
(e) No. Layer IV is the target of specific sensory projections; of all cortical areas, it is best developed in the primary visual cortex.

Answer 14.2

(a) Yes.
(b) No. Cones are associated with colour vision.
(c) No. Sympathetic activity causes pupillary dilatation (part of the 'flight or fight' reaction).
(d) No. It lies in the occipital lobe.
(e) No. Visual input from an object in the right visual field projects to the left hemisphere.

Answer 14.3

(a) No. The temporal half of the eye projects to the ipsilateral hemisphere and the nasal half to the contralateral hemisphere.
(b) No. The decussation is in the optic chiasma.
(c) No. It is from the frontal eye fields.
(d) No. Volitionally, we can only make saccadic movements.
(e) Yes. This avoids difficulties associated with cortical interpretation of a rapidly-moving input.

Answer 15.1

(a) Yes, if there were no γ action, the shortening of the muscle would unload the spindle receptors and reduce the afferent activity from the stretch receptors.
(b) Yes, via the γ loop.
(c) Yes, by direct activation of the extrafusal muscle fibres.
(d) No, a minimum value would be 100 ms; for many voluntary movements the reaction time is typically 250 ms. 30 ms would be correct for the reflex time of attendant jerk reflex.
(e) No, the dorsal columns consist entirely of afferent nerve fibres.

Answer 16.1

(a) Yes. Agnosia is an inability to recognize sensory images although detection of the occurrence of stimuli is unimpaired. It is characteristic of parietal lobe lesions.
(b) No. The patient can detect the occurrence of stimuli.
(c) Yes. Apraxia is an inability to perform motor acts, without impairment of primary motor function. It is characteristic of parietal lobe lesions.
(d) No. This would be true for lesions of the extrapyramidal system.

(e) No. Dysarthria is difficulty in performing the movements involved in speech. It is caused by pathology in the motor nuclei controlling speech (in the hindbrain) or by pathology in the peripheral nerves and muscles concerned with speech.

Answer 16.2

(a) No. The three speech areas lie in the frontal, parietal and temporal lobes of the left hemisphere.
(b) No. Disordered function of the speech areas results in difficulty with the structure of language, not with articulation.
(c) No. Just anterior.
(d) No. The motor speech area is released from control and characteristically subjects with such a lesion are verbose.
(e) No. This is true of young children but not of adults

Answer 16.3

(a) Yes.
(b) No. Mathematical ability and logic are attributes of the left ('dominant') hemisphere in most subjects.
(c) Yes
(d) Yes. Such an object projects to the speaking hemisphere.
(e) No. The opposite. The right (non-speaking) hemisphere is better at recognition than the speaking hemisphere.

Answer 17.1

(a) No. It is the principal excitatory transmitter for fast pathways.
(b) Yes.
(c) Yes, by acting on the NMDA subpopulation of glutamate receptors.
(d) Yes. by acting on the metabotropic glutamate receptors.
(e) No. Hypoxia results in excessive release of glutamate.

Answer 17.2

(a) Yes.
(b) No. The amino acids involved in the general metabolism of the body are the α amino acids.
(c) Yes.
(d) No. Synaptic vesicular membrane is synthesized in the cell body and transported to the nerve terminal by axonal transport.
(e) No. GABA-operated channels are voltage-insensitive.

Answer 17.3

(a) No. Repetitive activity of GABA-operated channels tends to decrease in synaptic efficacy.
(b) Yes. This is post-tetanic potentiation.
(c) No. Presynaptic inhibition has not been demonstrated in the mammalian cerebral cortex.
(d) No. They contribute to modulation of synaptic efficacy rather than as neurotransmitters in their own right and their time course of action is slow.
(e) No. They are synthesized in the cell body.

Answer 18.1

(a) Yes, intestine is very sensitive to stretch, insensitive to cutting.
(b) No, activity in thick afferent fibres reduces transmissivity along the pathways for pain.
(c) Yes, as in the control of pain by neurones activated from the Raphe nuclei of the brain stem.
(d) Yes, in the cornea the only type of sensory receptor is the nociceptor.
(e) Yes, it is common experience that a gentle touch to inflamed skin can be very painful.

Answer 18.2

(a) Yes.
(b) Yes, they are the only type of nerve ending to penetrate the epidermis.
(c) Yes, this occurs in the thalamic syndrome. The partial destruction of the thalamus results in an unusual pattern of afferent input to the higher centres and this is interpreted as pain.
(d) No, there are two groups of fibres signalling pain. The group with thicker axons are the smallest of the Group A fibres (Group A δ).
(e) No, cutting parts of the central nervous system itself does not cause pain. Cutting of the meninges is painful.

Answer 18.3

(a) Yes, this is the principal route for slow pain.
(b) Yes, the dorsal columns are the central afferent pathways for fine tactile discrimination and vibration sense, not for pain.
(c) Yes, normally noxious stimuli evoke protective reflexes and if absent, excessive wear and tear of joints with consequent damage is a common result.
(d) Yes, a very severe pain. In a person with a partial occlusion to the arterial supply to a limb, a little exercise brings on an excruciating pain. The condition is called ‘intermittent claudication’.

(e) Yes, embryologically the appendix is a midline structure so early in appendicitis, the pain may be felt as if it were in the midline. As the inflammation spreads to involve the peritoneum, the pain moves to the right iliac fossa.

Answer 18.4

(a) No, there are nociceptors in pleura, as evidenced by the extreme pain of pleurisy. The rubbing together of the inflamed layers of pleura make the movements of respiration very painful.
(b) Yes, since the gall bladder and the tip of the right shoulder are innervated by the same nerve root (C4).
(c) Yes, because the sensation of pain is mediated by the thinner fibres.
(d) Yes, the physiological basis is activity in thick nerve fibres which reduces transmissivity along the nociceptive pathway.
(e) No, it causes a flexor withdrawal reflex.

Answer 18.5

(a) Yes, pain is mediated by two groups of nerve fibres, C fibres (unmyelinated) and Group A delta fibres (myelinated).
(b) Yes, see (a)
(c) No, vibration is mediated by thick myelinated fibres which conduct quickly, nociception by thin myelinated fibres which conduct more slowly.
(d) Yes, for instance, after avulsion of a peripheral nerve, pain may develop and persist to become very disabling.
(e) Yes, they are free nerve endings i.e. the nerve endings for pain have no receptor structure which can be visualized with the light or electron microscope.

Answer 18.6

(a) Yes, antidromic impulses initiated at the site of stimulation pass down the dorsal columns along thick myelinated nerve fibres to the dorsal horn and reduce transmissivity along the nociceptive pathway.
(b) Yes, this increases activity in the descending pathways controlling pain.
(c) No, they decrease the firing rate of these neurones, thus alleviating pain.
(d) Yes, this projection allows the subject to localize pain.
(e) Yes, this projection is involved in the stress associated with pain.

Answer 19.1

(a) No, the skin receives no parasympathetic nerve supply.
(b) Yes.
(c) No, they lie in the lateral column in the grey matter within the spinal cord.
(d) Yes, preganglionic nerve fibres are thinly myelinated whereas postganglionic fibres are unmyelinated.
(e) Yes.

Answer 20.1

(a) Yes.
(b) No. The lower limit is around 60 mm Hg
(c) Yes. This may exacerbate cerebral hypoperfusion.
(d) Yes. This is a compensatory mechanism.
(e) No. When the blood pressure rises, autoregulation requires vasoconstriction.

Glossary

This is a glossary of the specialist vocabulary used by neurophysiologists. Some of the words are used in everyday speech but have a special meaning in neurophysiology. Most of the words are used in this book. The remainder are included here to allow the reader to look up unfamiliar words found in other textbooks of neurophysiology.

Acetylcholine An acetic ester of choline. In mammals, it is the chemical transmitter at a) the neuromuscular junction of skeletal muscles, b) autonomic synapses between pre- and postganglionic neurones, c) the neuroeffector junctions in the parasympathetic system, and d) at the sympathetic nerve endings innervating sweat glands and certain arterioles in skeletal muscles.

Acuity Visual acuity is the ability to discriminate between two points of light which are close together.

Adiadochokinesia a, neg. + Gr. diadochos, succeeding + kinesis, movement. Inability to perform rapidly alternating movements such as alternating pronation and supination of the hand. Also called dysdiadochokinesia.

Adrenaline Chemically, this is a catecholamine. It is secreted by the adrenal medulla into the blood stream whence it is distributed throughout the body. It acts on target organs such as the heart, where it causes an increase in both heart rate and myocardial contractility. Adrenaline is also called epinephrine.

Agnosia a, neg. + Gr. gnosis, knowledge. Lack of ability to recognize the meaning of sensory stimuli, although the threshold for detection of the existence of a stimulus is normal.

Agraphia a, neg, + Gr. grapho, to write. Inability to write even though voluntary control fo the skeletal musculature is intact. Agraphia occurs as a result of certain lesions in association cortex.

Alexia a, neg. + Gr. lexis, word. Inability to read the written word, although there is no defect in visual acuity or sensitivity. Alexia occurs with certain lesions in association cortex.

Allocortex Gr. allos, other + L. cortex, bark. The phylogenetically older cerebral cortex, e.g. hippocampus, usually consisting of three layers.

Amacrine a, neg. + Gr. makros, long + is, inos, fibre. Neurones without axons. They are found in the retina and thalamus. They often function without action potentials. They are phylogenetically precursors of neurones with axons.

Ammon's horn Ammon was an Egyptian deity with a ram's head. Ammon's horn is the hippocampus, which is in cross section suggestive of a ram's horn.

Amygdala L. from Gr. amygdale, almond. The amygdala is a nucleus in the temporal lobe of the cerebral hemisphere.

Anopsia an, neg. + Gr. opsis, vision. An inability to recognize visually presented objects despite normal acuity and sensitivity of vision. Anopsia occurs in individuals with certain lesions of the association cortex.

Anterior Adj. In front of. Synonymous with ventral in human anatomy.

Antidromic Gr. anti, against + dromos, running. An antidromic nerve impulse is one that is travelling in the direction which is the reverse of normal, e.g. an impulse from muscle to spinal cord in a motor nerve fibre.

Aphasia a, neg, + Gr. phasis, speech. An inability to express oneself in speech despite normal voluntary control of the skeletal musculature. Aphasia occurs in individuals with certain lesions of the association cortex.

Apraxia a, neg. + Gr. pratto, to do. Inability to perform an act in the absence of paralysis, sensory loss or deficity in comprehension. It occurs in individuals with lesions of the parietal region of the cerebral cortex.

Arachnoid Gr. arachne, spider's web + eidos, resemblance. The middle meningeal layer.

Archi Gr. Archaeus chief, leader hence archaic – ancient, old.

Archicerebellum Gr. archi, ancient + diminutive of cerebrum. The small, phylogenetically oldest portion of cerebellum. Also called the vestibulocerebellum because of its reciprocal connection with the vestibular nuclei. In mammals it comprises the nodule, uvula, vermis, flocculus and lingula. It is also called the paleocerebellum.

Archicortex Gr. archi, ancient + L. cortex, bark. This is phylogenetically old cortex, three-layered, located in the hippocampus and dentate gyrus of the temporal lobe.

Astereognosis a, neg. + stereos, solid + gnosis, knowledge. Loss of the ability to recognize objects by touching or feeling them, although the threshold for detection of the existence of a stimulus is normal.

Astrocyte Gr. astron, star + kylos, hollow (cell). An interneurone in the central nervous system.

Asynergy a, neg. + Gr. syn, with + ergon, work. Disturbance of the sequencing of the contraction of muscles in a voluntary movement.

Ataxia a, neg, + Gr. taxis, order. Incoordination of voluntary movement, with no loss of muscle power.

Athetosis a, neg. + Gr. thetos, position or place. Writhing movements of limbs and digits, occurring as a result of degeneration in the basal ganglia.

Autonomic Gr. autos, self + nomos, law. Self-regulating, as in autonomic nervous system.

Autonomic nervous system The efferent nerve fibres innervating all excitable tissues except skeletal muscle. It consists of two parts, the sympathetic nervous system, primarily concerned with fight and flight, and the parasympathetic nervous system, concerned with the integration of the vegetative functions of the body, such as digestion.

Autoradiography Gr. autos, self + L. radius, ray + Gr. grapho, to write. A technique in which radioactive isotopes are introduced into a tissue and the concentration of the isotope is detected by placing the tissue in contact with a photographic emulsion, which is subsequently developed. Also called radioautography.

Axolemma Gr. axo, axis + lemma, husk. The cell membrane of an axon.

Axon Gr. axo, axis. Nerve fibre.

Axon hillock The small hillock on the nerve cell body from which the axon arises.

Axoplasm Gr. axo, axis + plasm, formed. The contents of the axon.

Baroreceptor Gr. baros, pressure + receptor. A sensor found in the carotid sinus and aortic arch. It responds to changes in transmural pressure.

Betz cells These are the largest pyramidal cells of the cerebral cortex and are found in the deep layers of the primary motor strip. Their axons project down to the spinal motoneurones via the pyramids.

Bradykinesia Gr. brady, slow + kinesis, movement. Slowness of voluntary movements. It occurs as a consequence of disorder of function of the basal ganglia.

Brain stem The medulla, pons, and midbrain.

Bulb An old name for the medulla oblongata.

Catecholamines A group of dihydroxy phenols derived from catechol, a phenolic substance occurring naturally in certain plant tissue. The catecholamines occur naturally in the bodies of mammals; the catecholamines include adrenaline and noradrenaline.

Cauda equina A horse's tail. The lumbar and sacral spinal nerve roots in the lower part of the spinal canal at the level below the spinal cord.

Caudal L. tail. Adj. In the direction of the tail. Also spelt caudad.

Caudate nucleus Part of the corpus striatum. It is so named because it has a long extension which was likened to a tail by early neuroanatomists.

Central nervous system A term to describe the brain and spinal cord.

Cerebellum L. diminutive of cerebrum, brain. A fist-shaped structure with motor functions, situated behind the brain stem in the posterior cranial fossa.

Cerebrum L. brain. The cerebral hemispheres and diencephalon.

Chordotomy Gr. chord, cord + tome, a cutting. Surgical interruption of the spinothalamic tract for intractable pain. Also spelt cordotomy.

Chorea L. from Gr. choros, a dance. Involuntary dancing movements occurring in disorders of the basal ganglia.

Choroid Gr. chorion, a delicate membrane + eidos, form. The choroid plexuses in the ventricles of the brain are capillary networks from which cerebrospinal fluid is formed.

Chromatolysis Gr. chroma, colour + lysis, dispersal. Dispersal of the Nissl material in the cell body of a neurone as a result of cutting the axon.

Cingulum L. girdle. A nerve fibre tract in the white matter of the cingulate gyrus on the medial surface of the cerebral hemisphere.

Claustrum L. a barrier. A thin sheet of cell bodies, of unknown function, situated just lateral to the lentiform nucleus.

Colliculus L. a small elevation or mound. The superior and inferior colliculi are small mounds on the dorsal surface of the midbrain. The superior colliculi are concerned with visual orientation reactions whilst the inferior colliculi serve a similar function for audition.

Commissure L. a joining together. A bundle of nerve fibres passing from one side of the brain or spinal cord to the other.

Coning The forcing of the hindbrain down into the cone formed by the posterior fossa of the skull. The condition results from pressure applied to the brain stem.

Contralateral L. contra opposite + lateral side. On the opposite side of the body.

Corona L. from Gr. korone, a crown. The corona radiata is a prominent bundle of nerve fibres radiating from the thalamus via the internal capsule to the cerebral cortex.

Corpus callosum L. corpus, body + callosus, hard. A prominent tract of nerve fibres joining the two cerebral hemispheres.

Corpus striatum L. corpus, body + striatus, furrowed or striped. A series of nuclei traversed by bands of white matter lying deep within each cerebral hemisphere. The corpus striatum is a component of the basal ganglia.

Cortex L. bark. The outer layer matter of the cerebral and cerebellar hemispheres.

Cytoskeleton Gr. kutos, vessel and hence cell. The structural proteinous skeleton within cells.

Decussation L. decussatio, from decussis, the number X. The crossing of tracts from opposite sides of the body. The decussation of the pyramids in the medulla is an example.

Dendrite Gr. dendrites, tree. An extension of a nerve cell on which synapses from axons of other neurones terminate.

Dentate L. dentatus, toothed. The dentate nucleus is one of the cerebellar nuclei.

Diencephalon Gr. dia, through + enkephalos, brain. This name is applied to the thalamus and hypothalamus. The diencephalon arises from the posterior of the two brain vesicles formed from the prosencephalon during embryonic development.

Diplopia Gr. diplous, double + ops, eye. Double vision. This occurs when the visual axes of the two eyes are not aligned.

Dopamine A sympathomimetic agent released probably as a neurotransmitter by certain neurones, notably those comprising the nigro-striatal projection.

Dorsal L. back. Adj. towards the back. In human anatomy, it is synonymous with posterior.

Dura L. durus, hard. Dura mater, the thick outermost layer of the meninges.

Dyskinesia Gr. dys, difficult or disordered + kinesis, movement. Disorder of voluntary movement characterized by involuntary, purposeless movements.

Dysmetria Gr. dys, difficult or disordered + metron, measure. Disorder of the control of voluntary movement, in which the subject moves a finger in the wrong direction or too far.

Electroencephalogram Electrical brain activity recorded from electrodes attached to the scalp.

Emboliform Gr. embolos, plug + L.forma, form. The emboliform nucleus is one of the cerebellar nuclei.

Endoneurium Gr. endon, within + neuron, nerve. The delicate connective tissue sheath surrounding an individual nerve fibre of a peripheral nerve. It is formed by the Schwann cells.

Endorphins Endogenous chemicals resembling morphine both in their general pharmacological effects and in their effects on the excitability of central neurones involved in the perception of pain.

Entorhinal Gr. entos, within + rhis, nose. The entorhinal cortex is the region of cortex between the neocortex and the hippocampus.

Ependyma Gr. ependyma, garment. An epithelium lining the ventricles of the brain and central canal of the spinal cord.

Epicritic Gr. epi, upon, critos, judge. Discriminative sensation, contrasting with protopathic sensation, the primordial non-discriminative sensitivity of all animals, no matter how lowly.

Epineurium Gr. epi, upon + neuron. nerve. The connective tissue sheath surrounding the nerve fibres of a peripheral nerve.

Exteroceptor L. exterus, external + receptor, receiver. A sensory receptor sensing changes in the external environment (e.g. temperature of the skin).

Extrapyramidal system An obsolescent term which used to be used to describe of all motor parts of the central nervous system except the pyramidal motor system.

Fastigial L. fastigium, roof. The fastigial nucleus is one of the cerebellar nuclei.

Fimbria L. fimbriae, fringe. The fimbria is a band of nerve fibres running along the medial edge of the hippocampus and continuing into the fornix.

Foramen Magnum L. foramen, hole + magnum, large. The large hole in the base of the skull through which the brain and spinal cord are in continuity.

Fornix L. arch. The efferent tract from the hippocampal formation which courses over the thalamus and terminates in the mamillary body of the hypothalamus.

Fovea L. a pit. The fovea (fovea centralis) is a pit in the centre of the macula lutea of the retina.

Ganglion Gr. knot. A swelling on sympathetic nerves due to nerve cells. Also used for aggregations of nerve cells in the brain (e.g. basal ganglia).

Glia Gr. glue. The supporting cells of the nervous system.

Globus pallidus L. a ball + pale. Medial part of lentiform nucleus of corpus striatum. Phylogenetically the oldest part of the basal ganglia.

Gracilis L. slender. The nucleus gracilis is a nucleus on the dorsal aspect of the medulla which, together with the nearby cuneate nucleus, constitutes the relay station along the specific afferent pathway from the fibres of the dorsal column, which are first-order neurones. The nuclei project to the thalamus and constitute the second-order neurones of this pathway.

Granule L. granulum, diminutive of granum, grain. The term is used to describe small neurones, such as granule cells of cerebellar cortex and stellate cells of cerebral cortex.

Hemiballismus Gr. hemi, half + ballismos, jumping. A violent form of involuntary movement involving one side of the body, resulting from destruction of the subthalamic nucleus.

Hemiplegia Gr. hemi, half + plege, a stroke. Paralysis of one side of the body.

Hippocampus Gr. hippocampus, sea horse. A region of cerebral cortex which is phylogenetically ancient. The hippocampus is a gyrus and it constitutes an important part of the limbic system. Anatomically, it produces an elevation on the floor of the temporal horn of the lateral ventricle.

Homeostasis Gr. homois, like + stasis, standing. The tendency for the internal environment of the organism to be maintained constant in composition, temperature etc. Homeostasis depends on the operation of physiological control mechanisms.

Hydrocephalus Gr. hydro, water + kephale, head. Excessive accumulation of cerebrospinal fluid due to failure of drainage through the aqueduct or the foramina in the roof of the fourth ventricle. The brain is very thin and fails to function.

Hypothalamus Gr. hypo, under + thalamos, inner chamber. An anatomically tiny by physiologically vital region of the diencephalon that serves, together with the medulla oblongata, as the controlling centre of the autonomic nervous system.

Indusium L. induo, to put on. The indusium griseum. a thin layer of grey matter on the dorsal surface of the corpus callosum.

Insula L. island. The insula is a region of the temporal lobe of the cerebral cortex which, because it is infolded, cannot be seen when the surface of the brain is exposed.

Integument The skin surface.

Interoceptor L. inter, between + receptor, receiver. An interoceptor is a sensory receptor which senses the interior of the organism e.g. visceral receptors. Also spelt 'enteroceptor'.

Ipsilateral L. ipse, same + lateral, side. Adj. on the same side of the body.

Isocortex Gr. isos, equal + L. cortex, bark. Isocortex is synonymous with neocortex. It is cerebral cortex which is phylogenetically recent and has six layers.

Kinesthesia Gr. kinesis, movement + aisthesis, sensation. The perception of joint position and movement.

Lemniscus Gr. lemniskos, fillet (a ribbon or band). The term is used as a name for certain bundles of nerve fibres in the central nervous system (e.g. medial lemniscus and lateral lemniscus).

Lentiform L. lentil, lens + forma, shape. The lentiform nucleus is a component of the corpus striatum.

Leptomeninges Gr. leptos, slender + meninx, membrane. The delicate membranes of the meninges viz. the arachnoid and pia mater.

Limbus L. a border. Limbic lobe: a region of phylogenetically ancient cortex on the medial surface of the cerebral hemisphere. The limbic system is the limbic lobe, hippocampal formation, the mammillary body and anterior thalamic nuclei.

Ligament Connects bone to bone. e.g. ligaments supporting a joint.

Limen L. threshold.

Locus ceruleus L. place + caerulus, dark blue. A small faintly blue spot on each side of the floor of the fourth ventricle produced by a group of pigmented nerve cells.

Macroglia Gr. makros, large + glia, glue. The larger neuroglial cells i.e. astrocytes, oligodendrocytes, and ependymal cells.

Macroscopic Gr. macros, large + scopos, sight. Adjective to describe the appearance of a structure with the unaided eye; this contrasts with its microscopic appearance.

Macula lutea L. macula, a spot + lutea, yellow. A yellow area at the centre of the retina. The maculae sacculi and utriculi are sensory areas in the vestibular portion of the membranous labyrinth.

Mammillary L. mammilla, diminutive of mamma, breast (shaped like a nipple). The mammillary bodies are a pair of small swellings on the ventral surface of the hypothalamus. They are part of the diencephalic reticular formation.

Massa intermedia This is bridge of grey matter connecting the thalami of the two sides across the third ventricle, present in 70% of human brains.

Medulla L. medius, middle. The medulla, or medulla oblongata is the caudal portion of the brain stem.

Meninges L. meninx, membrane. The three membranes which cover and protect the surface of the brain. They are dura mater, arachnoid mater and pia mater, from outside inwards.

Mesencephalon Gr. mesos. middle + enkephalos, brain. The mesencephalon means the midbrain. It arises from the second of the three primary brain vesicles.

Metencephalon Gr. meta, after + enkephalos, brain. The region of the brain coming after the forebrain. It comprises the pons and cerebellum. It is the anterior of the two divisions of the rhombencephalon which arises from the posterior primary brain vesicle.

Microglia Gr. mikros, small + glia, glue. A type of neuroglial cell which is very small.

Mnemonic Gr. mneme, memory. An aide-memoire.

Molecular L. molecula, diminutive of moles, mass. Used in neurohistology to denote tissue containing large numbers of fine nerve fibres which endow the tissue with a punctate appearance in silver-stained sections. There are layers of cerebral and cerebellar cortices which are called molecular layers.

Motoneurone The motor nerve cells of the spinal cord and motor nuclei of the cranial nerves (Eccles, 1968, p.2).

Myelin Gr. myelos, marrow. An insulating sheath around many nerve fibres. It is a protein–lipid complex made up of spiral layers of cell membranes of glial cells. These glial cells are called Schwann cells in the peripheral nervous system and oligodendrocytes in the central nervous system.

Neocerebellum Gr. neos, new + diminutive of cerebrum. The phylogenetically most recent part of the cerebellum.

Neocortex Gr. neos, new + L. cortex, bark. Phylogenetically the youngest part of the cerebral cortex. It is six-layered and constitutes most of the cerebral cortex in humans.

Neostriatum Gr. neos, new + L. striatus, striped or grooved. The phylogenetically recent part of the corpus striatum consisting of the caudate nucleus and putamen. It is also called the striatum.

Net (movement of a chemical) The net movement of a chemical between two compartments A and B is the difference between the total, or gross, movement from A to B and that in the opposite direction, from B to A. The net movement is thus the change in amount of the chemical in a compartment.

Neurite Gr. neurites, of a nerve. The cytoplasmic extensions of neurones forming dendrites and axons.

Neurofibril Gr. neuron, nerve + L fibrilla, dimunitive of fibra, a fibre. The delicate structural filaments in the cytoplasm of neurones.

Neuroglia Gr. neuron, nerve + glia, glue. The cells of the central nervous system which provide structural support for the neurones. The neuroglia (literally nerve glue) form a supporting framework for the neurones of the central nervous system. They also contribute to controlling the chemical composition of the neurones' environment. The neuroglia of the central nervous system include astrocytes, oligodendrocytes, microglial cells, and ependymal cells.

Neurolemma Gr. neuron, nerve + lemma, husk. The delicate sheath surrounding a peripheral nerve fibre. It consists of a series of Schwann cells.

Neurone Gr. a nerve. Neurone means nerve cell. The neurone comprises the nerve cell body, the dendrites and the axon.

Neuropil Gr. neuron, nerve + pilos, felt. The complex net of dendrites and axons between cell bodies in grey matter.

Nociceptive L. noceo, to injury + capio, to take. Sensitive to injurious stimuli.

Noradrenaline Chemically it is a catecholamine. In the body, it is released from most peripheral sympathetic nerve endings to act locally as the neurotransmitter chemical. It also enters the blood and is carried away to act on distant target tissues and organs such as the heart.

Nystagmus Gr. nystagmos, a nodding, from nystazo, to be sleepy. An involuntary movement of the eyes occurring physiologically and in some pathological conditions. Nystagmus has two phases, a slow drift which, when the nystagmus is physiological, serves to stabilize the retinal image, and a fast phase, which brings the direction of gaze back to the midline when the slow phase has reached its limit.

Obex L. barrier. A small transverse fold overhanging the fourth ventricle.

Oligodendrocyte Gr. oligos, few + dendron, tree + kytos, cell. A neuroglial cell of ectodermal origin which forms the myelin sheath in the central nervous system.

Olive L. oliva. An olive-shaped nucleus in the medulla on either side.

Paleocerebellum Gr. palaios, ancient + diminutive of cerebrum. The phylogenetically old part of the cerebellum.

Paleocortex Gr. palaios, ancient + L. cortex, bark. The olfactory region of the cerebral cortex. It is simpler in structure than neocortex.

Paleostriatum Gr. palaios. ancient + L. striatus, striped or grooved. The globus pallidus or pallidum, which is the phylogenetically oldest part of the basal ganglia.

Pallidum L. pallidus, pale. See globus pallidus.

Pallium L. cloak. Pallium is used synonymously with cortex.

Paralysis Gr. paralyein. To loosen, dissolve, or weaken. Loss of voluntary movement.

Paraplegia Gr. para, beside + plege, a stroke. Paralysis of both legs.

Parasympathetic nervous system The parasympathetic nervous system and the sympathetic system together comprise the autonomic nervous system.
Parasympathomimetic An adjective to describe chemicals whose effects are similar to the effects of increased activity of the parasympathetic nervous system.
Paresis Gr. parienai, to relax. Loss of fine movements with no true paralysis. It occurs in individuals with lesions of the parietal region of the cerebral hemisphere.
Percutaneous This means 'through the skin'. Percutaneous stimulation of a nerve means stimulation by means of passing current through the skin overlying the nerve.
Perineurium Gr. peri, around + neuron, nerve. The connective tissue sheath around a bundle of nerve fibres in a peripheral nerve.
Peripheral nervous system The portions of neurones outside the cord and the brain.
Phenol Carbolic acid C_6H_5OH (hydroxybenzene).
Pia mater L. tender mother. The thin innermost layer of the meninges, continuous with the neural tissue of the brain and spinal cord. It forms the inner boundary of the subarachnoid space.
Pineal L. pineus, relating to the pine. The pineal body or pineal gland is a small appendix, of unknown function, lying posteriorly over the corpora quadrigemina.
Plexus L. plectere, to knit. A network of nerve fibres or blood vessels.
Pons L. bridge. The part of the brain stem that lies between the medulla and the midbrain. It contains relay nuclei from the cerebral cortex to the cerebellum and constitutes a bridge between the right and left halves of the cerebellum.
Posterior L. behind. Synonymous with dorsal in human anatomy.
Posterior fossa The region within the skull below the tentorium cerebelli.
Pronate To move to the prone position. It is applied to rotation of the forearm in the direction of the thumb moving inward. Opp. supinate.
Proprioceptor, proprioception L. proprius, one's own + receptor, receiver. A sensory ending in muscles, tendons, ligaments and joints providing information concerning movement and position of body parts in relation to each other.
Prosencephalon Gr. pros, before + enkephalos, brain. The forebrain, consisting of the telencephalon (cerebral hemispheres) and diencephalon. It arises from the anterior primary brain vesicle.
Protopathic Gr. proto, first + pathos, pain. The phylogenetically ancient sensation corresponding to the irritability of the lowliest animals. In humans, the poorly localized pain mediated by non-myelinated afferent nerve fibres is protopathic sensation. It contrasts with epicritic sensations.
Ptosis Gr. ptosis, fall. Drooping of the upper eyelid, usually as a result of interruption of the sympathetic nerve supply to smooth muscle in the levator palpebrae superioris muscle.

Putamen L. shell. The lateral part of the lentiform nucleus of the corpus striatum.

Pyramidal tract A large descending bundle of nerve fibres, partly made up of the axons of Betz cells in the primary motor cortex, which decussates in the medulla. It is so called because the decussation occupies a pyramid-shaped area on the ventral surface of the medulla.

Quadriplegia L. quadri, four + Gr. plege, stroke. Paralysis of the four limbs. Also called tetraplegia.

Raphe Gr. seam. The raphe nuclei are nuclei of the reticular formation lying in the midline of the medulla, pons, and midbrain.

Reticular L. reticularis, resembling a net. The reticular formation of the brain stem is so called because of its appearance. It consists of ill-defined nuclei crossed and interwoven with fibre tracts.

Rhinencephalon Gr. rhis, nose + enkephalos, brain. Rhinencephalon means literally nose-brain. It refers to the parts of the brain which developed from regions which were originally concerned with olfaction; it is often used synonymously with limbic system.

Rostral L. beak. Adj. towards the beak or, in the case of mammals, towards the nose.

Rubro L. ruber, red. Pertaining to the red nucleus as in 'rubrospinal'.

Saccadic Fr. saccader, to jerk. Small quick involuntary movements of the eyes.

Schwann cells Glial cells along peripheral nerves.

Sensitivity Applied to sensation such as vision or touch, it means the minimum stimulus which excites the sensory receptor.

Septal area An area on the medial aspect of the frontal lobe of the cerebral cortex. It is the site of the septal nuclei.

Septum pellucidum L. partition + transparent. A triangular double membrane separating the frontal horns of the lateral ventricles. It is situated in the median plane and fills in the space between the corpus callosum and the fornix.

Somatic Gr. somatikos, bodily. Used in neurophysiology to denote the body excluding the viscera. Hence the somatic nervous system is the part of the nervous system which serves the skeletal musculature.

Somaesthetic Gr. soma, body + aisthesis, perception. The somaesthetic senses are the general body senses of pain, temperature, touch, pressure, position, movement and vibration, but excluding the special senses.

Striatum L. striatus, furrowed. The phylogenetically recent part of the corpus striatum (neostriatum) consisting of the caudate nucleus and the putamen or lateral portion of the lentiform nucleus.

Subiculum L. diminutive of subex (subic-), a layer. Cortex intermediate in structure between neocortex and the hippocampus. Anatomically, also, it is sited between the neocortex and archicortex.

Substantia gelatinosa L. jelly-like substance. It is a lamina of small neurones at the apex of the dorsal grey horn throughout the spinal cord.

Substantia nigra A large nucleus in the midbrain. It is black becauses its constituent cells contain melanin.

Subthalamus L. under + Gr. thalamos, inner chamber. It is a region of the diencephalon beneath the thalamus, containing fibre tracts and the subthalamic nucleus.

Supinate To move to the supine position. Applied to rotation of the forearm, it is rotation in the direction such that the thumb moves outwards. Opp. pronate.

Sympathetic nervous system The parasympathetic nervous system and the sympathetic system together comprise the autonomic nervous system.

Sympathomimetic An adjective to describe chemicals whose effects are similar to the effects of increased activity of the sympathetic nervous system.

Synapse Gr. synapto, to join. The word was introduced by Sherrington in 1897 to describe the site of contact between neurones.

Syringomyelia Gr. syrinx, pipe, tube + myelos, marrow. A condition characterized by the formation of cavities in the centre of the spinal cord and gliosis around the cavities.

Tectum L. roof. The tectum is the roof of the midbrain consisting of the paired superior and inferior colliculi.

Telencephalon Gr. telos, end + enkephalos, brain. The telencephalon is the front end of the brain i.e. the cerebral hemispheres.

Tendon Connects muscle to its insertion, usually bone (patellar tendon is a misnomer, it is a ligament).

Tentorium L. tent. The tentorium cerebelli is a dural partition between the occipital lobes of the cerebral hemispheres and the cerebellum.

Tetraplegia Gr. tetra-, four + plege, a stroke. See quadriplegia.

Thalamus Gr. thalamos, inner chamber. A large mass of nuclei in the diencephalon. One group of thalamic nuclei act as a relay station on the specific afferent pathways to the cerebral cortex. Others relay between the major motor centres in the brain. Yet others are part of the reticular formation.

Tomography G. tomos, cutting + grapho, to write. It is a technique whereby it is possible to X-ray a section through an individual. This is achieved by a coupled movement of the X-ray source and the X-ray film during exposure. A modern development involves the use of a computer to calculate the X-ray density at any desired plane through a subject. This technique is called computerized tomography, or CT scan.

Transducer L. transducere, to lead across. A device for converting one form of energy into another. A sensory receptor transduces the energy to which it is sensitive into electrical energy, the latter being the currency of the nervous system.

Uvula L. little grape. A region of the inferior vermis of the cerebellum.

Velum L. sail, curtain, veil. The superior and inferior medullary vela are fine connective tissue sheaths forming the roof of the fourth ventricle.

Ventral L. venter, belly. Belly side. Synonymous with anterior in human anatomy.

Ventricle L. ventriculus, diminutive of venter, belly. The ventricles form a system of cavities within the brain. The ventricles are filled with cerebrospinal fluid. There are lateral, third, and fourth ventricles of the brain, the lateral ventricles being paired.

Vermis L. worm. The midline portion of the cerebellum.

References

Bard, P. (1928) Diencephalic mechanisms for expression of rage with special reference to sympathetic nervous system. *American Journal of Physiology*, **84**, 490–515.

Bard, P. and Rioch, D. McK. (1937) A study of four cats deprived of neocortex and additional parts of the forebrain. *Bulletin of the Johns Hopkins Hospital*, **60**, 65–146.

Barr, M.L. and Kiernan, J.A. (1989) *The Human Nervous System. An Anatomical Viewpoint*, 5th Edn, Lippincott, London.

Basbaum, A.I. (1973) Conduction of the effects of noxious stimulation by short-fibre multisynaptic system of the spinal cord. *Experimental Neurology*, **40**, 699–716.

Basbaum, A.I. (1992) The use of Fos immunocytochemistry to gain insights into mechanisms underlying the generation of nociceptive and neuropathic pain. *Neuroscience Letters*, **Suppl. A2**, S3.

Basbaum, A.I. and Fields, H.L (1984) Endogenous pain control systems. *Annual Review of Neuroscience*, **7**, 309–338.

Betz, V. (1874) Anatomischer Nachweiss zweier Gehirncentra. *Zentralblatte Medizinische Wissenschaft*, **12**, 578–580. 595–599.

Bockaert, J. (1992) Role of intra- and extracellular messengers in synaptic plasticity. *Neuroscience Letters*, **42**, S4.

Bond, M. (1979) *Pain, its Nature, Analysis and Treatment*. Churchill Livingstone, London.

Bullock, R. and Fujisawa, H. (1992) The role of glutamate antagonists for the treatment of CNS injury. *Journal of Neurotrauma*, **9** (Suppl, 2), 5443–5462.

Carpenter, R.H.S. (1990) *Neurophysiology*, 2nd Edn. London, Edward Arnold.

Carrea, R.M., Reissig, E.M. and Mettler, F.A. (1947) The climbing fibres of the simian and feline cerebellum. Experimental enquiry into their origin by lesions of the inferior olives and deep cerebellar nuclei. *Journal of Comparative Neurology*, **87**, 321–365.

Cajal, S.R. (1911) *Histologie du Systeme nerveux de l'homme et des Vertebres. Vol. 2*. A. Maloine, Paris.

Cajal, S.R. (1954) *Neuron Theory or Reticular Theory? Objective Evidence of the Anatomical Unity of Nerve Cells*. Madrid: Consejo Superior de Investigaciones Cientificas.

Chabre, M. (1987) The G protein connection: is it in the membrane or the cytoplasm? *Trends Biochemical Science*, **12**, 213–215.

Cowan, W.M., Gottlieb, D.I., Hendrickson, A.E., Price, J.L. and Woolsey, T.A. (1971) The autoradiographic demonstration of axonal connections in the central nervous system. *Brain Research*, **37**, 21–51.

Creed, R.S., Denny-Brown, D., Eccles, J.C., Liddell, E.G.T. and Sherrington, C.S. (1932) *Reflex Activity of the Spinal Cord*. Oxford University Press, Oxford.

Dale, H.H (1935) Pharmacology of nerve endings. *Proceedings of the Royal Society of Medicine*, **28**, 319–332.

Davson, H. and Eggleton, M., eds. (1962) *Starling and Lovatt Evans' Principles of Human Physiology*, 13th Edn, Churchill, London.

Desmedt, J.E. and Godaux, E. (1978) Ballistic skilled movements. *Progress in Clinical Neurophysiology*, **4**, 21–55.

Desmedt, J.E. and Monaco, P. (1960) Suppression par la stryhnine de l'effet inhibiteur centrifique exerce par le faisceau olivo-cocheaire. *Archives Internationale Pharmacodynamics*, **3**, 244–248.

Dockray, G.J., Gayton, R.J. and Williams, R.G. (1983) Enkephalins and endorphins: the physiology and pharmacology of pain. In: *The Neurobiology of Pain*, (eds. A.V. Holden and W. Winlow) Manchester University Press, Manchester, pp. 284–296.

Dudel, J. and Kuffler, S.W. (1961) Presynaptic inhibition at the crayfish neuromuscular junction. *Journal of Physiology*, **155**, 543–562.

Dunnett, S.B. and Richards, S.J. eds. (1990) *Neural Transplantation From Molecular Basis to Clinical Applications*. Elsevier, Amsterdam, New York, Oxford.

Eccles, J.C. (1955) The central action of antidromic impulses in motor nerve fibres. *Archiv Gesamte Physiologie*, **260**, 385–415.

Eccles, J.C. (1964) Presynaptic inhibition in the spinal cord. *Progress in Brain Research*, **12**, 65–91.

Eccles, J.C. (1964) *The Physiology of Synapses*. Springer-Verlag, Berlin.

Eccles, J.C. (1968) *The Physiology of Nerve Cells*. Johns Hopkins Press, Baltimore.

Eccles, J.C. and Sherrington, C.S. (1930) Numbers and contraction values of individual motor units examined in some muscles of the limb. *Proceedings of the Royal Society of London B*, **106** 326–357.

Erlanger, J. and Gasser, H.S. (1937) *Electrical Signs of Nervous Activity*. University of Pennsylvania Press.

Evarts, E.V. and Vaughn, W.J. (1978) Intended arm movements in response to externally produced arm displacements in man. *Progress in Clinical Neurophysiology*, **4**, 178–192.

Ferrieira, S.H. (1979) Prostaglandins. In *Handbook of Inflammation, Vol, 1, Chemical Messengers of the Inflammatory Process*. (ed. J.C. Houck) North-Holland, Amsterdam, pp. 113–157.

Fibinger, H.C. (1982) The organisation and some projections of cholinergic neurons of the mammilian forebrain. *Brain Research Reviews*, **4**, 327–388.

Fritsch, G. and Hitzig, E. (1870) Ueber die elektrische Erregbarkeit des Grosshirns. *Archive Anatome Physiologie Wissenschaff Medizin*, **10**, 300–332.

Fulton, J.F. (1926) *Muscular Contraction and the Reflex Control of Movement*. Bailliere, Tindall and Cox, London.

Gaffan, D. and Gaffan, E.A. (1991) Amnesia in man following transection of the fornix. *Brain*, **114**, 2611-2618.

Geschwind, N. (1970) The organization of languages and the brain. *Science*, **170**, 940–4.

Goodman Gilman, A. (1990) *The Pharmacological Basis of Therapeutics*, 8th edn. Pergamon Press, New York.

Gray, E.G. (1961) Ultrastructure of synapses of the cerebral cortex and certain specialisations of neuroglial membranes. In *Electron Microscopy in Anatomy*, ed. J.D. Boyd, F.R. Johnson and J.D. Lever, Edward Arnold, London, pp 54–73.

Gray, E.G. (1969) Electron microscopy of excitatory and inhibitory synapses: a brief review. *Progress in Brain Research*, **31** 141–155.

Harlow, J.M. (1848) Passage of an iron rod through the head. *Boston Medical and Surgical Journal*, **39**, 389–393.

Harlow, J.M. (1868) Recovery from the passage of an iron bar through the head. *Massachussetts Medical Society Publications*, **2**, 327–346.

Hebb, D.O. (1949) *The Organization of Behavior. A Neuropsychological Theory*. Wiley and Sons, New York.

Henneman, E., Somjen, G. and Carpenter, D.O. (1965) Functional significance of cell size in spinal motoneurones. *Journal of Neurophysiology*, **28**, 560–580.

Hodgkin, A.L. (1965) *The Conduction of the Nervous Impulse. The Sherrington lectures VI*. Liverpool University Press, Liverpool.

Holden, A.V. and Winlow, W., eds. (1983) *The Neurobiology of Pain. Symposium of the Northern Neurobiology Group*. Manchester University Press, Manchester.

Hubel, D.H. and Weisel, T.N. (1977) Functional architecture of macaque monkey visual cortex. *Proc. Roy. Soc.*, **B 198**, 1–59.

Hughes, J., Smith, T.W., Kosterlitz, H.W., Fothergill, L.A., Morgan, B.A. and Morris, H.R. (1975) Identification of two related pentapeptides from the brain with potent opiate agonist activity. *Nature*, **258**, 577–579.

Hultborn, H. (1992) Organisation of the spinal rhythm generator in mammals. *Neuroscience Letters*, **Suppl. 42**, S5.

Huxley, A.F. and Stampfli, R. (1949) Evidence for saltatory conduction in peripheral myelinated nerve fibres. *Journal of Physiology*, **108**, 315–339.

Iggo, A. (1972) The case for pain receptors, and critical remarks on the gate control theory. In *Pain* (eds R. Janzen, W.D. Keidel, A. Hertz, C. Steichele, J.P. Payne and R.A.P. Burt) Georg Thieme, Stuttgart and London, pp. 60, 127.

Iggo, A. 1982. Cutaneous sensory mechanisms. In *The senses*, (eds H.B. Barlow and J.D. Mollon), Cambridge University Press, Cambridge, pp. 369–408.

Isaacson, R.L. (1982) *The Limbic System*. Plenum Press, New York.

Jankowska, E., Jukes M.G.M., Lund, S. and Lundberg, A. (1967) The effect of DOPA on the spinal cord. *Acta Physiologica Scandinavica*, **70**, 369–402.

Jennett, B. (1977) Aspects of coma after severe head injury. *Lancet*, **i**, 878–81.

Jennett, B. (1980) Research in brain trauma. *Trends in Neurosciences*, **3(10)**, I–V.

Jennett, B. and Plum, F. (1972) Persistent vegetative state after brain damage. *Lancet*, **i**, 734–7.

Jennett, B. and Teasdale, G. (1977) Aspects of coma after severe head injury. *Lancet*, **i**, 878–81.

Jennett, S.J. and Holmes, O. (1983) *Multiple choice Questions in Physiology*. Pitman, London.

Jolly, W.A. (1911) On the time relations of the knee-jerk and simple reflexes. *Quarterly Journal of Experimental Physiology*, **4**, 67–87.

Jones, E.G. and Powell, T.P.S. (1970) Connexions of the somatic sensory cortex of the rhesus monkey. I. Ipsilateral cortical connexions. *Brain*, **92**, 477–505.

Jouvet, M. (1972) The role of monoamine and acetylcholine containing neurons in the regulation of the sleep-waking cycle. *Ergebnisse Physiologie*, **64**, 166–307.

Kandel, E.R., Schwartz, J.H. and Jessel, T.M. (1991) *Principles of Neuroscience*, 3rd edn. Edward Arnold, London.

Katz, B. (1950) Depolarisation of sensory terminals and the initiation of impulses in the muscle spindle. *Journal of Physiology*, **111**, 248–260.

Katz, B. (1966) *Nerve, Muscle and Synapse*. McGraw Hill, New York.

Katz, B. (1969) *The Release of Neural Transmitter Substances. The Sherrington lectures X*. Liverpool University Press.

Kemp, J.M. and Powell, T.P.S. (1970) The corticostriate pathway in the monkey. *Brain*, **93**, 525–546.

Ketty, S.S. and Schmidt, C.F. (1945) The determination of cerebral blood flow in man by the use of nitrous oxide in low concentrations. *American Journal of Physiology*, **143**, 53–66.

Kimura, M., Aosaki, T., Hu, A., Ishida, A. and Watanabe, K. (1992) Activity of primate putamen neurones is selective to the mode of voluntary movement: visually guided, self-initiated or memory-guided. *Experimental Brain Research*, **89**, 473–477.

Kornhuber, H.H. (1974) Cerebral cortex, cerebellum and basal ganglia: an introduction to their motor functions. In *The Neurosciences, 3rd Study Program*, (eds. F.O. Schmitt and F.G. Warden), MIT Press, Massachusetts, pp. 267–280.

Krnjevic, K., Pumain, R. and Renaud, L. (1971) The mechanism of excitation by actylcholine in the cerebral cortex. *Journal of Physiology*, **215**, 247–68.

Krogsgaard-Larsen, P., Nielsen, L. and Falch, E. (1986) The active site of the GABA receptor. In *Benzodiazepine/GABA Receptors and Chloride Channels: Structural and Functional Properties*, (eds R.W. Olsen, and J.C. Venter), Alan R. Liss, Inc, New York, pp. 73–94.

Kuypers, H.G.J.M., Svarcbast, M., Mishkin, M. and Rosvold, H. (1965) Occipitotemporal corticospinal connexions in rhesus monkeys. *Experimental Neurology*, **11**, 245–262.

Levy, J., Trevarthen, C. and Sperry, R.W. (1972) Perception of bilateral chimeric figures following hemispheric deconnexion. *Brain*, **95**, 61–78.

Lewis, P.R. and Shute, C.C.D (1963) Tracing presumed cholinergic fibres in rat forebrain. *Journal of Physiology*, **168**, 33–35P.

Lieberman, A.R. (1971) The axon reaction: a review of the principal features of perikaryal responses to axon injury. *International Review of Neurobiology*, **14**, 49–124.

Lloyd, D.P.C. and Chang, H.T. (1948) Afferent fibres in muscle nerves. *Journal of Neurophysiology*, **11**, 488–518.

Loewenstein, W.R. (1962) Excitation processes in a receptor membrane. *Acta Neurovegetativa*, **24**, 184–207.

Loewenstein, W.R., Terzuolo, C.A. and Washizu, Y. (1964) Separation of transducer and impulse-generating processes in sensory receptors. *Science*, **142**, 1180–1181.

Marsden, C.D., Merton, P.A., Morton, H.B., Adam, J.E.R. and Hallett, M. (1978) Automatic and voluntary responses to muscle stretch in man. *Progress in Clinical Neurophysiology*, **4**, 167–177.

Matthews, P.E. (1972) *Mammalian Muscle Receptors and Their Central Connections*. Edward Arnold, London.

Matthews, P.B.C. (1991) The human stretch reflex and the motor cortex. *Trends in Neuroscience*, **14**, 87–91.

Melzack, R. and Wall, P.D. (1965) Pain mechanisms: a new theory. *Science*, **150**, 971–979.

Melzack, R. and Wall, P.D. (1982) *The Challenge of Pain*. Middlesex, Penguin.

Moruzzi, G. and Magoun, H.W. (1949) Brain stem reticular formation and activation of the EEG. *Electroencephalography and Clinical Neurophysiology*, **1**, 455–473.

Mountcastle, V.B. (1957) Modality and topographic properties of single neurons of cat's somatic sensory cortex. *Journal of Neurophysiology*, **20**, 408–434.

Mountcastle, V.B. ed. (1974) *Medical Physiology Vol. 1, 13th Edn*. C.V. Mosby, Saint Louis.

Mountcastle, V.B., Davies P.W. and Berman, A.L. (1957) Neurons of cat somatosensory cortex to peripheral stimulation. *Journal of Neurophysiology*, **20**, 374–407.

Mountcastle, V.B., Lynch, J.C. and Georgopoulos, A. (1975) Posterior parietal association cortex of the monkey: Command functions for operations within extrapersonal space. *Journal of Neurophysiology*, **38**, 871–908.

Mountcastle, V.B. and Power, T.P.S. (1959) Position sense and Kinesthesis. *Bulletin of the Johns Hopkins Hospital*, **105**, 173–200.

Mountcastle, V.B. and Powell, T.P.S. (1959) Neural mechanisms subserving cutaneous sensibility. *Bulletin of the Johns Hopkins Hospital*, **108**, 201–32.

Nathan, P.W. (1976) The gate control theory of pain: a critical review. *Brain*, **99**, 123.

Nathan, P.W. (1983) Pain and the sympathetic nervous system. In *Current Topics in Pain Research and Therapy*. (eds T. Yokota, R. Dubner), Excerpta Medica, Amsterdam, pp. 241–7.

Noga, B.R., Bras, H. and Jankowska, E. (1992) Transmission from group II muscle afferents is depressed by stimulation of locus coeruleus/subcoeruleus, Kolliker-Fuse and raphe nuclei in the cat. *Experimental Brain Research*, **88**, 502–516.

O'Keefe, J. and Recce., M.L. (1992) Temporal firing patterns of hippocampal neurons and spatial memory. *Neuroscience Letters*, **suppl. 42**, 545.

Ommaya, A.K. and Gennarelli, GT.A. (1974) Cerebral concussion and transmotor unconsciousness. *Brain*, **97**, 633–654.

Penfield, W. (1958) *The Excitable Cortex in Conscious Man. The Sherrington Lectures*. The University Press, Liverpool.

Perrett, D.I., Hietanen, J.K., Oram, M.W. and Benson, P.J. (1992) Organisation and functions of cells responsive to faces in the temporal cortex. *Philosophical Transactions of the Royal Society of London* B, **335**, 23–30.

Plum, F. and Posner, J.B. (1980) *Stupor and Coma*, 3rd Edn. Davis, Philadelphia.

Porter, K. and Tucker, J. (1981) *The Ground Substance of the Living Cell*. Scientific American.

Purdon-Martin, J. (1967) *The Basal Ganglia and Posture Publ.* Pitman Medical Publ. Co., London.

Ranson, S.W.R. and Clark, S.L. (1959) *The Anatomy of the Nervous System*, 10th edn. W.B. Saunders Co., Philadelphia.

Roberts, T.D.M. (1978) *Neurophysiology of Postural Mechanisms*, 2nd edn. Butterworths, London.

Ruch, T. and Patton, H.D., eds. (1979) *Physiology and Biophysics: the Brain and Neural Function*. Saunders, Philadelphia.

Rushton, W.A.H. (1951) A theory of the effects of fibre size in medullated nerve. *Journal of Physiology*, **115**, 101–122.

Schwartzkroin, P.A. and Wheal, H.V. (1984) *Electrophysiology of Epilepsy*. Academic Press, London.

Sherrington, C.S. (1906) *The Integrative Action of the Nervous System*. Yale University Press, New Haven.

Sherrington, C.S. (1929) Some functional problems attaching to convergence. The Ferrier Lecture. *Proceedings of the Royal Society, B*, **105**, 332–62.

Shinohara, M. (1977) The (14C) deoxyglucose method for the measurement of local cerebral glucose utilisation: theory, procedure, and normal values in the conscious and anaesthetised albino rat. *Journal of Neurochemistry*, **28**, 897–916.

Sholl, D.A. (1956) *The Organisation of the Cerebral Cortex*. Methuen, London.

Shute, C.C.D. and Lewis, P.R. (1967). The ascending cholinergic reticular system: neocortical, olfactory and subcortical projections. *Brain*, **90**, 497–522.

Snider, R.S. (1972) The cerebellum. In *Scientific Foundations of Neurology* (eds M. Critchley, J.L. O'Leary, and B. Jennett) Heineman, London, pp. 71–4.

Sokoloff, L., Reivich, M., Kennedy, C. *et al.* (1977) The (^{14}C) deoxyglucose method for the measurement of local cerebral glucose utilisation: theory, procedure, and normal values in the conscious and anaesthetised albino rat. *Journal of Neurochemistry*, **28**, 897–916.

Springer, S.P. and Deutsch, G. (1985) *Left Brain, Right Brain*. W.H. Freeman and Co., New York.

Stein, J. (1978) Long-loop motor control in monkeys. *Progress in Clinical Neurophysiology*, **4**, 107–22.

Szenthagothai, J. (1978) The neuron network of the cerebral cortex: a functional interpretation. The Ferrier Lecture. *Proceedings of the Royal Society of London*, B, **201**, 219–48.

Travis, A.M. and Woolsey, C.N. (1965) Motor performance of monkeys after bilateral partial and total decortication. *American Journal of Physiology and Medicine*, **35**, 273–80.

Tusa, R.J., Zee, D.S. and Herdman, S.J. (1986) Effect of unilateral cerebral cortical lesions on ocular motor behaviour in monkeys: saccades and quick phases. *Journal of Neurophysiology*, **56**, 1590–1625.

Uchizono, K. (1967) Synaptic organisation of the Purkinje cells in the cerebellum of the cat. *Experimental Brain Research*, **4**, 97–113.

Vallbo, A.B. (1986) Proprioceptive activity from human finger muscles. In *Feedback and Motor Control in Invertebrates and Vertebrates* (eds W.J.P. Barnes and M.H. Gladden) Croom Helm, London, pp. 411–30.

Voogd, J.V. (1964) *The Cerebellum of the Cat. Structure and Fibre Connections*. Royal Van Gorcum, Netherlands.

Walberg, F. (1956) Descending connections to the inferior olive. An experimental study in the cat. *Journal of Comparative Neurology*, **104**, 77–173.

Wall, P.D. and McMahon, S.B. (1986) The relationship of perceived pain to afferent nerve impulses. *Trends in Neuroscience*, **96**, 254–5.

Walton, J. (985) *Brain's Diseases of the Nervous System*. 9th edn. Oxford University Press, Oxford.
Watkins, E.S. (1972) The basal ganglia. In *Scientific Foundations of Neurology*, (eds. M. Critchley, J.L. O'Leary, and B. Jennett). Heineman, London, pp. 75–82.
Weinberger, D.R. (1987) Implications of normal brain development for the pathogenesis of schizophrenia. *Archives of General Psychiatry*, **44**, 660–69.
Zimmerman, M. and Koschorke, G. M. (1987) Chemosensitivity of nerve sprouts in experimental neuroma. In *Fine Afferent Nerve Fibres and Pain*, (eds R.F. Schmidt, H.G. Schaible and C. Vahle-Hinz) VCH publishers, Weinheim, Germany, pp. 107–113.

Index

Page numbers given in italic refer to tables

α activation 122
Abdominal guarding 143
Abdominal and pelvic viscera 336
Abdominal reflexes 107
Abducens nerve 158, 362
Abduction 225
Aborting a Jacksonian seizure 250
Absence of pinocytotic vesicles 347
Accessory nerve 344
Accommodation 21, 58, 166
  for near vision 344
  in nerve 21
  reflex (pupil) 166
Acetylcholine 56, 335, 437
  and the cortex 221
  esterase 58
  evidence for acetylcholine as the transmitter at mammalian nerve–muscle junction 57
Action potentials 3, 10, 20
  all-or-none 3
  anti-dromically conducted 79
  climbing fibres, mossy fibres and Purkinje cells 235
  frequency of in joint movement 116
  membrane (voltage-gated channels) 68
  modification in presynaptic inhibition 83
  monophasic or diphasic 28
  in myelinated nerve 33
  originating in alpha motoneurones in the cord 66
  in Renshaw cells 78
Activation of gamma motoneurones 123
Activity in spinal neurones during nociception 330
Acuity of vision 159, 437
A δ group 52
Adaptation 4
  due to the capsule of Pacinian corpuscle 45
Adduction of the thumb, a characteristic of primates 225
Adenosine triphosphate (ATP), purine 340
Adiadochokinesis 437
Adrenal medulla 336
Adrenaline 328, 437
Adrenergic junctions 60
Adrenergic receptors 60
Afferent nerve fibres 3, 4, 39
Afferent projections 221
Agnosia 437
Agonist 336
Agonists (muscles) 12
Agranular cortex 249
Agraphia 437
Akinesia 224
Alexia 437
Algogen 322
Alimentary tract 335
Alkaloid 326
All-or-none, action potential 3, 26, 83
Allocortex 438
Allosteric relationship 307
Alzheimer's disease 211
  and the cytoskeleton 55

Amacrine cells 165, 438
  passive conduction 165
Amino acids 211, 304
Ammon's horn 438
Amoeboid motion, lack of 347
Ampulla 173
Amputation 318, 329
Amygdala 438
Amygdaloid body 363
Amygdaloid nucleus 219, 245, 255
Anaemic decerebration 241
Anaesthesia, below lesion in spinal transection 101
Anaesthetic 308
Anaesthetists 58
Analogue nature of receptor potential 41
Anatomical relationship of receptors in skeletal muscle to the muscle fibres 131
Anatomy of autonomic nervous system 334
Anatomy of the retina 159
Anatomy of the cranial parasympathetic outflow 344
Angina of effort (angina pectoris) 320
Angular gyrus (visual speech area) 288, 291, 364
Annulo-spiral (primary) sensory endings 118
Anopsia 438
Antagonist muscles 12, 147
Anterior 438
Anterior commissure 296
Anterior cortiscospinal tracts 112
Anterior crural muscles 180, 181
Anterior frontal 246
Anterior horn cell 9
Anterior pituitary 344
Anterior root 360
Anti-cholinesterase 58, 59
Antidromic 438
Anticipatory behaviour 343
Antidromic impulse 79
Antidromic inhibitory pathway 79
Antigravity muscles 206, 226
Antisera to VIP 340
Anxiety 286
Aortic arch 7, 342
Aortic bodies 7
Aphasia 288, 438
Apical dendrite 247
Apparent inertia 182
Appendicitis, pain of 321
Apposition, control of 252
Appreciation of music 292
Apraxia 287, 438
Aqueduct of Sylvius (midbrain) 188, 359
Arachidonic acid 314, 315
Arachnoid 438
Arachnoid granulation 359
Arachnoid mater 359
Arachnoid villi 359
Archi 438
Archicerebellum 438
  see also Paleocerebellum
Archicortex 438
Areflexia 103
Arginine 314
Argyll Robertson pupil 167
Arterial baroreceptors 342
Arterial blood pressure in spinal shock 103
Arterioles 336
  dilatation of 338
Artery 7
Asparate 306
Articular surfaces 318
Articulation 288
Ascending reticular formation 363
Aseptic precautions 104
Asphyxia 323
Aspirin 322
Association areas 285
  myelination of 247
Association cortex 1, 221, 243, 246
  hemispheric asymmetry 285
  parietal 246
  prefrontal area 285
  temporal 286
Astereognosis 287, 438
Astrocyte 438
Asynergy 438
Ataxia 124, 438
Athetosis 223, 439
Atlanto-occipital joint 186
Atria of the heart 7
Atropine 336
Atrophy of muscles in lower motoneurone lesion 108
Auditory area of cerebral cortex 244, 272

Auditory evoked potentials 259
Auditory cortex 272
Auditory nerve 362
Auditory speech area 290
Auditory ossicles 168
  control of vibrational energy reaching inner ear 169
  transmission of movement 168
Auerbach's plexus 335, 340
  parasympathetic relays in 340
Aura, epileptic 250
Autogenetic inhibition 153
Automatically 333
Automatic bladder 104
Autonomic 439
Autonomic and associated effects 103
Autonomic motor nuclei 334
Autonomic nerve fibres 387
Autonomic nervous system 333, 439
  central autonomic control 341
  compared with somatic nervous system 334
  feedback loops 343
  functions of 337
  history of discovery of 339
  modern modifications of classical view 340
  muscarinic and nicotinic receptors 336
  outline of anatomy 334
  peripheral 333
  physiological effects of activity in 337
  self-test 406
  sympathetic and parasympathetic systems opposing or working together 337
Autonomic preganglionic nerve fibres 334
Autonomic reflexes 105
Autoradiography 303, 439
Autoregulation, chemical and mechanical 349
Avulsion of the nerves 329
Axial musculature 275
Axolemma 439
Axon 439
Axon diameters, not uniformly distributed 29
Axon hillock 53, 439
Axon reflex 147
Axonal composition of cutaneous nerve 48
Axonal transport 54
Axons
  cut, living ends sprout 96
  primary afferent 135
Axoplasm 439
Axoplasmic current 24
Axoplasmic transport systems 54

Babies 107
Babinski responses 106, 107, 246
'Bag' fibres 132
Balance 39
  and the cerebellum 239
  self-test 399
Balance receptors 180
Balance reflexes 239
  neck reflexes 185
  standing and walking 179
  synthesis of vestibular and neck reflexes 189
  vestibular reflexes 181
Ballismus 223
Ballistic movements 218
Barbiturates 308
Baroreceptors 7, 343, 439
  arterial, discharge of 342
  in the right atrium 342
Basal dendrites 247
Basal ganglia 211, 218, 219, 245, 363
  and cerebellum 217
  connections of 230
  disorders of as positive signs of central nervous deficit 228
  functions of 223
  hemiballismus 228
  Parkinson's disease 223
Basilar membrane 170
Basis pedunculi 362, 366
BBB, *see* Blood–brain barrier
Bees 4
Behavioural disorders, and the limbic system 257
Benzodiazepine drugs 307
  receptor complex 308
Betz cells 273, 275, 439
  and activity in Broca's area 290
  ending of axons from 274
B fibres 335
Biliary colic 320
Binocular vision 253

Biological rhythms, depend on hypothalamus 343
Bipolar cells 165
Birds 4
Bladder 105, 335
  automatic bladder 104
  retention with overflow 104
Bladder reflex 103
Bleaching of visual pigment 166
Blind sight 267
Blind spot 164
Blocking agents, neuromuscular 58
Blood–brain barrier 228, 345
  and cerebrospinal fluid (CSF) 345
Blood, survival of stored blood 17
Blood flow, cerebral 348
Blood pressure
  after spinal transection 103
  in sleep 212, 213
Blood platelets 322
Blood supply, autoregulation 349
Blood vessels 335
  innervation of 336
Body map 251
Body senses, general, *109*
Boiler-maker's deafness 171
Bony labyrinth 169
Bradykinesia 439
Bradykinin 322, 328
Brain
  blood vessels in 346
  constant oxygen supply needed 350
  and its coverings 357
  embryological development of 360
  inertia of 200
  liability to inertial damage 200
  operations under local anaesthetic, reasons behind 252
  recovery of function in neural tissue 99
  see also Brain damage; Brain stem
Brain cell death 247
Brain damage 198
  brain-stem function 202
  irreversible 200
  mechanical considerations 200
  no brain stem function 200
  no cortical function 199
  space-occupying lesions 202
Brain stem 197, 357, 439
  consciousness, an alerting system for 204
  development of 246
  evoked potentials 205
  function 202
  housing respiratory centres 202
  motor nuclei 275
  reflexes 203
  rise in intracranial pressure 202
  self-test 396
Brain stem mechanisms 197
  and reflexes *204*
Brain stem reflexes 204
  corneal reflex 205
  facial nerve reflex 205
  gag reflex 205
  pupillary reaction to light 203
  vestibulo–ocular reflexes 197
Brain-stem reticular formation 209
  chemical transmission and the reticular formation 211
  consciousness and sleep 212
  sleep 212
  sleep centres 213
Brain tissue 6
Breathing 333
Broca's area (motor speech area) 288, 289
Broca's dysphasia 290
Bronchi 336
Brown–Sequard syndrome, *see* Spinal hemisection
Bulb 439
Bulk solution 15
Bursting of cells when sodium pump is poisoned 36

Calcium 26
  and excitotoxicity 315
Calcium and NMDA-operated channels 311
Calcium carbonate 175
Calcium channels 57
Calcium ions 62, 302
  effect of on nerve membranes in hypocalcaemic tetany 26
  released into axoplasm 62
Calf muscle (gastrocnemius) 10
Caloric testing 204
CAMP 302
Capillaries, with fenestrations 346
Capillary wall, leaky 346
Carbon dioxide 337
Carcinoma, intractable pain of 329

Cardiac muscle, fight or flight reaction 337
Cardiac plexus, and sympathetic system 336
Cardiovascular centres 103, 342
Carnosine 313
Carotid sheath 335
Carotid sinus 7, 342
Casualties in World War I 266
Cat(s) 246, 274
Catching a ball 274
Catecholamines 439
  and the cortex 214
Categorical hemisphere 293
Cauda equina 360, 439
Caudate nucleus 219, 363, 439
Causalgia 328
Causes of brain stem damage 200
Cell body 1, 13
  soma 1
Cell membrane 15, 35
Cells of origin of the spinothalamic tract 137
Cell volume, and the sodium pump 35
Central afferent pathways 108, 365
  pain and temperature 110
  position and vibration 110
  touch and pressure 110
Central autonomic control 341
  hypothalamus 342
  medullary and pontine centres 342
Central conduction time 260
Central delay (conduction through the cord) 142
Central efferent pathways 111, 365
Central excitatory state 129
Central fissure, cerebral hemispheres 243
Central grey matter of the brain stem 210
Centra gyrus 364
Central motor pathways, excitation of 261
Central nervous lesions: release of lower centres 107
Central nervous nuclei, spontaneous electrical activity 207
Central nervous structures, classification 357
Central nervous system 1, 13, 439
  decussation in 253
  lesions of; positive and negative signs of 107
  muscular contraction by coactivation of α – and γ – motoneurones 242
  switching off proprioceptive reflex 219
  temperature and conduction 35
Central neurones, inability to regenerate 98
Central pathways for nociception (injury messages) 323
Central reflex time 142
Central sulcus 364
Central summation of input 324
Centre of gravity 179, 191
Cephalo–ocular reflexes 203
Cerebellar cortex 77
  blocking the activity of
  electrical stimulation and decerebrate rigidity 241
  structure of 234
Cerebellar lesions and muscle tonus 240
Cerebellar nuclei 77, 229
Cerebellar peduncles 229
Cerebellar roof nuclei 233, 240
Cerebellum 218, 228, 238, 439
  an anti-hunting device 238
  control of ballistic movements 218
  and decerebrate rigidity 241
  effects of deficiency of the flocculo-nodular lobe 239
  effects of on tonus in skeletal musculature 239
  evolution 230
  functions and connections 229, 230, 237
  'Head ganglion of the proprioceptive system' 229
  integrative role 230
  localization 230
  possible mode of action 236, 282
  principal connections 230
  proprioceptive information relayed to structure of the cerebellar cortex 234
  supraspinal control of skeletal muscle 242
Cerebral arterial hypoxia 350
Cerebral arterioles 349
Cerebral arteriosclerosis 247

Cerebral blood flow 348
  autoregulation of blood supply 348
  cerebral metabolism 350
  mean 352
  and metabolism, measurement of: brain imaging 352
  regional 354
Cerebral capillaries 98
Cerebral cortex 75, 211, 221, 257
  association areas of 1
  cool the 241
  embryological development of 246
  evolutionary considerations 244
  and lack of oxygen 351
  limbic lobe 254
  motor functions of 273
  and movement 282
  parallel processing of visual information 267
  plasticity in 100
  reflex and reaction time 273
  self-test 400
  volitional control
Cerebral environment 345
Cerebral environment, constancy of 345
  blood–brain barrier and cerebrospinal fluid 345
  glia 348
Cerebral hemispheres: association areas 285
  hemispheric asymmetries 287
  memory 285
  speech 288
  split brain 293
Cerebral hemispheres 243, 363
  cerebral cortex: evolutionary considerations 244
  connections 230, 249
  cortical lamination 248
  cortical regions, gross histological differences between 249
  general features 243
  human cerebral cortex, embyrological development 246
  microscopic structure 247
  modular organization 249
  motor functions 273
  structure of 243
  subdivisions of the lobes of the 364
Cerebral ischaemia
  pharmacological protection against 356
Cerebral metabolism 350
  measurement of 354
Cerebral oedema 99
Cerebral perfusion pressure 349
Cerebrospinal fluid (CSF) 200, 359
Cerebrum 439
Cervical ganglia 335
Cervical roots 359
Cervical sympathetic trunk 335
C fibres 52, 323
C-fos 330
Chain fibre 132
  see also Nuclear chain fibres
Channel binding site for phencyclidines 313
Characteristics of reflexes 143
Characteristics of the neurones in the inferotemporal region of the cortex 270
Characteristics of the neurones in the posterior parietal area of the cortex 271
Charcot joint 318
Chemical composition 7
Chemical gradient 14
  see also Concentration gradient
Chemical neurotransmitters in the spinal cord 90
Chemical receptors 7
Chemical transmission and the reticular formation 211
Chemical transmitters 39
  and Parkinsonian rigidity 227
  released by neurones 60
  in the spinal cord 90
Chemicals
  endogenous, stimulating nociceptive nerve endings 322
  as modulators 340
Chemoreceptors 7, 343
Chloride 14, 15, 35, 66
  conductance 19
  equilibrium potential 19
  permeability 19
  synaptic inhibition 67
Chlorpromazine 122
Choline, reabsorbed by presynaptic terminal 62
Choline acetylase 59
Cholinergic 334
Cholinesterase 58, 303
Chorda tympani 344

Chordotomy 439
Chorea 440
Choreic movements 223
Choreiform movements 228
Choroid 440
Choroid plexus 359
Chromatolysis 95, 114, 440
Circulatory collapse 339
Circumlocutions 290
Ciliary body 166
Ciliary ganglion 344
Ciliary muscle 335, 344
Ciliary nerve 344
Cingulum 440
Clashing of nerve impulses 128
Clasp-knife reflex 117
Classification of nerve fibres 31
  of receptors 7
Claudication, intermittent 320
Claustrum 219, 440
Climbing fibres 235
Clitoris 335
Clonus 105
Closing of ion channels 166
Closing of membrane channels as a synpatic mechanism 70
Clumsiness 124
Cocaine 323
Co-ordination disorders, *see* Cerebellum
Cochlea 168
Coeliac plexus, and sympathetic system 336
Cog-wheel rigidity 226
Cold receptor 8, 39, 40, 52
Cold water 204
Colic 320
Collateral 75
Collateral circulation, opening of 99
Colliculus 272, 362, 440
  inferior 272, 362
  superior 362
Colour blindness 164
Colour triangle 163
Colour vision 161
Columns of cortical cells
  in the primary somatosensory areas 249
  in the primary visual cortex 269
Coma and stupor 199, 212
Complex cells 269
Commissure 440
Communication 13
Comparison of inhibitory neurotransmitters in the cord and in higher centres 306
Comparison of the anatomical layout of the somatic and autonomic nervous system 334
Competition of reflexes 147
Compound action potential 27, 28
Concentration gradient 14
  and diffusion 13
Conditioned reflexes, memory and learning 296
Conductance, membrane 19
Conduction of nerve impulse
  continuous 34
  saltatory 33
  and transmission, comparison of, *43*
Conduction velocity in nerve 34
Cones and colour vision 161
Conflicting information from the sense organs 194
Confusion 200
Confusional state 290
Congenitally absent 293
Coning 202, 440
Conjugate deviation, of the eyes 204
Connections of the basal ganglia 220
Conscious perception of proprioceptive activity 129
Consciousness 102, 197
  brain stem as the alerting system for 197
  and the reticular formation 210
  and surgery 210
Consensual light reflex 203
Consensual response 167
Constrictor pupillae muscle 335, 344
Continuous conduction 34
Contractions 250
Contralateral 440
Contrast sensitivity 161
Contribution of the spinal cord to the organization of complex movements 152
Control due to descending influences from higher centres: opioid peptides 326
Control due to input from the periphery 324
Control of eye movements 263

Covergence 145
and bipolar cells 165
and divergence 145
prejudices acuity 159
Convoluted 246
Coordination 218, 229
Core temperature 35
Corona 440
Cornea
sensation in 6
Corneal reflex 205
Coronary thrombosis 320
Corpora quadrigemina 267, 362
Corpus callosum 243, 293, 440
Corpus luteum 344
Corpus striatum 219, 440
Cortex 440
cerebellar 228
distribution of sensory and motor functions 243
generalized activation of 197
precentral 273
Cortical activity, normal 197
Cortical function, evidence for localization of 250
Cortical layers (lamination) 248
Corticofugal modulation 141
Corticospinal projection 112, 273
Cortical sensory areas, primary and secondary 243
Corticospinal tracts 106, 246, 274
lateral and anterior 108
myelination of 246
slowly conducting 273
Corticotopic projection 222
Consciousness 197
Co-transmission 90, 340
Convection, heat loss 338
Convergence along the visual pathway 161
Convulsive motions 286
Cortical and primary afferent modulation of cutaneous sensory information in the cord 135
Corticospinal projection 137
Counter irritation 324
Cranial nerves 9
Cranial parasympathetic nerves, structures innervated 335
Cremaster reflex 107
Crista 172
Cristae 173
Crossed extensor reflex 147
Crushed 97
CSF, *see* Cerebrospinal fluid (CSF)
Cuneate nucleus 138, 365
Cupula 173
Curare, long-term relaxant 58, 59
Cutaneous mechanical receptors 138
Cutaneous nerve, composition of 48
loss of sensation when cut 93
Cutaneous sensation 48
Cutaneous pressure receptors 7, 181
Cutaneous receptors 48
Cytochrome oxidase 269
Cytoplasmic protein synthesis 95
Cytoskeleton 55, 440
Cytosolic enzymes 313

Dale, Sir Henry 92
Dale's principle 90, 92
Damping of oscillation of the otolith system 176
Dark adaptation 6, 164
Decerebrate attacks 207
Decerebrate rigidity 205, 206, 226
alpha of gamma 207
and the cerebellum 241
mechanism of 206
mesencephalic fits 207
Decussation 110, 253, 440
in the central nervous system 253
double 232
in the medulla 366
of the optic nerve 166
of the pyramids 362
Defecation 33, 337
Definition of reflex response time 155
Deformation receptors 8
Degeneration of the substantia nigra in Parkinson's disease 223
Delirium 200
Dendrites 1, 440
apical 247
basal 247
pyramidal cells 247
stellate cells 247
Dentate 440
Deoxyglucose and the mapping of metabolic activity 355
Depolarization
and excitatory postsynaptic potential 63
fast and slow 70

Depolarization block 58
Depolarizing electrical stimulus 20
Dermal papillae 44
Dermatomes, *see* Spinal segments
Descending colon 335
Descending pathways 111
Determinants of the membrane potential 36
Diaphragmatic 101
Diaschisis, *see* Neural shock
Diencephalon 357, 363, 440
Different spectral sensitivities of rods and cones 163
Different types of spindle receptor 130
Differential nerve block 122
Difficulty with saying short words 290
Diffuse flair, *see* Triple response
Diffusion 13, 15
  of oxygen 9
Digestion 337
Digital nerves 129
Dilatation of arterioles 338
Dipeptides 313
Diphasic action potential 29
Diplopia 441
Direct and indirect pathways, differentiation between 120
Direct electrical stimulation of the precentral cortex 250
Direct light reflex 167, 203
Discrimination, two-point 48, 110
Disdiadochokinesis 238
Disorders of the basal ganglia as release phenomena 228
Disoriented 213
Displacement of the nucleus of an axotomized cell 114
Dissociated sensory loss 112
Distension of viscus, producing pain 320
Disturbances of body image 292
Disynaptic reflex 76
Divergence 145
Diversity of receptors 7
Dizziness, a cause of 194
Doll's head phenomenon 203
Domestic power supply 23
Dominant hemisphere 288
Dopa (L-Dopa) 228
  does not cross blood–brain barrier 228
  as a synaptic transmitter 227
Dopamine 228, 441
Dorsal 441
Dorsal columns 108
  nuclei 138, 365
Dorsal horn neurones 326
Dorsal root ganglion 1, 360
Dorsal roots 124
  section of and limb flaccidity 120
Drive 133
Drowsiness 210
Drugs, and the blood–brain barrier 348
Drunkenness and incoordination 274
Dura 441
Dura mater 357
Duration of shock and recovery from shock 104
Dye 13
Dynamic bag fibres 132
Dynamic component 116
Dynamic polarization 53
Dynorphins 326
Dysarthria 288
Dysdiadochokinesis 238
Dyskinesia 441
Dysmetria 238, 441
Dysphasia 288
Dyssynergia 238

Ear 163
  the cochlea 170
  inner 169
  lobe, skin of the 39
  middle 168
  outer 168
  receptors in the vestibules and semicircular canals 172
  semicircular canals 170
  the utricle and saccule – balance 174
Ear-drum 168
  vibration of 169
Edinger–Westphal nucleus 203
  see also Autonomic motor nucleus
EEG (electroencephalogram) 199
Effect of transection of the midbrain on movements in limbs 208
Efferent fibre 9
Effects of fusimotor and skeletomotor activity on spindle stretch receptors 122
Effects of membrane potential on NMDA-operated channels 309

Efferent projections 222
Elbow joint 130
Elbow joint, conscious perception of proprioceptive activity 129
Electrical energy 13
Electrical forces 13
Electrical neutrality, principle of 15
Electrical voltage, development across selectively permeable membranes 13
Electric eels 4
Elective operation on the brain 251
Electrochemical gradient 18, 19
  potassium 19
  sodium 19
Electroencephalogram 199, 212, 441
Electromyogram (EMG) 126
Electromyographic activity during voluntary movement 125
Electronic (passive) conduction 39
Electrophysiological evidence for the pathway 78
Emboliform nucleus 441
Embryo 246
EMG, *see* Electromyographic activity during voluntary movement
Emotion and the limbic system 256
Encapsulated endings 44, 45
Encapsulated receptors 44
Encephalins 326
Ending, unmyelinated 39
Endocrine system, positive feedback loop 343
Endogenous
  chemicals 322
  excitotoxins 315
endolymph 170, 203
Endoneurial tubes 95, 96, 97
Endoneurium 441
Endoplasmic reticulum 95
End-plate potential 57
End-plates 55
Endorphins 326, 441
Endothelial cells, cerebral 346, 347
  lack of contractility 347
Engram, *see* Memory image, formation of
Enhancement of inhibitory GABAergic transmission by benzodiazepines 308
Entorrhinal cortex 441
Enzymes 340
Ependyma 441
Epicritic 441
Epicritic sensibility 96
Epidermis 322
Epilepsy 293, 316
  and inhibition disturbance 75
Epileptic cerebral dysrhythmias 351
Epileptic fits 250
Epineurium 441
Equilibrium potential 17, 18
  for excitatory postsynaptic potential 65
  for inhibitory postsynaptic potential 67
Erectile tissue 335, 336
Erection and ejaculation 338
Erlanger's classification of nerve fibres 31
Eustachian tube 169
Evoked potentials 259
  brain stem 205
Excitable cortex in conscious man 251
Excitatory and inhibitory synapses 62
Excitatory postsynaptic potential (EPSP) 63, 82
Excitatory reticular formation 210
Excitatory synapses 65
Excitotoxins 315
Exogenous 315
Experimental neurology 98
Exploratory behaviour 343
Explore the environment 274
Exposed end of the tendon 129
Extensor muscles 206
Extensor thrust (Babinski sign) 106
Exteroceptor 441
External auditory meatus 168, 204
External genitalia 336
External sex organs 336
Extracellular currents 28
Extracelleular fluid 345
Extracellular space 55
Extrafusal muscle fibre 115, 118
Extrapyramidal pathways 217
  rapidly conducting nerve fibres 273
Extrapyramidal system 441
Extrinisic ocular muscles 11, 158
Extrinsic influences on the dorsal column nuclei 140
Eye 6, 157, 335
  acuity of vision 159
  Argyll Robertson pupil 167

blind sight 267
the blind spot 164
colour blindness 164
colour vision 161
dilator pupillae muscle 335
movements; control of 263
movements; vision and 263
pupillary reactions to light 203
retina 159
rods and cones, different spectral sensitivities of 163
the visual pathway 166
Eyeball 8, 157
smooth muscle in 165
Eyes, conjugate deviation of 203

Facial nerve 344, 362
Facial nerve nucleus 205
Facial nerve reflex 205
Facilitation 146
Falx cerebri 200
Familial 317
Fascicles 96
Fasciculation 58
Fascicles of nerve fibres 96
Fast and slow information from receptors 51
Fast and slow muscle fibres 149
Fastigial 441
Fatigue in muscle 149
Features of junctional transmission 67
Feedback, negative and positive 343
Feedback inhibition 78
Feedback loops 343
Ferry boats 54
Fick principle 352
Fight or flight reaction 337
Fimbria 441
Final common path 12, 217, 280
Fine tactile discrimination in different areas of the body 47
Finger and toe joints, conscious perception of 129
proprioception 115
First-order nerve fibres 110
Fissure of Rolando, *see* Central fissure, cerebral hemispheres
Fits, mesencephalic 207
Five-hydroxytryptamine 322
Flaccid paralysis 103
Flaccidity 103
Flexion and extension movements of the neck 186
Flexor reflexes 106
Flexor withdrawal reflex 106
Flexure 335
Flocculo–nodular lobe 229
connections with vestibular system 229
effects of deficiency of 239
Flower spray (secondary) sensory endings 118
Fluid movement, semicircular canals 173
Folium (folia) 229
Food and energy balance 342
Foramen magnum 361, 442
Foramen of Monro 255
Foramen ovale, *see* Oval window
Foramina of Lushka and Magendie 188
Forebrain 357
Formation of urine 337
Fornix 255, 442
Fourth ventricle 188, 359, 362
Fovea 159, 442
Frequency filtering
ear 171
eye 162
Fritsch, G. and Hitzig, E. (neurologists) 250
Frog, spinal shock in 104
Frog nerve
compound action potential 28
Frontal eye field 263
Frontal lobe 243, 364
Frontal lobotomy 286
Funnel 12
'Funny bone' 6
Fusimotor fibres 118

γ activation 120, 122
γ-amino-butyric acid (GABA)
in presynaptic inhibition in the cord 90
GABA 304
binding sites 308
postsynaptic receptor for 307
Gag reflex 205
Gain of the receptors 124
Gait, ungainly 238
Gall-bladder 321
as principal inhibitory neurotransmitter in higher brain centres 304

Gamma rigidity 207
Ganglion 334, 442
Ganglion cells in retina 165
Ganglionic receptors, nicotinic 336
Gastrocnemius-soleus muscle group 10, 118, 127, 180
Gate theory of pain 331
General body senses 39, 108
Generator potential, *see* Receptor potential
Geniculo–cortical afferent nerve axons 264
Giant axon, resting potential 19
Glabrous skin 51
Glasgow coma scale 200, 214
Glia, glial cells 348, 442
 stabilizing effect of 348
Globus pallidus 219, 442
Glossary 437–49
Glossopharyngeal nerve 344, 362
Glucose 337
 aiding survival of stored blood 17
L-glutamate 315
 as excitatory neurotransmitter 303
 toxicity 316
Glutamic acid decarboxylase (GAD) 304
Glycine 305
 binding site 313
 as inhibitory neurotransmitter 305
Glycogen 351
Goats 59
G proteins 302, 313
Gracile and cuneate nuclei 110, 365
Gracile nucleus 110
Gracilis 442
Graded increase of muscular effort 10
Granular cortex 249
Granular layer 235
Granule cells 235, 442
Greater superficial petrosal nerve 344
Grey 335
Grey matter 9
 central 210
 cerebellar cortex 228
Grimacing reflex, *see* Facial nerve reflex
Groom 274
Group A S fibres 323
Group Ia afferent nerve fibres 75
Guarding, abdominal 143
Gut 337
Gyri 245
 cingulate 255
 hippocampal 255
 of neocortex 364
 and sulci 245

H response 127
Haematoma 351
Haemorrhgae, haemorrhagic shock 339
Hair cells
 cochlear 172
 orientation of in maculae 175
 vestibules and semicircular canals 172
Half-centres 153
Head flexing, involvement of neck and vestibular reflexes 190
'Head ganglion of the proprioceptive system' 29
Hearing (inferior colliculi) 267
Hearing receptors 39
Heart 336, 337, 344
 innervation of 335
Heart rate 324
 in fight or flight reaction 337
 in sleep 213
Heat 337
 loss of 338
Hebbian synpases 298
Hemiballismus 228, 442
Hemiplegia 442
Hemiplegic cerebral palsy 275
Hemispheric asymmetry 287
 angular gyrus (visual speech area) 288
 Broca's area (motor speech area) 288
 the right hemisphere 292
 speech 288
 speech areas and age 288
 Wernicke's area (auditory speech area) 288, 290
Hennemen size principle 125
Hereditary material of neurones 1
High-threshold mechanical 139
Hindbrain
 and coning 202
 and gag reflex 205
Hippocampal commissure 296
Hippocampal neurones signal spatial information 258

Hippocampus 245, 255, 442
and memory 257
result of destruction 257
Histamine 322, 328, 347
Histochemical 59
Histological changes in the cell body 95
'Hitting the funny bone' 93
Hitzig, E. 250
Hodgkin and Huxley 19
Hoffman 127
Homeostasis 442
Homeostatic mechanisms
negative and positive feedback 343
role of hypothalamus 342
Homologous motoneurone pools 276
Homo-typical cortex (association cortex) 249
Homunculus 252
Hopping reflex 179
Horizontal cells 165
Hormones, as transmitters 60
Horseradish peroxidase 55
Hot 39, 51
Hunting (oscillation) 238
Hydrocephalus 442
neck reflexes in 188
Hydrogen ion concentration 7
buffering of 350
Hydrostatic pressure of blood 7
Hydrostatic pressure difference 35
Hyper-reflexia 105
in lesions of cerebellar cortex 240
Hyperalgesia 321, 322
primary and secondary 331
Hypercomplex cells 269
Hyperexcitability, hypocalcaemic 26
Hyperpolarization 166
Hyper-reflexia 105
Hyperventilation and hypocalcaemic tetany 26
Hypocalcaemic tetany (hypocalcaemia) 26
Hypogastric plexus, and sympathetic system 336
Hypoglossal nerve 362
Hypoglycaemia 316, 351
Hyporeflexia 237
Hypothalamic centres 339
Hypothalamus 8, 221, 255, 341, 342, 363, 424
input to neostriatum 221
mammillary body and 255
osmoreceptors 343
Hypotonia 237, 240
Hypoxia 316
Hypoxic damage 98
Hypoxic neural tissue 316

Image on the retina 8
Immediate and long-term effects of neocerebellar insufficiency 239
Impairment of consciousness 210
Improvement by substitution of function 100
Incus 169
Identification of neurotransmitter chemicals 302
Indomethacin 322
Indusium 443
Indwelling catheter 104
Inertia 174
Infection 318
Inferior frontal gyrus 364
Inferior parietal lobule 364
Inferior temporal cortex 271
Inflammation 322
Infra-red 3, 4
Inhibition 67
role of 74
Inhibitory influences 12
Inhibitory pathways
orthodromic 75
recurrent 77
role of in spinal cord 75
Inhibitory postsynaptic potential (IPSP) 65
Inhibitory reticular formation 206, 207, 210
Inhibitory synapses 73
Initial unmyelinated segment of the axon 53
Initiation of nerve impulses 65
Initiation of voluntary movement 218
Initiation rites and rituals 318
Initial unmyelinated segment 53
Inner-ear receptors 169
Innervation
of blood vessels 336
reciprocal; of α and γ motoneurones, pathway not known 242
segmental 361
of skeletal muscle

Input 231, 232
Insula 443
Insulating 33
Integrates synaptic potentials 65
Integration system 13
Integration with the retina 164
Integument 443
Intellectual facilities 286
Intention tremor 238
Interaction of vestibular and neck reflexes 190
Internal capsule 342, 366
Internal environment 221
Internal intercostal muscles 101
Interneurones 75
  inhibitory 78
  Renshaw cells 78
Internodes 33
Intermittent claudication 320
Intermittent pain 320
Interoceptor 443
Interosseous membranes 7
Interphalangeal joints 129
Interstitial fluid 9
Interruption of peripheral nerve and of the spinal cord 94
Intervertebral foramina 357
Intestinal colic 320
Intracellular calcium ion concentration 313
Intracellular fluid 14
Intracellular messenger 302
Intracellular organelles 53, 114
Intracellular stores 313
Intracranial pressure 351
Intrafusal muscle fibre 118, 123
  sensory endings 117
Intraparietal sulcus 364
Intrinsic ocular muscles 158
Invertebrate giant axons 19
Invertebrate receptors 41
Involuntary contractions 250
Involuntary movement 223, 225, 281
  and lesions of the basal ganglia 223
  in Parkinson's disease 225
Ionic channels, opening of 64
Ionic mechanism
  of excitatory postsynaptic potential 64
  of inhibitory postsynaptic potential 67
  of presynaptic inhibition 82
Ionic mechanism of the receptor potential 43
Ipselateral 443
Iris 157
Ischaemia 319, 323
Ischaemic damage 320
Ischaemic muscle pain 319
Isocortex 443
Isometric contraction 122

Jackons, Hughlings 107, 250
Joint capsules 7, 115
  intervertebral joints 186
Joint position sense 129
Joint receptors, stimulation of 115
Joint and tendon receptors 115
Junctional transmission 55
  acetylcholine, transmitter at nerve–muscle junction 57
  features of 55
  neuromuscular transmission 55
  transmission at junctions other than end-plate 59

Kinaesthetic sense 108
Kidney 337
Kinesthesia 443
Kinocilium 172

Labelled line 48
Lack of functioning of the spinal cord 103
Lachrymal glands 335, 344
Lactic acid 337
Lamellae 45
Large intestine 344
Larynx 336
Last order interneurones 152
Lateral 112
Lateral fissure, cerebral hemispheres 243
Lateral geniculate body 166
Lateral horn, lateral horn cells 335
Lateral spinothalamic tract 329
Lateral ventricles 449
Lateral vestibular nucleus 206
  role of in decerebrate rigidity 206
Layers (laminae) of cerebral cortex 248
  layers V and VI 273
L-dopa 228, 348

Lead pipe rigidity 226
Learning 314
Lemniscus 443
Lens 157
Lentiform nucleus 219
Leptomeninges 443
Lesions of the central nervous system 98
Lesions, self-test 382
Lesser superficial petrosal nerve 344
Lentiform nucleus 443
Leucotomy 286
Levels of transection compatible with survival 100
Levodopa (L-dopa) 228
Ligaments 7, 443
  stretched 318
Ligand 308
Light
  colour determined by spectral composition 163
  pupillary reactions 167
Light energy 3
  transduction of 3
Limb muscles, as effectors 183
Limbic lobe 254
Limbic system 219, 245, 254,271, 285, 343
  components of 255
  functions of 256
  the hippocampus and memory 257
  and psychosis 257
Limen 443
Linear acceleration 191
Lissencephalic 244
Lloyd's classification of nerve fibres 31
Local anaesthetic 129
Local electrical stimulation 251
Locus coeruleus 152, 213, 443
  and sleep 213
Long-loop reflex 279
'Long-loop' or transcortical reflexes and voluntary movement 277
Long propositional neurones 137
Long-term memory 297
Long-term potentiation 297, 314
Loss of heat 338
Low back pain 319
Lower motoneurones 217
Lumbar puncture 360
Lungs 344
  innervation of 335, 336
Luteinizing hormone 344
Lymph 169

M and H responses 126
M1 and M2 reflexes 279
M (motor) response 126
MacEwen, Sir William, first elective brain operation 251
Macroglia 443
Macromolecular proteins 20
Macroscopic 443
Macula 172
  lutea (yellow spot) 159, 443
Maculae 172, 182
  orientation of hair cells 715
Magnesium 313
Magnetic field 4
Malapropisms 290
Malleus, handle of 168
Mammillary bodies 255 443
Map of the skin 97
Mapping 251
Massa intermedia 444
Mass reflex 105
Mechanical deformation, and receptors 3, 7
Mechanical sensitivity 96
Mechanism of adaptation 45
Mechanism of decrerebrate rigidity 206
Mechanisms in synaptic transmission 60
Mechanoreceptors 7, 49, 50, 115, 322
  muscle 326
  visceral 45
Medial lemniscus 110, 365
Medulla 103, 229, 341, 443
Medulla oblongata 110, 361
Medullary cardiovascular centres 103, 342
Medullary and pontine centres 341
Medullary pyramids 113, 366
Medulloblastoma 239
Meissner's corpuscles 44, 51
Membrane 15
Membrane channels, closing of as a synaptic mechanism 70
Membrane conductance 19
Membrane permeability 19

Membrane potential 13
  determinants of 15
  giant axon 19
  Pacinian corpuscle 41
  threshold 21
Membraneous labyrinth 169
Memory 257, 314
  and the hippocampus 257
Memory image, formation of 257
Meninges 357, 444
Meningitis, pin-point pupil 203
Mental prowess, myelination and neuronal death 247
Mesencephalic fits 207
Mesencephalon 357, 444
  roof of the 267
  see also Midbrain
Messenger RNA 330
Metabolic activity 53, 95
  and cell survival 17
  mapping of, use of deoxyglucose 355
Metabolic energy 35
Metabolic factory 53
Metabolism, products of 9
Metabotropic receptors 313
Microelectrode 19, 46
Microglia 444
Microneurography 46, 51, 324
Micturition 103, 333, 337
  indirect control of 105
  reflex 103
Midbrain 166, 206, 357, 362
  nuclei 166
  site of sleep centres 213
Migration of birds 4
Mind, reducing pain 318
Miniature end-plate potential 57
Mirror movements 275
Mitochondria 61
Mitosis 93
Mitral cells 177
Mixed peripheral nerve 93, 335
Mnemonic 444
Modular organization 249
Modulators 340
Molecular 444
Molecular layer 235
Monocytes 99
Monophasic 29
Monosynaptic 120
Morphine 326
Morphology of nerve cells 53
Mossy fibres of cerebellum 236
Motoneurone 444
Motoneurones 9, 12, 75
  α motoneurones 119, 120
  γ motoneurones 120
  in different sites in the cord 275
  pool 11
  spinal, equilibrium potential for chloride 67
  upper and lower 217
motor activity and reflexes 9
  innervation of skeletal muscle 9
  production of movement 9
Motor area 112, 221
  primary 288
Motor control
  co-ordination of 217
Motor control loops, summary *280*
Motor cortex 275
Motor end-plates 118
Motor functions of the precentral gyrus 250
Motor nerve fibres 360
  action potential in 9, 10, 12
Motor response 128
Motor tracts, stimulation of 260
Motor unit 9
Movement
  ballistic movements 218
  co-ordination of; and the cerebellum 218
  control of 217
  ramp movements 218
  supraspinal control of 217
Movement memory in speech, Broca's area 281
Muscarine 336
Muscarinic 336
Muscle cell 57
Muscle end plates 96
Muscle fatigue 149
  compensation in γ-commanded contraction 124
Muscle fibres, intrafusal and extrafusal 118
Muscle spindle 118
  initiating stretch reflex 120
  initiating tendon-jerk reflex 119
  innervation of 118
  mammalian 131
  selective activation of receptors 133

Muscle spindle receptors, primary and secondary 40, 118
Muscle tone, *see* Tonus (tone) in skeletal muscle
Muscle twitching 26
Muscle wasting 108
Muscular effort, mechanism for graded increase of 10
Musculature malfunction and cortical lesion 281
Music score, reading of 292
Myelin 444
Myelin sheath 33
Myelinated 48, 53, 246
Myelination 32
Myofibrils 57

Naloxan 326
Naso-pharynx, connection to middle ear 169
Neck muscles, as effectors 182
Neck reflexes 185, 186, 188
  rotation 188
Neck righting reflexes 181
Neocerebellum 230, 234, 444
  effects of deficiency of 237
  effects on voluntary movement 237
  ocular disturbances 238
Neocortex 245, 285, 444
  lamination of 248
  myelination of 247
Neologisms 290
Neostriatum 219, 221, 444
Nernst equation and equilibrium potential 18
Nerve axon 13, 15
Nerve, self-test 370
Nerve block
  differential by local anaesthetic 122
Nerve cells (neurones)
  axonal transport 54
  morphology of 53
Nerve fibre groupings 31
  plexus 335
  regeneration 96
  sensory 48, 337
  size of in central nervous system 35
  unmyelinated 335
Nerve impulse 3, 13
  initiation of 65
  see also Action potentials
Nerve roots 359
Nerve terminals 83
Nerve trunk 28
Nervous system 1
  autonomic 333
  parasympathetic 335
  sympathetic 335
Net movement of chemicals 15, 444
Neural circuitry, subserving pain, self-perpetuating 78
Neural control of skilled human movement 281
Neural function
  central conduction time 260
  physiological assessment of 259
Neural influences 12
Neural shock 100
Neuralgia, post-herpetic 329
Neuraxis, transection of at midbrain level and decerebrate rigidity 205
Neurite 444
Neuroactive peptides 313
Neuroanatomy 357
Neuro-effector junction, *see* Synapses
Neurofibril 445
Neuroglia 445
Neurolemma 445
Neuromas 96
Neuromodulation 211, 340
Neuromuscular junction 303
Neuromuscular blocking agents 58
Neuromuscular transmission 55
  as an exceptional synaptic mechanism 69
Neuronal circuits in volitional movement 277
Neuronal integration circuits 13
Neurone 445
Neuro-neuronal transmission, *see* Synaptic transmission
Neural grafting 98
Neural interactions, self-test 378
Neural trajectories, crossing in the cord 365, 366
Neurones 1
  of the cortex 247
  damage to in human adults 93
  denervation hypersensitivity of 329
  with short axons 77
  see also Interneurones; Motoneurones

Neurones which are sensors of the cerebral environment 341
Neuropathic pain 330
Neuropil 445
Neurophysiological basis of pathological pains 328
Neurosurgical operations, central conduction time 260
Neurotransmitters 301
  in the cerebral cortex 301
  chemicals in the brain 301
  excitatory 306
  for fast pathways 303
  in general 301
  inhibitory 304
  mediating sleep 213
Nicotine 336
Nicotinic receptors 336
Night vision 158
Nigral insufficiency and rigidity 241
Nigro–thalamic deficiency 226
Nigro–striatal pathway, degeneration of 227
Nitric acid 314
Nitrogen excretion 324
Nitrous oxide 352
NMDA antagonists 314
NMDA-operated channel as a mechanism for heterosynaptic interaction 311
NMDA receptor complex 312
NMDA receptors the hippocampus and memory 311
Nociception 137
Nociceptive 445
Nociceptive mechanisms 321
Nociceptive pathways, control of 324
Nociceptive receptors 322
Nociceptive reflexes 143
  see also Protective reflexes
Nociceptors 49, 52, 322
  mechanical 323
Nocifensive behaviour 138
Node of Ranvier 33, 43
Nominal aphasia 291
Noradrenaline 152, 328, 340, 445
Nose–brain 245
Noxious 140
Nuclear bag fibres 132
Nuclear chain fibres 132
Nucleus 1, 53, 95
Nucleus coerculeus, *see* Locus coeruleus
Nucleus of the raphe 152
Nystagmus 182, 204, 238, 445

Obstruction of the bowel 143
Obex 445
Occipital lobe 243, 357, 365
Occlusion 146
Ocular muscles
  extrinsic 158
  intrinsic 158
Oculomotor nerve 158, 344
Oesophagus 344
Oestrus cycle 343
Olfaction 39, 176
  and the limbic system 254
Olfactory epithelium 176
Olfactory receptors 176
  sensitivity of 6
Oligodendrocyte 445
Olivary nucleus 362
Olive 445
Olivo–cochlear bundle 172
Ontogeny 246
Open channel blocker 313
Opening of collateral circulation 99
Operant conditioning 297
Opioid peptides 318, 326
Ophthalmoscope 159
Optic chiasma 166, 343
Optic disc 164
Optic nerve 363
  decussation of 166
Optic nerve fibres 267
Optic tract 166
Optical illusions and the operation of lateral inhibition 86
Organization of spinal rhythm generators in mammals 153
Organic anions, in intracellular fluid 15, 35
Orientation of the receptor 173
Orientation sensitivity of hair cells 173
Origin of spinal shock 104
Orthodox sleep 212
  and the raphe system 152
Orthodromic inhibitory pathway 75, 147
  circuitry 75
  function of 75
Orthograde transport systems 54, 55
Orthostatic hypotension 105

Oscillatory movements in cerebellar disorders 237
Osmoreceptors 8, 343
Osmotic
  pressure 35
  strength 8
Otic ganglion 344
Otolith organs 174
Otolith reflexes 182
Otoliths 175, 182
Oval window 169
Oxygen 319
  debt 351

Pacinian corpuscle 45, 51
Packets of transmitter 57
Pain 96, 317
  of angina pectoris 321
  caused by cutting a nerve 95
  definition of 317
  different tissues; stimuli which elicit pain 319
  from different tissues 319
  and emotional stress 317
  first and second, mediated by different groups of nerve fibres 322
  functions of 317
  gate theory of 331
  ischaemic muscle pain 319
  lack of 317
  pathological 327
  peripheral mechanisms of 324
  and peripheral nerve regeneration 96
  referred 321
  self-test 404
  sensation of; absence of and its effects; from 317
  slow 322
  and temperature 110
  variation in perception of 318
  in viscera 319
  visceral 320, 321
Pain and temperature, contralateral loss of 112
Pain threshold 318
Pain-suppressing mechanisms 324, 325
Paleocerebellum 230, 234, 445
Paleocortex 245, 445
Paleostriatum 219, 445
Pallidum 445
Pallium 445
Papillae, dermal 44
Paradoxical breathing, after spinal transection 101
Paradoxical sleep 212
  disturbed by lesions of the locus coeruleus 213
Parallel fibres 236
Paralysis 445
Paralysis, and precentral stimulation 251
Paralysis agitans, *see* Parkinson's disease
Paraplegia 445
Parasympathetic nervous system 333, 335, 446
Parasympathomimetic 446
Parenchyma of the brain 347
Paresis 446
Parietal association cortex, disorders resulting from lesions of 287
Parietal lobe of cerebral hemisphere 243, 364
Parietal region, posterior, association areas 285
Parkinson's disease 223
  akinesia 223, 224
  involuntary movement 223, 225
  reduced dopamine level 228
  rigidity; and chemical transmitters 223, 227
Parotid gland 340
  salivary 344
Partial pressure of oxygen 7
Passive head rotation 203
Passively 39
Past-pointing 238
Patellar hammer 120
Patellar tendon 141
Pathological pains 327
Pathological synchronization of neuronal discharge 250
Peak of the action potential 21
Pelvic splanchnic nerves 335
Pendular tendon jerk 237
Penfield, Wilder, study of excitable cortex 251
Penis 335, 338
Peptides 326
Perception 3
Perception and sensation 135
Perceptive threshold 46
Percutaneous 446

Percutaneous stimulation of nerve 126
Perfumes 176
Perilymph 170
Perineurium 446
Periodic breathing 207
Peripheral nerve
  regeneration of 96
  section of 93
Peripheral nervous system 1, 446
Peristalsis 320
Perseveration 286
Personality changes 286
PET scanning, *see* Position emission tomography
Phantom limb pain 329
Pharmacological experiments 303
Pharmacological protection against cerebral ischaemia 356
Pharyngotympanic tube, *see* Eustachian tube
Phases in memory 297
Phasic 277
Phenol 446
Phineas Gage 285
Photic receptors 157
Photopigments 162
Photons 6
Photoreceptors (photosensitive cells) 3, 158
Phrenic nerve 100
Phylogenetically 108
Phylogenetic scale 104
Phylogeny 246
Physiological effects of activity in the autonomic nervous system 337
Physostigmine 59
Pia mater 446
Picrotoxin 308
Pigment molecules 3
Pigment rhodosin 158
Pill rolling 225
Pineal 446
Pineal body 362
Pinna (earflap) 168
Pinocytosis 176
Pinocytotic vesicles, lack of 347
Pin point pupil 203
Pit vipers 4
Placing reactions 107
Plane of polarized 4
Plantar reflex 106
Plasma membranes 13
Plasticity, in cerebral cortex 100
Plexus 446
Plexus of nerves 335, 336
Plumb line 175
Poisoned 35
Polar regions 118
Polypeptides 313
Pons 229, 362, 446
Pontine nuclei 362
Pontocerebellum 230
Posterior crural muscles 180
Position and vibration 110
  ipselateral loss of 112
Positive-feedback 75
Positron emission tomography 355
Post-herpetic neuralgia 329
Postcentral gyrus 364
  stimulation of 251
Postcentral sulcus 364
Posterior 364, 446
Posterior fossa 446
Posterior horn of spinal cord 11
Posterior parietal cortex 222, 287
Posterior parietal region 285
Posterior root 360
Postganglionic nerve fibres 334
Postsynaptic membrane 303
  receptors in 307
Postsynaptic potential 62, 303
Postsynaptic receptors for amino acid neurotransmitters 307
Postural adjustment 218, 219
Postural component, movement 12
Postural reflexes 143
  neck reflexes 185
Posture 180, 275
Potassium 35, 337
Potassium, extracellular rise in, buffered by glia 348
Potassium channels 211
  closure of 70
Potassium conductance 15
Potassium equilibrium potential 19
Potassium ions 15
Potential
  difference 13
  generator 40
  receptor 40
  spindle 40
Precentral gyrus 218, 288, 364
  and cortical epilepsy 250
  lesions of 281

in Parkinson's disease 226
as penultimate final common path 280
and speech 288
stimulation of 260
Precentral motor strip
lesions of the precentral gyrus 281
status of 280
Precentral sulcus 364
Prefrontal area of cortex 263, 285, 324
see also Association areas
Preganglionic nerve fibres 334
Prepotent reflexes 147
Pressure
intrancranial 98, 351
receptors 39
removal of (brain) 98
Presynaptic inhibition 81, 90
Primary afferent axons 135
Primary cortical evoked potentials 259
Primary endings 118, 130
Primary motor areas 243, 245
Primary sensory areas 243
Primary sensory endings 120
Primary somatosensory areas 245
Primary spectral wavelengths 162
Primary visual cortex 266, 269
modular organization of 269
Prime mover muscle 12
Pronation of wrist 446
Production of movement 11
Progesterone 344
Prone 204
Propagation, of the action potential 23
Proprioception, self-test 383
Proprioceptive receptors 115
Proprioceptors 7, 115, 229, 446
conscious awareness of 129
joint and tendon receptors 115
in muscle 117
spindle receptors, different types 130
voluntary movement 124
Propriospinal system of neurones 137
Prosencephalon 357, 446
Prostaglandins in nociception 322
Prostaglandin synthetase 322
Prostaglandins 322
Protective reflexes 143
Protein catabolism 324
Protein, in cerebral interstitial fluid 346
Protein macromolecules, as receptors 307
Protein synthesis 53
Protopathic sensation 96, 446
Protopathic sensibility 96
Psychic blindness 286
Psychosis and the limbic system 257
Psychosurgery 286
PTN, *see* Pyramidal tract, neurones
Ptosis 446
Punctate 47
Pupil 164
Pupillary reaction to light 167, 203
Purinergic transmission 337
Purkinje cells in the cerebellar cortex 77, 232, 234
Putamen 219, 447
Pyramidal cells 273, 366
Pyramidal decussation 112
Pyramidal system 217
Pyramidal tract 107, 273, 447
neurones 275

Quadriplegia 447
Quantum 57

Rabbit 246
Radiation 338
Radioactive indicators 354, 355
Radio waves 3
Ramp movement 218, 219
Raphe nuclei 152, 447
Rapid eye movement (REM) sleep 212
Rapidly adapting mechanoreceptors 140
Rapidly adapting receptors 51
Rat 246
Reaction of the neurone to axotomy 114
Reactions 318
Reaction time 156
Reading a music score 292
Reading and writing disorders 291
Rebound 238
Receiving areas, neocrotex 244
Receptive fields after regeneration 97
Receptive field, of the sensory unit 4
Receptor cells 172
Receptors 13, 39, 96, 336
adaptation in the Pacinian corpuscle 45

classification of 7
cold receptor 40
for excitatory neurotransmitters 308
mechano-thermal 323
muscarinic and nicotinic 336
polymodal 323
specificity 3
stretch 39
in the vestibules and semicircular canals 172
Receptor mechanisms, self-test 374
Receptor potential 40, 41
strength–response curve 40
Reciprocal innervation 146
Recovery of neural function 99
Recruitment, of nerve fibres 4
Rectum 335
Recurrent branch of motor nerve fibre 78
Recurrent collateral branches 247
Recurrent inhibitory pathway 77
electrophysiological evidence for 78
function of 77
Red nucleus 206, 211
Refractory period 24
Refractoriness, M and H responses 128
Referred pain 321
Reflex 12
Reflex effects of Group Ib afferents 153
Reflex relaxation 12
Reflex time 141, 276
and reaction time 276
Reflexes
abdominal 143
balance 179
some characteristics of 143
classification 143
crossed extensor reflex 147
introduction to 9
with otolith organs as receptors 182
postural 143
protective 143
reciprocal innervation 146
with semicircular canals as receptors 182
Refractive index 158
Refractory period, absolute or relative 21
Refractory power 158
Regenerating C fibre sprouts 328
Regenerating nerve tips, sensitivity of 96
Regeneration of peripheral nerve 96
Reissner's membrane 170
Relative positions of the different parts of the body 7
Relative refractory period 24
Relax the smooth muscle 316
Relay nuclei 84
Removal of the precentral area 250
Renal colic 320
Renshaw cells 78
Reorganization of corticospinal projection 275
Representational hemisphere 293
Reproduction 342
Resolution of the eye 158
Resolution of oedema 99
Respiration affected by pain 324
control of 350
and speech 288
Respiratory centres 350
Respiratory failure 207
Responses depending on connections of the cord with the cerebral cortex 106
Resting potential 13
of a neurone 63
origin of 15
Restoration of neural function 97
Retention with overflow 104
Reticular formation 209, 275, 447
and chemical transmission 211
excitatory, cholinergic 211
excitatory portion 210
inhibitory portion 210
Retina 3, 159
Retinal ganglion cells 267
Retinal receptors 3
Retinal sensitivity
greater off visual axis 159
regional differences to illumination 160
Retino–collicular mapping 267
Retino–cortical map 264
Retino–tectal projection 166
interruption of, Argyll Robertson pupil 167
Retractor muscle of the upper eyelid 335
Retrograde amnesia 298

Retrograde messenger 315
Retrograde transport system 55
Rheumatoid arthritis 322
Rhinencephalon 245, 447
Rhizotomy 328
Rhodopsin 158
Ribosomes 114
Right- and left-handedness 287
Right iliac fossa 321
Righting reactions 181
Rigid bony case 98
Rigidity 226
  decerebrate 205
  in Parkinson's disease 226
Rites and rituals 318
RNA synthesis 95
Rods and cones 158
  density of 160
  different spectral sensitivities of 163
Role of inhibition 74
Role of the lateral vestibular nucleus in the maintenance of normal posture 206
Rostral 447
Rotation of the head 188
Rough endoplasmic reticulum 2, 114
Rubro–reticulo–spinal tract 222

Saccades, saccadic eye movements 263, 447
Saccule 174
Sacral parasympathetic nerve fibres 335
Safety factor of nerve conduction 57
Saltatory 33
Salivary 335
  fluid 340
Saltatory conduction 33
Scala tympani 170
Scala vestibuli 170
Schwann cells 32, 33, 447
Scotoma 267
Scratching 12
Scratch reflex 143
Seat of the intellect 285
Secondary ending 131
Secondary motor areas 243
Secondary visual cortex 270
Second messenger 302
Second messengers and retrograde messengers 314
Second-order neurones and nerve fibres 107, 110
Section of a peripheral nerve 93
Seeing and hearing 263
Self-generated voluntary movements 224
Self-test questions and answers 369–436
Semen 338
Semicircular canals 173, 203
Seminal vesicles 338
Senile dementia 211
Sensation 2
  adaptation 4
  the sensory unit 4
Sensitivity 158, 447
Sensitivity of different parts of the body 6
Sensory cortex 242
Sensory discriminating task 135
Sensory modalities *109*
Sensory nerve fibres 3, 4, 337, 360
  classifications 31
Sensory neurones (nerve cells) 54
Sensory perception 46
Sensory receptors 2, 20
Sensory unit 4
Sepsis 96
Septal area 447
Septum pellucidum 447
Sequence of recovery of somatic reflexes as spinal shock wears off 105
Sequence of recruitment of motor unis 149
Serotonin 152
Sexual function 342
Sham rage 257
Shell temperature 35
Sherrington, Sir C.S. 230
Shivering 103
Short-term memory 297
Short-term survival of patients with brain stem death 202
Short words 290
Sick cells swell 17
Sigmoid shape of strength response curve 41
Significance of pre- as opposed to postsynaptic inhibition 83
Silent 285
Simple cells 269

Singing 292
Sinusoidal current 23
Sites, receptor 39
Size of the animal 13
Size of nerve fibres in the central nervous system 35
Skeletal muscle 12, 115, 340
Skilled movement, and the cerebellum 229
Skin 3, 6, 336
  earlobe, of the 39
  encapsulated receptors 44
  hand, of the 47
  hand, receptor groups 48
  sensory innervation of 48
  ulceration 318
  vessels are dilated 103
Skull, foramina 357
Sleep 211, 212
Sleep centres 213
Sleep deprivation 212
Slow conjugate deviation 204
Slowly adapting receptors 47
Small intestine 344
Smell 245
Smooth muscle 158, 166, 340
Snake venom 69
Social behaviour 343
Sodium 35
Sodium channels 20
Sodium concentration gradient 15
Sodium equilibrium potential 21
Sodium ions 16, 17
Sodium pump 17
  and cell volume 17
Soldiers and pain 318
Soma, *see* Cell body
Somaesthetic 447
Somaesthetic receptors 39
Somaesthetic senses 108, 157
  see also General body senses
Somatic 447
Somatic nervous system, cf. autonomic nervous system 334
Somato-cortical map 265
Somatosensory cortical evoked potentials 259
Somatosensory projection 226
Somatotopic mapping 138
  disorganization after nerve section and regeneration 97
  skin to dorsal grey somatotopic matter 97
Somatotopic mapping 251
Sound 168
Sound frequencies, recognition of 171
Space-occupying lesion 351
Spastic paralysis 105
Spasticity 105
Spatial awareness 292
Spatial discrimination 47
Spatial interaction 135
Special senses 8, 157
  self-test 391
  see also Balance; Ear; Eye; Olfaction
Specific afferent projection 222
Specificity of sensation 3
Specific receptor sites 326
Specific recognition sites 303
Specific relay nucleus 226
Speech 100, 288
  a function of the left hemisphere 287
  jerky and explosive in cerebellar disorder 238
Speech apparatus 288
speech areas
  and age 292
  characteristics of interference with, *291*
Speech centres 288
Spheno-palatine ganglion 344
Sphincter 337
Sphygmomanometer 319
Spinal hemisection (Brown-Sequard syndrome) 112
Spinal cord 1, 108, 218, 357, 359
  chemical transmitters in 90
  cutaneous distribution of spinal segments 361
  extent of (adult) 360
  influenced by cerebral cortex 182
Spinal nerves 360
Spinal reflex mechanisms, modified by corticopspinal tracts 128
Spinal reflexes 103, 141
  activity activated by Group II afferent fibres 151
  elicited by stimulation of receptors in skeletal muscle 150
  facilitation and inhibition of by higher centres 128
  self-test 382

Spinal rhythm generators 152
Spinal segments (dermatomes), cutaneous distribution of 361
Spinal shock 100, 107
  duration of 104
  origin of 104
  sequence of recovery of somatic reflexes 105
Spinal stretch reflex, over-activity of 206
Spinal transection 100, 102
  autonomic and associated effects 103
  autonomic reflexes 103
  Babinski sign 106
  effects of, *101*
  motor effects of 101
  plantar responses 106
  resposes returning in modified form or failing to return 106
  segmental levels of compatible with survival 100
  sensory effects of 101
Spindle afferent discharge, during voluntary movement 125
Spindle, muscle of the 40
Spindle potential 40
Spindle receptors
  primary ending 118
  secondary ending 118
Spino–cerebellar tracts 232
Spino–olivary tract 233
Spino–thalamic tract 329
Spinocerebellum 230
  connections of 232
Spiral ganglion, in bony labyrinth 170
Splenic flexure 335, 344
Split brain in man 293
Spoken language, after lesions in Wernicke's area 290
  Spontaneously active 340
Stability vs. need for change 343
Stabilizing the trunk 275
Staggering 239
Standing and walking 179
  receptors important in maintenance of posture 181
  and running 206
Stapedius 169
Stapes 169
Stars 159
Static bag fibre 132
Static component 116
Stellate cells 247
Stepping reflex 175
Stereognosis 222
Stereophonic sensitivity 272
Stereocilia 172
Stimuli eliciting pain 319
  specific, evoking given reflexes 143
  strength of, in sensation 4
  threshold, subthreshold and suprathreshold 41
Stimulus 3
  intensity of 4
  strength of 4
Stomach 344
Strategy for survival 318
Strength response curve for receptor potential 40
Stretching, an involuntary movement 281
Stretch receptors 39
Stretch reflex 120
Striatum 447
Strong coarse movements 10
Structural correlates of excitatory and inhibitory synapses 62
Structural proteins 2
Structure of the language 288
Strychnine 90
Stupor 212
Subarachnoid haemorrhage, recovery from 98
Subarachnoid space 359
Subcortical motor nuclei, self-test 398
Subdivisions of the lobes of the cerebral hemispheres 364
Subdural haemorrhage 359
Subdural space 357
Subiculum 447
Subliminal fringe 146
Submandibular and lingual salivary glands 344
Submandibular ganglion 344
Substantia gelatinosa 448
Substantia nigra 448
Subthalamic nucleus 219, 221
Subthalamus 363, 448
Subthreshold 4, 41
Subthreshold stimulus 41
Succinylcholine, short-term relaxant 58
Sudomotor nerves 339
Sulci 243

Sunburn 322
Substance P 320
Substantia nigra 219, 221
  degeneration of in Parkinson's disease 223
  input to neostriatum 227
Superior and inferior colliculi 206, 267
Superior cervical ganglion, as a synaptic relay 335
Superior colliculi
Superior, middle and inferior frontal gyri 364
Supinate 448
Supra-orbital notch 205
Supra-chiasmatic nuclei 342
Supra-marginal gyrus 364
Supraspinal control
  of movement 217
  of skeletal muscle 273
  of spinal reflexes 128
Supraspinal mechanisms 106, 135
Suprathreshold 43
Surround (lateral) inhibition 84
Swaying 239
Sweat glands 336
Sweating 339
  after spinal transection 103, 106
Sylvian fissure 288
  see also Lateral fissure, cerebral hemispheres
Sympathetic collateral ganglia 335
Sympathetic ganglia 335
Sympathetic nervous system 333, 335, 448
  fight or flight reaction 337
  reflex discharge of in severe haemorrhage 339
  subtle patterns of activity 339
Sympathomimetic 448
Synapses 53, 448
  excitatory and inhibitory 62, 63
  number on cord pathways for different reflexes 143
Synapsin I 62
Synaptic cleft 55
  differs at different junctions 60
Synaptic delay 142
Synaptic gap 55
Synaptic integration 73
  membrane mechanism of 69
  role of inhibition 74
Synaptic potentials 63
Synaptic relays 334
  delays in conduction velocity 142
  in the thalamus 109
Synaptic transmission 60
  closing of membrane channels as a synaptic mechanism 70
  excitatory and inhibitory synapses 62
  excitatory postsynaptic potential 63
  features of junctional transmission 65
  inhibitory postsynaptic potential 65
  initiation of nerve impulse 65
  synaptic delay 142
Synaptic vesicles 56
  fusion of membrane of 61
Syringomyelia 448
  intractable pain of 329
Systematically 122

Tactile discrimination 110
Task-related pattern of organization 275
Taste 178
Taste-buds in the tongue 7, 39
Tectorial membrane 170
Tectum 44
Telegram speech 290
Telencephalon 359, 448
Temperature and conduction velocity in nerve 35
Temperature receptors 3, 343
Temperature regulation 342
Temperature sensitivity 7
Temporal association cortex 246, 286
  role of 286
Temporal lobe 243, 257, 285, 365
  association areas 246, 286
  inferotemporal region 270
Temporal sequence of contractions 12
Tendon 448
Tendon organ, siting of 118
Tendon receptors 116
Tendons 115
Tendon-jerk reflex 75, 119
  components of reflex time 142
Tensor tympani 169
Tentorium 448
Tentorium cerebelli 200, 357
Tetanic contractions 9
Tetanus 90

Tetraplegia 448
Tetrodotoxin 69
Thalamic nuclei, non-specific 255
Thalamic relay nuclei 108, 363
Thalamic syndrome 329
Thalamo-striate artery, occlusion of 329
Thalamus 222, 363, 448
  dorsomedial nucleus of 285
  lateral nuclei 363
  medial nuclei 363
  as relay nucleus 222, 363
  ventral lateral nucleus 255
  ventral posterior nucleus 225
  ventral region, intermediate nucleus 221
Thalamotomy 226
Thermal stimuli 139
Thermometer 8
  information about 8
Thermoreceptors 49, 52, 322
Thick nerve fibres
  pain-reducing effect of activity in 324
Thickly myelinated nerve fibres 324
Thin striated muscle fibre 118
Third ventricle 342
Thoraco–lumbar region 333
Threshold 21, 43
  calcium 26
  propagation 23
  refractoriness 24
Thrust force 191
Tibialis anterior 181
Tight junctions 347
Timing and degree of contraction 218
Tissue metabolism, effect of increase in 333
Tomography 448
Tongue 6, 178
Tonic mesencephalic fits 207
Tonic neck reflexes 186
Tonotopic 272
Tonus (tone) in skeletal muscle 12, 120, 179
  and balance 206
  effects of cerebellum on 240
  extensor 206
Toppling, prevented by reflex adjustment 179
Touch and pressure 51, 110
Touch receptors 3
Toxins 104
Trachea 336
Transducer 448
Transduction 39
  of light energy 166
Transected spinal cord 206
Transmission at other junctions 59
Transmitter chemicals, *see* Chemical transmitters
Transmural passage of plasma proteins 98
Transport, axonal 54
Travel sickness 195
Temorogenic zone 226
Trepanation (African ritual) 318
Triple response 147
Trochlear nerve 158, 362
Trunk ataxia 239
Trunk muscles, as effectors 184
Twitch fibre 149
Twitch response 10
Twitch tension, maximal 10
Twitching, in extrafusal muscle fibres 26
Two-point discrimination 48, 110
Two types of neuronal membrane 67
Types of reflex 143

Ulnar nerve 6, 93
Ultraviolet 3, 319
Unconsciousness 199
Unmyelinated 34, 48, 335
Unmyelinated ending 39
Unmyelinated initial segment 65
Unmyelinated nociceptive afferent nerve fibres 148
Upper motoneurone lesion 108, 113
Uptake and release of glutamate by synaptic terminals 306
Ureter 320
Urethra 338
Uterus 335
Utricle and saccule – balance 174
Uvula 449

Vagus nerve 335, 344, 362
Variation in the perception of pain depending on circumstance 318
Vasculature 335
Vas deferens 326, 338
Vasoactive intestinal peptide (VIP) 340
Vasodilation, in brain blood vessels 338, 340

Vegetative functions 338
Vegetative nervous system 339
Vegetative physiological mechanisms 338
Vegetative wreck 199
Veins 336
Velum 449
Ventilation, artificial 202
Ventral 449
Ventricle 449
Vermis 229, 234, 449
Vertebrae, naming of 359
Vertebral canal 357
Vertebral column, in early months 360
Vestibular apparatus, for balance 180
Vestibular information, disregarded by pilots 194
Vestibular and neck reflexes, synthesis of 189
  linear acceleration 191
  sense organs, conflicting information from 194
Vestibular nuclei 275
Vestibular nucleus, flocculo-nodular lobe as 233
Vestibular reflexes 181, 189, 203
Vestibular righting reflexes 181
Vestibular system 218
Vestibules 169, 182
  as sensors, *181*
Vesibulo-ocular reflexes 203
Vestibulocerebellum 230
  connections of 233
Vestibulospinal tract 206
Vibration 51
VIP (vasoactive intestinal peptide) 340
Viscera 6, 333
  double innervation of 333
Visceral brain, limbic system as 255
Visceral mechanoreceptors 7
Visceral pain 320
  colic 320
  localization of 321
Viscero-somatic neurones 321
Vision 39
  and eye movements 263
  (superior colliculi) 267
Visual areas 269
  primary 269
  secondary 270
Visual aspects of speech 291
Visual cortex 264
  secondary 270
Visual field 266
Visual evoked potentials 259
Visual image, interpretation of 271
Visual information, processing of 267
Visual pathway 166, 264
  pigment 162
Visual system
  columnar organization in the visual cortex 269
  retino-cortical map 264, 267
  visual image, perception of 271
Volition, phasic effect 277
Volitional movement 282
  probable route of information flow 276
Voltage dependent 68
Voluntary movement 124, 223
  effects of deficiency of the cerebellum 238
  Hennemen size principle 125
  and 'long-loop' or transcortical reflexes 276
  loss of 281
  M and H response 126
  skilled 281
  supraspinal control of spinal reflexes 128
Voluntary muscles 12
Volunary ramp movements out of control 223

Walking 12, 180
Wasting of muscles in lower motoneurone lesion 108
Water content and osmolarity of the body 342
Wernicke's area (auditory speech area) 290
White matter 247
  cerebellar cortex 247
  result of damage to 199
Withdrawal reflex 106
Word blindness 291
Worm (vermis of cerebellum) 108
Wrong choice of words 290

X-rays 3, 319
X-cells (parvocellular) 268

Y cells (magnocellular) 268
Yellow disc 159
Young–Hemholz theory of colour vision 162